Lecture Notes in Statistics **186**

Edited by P. Bickel, P. Diggle, S. Fienberg, U. Gather, I. Olkin, S. Zeger

Lecture Notes in Statistics

For information about Volumes 1 to 132, please contact Springer-Verlag

133: Dipak Dey, Peter Müller, and Debajyoti Sinha (Editors), Practical Nonparametric and Semiparametric Bayesian Statistics. xv, 408 pp., 1998.

134: Yu. A. Kutoyants, Statistical Inference For Spatial Poisson Processes. vii, 284 pp., 1998.

135: Christian P. Robert, Discretization and MCMC Convergence Assessment. x, 192 pp., 1998.

136: Gregory C. Reinsel, Raja P. Velu, Multivariate Reduced-Rank Regression. xiii, 272 pp., 1998.

137: V. Seshadri, The Inverse Gaussian Distribution: Statistical Theory and Applications. xii, 360 pp., 1998.

138: Peter Hellekalek and Gerhard Larcher (Editors), Random and Quasi-Random Point Sets. xi, 352 pp., 1998.

139: Roger B. Nelsen, An Introduction to Copulas. xi, 232 pp., 1999.

140: Constantine Gatsonis, Robert E. Kass, Bradley Carlin, Alicia Carriquiry, Andrew Gelman, Isabella Verdinelli, and Mike West (Editors), Case Studies in Bayesian Statistics, Volume IV. xvi, 456 pp., 1999.

141: Peter Müller and Brani Vidakovic (Editors), Bayesian Inference in Wavelet Based Models. xiii, 394 pp., 1999.

142: György Terdik, Bilinear Stochastic Models and Related Problems of Nonlinear Time Series Analysis: A Frequency Domain Approach. xi, 258 pp., 1999.

143: Russell Barton, Graphical Methods for the Design of Experiments. x, 208 pp., 1999.

144: L. Mark Berliner, Douglas Nychka, and Timothy Hoar (Editors), Case Studies in Statistics and the Atmospheric Sciences. x, 208 pp., 2000.

145: James H. Matis and Thomas R. Kiffe, Stochastic Population Models. viii, 220 pp., 2000.

146: Wim Schoutens, Stochastic Processes and Orthogonal Polynomials. xiv, 163 pp., 2000.

147: Jürgen Franke, Wolfgang Härdle, and Gerhard Stahl, Measuring Risk in Complex Stochastic Systems. xvi, 272 pp., 2000.

148: S.E. Ahmed and Nancy Reid, Empirical Bayes and Likelihood Inference. x, 200 pp., 2000.

149: D. Bosq, Linear Processes in Function Spaces: Theory and Applications. xv, 296 pp., 2000.

150: Tadeusz Caliński and Sanpei Kageyama, Block Designs: A Randomization Approach, Volume I: Analysis. ix, 313 pp., 2000.

151: Håkan Andersson and Tom Britton, Stochastic Epidemic Models and Their Statistical Analysis. ix, 152 pp., 2000.

152: David Ríos Insua and Fabrizio Ruggeri, Robust Bayesian Analysis. xiii, 435 pp., 2000.

153: Parimal Mukhopadhyay, Topics in Survey Sampling. x, 303 pp., 2000.

154: Regina Kaiser and Agustin Maravall, Measuring Business Cycles in Economic Time Series. vi, 190 pp., 2000.

155: Leon Willenborg and Ton de Waal, Elements of Statistical Disclosure Control. xvii, 289 pp., 2000.

156: Gordon Willmot and X. Sheldon Lin, Lundberg Approximations for Compound Distributions with Insurance Applications. xi, 272 pp., 2000.

157: Anne Boomsma, Marijtje A. J. van Duijn, and Tom A.B. Snijders (Editors), Essays on Item Response Theory. xv, 448 pp., 2000.

158: Dominique Ladiray and Benoît Quenneville, Seasonal Adjustment with the X-11 Method. xxii, 220 pp., 2001.

159: Marc Moore (Editor), Spatial Statistics: Methodological Aspects and Some Applications. xvi, 282 pp., 2001.

160: Tomasz Rychlik, Projecting Statistical Functionals. viii, 184 pp., 2001.

161: Maarten Jansen, Noise Reduction by Wavelet Thresholding, xxii, 224 pp., 2001.

(*Continued after index*)

Estela Bee Dagum
Pierre A. Cholette

Benchmarking, Temporal Distribution, and Reconciliation Methods for Time Series

Springer

Estela Bee Dagum
University of Bologna
Faculty of Statistical Sciences
via delle Belle Arti 41
Bologna, Italy
beedagum@stat.unibo.it

Pierre A. Cholette
Statistics Canada
Time Series Research
and Analysis Centre
Ottawa, Canada K1A 0T6
Pierre_A.Cholette@Sympatico.ca

Library of Congress Control Number: 2005938354

ISBN-10: 0-387-31102-5
ISBN-13: 978-0387-31102-9

Printed on acid-free paper.

Printed in the United States of America. (SBI)

9 8 7 6 5 4 3 2 1

springer.com

Dedication

We would like to dedicate this book to the memory of Camilo Dagum, a "true Renaissance man" who possessed a deep knowledge on a vast array of Sciences, from philosophy to history, political sciences, economics, sociology, mathematics, and statistics. His versatile scientific knowledge, combined with an inquisitive mind, enabled him to carry out a remarkable scientific work in economics, statistics, econometrics, and philosophy of science. His rigorous scientific contributions have stood the passing of time, and many among them belong to today's scientific paradigm. He was privileged by a long working experience and friendship with other famous men of science: Corrado Gini, Francois Perroux, Oskar Morgenstern, Maurice Allais, Herman Wold, and Nicholas Georgescu Roegen.

Camilo Dagum, Emeritus Professor of the University of Ottawa, Canada, and Professor of the University of Bologna, Italy, was born August 11, 1925, in Rosario de Lerma, Salta, Argentina. He passed away in Ottawa on November 5, 2005. He obtained his Ph.D. degree from the National University of Cordoba with a doctoral dissertation on the "Theory of Transvariation and Its Application to Economics," which was approved with the highest honors and recommended by the jury to be published as a book. His dissertation was a major contribution to the transvariation theory that gave birth to a series of papers later published in Spanish, Italian, English, French, and German. He graduated Summa Cum Laude and was awarded the University Gold Medal Prize as the best student of his class.

Camilo Dagum's long and outstanding career in teaching and research started in 1949 at the Faculty of Economic Sciences of the National University of Cordoba, where he became Full Professor in 1956. Having been awarded a fellowship by the Argentine Research Council, in 1960, he went to the London School of Economics as a Visiting Research Economist and worked with Professors M.G. Kendall and A.W. Phillips. Upon his return in 1962, he was elected Dean of the Faculty of Economic Sciences, a position he renounced in 1966 due to an Army Coup d'Etat that eliminated university autonomy and destroyed the Argentinean constitutional and democratic order for 17 years. Forced to leave the country, he accepted an invitation by Professor Oskar Morgenstern, and went to Princeton University.

Camilo Dagum's outstanding scientific contributions have been recognized by his peers over the years. He was awarded the College of France Medal and the degree of Doctor Honoris Causa by the Universities of Bologna, Cordoba (Argentina), Montpellier (France), and Naples "Parthenope." He is an elected member of the Academy of Sciences of the Institute of Bologna and of the Lombard Institute, elected member of the International Statistical Institute and the International Institute of Sociology.

In 1988, he was invited by the Alfred Nobel Foundation and by the Nobel Prize Laureate in Economics of that year, Maurice Allais, to be one of his 10 companions for the Nobel Week Celebration and the Nobel Laureate convocation in Stockholm. Later, in 1993, Camilo Dagum was invited by Maurice Allais to become a member of the "Comité d'honneur en vue de la remise de l'épée d'académicien au Professeur Maurice Allais par l'Académie des Sciences Morales et Politiques de France."

During his long research career initiated at the young age of 24 years with an econometric study on the demand theory, Camilo Dagum published 12 books and more than 200 peer review articles. He is one of the rare economists who has worked and written in an overwhelming number of areas of diverse interests.

The concepts and analysis of national structures are basic knowledge to understand the dynamics of economic processes and the leading forces that generate the dynamics of structural change. The contributions done to this field of research allowed Camilo Dagum to identify the roles of education, health, finance, and social security systems as the prime movers of economic development. As a consequence, he pursued new research on functional and personal income distributions, inequality within and between income distributions, wealth distribution, and personal and national human capital and poverty. He wrote pioneering and seminal papers in all the above topics, creating new research paths that made him a world leader on income distribution. He also made breakthrough contributions to income inequality. He introduced a directional economic distance ratio, which provides a rigorous and clear representation of the inequality structure between income groups, hence complementing the intra-income inequality measure developed by Gini.

In his last years Camilo Dagum got involved with two important socioeconomic issues, poverty and human capital. Concerning poverty, he developed a multivariate measure that enables the identification of its main causes, thus offering essential information to political decision makers to advance structural socioeconomic policies. These latter are capable of breaking the vicious circle of reproducing poverty from generation to generation, as it occurs with policy limited only to income transfers.

Camilo Dagum was a brilliant scientist and a man of the utmost integrity, impeccable character, and unparalleled kindness. He has always been almost revered by his students in all parts of the world and admired by the peers for his creative intelligence. He was much loved by his wife, Estela, and his three sons, Alexander, Paul, and Leonardo.

Preface

In modern economies, time series play a crucial role at all levels of activity. They are used by decision makers to plan for a better future, namely by governments to promote prosperity, by central banks to control inflation, by trade unions to bargain for higher wages, by hospitals, school boards, builders, manufacturers, transportation companies, and consumers in general.

A common misconception among researchers, practitioners, and other users is that time series data are direct and straightforward compilations of sample survey data, census data, or administrative data. On the contrary, before publication, time series data are subject to several statistical adjustments intended to increase efficiency, reduce bias, replace missing values, correct errors, facilitate analysis, and fulfill cross-sectional additivity constraints. Some of the most common adjustments are benchmarking, interpolation, temporal distribution, calendarization, and reconciliation.

This book brings together the scattered literature on some of these topics, presents them under a consistent notation and unifying view, and discusses their properties and limitations. The book will assist large producers of time series, e.g., statistical agencies and central banks, in choosing the best procedures, and enable decision makers, whether governments, businesses, or individuals, to perform better planning, modeling, prediction, and analysis.

One widely applied transformation is benchmarking, which arises when time series data for the same target variable are measured at different frequencies with different levels of accuracy. To simplify the presentation, we label the more frequent series as "sub-annual" and the less frequent as "annual," unless it is relevant to be more specific. The sub-annual series is usually less reliable than the annual series, which is therefore considered as benchmarks. Annual discrepancies are observed between the annual benchmarks and the corresponding annual sums of the sub-annual series. Historically, benchmarking has mainly consisted of imposing the level of the annual benchmarks, which eliminates the annual discrepancies; and of adopting the observed sub-annual movement as much as possible, which is the only one available.

Another commonly performed adjustment is temporal distribution or interpolation, which occurs when annual data are available and sub-annual data are required. This is the case for many of the quarterly National Accounts series with respect to yearly benchmarks typically originating from the yearly Input–Output Accounts. The adjustment consists of estimating sub-annual values from the annual data and from sub-annual indicator series or sub-annual related series deemed to behave like the target variable.

In recent years, statistical agencies have performed calendarization, which consists of transforming data covering fiscal periods into estimates covering calendar periods, namely transforming fiscal year data into calendar year values, fiscal quarter data into calendar quarter values, multiweekly data covering four or five weeks into monthly values.

For systems of time series, the series must cross-sectionally (contemporaneously) add to the marginal totals. For example, the National Accounts series must add up to the Commodity totals at the national level and to the Provincial totals and to the Grand Canada total for each quarter, and also temporally sum to the yearly totals. This is also the case for Labour Force series, Retail Trade, Exports, Imports, etc. The imposition of cross-sectional constraints is referred to as reconciliation, balancing, or "raking," depending on the context.

We have been involved with these statistical transformations over 30 years, as either teachers, consultants, or practitioners. The book contains 14 chapters and 5 appendixes. The first two chapters are introductory and important to understand the effects of the adjustments on the components of time series, namely on the business cycle, seasonality, and so on. The rest of the book can be divided in two parts. The first part, consisting of Chapters 3 to 10, is devoted to adjustments for single time series. The second part, Chapters 11 to 14, deals with adjustments for systems of time series, namely reconciliation or balancing. Each chapter is written as self-contained as possible, to accommodate readers interested in one specific type of adjustment; this entails some repetition. Appendixes A, B, and C prove important theorems and matrix results not readily accessible. Appendixes D and E discuss various seasonal and trading-day regressors for benchmarking, interpolation, temporal distribution, and calendarization models.

This book will prove useful as a text for graduate and final-year undergraduate courses in econometrics and time series and as a reference book for researchers and practitioners in governments and business. The prerequisites are a good knowledge of matrix algebra and linear regression and some acquaintance with ARIMA modelling.

We would like to thank Mr. Guy Leclerc, Assistant Chief Statistician, who brought to our attention the spurious fluctuations introduced in the System of National Accounts series by inappropriate benchmarking, as soon as 1974. We are also very thankful to Dr. Ivan Fellegi, Chief Statistician of Canada, who several years later requested us to address the problem of lack of additivity in systems of seasonally adjusted series, namely in the system of Labour Force series, in order to facilitate interpretation.

We are indebted to participants and students who during many seminars and presentations raised valuable questions answered in this book. We would also like to thank several client divisions at Statistics Canada and other statistical agencies for bringing to our attention some of the real-life cases discussed. Our most sincere gratitude goes to our colleagues who encouraged us to write the book and all those who provided many valuable and useful suggestions through lively discussions or written comments. Particularly, we are thankful to Italo Scardovi, Paola Monari, James Durbin, Steve Feinberg, Tomasso Di Fonzo, Zhao-Guo Chen, Benoit Quenneville, Bernard Lefrançois, Norma Chhab, Marietta Morry, and Irfan Hashmi. Our thanks also go to John Kimmel, Statistics Executive Editor of Springer, for his sustained and valuable support while writing the book.

We are solely responsible for any errors and omissions.

Estela Bee Dagum
Pierre A. Cholette
November 5, 2005

Ottawa, Canada

Table of Contents

1 Introduction

Time series analysis plays a very important role in statistical methodology because one of its main objective is the study of the temporal evolution of dynamic phenomena of socio-economic and environmental character.

A time series is a sequence of observations often measured at a constant frequency, e.g. daily, weekly, monthly, quarterly and yearly. The observations are time dependent and the nature of this dependence is of interest in itself. The measurements can originate from administrative records, repeated surveys and censuses with different levels of reliability. A commonly held view among time series users is that the data result from straight compilations of the measurements from one of these sources. In fact, before their official release by statistical agencies, time series data are subject to several adjustments to increase efficiency, reduce bias, replace missing values, facilitate analysis and change the frequency (e.g. from yearly to quarterly). The most common such adjustments are benchmarking, signal extraction, interpolation and temporal distribution and calendarization.

A great majority of time series also belong to a system of series classified by attributes. For example, labour force series are classified by province, age, sex, part-time and full-time employment. In such cases, the series of the system must be reconciled to satisfy cross-sectional (contemporaneous) aggregation constraints. The latter require that the values of the component *elementary series* add up to the marginal totals for each period of time. For example, if the system is classified by N Industries (or sectors) and P Provinces, the system must satisfy N sets of industrial cross-sectional constraints over the Provinces in each Industry, and P sets of provincial cross-sectional aggregation constraints over the Industries in each Province.

Under various names, the reconciliation problem has pre-occupied economists and statisticians for more than sixty years. The problem was originally posed in the context of systems of National Accounts, where it was referred to as "balancing"; and, in the context of survey estimation, as "Iterative Proportional Fitting" or "raking".

This book discusses statistical methods for benchmarking, interpolation, temporal distribution and calendarization, which are widely applied by statistical agencies. Our purpose is to point out the properties and limitations of existing procedures ranging from ad hoc to regression-based models, to signal extraction based models. We bring together the scattered literature on these topics and present them using a consistent notation and a unified view.

We put the emphasis on regression based models because of their clarity, ease of application and excellent results for real life cases. Although seldom used, we also discuss other methods based on ARIMA and structural time series models. We illustrate each method with many examples of real and simulated data. To facilitate understanding as well as comparisons of several methods, we present a real data example, the Canadian Total Retail Trade series throughout the book.

Knowledge about the nature and the impact of these adjustments on time series data - namely on the trend, the business cycle and seasonality - will be beneficial to the users as well as statistical agencies engaged in their production. Decision makers, whether governments, businesses or individuals, will be aware that wrongly applied procedures adversely affect modelling, prediction, planning and analysis.

Statistical agencies and central banks, which are large producers of time series, will benefit by using the most appropriate techniques in each situation, in order to avoid the introduction of serious distortions in time series data.

1.1 Benchmarking

The benchmarking problem arises when time series data for the same target variable are measured at different frequencies with different levels of accuracy. Typically, one source of data is more frequent than the other: for example, monthly series on one hand and yearly series on the other, monthly and quarterly series, quarterly and yearly, daily and multi-weekly (e.g. 4- or 5-week). The more frequently observed series is usually less reliable than the less frequently observed. For this reason, the latter is generally considered as benchmark. Typically, discrepancies are observed between the yearly benchmarks (say) and the corresponding yearly sums of the more frequent series. Broadly speaking, benchmarking consists of combining the relative strengths of the two sources of data. More specifically, this has essentially consisted of adopting the movement of the more frequent series and the level of the benchmarks.

In order to simplify the exposition, we will refer to the high frequency series as the *sub-annual* series and to the low frequency series as the *annual* series, unless it is necessary to be more specific.

Situations requiring benchmarking are very common in statistical agencies. In some cases, both the sub-annual and annual series originate from surveys or censuses; in others, none of the two series originate from surveys or only one does. The following examples illustrates some cases.

Until the end of the 1990s, Statistics Canada obtained the monthly Sales from the Monthly Wholesale and Retail Trade Survey and the yearly benchmarks from the Annual Census of Manufacturing. Similarly, Statistics Canada obtained the monthly values of Wages and Salaries from a survey and the yearly benchmarks from administrative data, namely the tax data file anonymously supplied by Revenue Canada.

In Canada, the Census of the Population takes place every fifth year (1996, 2001, etc.). The yearly estimates of the Population are derived from the last quinquennial census value and from the yearly flows, adding to the stock of Population (Births and Immigration) and subtracting from the Population (Deaths and Emigration). These yearly flows are more accurate at the national level, but not so accurate at the provincial and sub-provincial levels. When the quinquennial census data become available, demographers re-estimate the flows for the last five years under the light of the new provincial, sub-provincial and inter-provincial information. Conceptually, this is a form of benchmarking although no formal benchmarking method is used.

In many situations none of the data originate from surveys. This is often the case of the System of National Accounts of many countries. The time series of the quarterly Accounts are interpolated from many sources to fit the concepts and definitions of the Accounts. The quarterly Accounts are benchmarked to the yearly values provided by the Input-Output Accounts. The quarterly Accounts must also satisfy cross-sectional additive constraints, which is a case of reconciliation. The Provincial Accounts are obtained by similar methods. Furthermore, they must cross-sectionally add up to the National Accounts. Benchmarking and reconciliation are critical processes used in building the National Accounts.

Benchmarking also occurs in the context of seasonal adjustment. Seasonally adjusting a monthly or quarterly time series causes discrepancies between the yearly sums of the raw (seasonally *un*adjusted) series and the corresponding yearly sums of the seasonally adjusted series. Such seasonally adjusted series are then benchmarked to the yearly sums of the raw series. The benchmarking is usually performed within the seasonal adjustment method, such as the X-11-ARIMA method (Dagum 1980 and 1988) and X-12 ARIMA method (Findley et al. 1998), using formal benchmarking methods.

The need for benchmarking arises because the annual sums of the sub-annual (high frequency) series are not equal to the corresponding annual (low frequency) values. In other words, there are *annual discrepancies*, d_m, between the annual benchmarks and the sub-annual values:

$$d_m = a_m - \Sigma_{t=t_{1m}}^{t_{Lm}} j_{mt}\, s_t\,,\; m=1,...,M, \tag{1.1a}$$

where t_{1m} and t_{Lm} are respectively first and last sub-annual periods covered by the m-th benchmark, e.g. quarters 1 to 4 for the first benchmark, 5 to 8 for the second, and so forth. The quantities j_{mt} are the coverage fractions, here assumed equal to 1. Note that the quantity $j_{m\bullet} = \Sigma_t\, j_{mt}$ is the number of sub-annual periods covered by the m-th benchmark.

The annual discrepancies are more often expressed in terms of *proportional annual discrepancies*:

$$d_m^{(p)} = a_m \,/\, (\Sigma_{t=t_{1m}}^{t_{Lm}} j_{mt}\, s_t),\; m=1,...,M, \tag{1.1b}$$

which may be less, equal or greater than 1.0.

Benchmarking usually consists of imposing the annual values onto the sub-annual values. In other words, the sub-annual series is modified so that the annual sums of the sub-annual series be equal to the corresponding benchmark. That is

$$a_m - \Sigma_{t=t_{1m}}^{t_{Lm}} j_{mt}\, \hat{\theta}_t = 0,\; m=1,...,M, \tag{1.2}$$

where $\hat{\theta}_t$ is the benchmarked series.

Classical benchmarking assumes that the annual benchmarks are fully reliable and therefore *binding*, i.e. they satisfy (1.2). A broader definition of benchmarking recognizes the fact the benchmarks may be observed with error. Such benchmarks are called *non-binding*, because the benchmarked series may not necessarily satisfy the constraints (1.2). As a result, benchmarking consists of optimally combining both the more frequent and less frequent data to increase their reliability. An improved set of annual values may then be obtained by taking the annual sums of the benchmarked series:

$$\hat{a}_m = \Sigma_{t=t_{1m}}^{t_{Lm}} j_{mt}\, \hat{\theta}_t,\; m=1,...,M. \tag{1.3}$$

We discuss extensively the additive and multiplicative versions of the regression model based method developed by Cholette and Dagum (1994) and the classical Denton (1971) method with its variants. It is shown that the latter is a particular case of the former method. We also present signal extraction and benchmarking methods based on ARIMA models (Hillmer and Trabelsi 1987) and structural time series models cast in state space form (Durbin and Quenneville 1997).

1.2 Interpolation and Temporal Distribution

Low frequency (annual) data are usually detailed and of high precision but not very timely. On the other hand, high frequency (sub-annual) data are both less detailed and precise but more timely. Indeed producing high frequency data at the same level of detail and precision would typically require more resources and impose a heavier response burden to businesses and individuals.

At the international level, the problem is compounded by the fact that different countries produce data at different frequencies which complicates comparisons. In some cases, different countries produce at the same frequency but with different timings, for example every fifth year but not in the *same* years. Moreover, in some cases, the data are irregularly spaced in time or displays gaps.

Current socio-economic and environmental analyses require an uninterrupted history of frequent and recent data about the variables of interest. Studies pertaining to long term phenomena, e.g. green house gazes and population growth, need a very long history of data. On the contrary, studies pertaining to fast growing leading edge industries, like bio-engineering, nano-technologies, require timely data but no more than two decades.

Researchers and academics often face situations of data with different frequencies and with temporal gaps. This complicates or impedes the development of a quarterly models (say) providing frequent and timely predictions. Benchmarking, interpolation and temporal distribution address these issues by estimating high frequency values from low frequency data and related high frequency data; and by providing frequent extrapolations where needed.

Interpolation and temporal distribution problem are related to each other and to benchmarking. Indeed, it is possible to produce (say) quarterly interpolations and temporal distributions from low frequency data specified as binding benchmarks. For example, benchmarking a quarterly *indicator series*, e.g. a seasonal pattern or proxy variable, to yearly data generates quarterly estimates. This practice is very widely used in statistical agencies. As a result, many of the concepts and notation used in benchmarking also apply to interpolation and temporal distribution.

The problem of interpolation usually arises in the context of stock series, where the yearly value should typically be equal to that of the fourth quarter or twelfth month. The purpose is to obtain estimates for the other quarters or months. This is a problem of producing missing values within the temporal range of the data.

On the contrary, the problem of temporal distribution, also referred to as temporal disaggregation, is usually associated with flow series, where low frequency annual data correspond to the annual sums of the corresponding higher frequency data. Temporal distribution is a standard practice in the system of National Accounts, because of the excessive cost of frequent data collection. High frequency data is therefore obtained from temporal distribution of the annual data. This process usually involves indicators which are available on a high frequency basis and deemed to behave like the target variable.

Extrapolation refers to the generation of values outside the temporal range of the data, whether the series under consideration is a flow or a stock. The extrapolation process may be backward or forward.

Despite the conceptual differences between interpolation, temporal distribution and extrapolation, they are considered similar from the viewpoint of regression-based estimation. In this regard, Chow and Lin (1971 p 374) themselves wrote:

> *" ... our methods treats interpolation and distribution identically.* [...] *This distinction is unjustified from the view point of the theory of estimation ...".*

Our experience with temporal distribution suggests that the distinction between interpolation and distribution breaks down in practice. Indeed, monthly values (*sic*) and yearly values are often specified as benchmarks for monthly flow series; and similarly with quarterly flow series. In fact, interpolation can also occur for flow variables. For example, immigration to a province may be measured every fifth year, and the intervening years or quarters need to be interpolated.

Several procedures have been developed for interpolation and temporal distribution. In this book, we discuss those methods more widely applied by statistical agencies The methods analysed can be grouped as (1) ad hoc, (2) the Cholette-Dagum regression-based methods, (3) the Chow-Lin method and its variants and (4) the ARIMA model-based methods. Methods (2) to (4) produce smooth values, whereas the ad hoc techniques generate kinks in the estimated series.

1.3 Signal Extraction and Benchmarking

In time series analysis, it is often assumed that the observations y_t are contaminated with errors e_t such that

$$y_t = \eta_t + e_t, \ t=1,...,n,$$

where η_t denotes the true un-observable values. Following from electrical engineering, η_t is called the "signal" and e_t the "noise". The signal extraction problem is to find the "best" estimates of η_t where best is usually defined as minimizing the mean square error.

Signal extraction can be made by means of parametric models or non-parametric procedures. The latter has a long standing and was used by actuaries at the beginning of the 1900's. The main assumption in non-parametric procedures is that η_t is a smooth function of time. The most common smoothers are the cubic splines originally applied by Whittaker (1923) and Whittaker and Robinson (1924) to smooth mortality tables. Other smoother are moving averages and high order kernels used in the context of seasonal adjustment and form the basis of methods such as Census X-11 (Shiskin et al. 1967), X-11-ARIMA (Dagum 1980 and 1988), X-12-ARIMA (Findley et al. 1998), STL (Cleveland et al. 1990).

On the other hand, signal extraction by means of explicit models arrived much later. Under the assumption that the entire realization of y_t is observed from $-\infty$ to ∞ and η_t and e_t are both mutually independent and stationary, Kolmogorov (1939 and 1941) and Wiener (1949) independently proved that the minimum mean square estimator of the signal η_t is the conditional mean given the observations y_t, that is $\hat{\eta}_t = E(\eta_t | y_t, y_{t-1}, ...)$. This fundamental result was extended by several authors who provided approximative solutions to the non-stationary signal extraction, particularly Hannan (1967), Sobel (1967) and Cleveland and Tiao (1976). Bell (1984) provided exact solutions for the conditional mean and conditonal variance of vector $\boldsymbol{\eta}$ when non-stationarity can be removed by applying differences of a finite order. Signal extraction is a data adjustment independent of benchmarking. When modelling is feasible, it provides a genuine covariance function to be used in the benchmarking process, which is a major advantage.

This book discusses two signal extraction and benchmarking methods, one due to Hillmer and Trabelsi (1987), which is based on ARIMA modelling; and the other due to Durbin and Quenneville (1997), based on structural time series modelling cast in state space form.

1.4 A Unified View

We present a generalized dynamic stochastic regression model developed by Dagum, Cholette and Chen in 1998, which encompasses the previous statistical adjustments: signal extraction, benchmarking, interpolation, temporal distribution and extrapolation. This dynamic model provides a unified regression-based framework for all the adjustments often treated separately. ARIMA model-based techniques, e.g. benchmarking with ARIMA model signal extraction by Hillmer and Trabelsi (1987), are shown to be special cases of the generalized regression model. Other benchmarking methods, e.g. those by Denton (1971) and its variants, are also included. Several interpolation and temporal distribution are shown to be particular cases of the generalized model, e.g. the Boot, Feibes and Lisman (1967) method, the Chow-Lin (1971, 1976) method and several of its variants. Restricted ARIMA forecasting methods are also comprised, e.g. Cholette (1982), Pankratz (1989), Guerrero (1989) Trabelsi and Hillmer (1989).

1.5 Calendarization

Time series data do not always coincide with calendar periods. To reduce response burden to surveys and censuses, statistical agencies often accept data with fiscal reporting periods. For example, the Canadian Federal and Provincial governments have a common fiscal year ranging from April 1 to March 31 of the following year; banks have a common fiscal year ranging from November 1 to October 31; School Boards and academic institutions, from September 1 to August 31 of the following year.

The timing and duration of a reported value covering fiscal or calendar periods depend on the starting and ending dates of the reporting period. The *duration* is the number of days in the reporting period. The *timing* can be broadly defined as the middle date of the reporting period.

In cases of homogenous fiscal years within an industry or sector, e.g. banking, taking the cross-sectional sums of the fiscal year data produces valid *tabulated* yearly values. These may be used as such in socio-economic analysis provided awareness of the timing of the data; and in benchmarking, as long as the coverage fractions reflect the correct duration and timing.

In many industries or sectors, however, the fiscal years are *not* homogeneous and vary from unit to unit (respondent). As a result, the timing and the duration of the reported values differ from unit to unit and even for a given unit. Hence, the fiscal data is not very useful. It is not possible to cross-sectionaly sum over the various fiscal years for a given calendar year because they do not have the same timing. One crude approach to calendariztion is the assignment procedure, which allocates fiscal data to a specific year (or other calendar period) according to an arbitrary rule. This technique systematically distorts the timing and duration of the estimates differently in different industries, and under-estimates the amplitude of business cycles. The technique mis-represent the targeted calendar periods, blurs causality between socio-economic variables and leads to tainted analysis.

Another ad hoc procedure widely applied is the fractional "method", which sets the calendarized values $\hat{a}_t$ equal to a weighted sum of the two overlapping fiscal values, e.g.

$$\hat{a}_{2003} = (1/4)\, a_{2002-03} + (3/4)\, a_{2003-04} ,$$

in the case of the Canadian Government fiscal year. The limitations of these two approaches are extensively discussed in Chapter 9.

Calendarization can also be seen as a benchmarking-interpolation problem. The approach consists of benchmarking a sub-annual series to fiscal data and of taking the temporal sums of the benchmarked series over the desired calendar periods (Cholette 1989). For example, a monthly series is benchmarked to fiscal year values and the calendar year values are set equal to the sums of the benchmarked series over calendar years. Casting calendarization as a benchmarking problem provides a more general framework to address the issue. Calendarization can be done using the Cholette-Dagum regression-based method, as well as the modified Denton versions.

In some cases, businesses report on a multi-weekly basis, i.e. every four or five weeks. The book discusses two ways to calendarize reported values of short durations: one by means of daily interpolations (temporal distributions) and the other by directly generating monthly interpolations.

1.6 Data Requirements for Benchmarking and Calendarization

The success of the methods discussed strongly depend on the availability of adequate data. Each benchmark must be dated exactly, by means of an actual starting date and of an actual ending date, both defined in terms of day, month and year, e.g. April 15 2003 and April 14 2004. This is especially critical in the case of calendarization. Because calendarization is a relatively new adjustment made by statistical agencies, the starting and ending dates of the reported data may be inexact or absent. Sometimes the only date on the record is that intended for the record by the survey design, for instance year 2004, instead of April 15 2003 and April 14 2004.

Temporal gaps in the benchmarks are acceptable, because the benchmarks may become available every second year, or every fifth year, or even irregularly. On the other hand, such gaps are unacceptable in the sub-annual series to be benchmarked. More specifically, the original monthly series (say) must provide observations for all the months (say) in the interval of months covered by the benchmarks. If the first benchmark has starting date equal to Jan. 15 1991 and the last benchmark has ending date equal to June 14 2005, the original unbenchmarked series must cover at least from Jan. 1 1991 to June 30 2005.

When new data become available, benchmarking must be applied to all the available data, namely the past un-benchmarked series and the corresponding benchmarks, followed by the current un-benchmarked series and the newly available benchmark(s). Benchmarking must never be applied to an already benchmarked series followed by a segment of new data. If this requirement is not met, the only applicable method is prorating, despite the resulting inter-annual discontinuities (steps).

In the case of administrative records, the starting dates may be absent from the file. It is possible to retrieve it from the previous record, by setting the starting date equal to previous ending date plus one day, under the assumption that no record is missing between the current and the previous records. In the case of third party administrative file, the choice of variables coded may be unsuited for the purpose of a statistical agency. Furthermore, the industrial classification may be absent or tentative. When a unit changes industrial classification, a "death record" (with reported values equal to zero and corresponding flags) should be inserted in the former class with appropriate reporting periods; and a "birth record", inserted in the new industrial class.

In the case of surveys, both the starting and ending dates must appear on the record, if calendarization is to be applied. Sometimes, the only date on record is that intended by the survey design, for example June 2005 when in fact the reported covered the 35 days from May 18 to June 21 2005.

The above requirements also apply to interpolation and temporal dis-aggregation.

1.7 Reconciliation or Balancing Systems of Time Series

The vast majority of time series data produced by statistical agencies are typically part of a system of series classified by attributes. For example, the Canadian Retail Trade series are classified by Trade Group (type of store) and Province. In such cases the series of the system must satisfy *cross-sectional aggregation constraints*. In many cases, each series must also temporally add up to its annual benchmarks; these requirements are referred to as *temporal aggregation constraints*.

The problem of reconciling systems of data has preoccupied economists and statisticians for a long time. Deming and Stephan (1940) suggested the Iterative Proportional Fitting approach as an approximation to their "least square" numerical optimisation, probably justified by the lack of computing power at the time. According to Byron (1978), van der Ploeg (1982) and Weale (1988), the method of least squares was initially suggested by Stone et al. (1942). A major problem proved to be the amount of computations involved. Hence for the purpose of balancing input-output tables, the RAS method developed by Bacharach (1965 1971) became very popular. However, it had the following shortcomings: no allowance for varying degrees of uncertainty about the initial estimates and the constraints, dubious economic interpretation of pro rata adjustments to the accounts, and a very large number of iterations before convergence.

The attention reverted to the least square approach thanks to the work of Byron (1978), following the approach proposed by Stone et al. (1942). Byron developed alternative procedures for minimizing a constrained quadratic loss function based on the conjugate gradient algorithm.

The feasibility of balancing a large set of accounting matrices using least squares was demonstrated by van der Ploeg (1982) and Barker et al. (1984). Weale (1988) extended the approach to deal with the simultaneous adjustment of values, volumes and prices, by introducing non-linear constraints linearly approximated by means of logarithmic transformations.

Byron, van der Ploeg, Weale and others, observed that the balanced estimates based on the least squares approach are best linear unbiased. In his appendix, van der Ploeg (1982) claimed that

> *"the optimal* [least squares] *algorithm is not necessarily more expensive than the RAS approach " and that "the advantages of a more general treatment, allowing for uncertain row and column totals and relative degrees of reliability, are easily obtained ".*

Theil (1967), Kapur et al. (1992) and Golan et al. (1994, 1996) proposed methods based on maximum entropy. MacGill (1977) showed that the entropy and the RAS methods converge to the same solution.

In this book, we present models for one-way and two-way classified systems of series; and for the marginal two-way model, which involves only the row and column totals of the two-way model.

1.8 Book Outline

This book contains fourteen chapters and five appendices. Chapter 1 gives a general introduction of the topics discussed. Chapter 2 deals with time series decomposition models and the concepts of latent components which affect time series dynamics, namely the trend, business cycle, seasonality, trading-day variations, moving-holiday effects, irregulars and outliers. This chapter is useful to readers not familiarized with time series analysis for all the statistical adjustments made to raw data inevitably affect the time series components.

Chapters 3, 4, 5, 6 and 8 are exclusively devoted to benchmarking. The first three discuss the additive and multiplicative regression-based model developed by Cholette and Dagum (1994) widely applied by statistical agencies. Chapter 6 deals with the original classical Denton (1971) benchmarking method and all its variants. Chapter 8 presents two signal extraction benchmarking methods, one based on ARIMA modelling developed by Hillmer and Trabelsi (1987) and the other based on structural time series models cast in state space form, developed by Durbin and Quenneville (1997).

Chapter 7 focuses on interpolation, temporal distribution and extrapolation. Some ad hoc procedures are presented together with other more formal methods: the regression-model based model, ARIMA model based and Chow-Lin and its dynamic extensions.

Chapter 9 examines the problem of calendarization, which basically consists of transforming fiscal year data into calendar year estimates, fiscal quarter data into calendar quarter estimates and multi-weekly data into calendar month estimates.

Chapter 10 completes the various statistical adjustments made to single time series with a unified regression-based framework for signal extraction, benchmarking and interpolation.

The last chapters 11 to 14 are devoted to the problem of reconciliation or balancing of systems of time series. These chapters provide analytical solutions which enable implementation for large systems of series.

Chapter 11 introduces the problem and discusses the Iterative Proportional Fitting approach and a general regression-based method.

Chapter 12 deals with one-way classified time series systems and presents different cases often encountered in real life situations.

Chapter 13 focusses on the marginal totals of two-way classified systems. The marginal two-way model enables the earlier publication of the more important marginal totals, and the publication of the elementary series at a later date.

Finally, Chapter 14 presents the reconciliation model for two-way classified systems of time series and a variant applicable to large Input-Output models.

Appendices A, B and C contain proofs of important theorems and a number of matrix algebra results, not readily accessible. Appendix D and E respectively describe various kinds of seasonal and trading-day regressors to be used in the context of the regression-based models.

Each chapter has been written as self-contained as possible, to accommodate readers interested on specific adjustments. To achieve such a goal, a few key formulae and equations are repeated in some of the chapters. Each chapter illustrates the methods discussed with a large number of real life data examples. Furthermore, a real data series, the Canada Total Retail Trade, is used throughout the book to illustrate the adjustments made and facilitate comparisons among the various statistical adjustment procedures.

The basic prerequisites to understand the book are a good knowledge of matrix algebra, regression analysis and some acquaintance with ARIMA modelling.

2 The Components of Time Series

2.1 Introduction

This chapter is intended as a general introduction to time series analysis. A time series consists of a set of observations order in time, on a given phenomenon (target variable). Usually the measurements are equally spaced, e.g. by year, quarter, month, week, day. The most important property of a time series is that the ordered observations are dependent through time, and the nature of this dependence is of interest in itself.

Formally, a time series is defined as a set of random variables indexed in time, $\{X_1,...,X_T\}$. In this regard, an observed time series is denoted by $\{x_1,...,x_T\}$, where the sub-index indicates the time to which the observation x_t pertains. The first observed value x_1 can be interpreted as the realization of the random variable X_1, which can also be written as $X(t=1,\omega)$ where ω denotes the event belonging to the sample space. Similarly, x_2 is the realization of X_2 and so on. The T-dimensional vector of random variable can be characterized by different probability distribution.

For socio-economic time series the probability space is continuous, and the time measurements are discrete. The frequency of measurements is said to be high when it is daily, weekly or monthly and to be low when the observations are quarterly or yearly.

The target variable or phenomenon under question can be classified as flow or stock.

Flow variables are associated with the speed of a phenomenon, for example: sales *per* month; immigration, emigration, births, deaths *per* year; government deficit or surplus *per* year. As a consequence, the yearly values of flows (e.g. sales) correspond to the yearly sums of monthly values in the year, and the quarterly values correspond to the sum of the monthly values in the quarter, etc.[1]

Stock variables, on the other hand, pertains to the level of a phenomenon *at* a very specific date (e.g. 12:00pm Dec 31 2001). Example of stock

[1] Instead of kilometres per hour, we have dollars, persons, etc. per month, quarter, year and so forth. A motorist driving at one kilometre per minute covers 60 kilometres in one hour.

variables are inventories, assets, liabilities, prices, population, employment, unemployment, the national debt.

The distinction between flow and stock variables becomes more clear if we look at the relationship between them. Indeed, some flows contribute to stocks, for example, purchases of a good by a store increase its inventories, whereas sales decrease them. Similarly, government deficits add to the national debt of a country, and surpluses subtract from it. Births and immigration add to the stock of population, whereas deaths and emigration subtract from it.

2.2 Time Series Decomposition Models

An important objective in time series analysis is the decomposition of a series into a set of non-observable (latent) components that can be associated to different types of temporal variations. The idea of time series decomposition is very old and was used for the calculation of planetary orbits by seventeenth century astronomers. Persons (1919) was the first to state explicitly the assumptions of unobserved components. As Persons saw it, time series were composed of four types of fluctuations:

(1) A long-term tendency or secular trend.
(2) Cyclical movements super-imposed upon the long-term trend. These cycles appear to reach their peaks during periods of industrial prosperity and their troughs during periods of depressions, their rise and fall constituting the business cycle.
(3) A seasonal movement within each year, the shape of which depends on the nature of the series.
(4) Residual variations due to changes impacting individual variables or other major events such as wars and national catastrophes affecting a number of variables.

Traditionally, the four variations have been assumed to be mutually independent from one another and specified by means of an additive decomposition model:

$$X_t = T_t + C_t + S_t + I_t \,, \tag{2.1}$$

where X_t denotes the observed series, T_t the long-term trend, C_t the cycle, S_t seasonality and I_t the irregulars.

If there is dependence among the latent components, this relationship is specified through a multiplicative model

$$X_t = T_t \times C_t \times S_t \times I_t \,, \tag{2.2}$$

where now S_t and I_t are expressed in proportion to the trend-cycle $T_t \times C_t$.
In some cases, mixed additive-multiplicative models are used.

Whether a latent component is present or not in a given time series depends on the nature of the phenomenon and on the frequency of measurement. For example, seasonality is due to the fact that some months or quarters of a year are more important in terms of activity or level. Because this component is specified to cancel out over 12 consecutive months or 4 consecutive quarters, or more generally over 365.25 consecutive days, yearly series cannot contain seasonality.

Flow series can be affected by other variations which are associated to the composition of the calendar. The most important are the trading-day variations, which are due to the fact that some days of the week are more important than others. Months with five of the more important days register an excess of activity (*ceteris paribus*) in comparison to months with four such days. Conversely, months with five of the less important days register a short-fall of activity. The length-of-month variation is usually assigned to the seasonal component. The trading-day component is usually considered as negligible in quarterly series and even more so in yearly data.

Another important calendar variation is the *moving-holiday* or *moving-festival* component. That component is associated to holidays which change date from year to year, e.g. Easter and Labour Day, causing a displacement of activity from one month to the previous or the following month. For example, an early date of Easter in March or early April can cause an important excess of activity in March and a corresponding short-fall in April, in variables associated to imports, exports and tourism.

Under models (2.1) and (2.2), the trading-day and moving festival components (if present) are implicitly part of the irregular. In 1965, Young developed a procedure to estimate trading-day variations which was incorporated in the X-11 seasonal adjustment method (Shiskin et al. 1967) and its subsequent versions, the X-11-ARIMA (Dagum 1980 and 1988a) and X-12-ARIMA (Findley et al. 1998) methods. The later two versions also include models to estimate moving-holidays due to Easter.

Considering these new components, the additive decomposition model becomes

$$X_t = T_t + C_t + S_t + D_t + H_t + I_t\,, \tag{2.3}$$

where D_t and H_t respectively denote the trading-day and moving-holiday components. Similarly, the multiplicative decomposition model becomes

$$X_t = T_t \times C_t \times S_t \times D_t \times H_t \times I_t\,, \tag{2.4}$$

where the components S_t, D_t, H_t and I_t are proportional to the trend-cycle $T_t \times C_t$.

Decomposition models (2.3) and (2.4) are traditionally used by seasonal adjustment methods. Seasonal adjustment actually entails the estimation of all the time series components and the removal of seasonality, trading-day and holiday effects from the observed series. The rationale is that these components which are relatively predictable conceal the current stage of the business cycle which is critical for policy and decision making.

There is another time series decomposition often used for univariate time series modelling and forecasting:

$$X_t = \eta_t + e_t\,, \tag{2.5}$$

where η_t and e_t are referred to as the "signal" and the "noise", according to the electrical engineering terminology. The signal η_t comprises all the systematic components of models (2.1) to (2.4), i.e. $T_t \times C_t$, S_t, D_t and H_t.

Model (2.5) is classical in "signal extraction" where the problem is to find the "best" estimates of the signal $\{\eta_t\}$ given the observations $\{x_t\}$ corrupted by noise $\{e_t\}$. The "best" estimates are usually defined as minimizing the mean square error.

Finally, given its fundamental role in time series modelling, we summarize the famous decomposition theorem due to Wold (1938). Wold proved that a any second-order stationary stochastic process $\{X_t\}$ can be decomposed in two mutually uncorrelated processes $\{Z_t\}$ and $\{V_t\}$, such that

$$X_t = Z_t + V_t\,, \tag{2.6a}$$

where

$$Z_t = \Sigma_{j=0}^{\infty} \psi_j\, a_{t-j},\;\; \psi_0 \equiv 1,\;\; \Sigma_{j=1}^{\infty} \psi_j^2 < \infty, \tag{2.6b}$$

with $\{a_t\} \sim WN(0, \sigma_a^2)$.

Component $\{Z_t\}$ is a convergent infinite linear combination of the a_t's, assumed to follow a white noise (WN) process of zero mean, constant variance σ_a^2 and zero autocovariance. Model (2.6b) is known as an infinite moving average $MA(\infty)$ and the a_t's are the "innovations". The component $\{Z_t\}$ is called the non-deterministic or purely linear component since only one realization of the process is not sufficient to determine future values $Z_{t+\ell}$, $\ell > 0$, without error.

Component $\{V_t\}$ can be represented by

$$V_t = \mu + \Sigma_{j=1}^{\infty} [\alpha_j \sin(\lambda_j t) + \beta_j \cos(\lambda_j t)], \ -\pi < \lambda_j < \pi , \qquad (2.6c)$$

where μ is the constant mean of process $\{X_t\}$ and $\{\alpha_j\}$, $\{\beta_j\}$ are mutually uncorrelated white noise processes. The series $\{V_t\}$ is called the deterministic part because it can be predicted in the future without error from a single realization of the process by means of an infinite linear combination of past values.

Wold's theorem demonstrates that the property of stationarity is strongly related to that of linearity. It provides a justification for autoregressive moving average (ARMA) models (Box and Jenkins 1970) and some extensions, such as the autoregressive integrated moving average (ARIMA) and regression-ARIMA models (RegARIMA).

A stochastic process $\{X_t\}$ is second-order stationary or weakly stationary, if the first two moments are not time dependent, that is, the mean and the variance are constant, and the autocovariance function depends only on the time *lag* and not the time origin:

$$E(X_t) = \mu < \infty, \qquad (2.7.a)$$

$$E(X_t - \mu)^2 = \sigma_X^2 < \infty, \ E[(X_t - \mu)(X_{t-k} - \mu)] = \gamma(k) < \infty, \qquad (2.7b)$$

where $k = 0, 1, 2, \ldots$ denotes the time lag.

2.3 The Secular or Long-Term Trend

The concept of trend is used in economics and other sciences to represent long-term smooth variations. The causes of these variations are often associated with structural phenomena such as population growth, technological progress, capital accumulation, new practices of business and economic organization. For most economic time series, the trends evolve smoothly and gradually, whether in a deterministic or stochastic manner. When there is sudden change of level and/or slope this is referred to as a structural change. It should be noticed however that series at a higher levels of aggregation are less susceptible to structural changes. For example, a technological change is more likely to produce a structural change for some firms than for the whole industry.

The identification and estimation of the secular or long-term trend have posed serious challenges to statisticians. The problem is not of statistical or mathematical character but originates from the fact that the trend is a latent (non-observable) component and its definition as a long-term smooth movement is statistically vague. The concept of "long-period" is relative, since a trend estimated for a given series may turn out to be just a long business cycle as more years of data become available. To avoid this problem statisticians have used two simple solutions. One is to estimate the trend and the business cycles, calling it the trend-cycle. The other solution is to estimate the trend over the whole series, and to refer to it as the longest non-periodic variation.

It should be kept in mind that many systems of time series are redefined every fifteen years or so in order to maintain relevance. Hence, the concept of long-term trend loses importance. For example the system of Retail and Wholesale Trade series was redefined in 1989 to adopt the 1980 Standard Industrial Classification (SIC), and in 2003 to conform to the North American Industrial Classification System (NAICS), following the North American Free Trade Agreement. The following examples illustrate the necessity of such reclassifications. The 1970 Standard Industrial Classification (SIC) considered computers as business machines, e.g. cash registers, desk calculators. The 1980 SIC rectified the situation by creating a class for computers and other goods and services. The last few decades witnessed the birth of new industries involved in photonics (lasers), bio-engineering, nano-technology, electronic commerce. In the process, new professions emerged, and Classification systems had to keep up with these new realities.

There is a large number of deterministic and stochastic models which have been proposed for trend estimation (see Dagum and Dagum 1988). Deterministic models are based on the assumption that the trend can be well approximated by mathematical functions of time such as polynomials of low degree, cubic splines, logistic functions, Gompertz curves, modified exponentials. Stochastic trends models assume that the trend can be better modelled by differences of low order together with autoregressive and moving average errors.

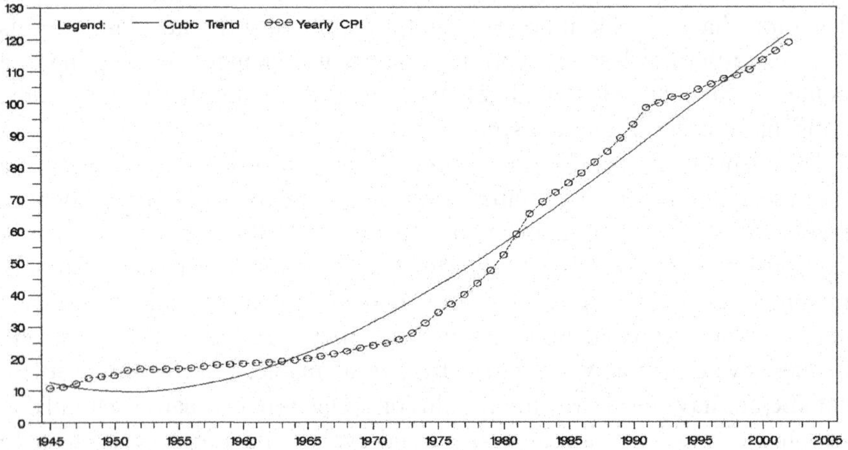

Fig. 2.1. Cubic trend fitted to the yearly averages of the Canadian Consumer Price Index

2.3.1 Deterministic Trend Models

The most common representation of a deterministic trend is by means of polynomial functions. The observed time series is assumed to have a deterministic non-stationary mean, i.e. a mean dependent on time. A classical model is the regression error model where the observed data is treated as the sum of the trend and a random component such that,

$$Y_t = T_t + u_t , \tag{2.8a}$$

where T_t denotes the trend and u_t is assumed to follow a stationary process, often white noise. The polynomial trend can be written as

$$T_t = \alpha_0 + \alpha_1 t + +\ldots + \alpha_n t^n , \tag{2.8b}$$

where generally $n \leq 3$. The trend is said to be of a deterministic character because the observed series is affected by random shocks which are assumed to be uncorrelated with the systematic part.

Fig 2.1 show a cubic trend fitted to the yearly averages of the Canadian Consumer Price Index.

Besides polynomial of time, three very well known growth functions have been widely applied in population and economic studies, namely the modified exponential, the Gompertz and the logistic. Historically, the first growth model for time series was proposed by Malthus (1798) in the context of population growth. He stated two time path processes, one for the supply of food and the other for population. According to Malthus, the supply of food followed an arithmetic progression and population a geometric one. Mathematically, the food supply is represented by a non-negative and increasing linear function of time t, that is

$$F_t = a + bt, \quad a>0, \; b>0, \tag{2.9}$$

at time $t=0$ the starting level of the food supply is equal to a.

For the population growth P_t, the model is

$$(dP_t / dt) / P_t = c, \quad c>0, \tag{2.10}$$

which means a constant *relative* rate of growth. Solving the differential equation (2.10) leads to an exponential trend

$$P_t = P_0 \, e^{ct}, \quad P_0>0, \; c>0, \tag{2.11}$$

The modified exponential trend is given by

$$P_t = a + P_0 \, e^{ct}, \; P_0>0, \; c>0, \tag{2.12}$$

where a is a real number. During periods of sustained growth or rapid inflation, several economic variables can be well approximated by models (2.11) and (2.12). But in the long run, socio-economic and demographic phenomena are subject to obstacles which slow their time paths.

Pearl and Reid (1920) observed that the relative rate of population growth was not constant but a decreasing function of the size of the population. Verhulst (1838) seem to have been the first to formalize it by deducing the logistic model given by

$$P_t = (b/a)(1 + e^{-a(t-\bar{t})})^{-1},\ 0<a<1,\ b>0, \tag{2.13}$$

where parameter a is a constant of integration. Model (2.13) is the solution of the differential equation,

$$(dP_t / dt)/P_t = a - bP_t,\ a>0,\ b>0.$$

The logistic function (2.13) is a non-negative and non-decreasing function of time. The logistic model belongs to the S-shaped (saturation) curves and often used to represent the time behaviour of new electronic goods entering the market.

The function (2.13) ranges from 0 to b/a;. The function reaches both its mid-course and inflection point at $t=\bar{t}$, where $P_t = (b/a)/2$. Parameter a controls the speed of the process: the larger the value of a, the faster the course from 0 to b/a about $t=\bar{t}$. The change ΔP_t is positive and symmetric about $t=\bar{t}$, i.e. $\Delta P_{\bar{t}-k} = \Delta P_{\bar{t}+k}$; ΔP_t reaches its maximum (inflection point) at $t=\bar{t}$ and declines afterward.

Model (2.13) can be extended by introducing two more parameters

$$P_t = c + (b/a)(1 + e^{d-a(t-\bar{t})})^{-1},\ a>0,\ b>0, \tag{2.14}$$

where parameter c controls the starting level and d control the timing of the inflection point. The function ranges from c to $c+(b/a)$. If $d=0$, the function reaches both its mid-course and inflection point at $t=\bar{t}$, where $P_t = c + (b/a)/2$. Paramater d moves that point to $t=\bar{t}-d/a$.

Fig. 2.2 displays the logistic trend of Eq. (2.13) with parameters $b=20$ and $a=0.10$; its inflection point occurs at $t=\bar{t}=60$. The figure also shows the modified logistic of Eq. (2.14) with parameters $b=15$, $a=0.10$, $c=50$ and $d=-2$. Parameter $c=50$ causes the curve to start at level 50, and parameter $d=-2$ moves the inflection point to $t=40$,

A good example of logistic trend is given by the proportion of households with television sets, which grew slowly when television was introduced in the 1950s, then boomed in the 1960s, and eventually levelled off. Another example is the participation rate of women to the labour force. In the first part of the 20-th century women had a relatively low participation rate to the labour force; working women consisted mainly of young women, teachers and nurses. Their participation rate grew rapidly in the 1960s and 70s, to reach a saturation level comparable to that of men in the 1980s.

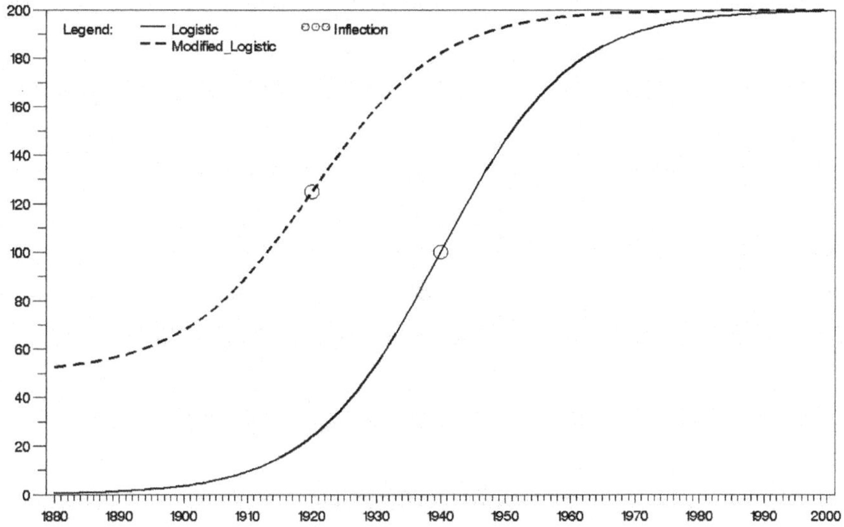

Fig. 2.2. Examples of logistic trends with and without starting level and timing parameters (*c* and *d*)

2.3.2 Stochastic Trends

Stochastic models are appropriate when the trend is assumed to follow a non-stationary stochastic process. The non-stationarity is modelled with finite differences of low order (cf. Harvey 1985, Maravall 1993).

A typical stochastic trend model often used in structural time series modelling, is the so-called random walk with constant drift. In the classical notation the model is

$$\mu_t = \mu_{t-1} + \beta + \xi_t,\ t=1,2,...n;\ \xi_t \sim N(0,\sigma_\xi^2)\ , \tag{2.15a}$$

$$\Delta\mu_t = \beta + \xi_t\ ,$$

where μ_t denotes the trend, β a constant drift and $\{\xi_t\}$ is a normal white noise process. Solving the difference equation (2.15a) and assuming $\xi_0 = 0$, we obtain

$$\mu_t = \beta t + \Delta^{-1}\xi_t \quad = \beta t + \Sigma_{j=0}^{\infty}\xi_{t-j}\,,\ t=1,...,n, \tag{2.15b}$$

which show that a random walk with constant drift consists of a linear deterministic trend plus a non-stationary infinite moving average.

Another type of stochastic trend belongs to the ARIMA (p,d,q) class, where p is the order of the autoregressive polynomial, q is the order of the moving average polynomial and d the order of the finite difference operator $\Delta = (1-B)$. The backshift operator B is such that $B^n z_t \equiv z_{t-n}$. The ARIMA (p,d,q) model is written as

$$\varphi_p(B)(1-B)^d z_t = \theta_q(B)\, a_t,\ a_t \sim N(0,\sigma_a^2), \tag{2.16}$$

where z_t now denotes the trend, $\varphi_p(B)$ the autoregressive polynomial in B of order p, $\theta_q(B)$ stands for the moving average polynomial in B of order q, and $\{a_t\}$ denotes the innovations assumed to follow a normal white noise process. For example, with p=1, d=2, q=0, model (2.16) becomes

$$(1-\varphi_1 B)(1-B)^2 z_t = a_t\ , \tag{2.17}$$

which means that after applying first order differences twice, the transformed series can be modelled by an autoregressive process of order one.

2.4 The Business Cycle

The business cycle is a quasi-periodic oscillation characterized by periods of expansion and contraction of the economy, lasting on average from three to five years. Because most time series are too short for the identification of a trend, the cycle and the trend are estimated jointly and referred to as the trend-cycle. As a result the concept of trend loses importance. The trend-cycle is considered a fundamental component, reflecting the underlying socio-economic conditions, as opposed to seasonal, trading-day and transient irregular fluctuations.

The proper identification of cycles in the economy requires a definition of contraction and expansion. The definition used in capitalistic countries to produce the chronology of cycles is based on fluctuations found in the aggregate economic activity. A cycle consists of an expansion phase simultaneously present in many economic activities, followed by a recession phase and by a recovery which develops into the next expansion phase. This sequence is recurrent but not strictly periodic. Business cycles vary in intensity and duration. In Canada for example, the 1981 recession was very acute but of short duration, whereas the 1991 recession was mild and of long duration. Business cycles can be as short as 13 months and as long as 10 years.

A turning point is called a *peak* or *downturn* when the next estimate of the trend-cycle indicates a decline in the level of activity; and a *trough* in the opposite situation.

According to the literature (e.g. Dagum and Luati 2000, Chhab et al. 1999, Zellner et al. 1991) a downturn is deemed to occur at time t in the trend-cycle of monthly series, if

$$c_{t-3} \leq \ldots \leq c_{t-1} > c_t \geq c_{t+1}; \tag{2.18a}$$

and an upturn, if

$$c_{t-3} \geq \ldots \geq c_{t-1} < c_t \leq c_{t+1}. \tag{2.18b}$$

Thus a single change to a lower level c_t, between $t-1$ and t, qualifies as a downturn, if $c_{t+1} \leq c_t$ and $c_{t-3} \leq c_{t-2} \leq c_{t-1}$; and conversely for an upturn.

The dating of downturns and upturns is based on a set of economic variables related to production, employment, income, trade and so on.

2.4.1 Deterministic and Stochastic Models for the Business Cycle

Similarly to the trend, the models for cyclical variations can be deterministic or stochastic. Deterministic models often consist of sine and cosine functions of different amplitude and periodicities. For example, denoting the cycle by c_t, a deterministic model is

$$c_t = \Sigma_{j=1}^{2} \, [\alpha_j \, cos(\lambda_j t) + \beta_j \, sin(\lambda_j t)], \tag{2.19}$$

where $\lambda_1 = 2\pi/60$ and $\lambda_2 = 2\pi/40$. Model (2.19) takes into consideration two dominant cycles found in the European and American economies, those of 60 and 40 months respectively.

Stochastic models of the ARIMA type, involving autoregressive models of order 2 with complex roots, have also been used to model the trend-cycle. For example,

$$c_t = \varphi_1 c_{t-1} + \varphi_2 c_{t-2} + a_t, \;\; a_t \sim N(0, \sigma^2), \tag{2.20}$$

where c_t denotes the cycle, a_t is assumed Normal white noise, and the following conditions apply to the parameters: $\varphi_1 + \varphi_2 < 1$, $\varphi_2 - \varphi_1 < 1$ and $-1 < \varphi_2 < 0$ (see Box and Jenkins 1970).

2.4.2 Limitations of Same-Month Comparisons

In the absence of seasonal adjustment, only the raw series is available. In such cases, it is customary to use same-month comparisons from year to year, $z_t - z_{t-12}$, to assess the phase of the business cycle. The rationale is that the seasonal effect in z_t is approximately the same as in z_{t-12}, under the assumption of slowly evolving seasonality. Same-month year ago comparisons can be expressed as the sum of the changes in the raw series between z_t and z_{t-12},

$$z_t - z_{t-12} \equiv (z_t - z_{t-1}) + (z_{t-1} - z_{t-2}) + (z_{t-2} - z_{t-3}) + \\ \dots + (z_{t-11} - z_{t-12}) \equiv \Sigma_{j=1}^{12} (z_{t-j+1} - z_{t-j}). \quad (2.21)$$

Eq. (2.21) shows that same-month comparison display an increase if increases dominate decreases over the 13 months involved, and conversely. The timing of $z_t - z_{t-12}$ is $t-6$, the average of t and $t-12$. This points out a limitation of this practise: the diagnosis provided is not timely with respect to t. Furthermore, z_t and z_{t-12} contain irregular variations which may affect one observation positively and the other negatively, hence conveying instability to the comparison. Moreover, for flow data the comparison is systematically distorted by trading-day variations if present.

As already mentioned, seasonal adjustment entails the removal of seasonality, trading-day variations and moving-holiday effects from the raw data, to produce the seasonally adjusted series which consists of the trend-cycle and the irregular components. The irregular fluctuations in the seasonally adjusted series can be reduced by smoothing, to isolate the trend-cycle and to enable month-to-month comparisons.

The official inflation rate in Canada is based on same-month comparisons of the Consumer Price index, probably for historical reasons and international comparisons. However, Statistics Canada does provide an analysis of current trend-cycle movements and projection thereof to complement the official rate.

Fig. 2.3a displays the raw Unemployment series along with the trend-cycle and seasonally adjusted estimates. In order to illustrate same-month comparisons, note that the raw Unemployment declined between Dec. 1978 and Dec. 1979 for instance. On the other hand, the trend-cycle over the last months of 1979 was clearly rising. The difference in diagnosis results from the timing: $t-6$ (June 1979) for $z_t - z_{t-12}$ and $t-0.5$ for $c_t - c_{t-1}$.

At the end of the series, the raw Unemployment series rose by about 50% between Dec. 1981 and Dec. 1982, which was a staggering and unprecedented growth in such a short time. On that basis, most observers anticipated even higher levels for 1983. By looking at the trend-cycle values however, there are signs that Unemployment is slowing down between Nov. and Dec. 1982. This suggests that unemployment *could* stabilize in early 1983. As shown by Fig. 2.3b, Unemployement did stabilize and started to decline in 1983. However, it is important to keep in mind that the most recent trend-cycle estimates are often subject to revisions. Hence any proper prognosis would require one or two more observations. Fig. 2.3b also illustrates that business cycles vary in length and amplitude.

Fig. 2.3c displays the trend-cycle of Sales by Canadian Department Stores, which are less sensitive to the current economic situation than Unemployment exhibited in Fig. 2.3b. Indeed, in the short run the unemployed continue spending from their unemployment benefits and savings.

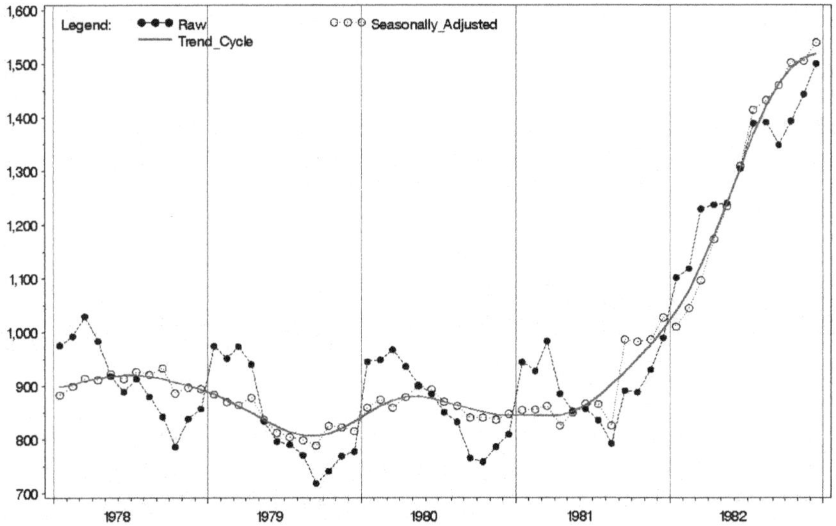

Fig. 2.3a. Observed raw Unemployment in Canada and corresponding seasonally adjusted series and trend-cycle estimate from 1978 to 1982

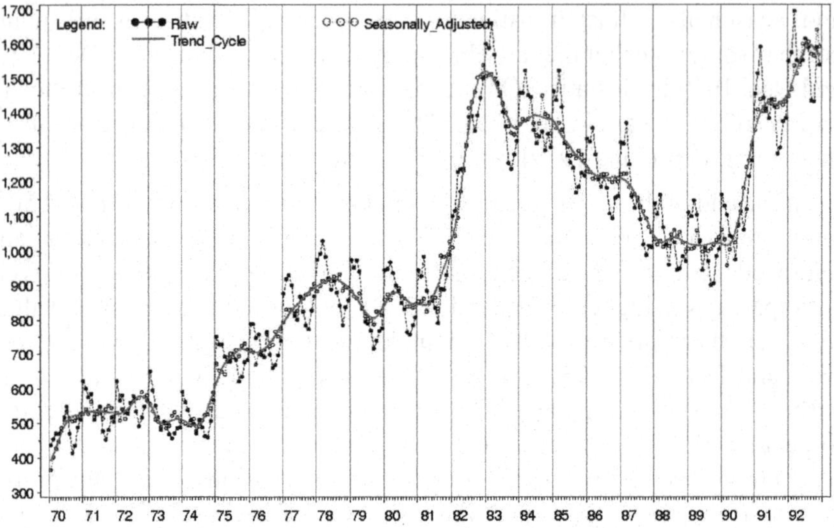

Fig. 2.3b. Observed raw Unemployment in Canada and corresponding seasonally adjusted series and trend-cycle estimates from 1970 to 1992

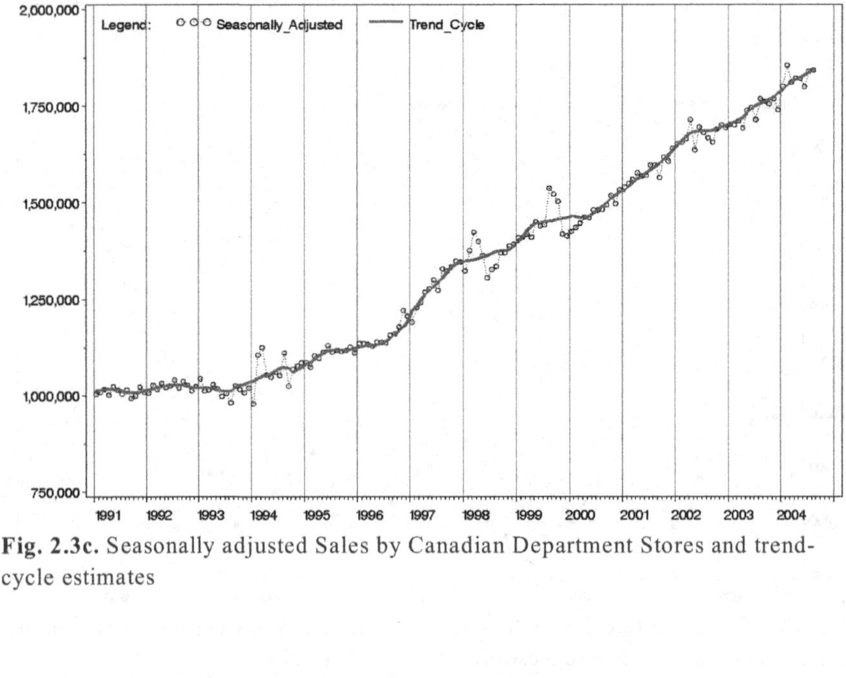

Fig. 2.3c. Seasonally adjusted Sales by Canadian Department Stores and trend-cycle estimates

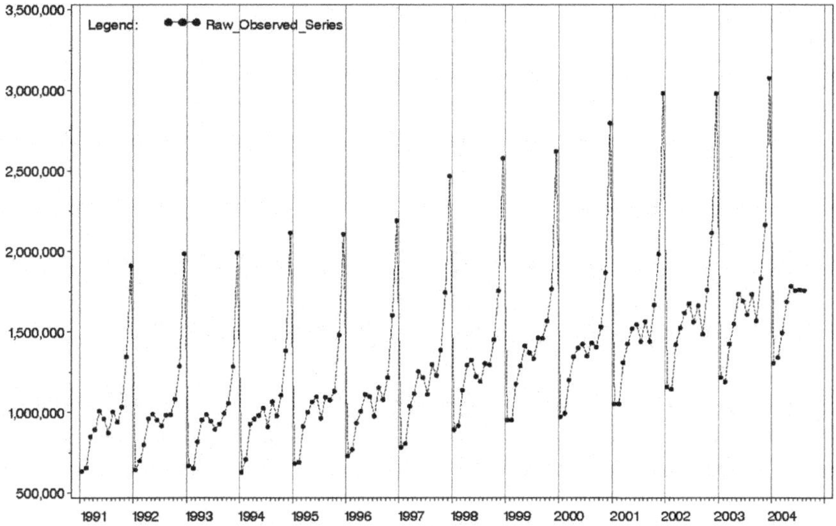

Fig. 2.4a. Raw Sales by Canadian Department Stores

2.5 Seasonality

Time series of sub-yearly observations, e.g. monthly, quarterly, weekly, are often affected by seasonal variations. The presence of such variations in socio-economic activities has been recognized for a long time. Indeed seasonality usually accounts for most of the total variation within the year.

Seasonality is due to the fact that some months, quarters of the year are more important in terms of activity or level. For example, the level of unemployment is generally higher during the winter and spring months and lower in the other months. Yearly series cannot contain seasonality, because the component is specified to cancel out over 12 consecutive months or 4 consecutive quarters.

2.5.1 The Causes and Costs of Seasonality

Seasonality originates from climate and conventional seasons, like religious, social and civic events, which repeat from year to year.

The climatic seasons influence trade, agriculture, the consumption patterns of energy, fishing, mining and related activities. For example, in North America the consumption of heating oil increases in winter, and the consumption of electricity increases in the summer months because of air conditioning.

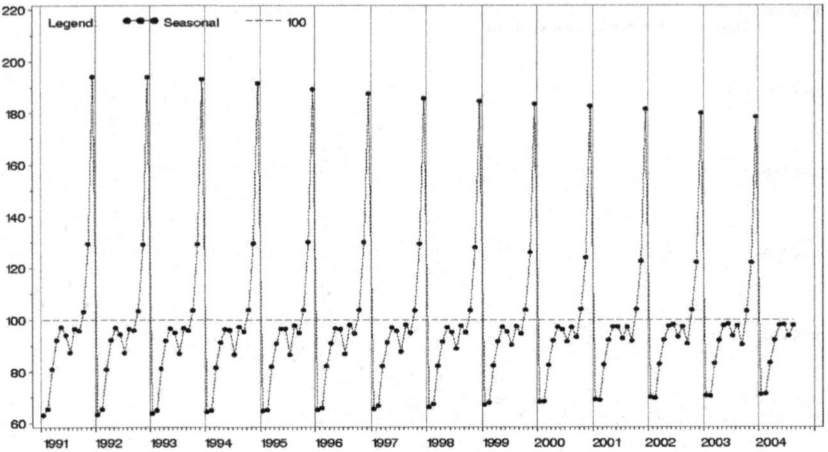

Fig. 2.4b. Seasonal component of Sales by Canadian Department Stores

Institutional seasons like Christmas, Easter, civic holidays, the school and academic year have a large impact on retail trade and on the consumption of certain goods and services, namely travel by plane, hotel occupancy, consumption of gasoline.

In order to determine whether a series contains seasonality, it is sufficient to identify at least one month (or quarter) which tends to be systematically higher or lower than other months. For example, the sales by Canadian Department Stores in Fig. 2.4a are much larger in December and much lower in January and February with respect to other months.

Fig. 2.4b exhibits the seasonal pattern of the series. The seasonal pattern measures the relative importance of the months of the year. The constant 100% represents an average month or a non-seasonal month. The *peak* month is December, with sales almost 100% larger than on an average month; the *trough* months are January and February with sales almost 40% lower than on an average month. The *seasonal amplitude*, the difference between the peak and trough months of the seasonal pattern, reaches almost 140%.

Seasonality entails large costs to society and businesses. One cost is the necessity to build warehouses to store inventories of goods to be sold as consumers require them, for example grain elevators. Another cost is the under-use and over-use of the factors of production: capital and labour.

Capital in the form of un-used equipment, buildings and land during part of the year has to be financed regardless. For example, this is the case in farming, food processing, tourism, electrical generation, accounting. The cold climate increases the cost of buildings and infrastructure, e.g. roads, transportation systems, water and sewage systems, schools, hospitals; not to mention the damage to the same caused by the action of ice.

The labour force is over-used during the peak seasons of agriculture and construction for example; and, under-used in trough seasons sometimes leading to social problems.

A more subtle unwanted effect is that seasonality complicates business decisions by concealing the fundamental trend-cycle movement of the variables of interest.

The four main causes of seasonality are attributed to the weather, composition of the calendar, major institutional deadlines and expectations. Seasonality is largely exogenous to the economic system but can be partially offset by human intervention. For example, seasonality in money supply can be controlled by central bank decisions on interest rates. In other cases, the effects can be offset by international and inter-regional trade. For example Hydro Québec, a major Canadian electrical supplier, sells much of it excess power during the summer seasonal trough months to the neighbouring Canadian provinces and U.S. states; and imports some of it during the winter seasonal peak months of electrical consumption in Québec. The scarcity of fresh fruits and vegetables in Canada is handled in a similar manner. Some workers and businesses manage their seasonal pattern with complementary occupations: for example landscaping in the summer and snow removal in winter.

To some extent seasonality can evolve through technological and institutional changes. For example the developments of appropriate construction materials and techniques made it possible to continue building in winter. The development of new crops, which better resist cold and dry weather, have influenced the seasonal pattern. The partial or total replacement of some crops by chemical substitutes, e.g. substitute of sugar, vanilla and other flavours, reduces seasonality in the economy.

As for institutional change, the extension of the academic year to the summer months in the 1970s affected the seasonal pattern of unemployment for the population of 15 to 25 years of age. Similarly the practice of spreading holidays over the whole year impacted on seasonality.

In the 1970s, December rapidly became a peak month for marriages in Canada. This was surprising because the month had always be a low month since the 1930s. Subject matter research determined that the Canadian law on income tax allowed the deduction of the spouse as a dependent for the whole year, even if the marriage had taken place in the last days of the year. In the 1980s, the Canadian government terminated that fiscal loophole, which resulted in a dramatic drop of marriages in December. The drop was so sudden that the series became difficult to seasonally adjust in the early 1980s. Indeed, seasonal adjustment methods assume gradual seasonal evolution. The same situation occurred in the United Kingdom. This kind of abrupt change in seasonal pattern is exceptional.

The changing industrial mix of an economy also transforms the seasonal pattern, because some industries are more seasonal than others. In particular, economies which diversify and depend less on "primary" industries (e.g. fishing, agriculture) typically become less seasonal.

In most situations, seasonality evolves slowly and gradually. Indeed the seasonal pattern basically repeats from year to year, as illustrated in Fig. 2.4b. Merely repeating the seasonal pattern of the last twelve months usually provides a reasonable forecast.

2.5.2 Models for Seasonality

The simplest seasonal model for monthly seasonality can be written as

$$S_t = \Sigma_{j=1}^{12} \alpha_j d_{jt} + u_t, \; d_{jt} = \begin{cases} 1, & j=t\pm 12k,\; k=0,1,2,...,11, \\ 0, & otherwise, \end{cases} \tag{2.22}$$

subject to $\Sigma_{j=1}^{12} \alpha_j = 0$, $\{u_t\}$ is assumed white noise. The α_j are the seasonal effects and the d_{jt}'s are dummy variables.

Model (2.22) can be equivalently written by means of sines and cosines

$$S_t = \Sigma_{j=1}^{6} [\alpha_j cos(\lambda_j t), + \beta_j sin(\lambda_j t)], \tag{2.23}$$

where $\lambda_j = 2\pi j/12$, $j=1,2,...,6$ and $\beta_6 = 0$. The λ_js are known as the seasonal frequencies, with j corresponding to cycles lasting 12, 6, 4, 3, 2.4 and 2 months respectively. This model is discussed in Appendix D.

In order to represent stochastic seasonality, the α_j of Eq. (2.22) are specified as random variables instead of constant coefficients (see Dagum 2001). Such a model model is

$$S_t = S_{t-12} + \omega_t, \tag{2.24a}$$

or

$$(1-B^{12})S_t = \omega_t, \tag{2.24b}$$

subject to constraints $\Sigma_{j=0}^{11} S_{t-j} = \omega_t$ where ω_t is assumed white noise.

Model (2.24) specifies seasonality as a non-stationary random walk process. Since $(1-B^s) \equiv (1-B)(1+B+...+B^{s-1})$, model-based seasonal adjustment method assigns $(1-B)$ to the trend and $S(B) = \Sigma_{j=0}^{s-1} B^j$ to the seasonal component. Hence, the corresponding seasonal model is

$$\Sigma_{j=0}^{s-1} S_{t-j} = \omega_t , \tag{2.25}$$

which entails a volatile seasonal behaviour, because the sum is not constrained to 0 but to the value of ω_t. Indeed, the spectrum of $\Sigma_{j=0}^{s-1} B^j$ (not shown here) displays broad bands at the high seasonal frequencies, i.e. corresponding to cycles of 4, 3, and 2.4 months.

Model (2.25) has been used in many structural time series models (see e.g. Harvey 1981, Kitagawa and Gersch 1984). A very important variant to model (2.25) was introduced by Hillmer and Tiao (1982) and largely discussed in Bell and Hillmer (1984), that is

$$\Sigma_{j=0}^{s-1} S_{t-j} = \eta_s(B)\, b_t ,$$

where $\eta_s(B)$ is a moving average of $s-1$ minimum order and $b_t \sim WN(0, \sigma_b^2)$. The moving average component enables seasonality to evolve gradually. Indeed, the moving average eliminates the aforementioned bands at the high seasonal frequencies.

Another stochastic seasonality model is based on trigonometric functions (see Harvey 1989) defined as

$$S_t = \Sigma_{j=1}^{[s/2]} \gamma_{jt} ,$$

where γ_{jt} denotes the seasonal effects generated by

$$\begin{bmatrix} \gamma_{jt} \\ \gamma_{jt}^* \end{bmatrix} = \begin{bmatrix} \cos\lambda_j & \sin\lambda_j \\ -\sin\lambda_j & \cos\lambda_j \end{bmatrix} \begin{bmatrix} \gamma_{j,t-1} \\ \gamma_{j,t-1}^* \end{bmatrix} + \begin{bmatrix} \omega_{jt} \\ \omega_{jt}^* \end{bmatrix},$$

and $\lambda_j = 2\pi j/s$, $j=1,...,[s/2]$ and $t=1,...,T$. The seasonal innovation ω_{jt} and ω_{jt}^* are mutually uncorrelated with zero means and common variance σ_ω^2.

2.6 The Moving-Holiday Component

The *moving-holiday* or *moving-festival* component is attributed to calendar variations, namely the fact that some holidays change date from year to year. For example, Easter can fall between March 23 to April 25, and Labour Day on the first Monday of September. The Chinese New Year date depends on the lunar calendar. Ramadan falls eleven days earlier from year to year. In the Moslem world, Israel and in the Far East, there are many such festivals. For example, Malaysia contends with as many as eleven moving festivals, due to its religious and ethnic diversity. These festivals affect flow and stock variables and may cause a displacement of activity from one month to the previous or the following month. For example, an early date of Easter in March or early April can cause an important excess of activity in March and a corresponding short-fall in April, in variables associated to imports, exports, tourism. When the Christian Easter falls late in April (e.g. beyond the 10-th), the effect is captured by the seasonal factor of April. In the long run, Easter falls in April 11 times out of 14.

Some of these festivals have a positive impact on certain variables, for examples air traffic, sales of gasoline, hotel occupancy, restaurant activity, sales of flowers and chocolate (in the case of Easter), sales of children clothing (Labour Day).[2] The impact may be negative on other industries or sectors which close or reduce their activity during these festivals.

The festival effect may affect only the day of the festival itself, or a number of days preceding and/or following the festival. In the case of Easter, travellers tend to leave a few days before and return after Easter, which affects air traffic and hotel occupancy, etc., for a number of days. Purchases of flowers and other highly perishable goods, on the other hand, are tightly clustered immediately before the Easter date.

The effect of moving festivals can be seen as a seasonal effect dependent on the date(s) of the festival. Fig. 2.4c displays the Easter effect on the sales by Canadian Department Stores. In this particular case, the Easter effect is rather mild. In some of the years, the effect is absent because Easter fell too late in April. The dates of Easter appear in Table 2.1.

In the case exemplified, the effect is felt seven days before Easter and on Easter Sunday but not after Easter. This is evidenced by years 1994, 1996 and 1999 where Easters falls early in April and impacts the month of March. Note that the later Easter falls in April, the smaller the displacement of

[2] In Canada and the United States, the school year typically starts the day after Labour Day (the first Monday of September).

activity to March; after a certain date the effect is entirely captured by the April seasonal factor. The effect is rather moderate for Department Stores. This may not be the case for other variables. For example, Imports and Exports are substantially affected by Easter, because Customs do not operate from Good Friday to Easter Monday. Easter can also significantly affect quarterly series, by displacing activity from the second to the first quarter.

Table 2.1. Dates of Easter and presence of effect in March

March 31 1991	April 12 1998, no effect
April 19 1992, no effect	April 4 1999
April 11 1993, no effect	April 23 2000, no effect
April 3 1994	April 15 2001, no effect
April 16 1995, no effect	March 31 2002
April 7 1996	April 20 2003, no effect
March 30 1997	April 11 in 2004, no effect.

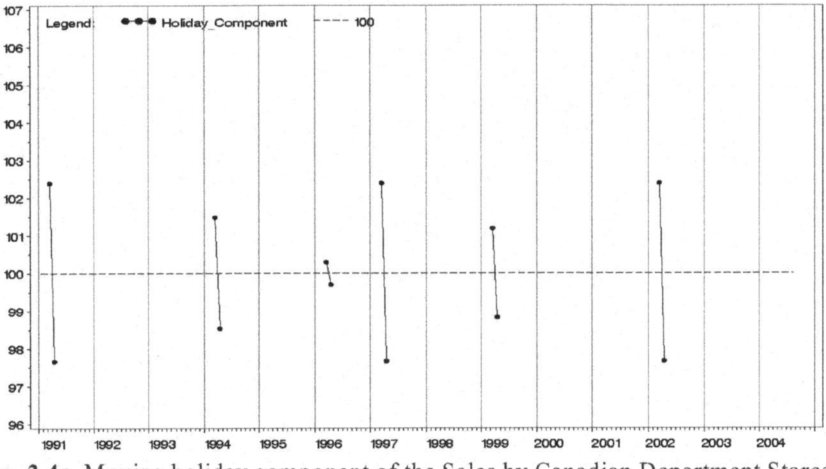

Fig. 2.4c. Moving holiday component of the Sales by Canadian Department Stores

Morris and Pfefferman (1984) proposed a method to estimate festival effects. They consider the "rectangular" effect of a festival, which assumes the activity spreads uniformly over the days of the festival period; and the "triangular" effect, which assumes the activity increases gradually (linearly) before the actual date of the festival and declines gradually thereafter.

The rectangular specification of the daily regressor is given by

$$\delta_\tau = \begin{cases} 0, & \tau<\tau_0-n,\ \tau>\tau_0+k, \\ 1/(n+k+1), & \tau_0-n \le \tau \le \tau_0+k, \end{cases} \tag{2.26}$$

where n and k are the number of days before and after the Easter festival for example, and where τ stands for the daily date and τ_0 for Easter Sunday. In order to mathematically process the daily dates τ, one integer value is assigned to each day of the 20-th and 21-st centuries in an ascending manner, e.g. 1 for Jan. 1 1900, 2 for Jan. 2 1900, etc.

The regressor (2.26) may be used as such for daily series, to capture the excess of activity due to Easter in March and April. For monthly series, the regressors are the sums of the daily regressors over the days of March and April:

$$h_t = \Sigma_{\tau\in t}\ \delta_\tau, \tag{2.27a}$$

$$h_{t+1} = \Sigma_{\tau\in(t+1)}\ \delta_\tau, \tag{2.27b}$$

where t stands for March and $t+1$ for April. The regressors of Eq. (2.27) measure the excess activity associated with Easter, but do not assign it to any of the two months.

One practice is to assign the full Easter effect to April because Easter occurs in April most of the time. This is achieved by redefining the regressor of April in Eq. (2.27b) as $h_{t+1} = -h_t$. This is the case in the example exhibited in Fig. 2.4c.

Another practice is to assign half of the monthly Easter effect to March and the other half to April. This is achieved by setting the regressors to

$$h_t = (\Sigma_{\tau\in t}\ \delta_\tau)/2, \tag{2.28a}$$

$$h_{t+1} = -(\Sigma_{\tau\in t}\ \delta_\tau)/2. \tag{2.28b}$$

The fractions could be 11/14 for March and 3/14 for April (instead of 1/2), which would leave 3/14 of the Easter effect in March and 11/14 in April.

The triangular festival effect assumes that activities associated to the festival start gradually, culminate on Easter Sunday and decline gradually afterwards. The triangular regressor for daily series is given by

$$\delta_\tau = \begin{cases} 0, & \tau < \tau_0 - n,\ \tau > \tau_0 + k, \\ [(n+1-|\tau-\tau_0|)/(n+1)]/\kappa, & \tau_0 - n < \tau \le \tau_0, \\ [(k+1-|\tau-\tau_0|)/(k+1)]/\kappa, & \tau_0 < \tau \le \tau_0 + k, \end{cases} \tag{2.29}$$

where the constant $\kappa = (n+2)/2 + (k/2)$ standardizes the regressor to sum one for each year. The corresponding monthly regressors are given by (2.27) or by (2.28).

Under both specifications, the Easter effect is given by $H_t = h_t\,\beta$ and $H_{t+1} = h_{t+1}\,\beta$ for March and April respectively.

It should be noted that the Morris and Pfefferman (1984) model can be tailored to particular festivals. For example, other shapes of activity could be chosen, namely the trapezoid with a flat portion over the more intense days of the festival. Festivals can affect more than two months, for example the Ramadan rotates through all months of the year in a cycle of 33 years. The approach can be easily adapted to quarterly series impacted by Easter and other festivals, by taking the quarterly sums of the daily regressors.

There has been cases of complete reversal on the timing of the Easter effect. For example, Marriages in Canada were performed mainly by the Church during the 1940s up to the 1960s. The Church did not celebrate marriages during the Lent period, i.e. the 40 days before Easter. Some marriages therefore were celebrated before the Lent period, potentially affecting February and March. However, if Easter fell too early, many of these marriages were postponed *after* Easter.

Generally, festival effects are difficult to estimate, because the nature and the shape of the effect are often not well known. Furthermore, there are few observations, i.e. one occurrence per year.

Easter adjustment models similar to (2.26) and (2.29) have been implemented in the seasonal adjustment software, X-11-ARIMA and X-12-ARIMA (see Dagum, Huot and Morry 1988).

2.7 The Trading-Day Component

Flow series may be affected by other variations associated with the composition of the calendar. The most important calendar variations are the trading-day variations, which are due to the fact that some days of the week are more important than others. Trading-day variations imply the existence of a daily pattern analogous to the seasonal pattern. However, these "daily factors" are usually referred to as *daily coefficients*.

2.7.1 Causes and Costs of Daily Patterns of Activity

Depending on the socio-economic variable considered, some days may be 60% more important than an average day and other days, 80% less important.

If the more important days of the week appear five times in a month (instead of four), the month registers an excess of activity *ceteris paribus*. If the less important days appear five times, the month records a short-fall. As a result, the monthly *trading-day component* can cause increase of +8% or -8% (say) between neighbouring months and also between same-months of neighbouring years. The trading-day component is usually considered as negligible and very difficult to estimate in quarterly series.

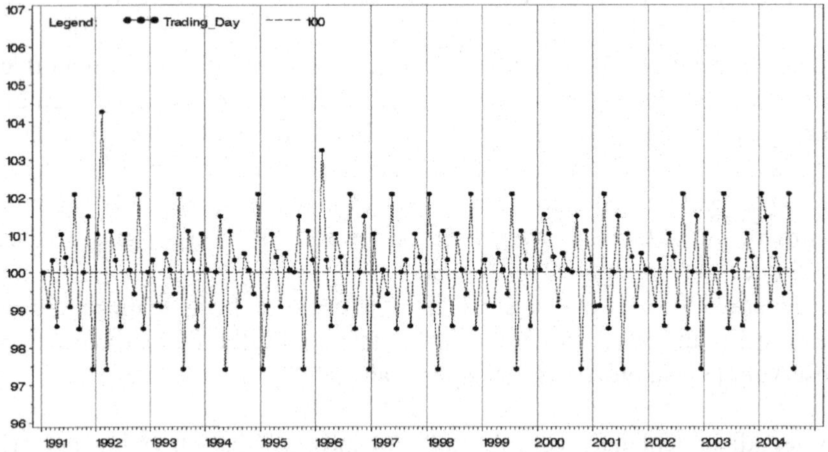

Fig. 2.4d. Trading-day estimates of the Sales by Canadian Department Stores

Fig. 2.4d displays the monthly trading-day component obtained from the following daily pattern: 90.2, 71.8, 117.1, 119.3, 97.6, 161.3 and 70.3 for Monday to Sunday (in percentage) respectively. The daily pattern indicates that Saturday is approximately 61% more important than an average day (100%); and that Tuesday and Sunday, 30% less important.

For the multiplicative, the log-additive and the additive time series decomposition models, the monthly trading-day component is respectively obtained in the following manner

$$D_t = \Sigma_{\tau \in t}\, d_\tau / n_t \;\equiv\; (2800 + \Sigma_{\tau \in t\ 5\,times}\, d_\tau) / n_t, \qquad (2.30a)$$

$$D_t = exp(\Sigma_{\tau \in t}\, d_\tau / n_t) \;\equiv\; exp((\Sigma_{\tau \in t\ 5\,times}\, d_\tau) / n_t), \qquad (2.30b)$$

$$D_t = \Sigma_{\tau \in t}\, d_\tau \;\equiv\; (\Sigma_{\tau \in t\ 5\,times}\, d_\tau), \qquad (2.30c)$$

where d_τ are the daily coefficients in the month. The preferred option regarding n_t is to set it equal to the number of days in month t, so that the length-of-month effect is captured by the multiplicative seasonal factors, except for Februaries.[3] The other option is to set n_t equal to 30.4375, so that the multiplicative trading-day component also accounts for the length-of-month effect. The number 2800 in Eq. (2.30a) is the sum of the first 28 days of the months expressed in percentage.

The monthly trading-day estimates of the Sales by Canadian Department Stores shown in Fig. 2.4d were obtained with the log-additive model (2.30b). They display a drop of 7% between Jan. and Feb. 1992 and a drop of 5% between Jan. 1992 and Jan. 1993. One can identify several instances where the change between same-months is significant. Indeed, same-month year-ago comparisons are never valid in the presence of trading-day variations, not even as a rule of thumb. Furthermore, it is apparent that the monthly trading-day factors in the figure are identical for quite a few months. Indeed for a given set of daily coefficients, there are only 22 different monthly values for the trading-day component, for a given set of daily coefficients: seven values for 31-day months (depending on which day the month starts), seven for 30-day months, seven for 29-day months and one for 28-day months. In other words, there are at most 22 possible arrangements of days in monthly data.

For certain variables, e.g. Income Tax, Superannuation and Employment Insurance premiums, the daily pattern is defined over two weeks instead of one. These variables are related to pay days and Payroll Deductions, which typically take place every second week. Appendix E provides regressors for weekly and bi-weekly daily patterns. Section 9.6 of Chapter 9 discusses the calendarization of such data.

[3] To adjust Februaries for the lengh-of-month, the seasonal factors of that month are multiplied by 29/28.25 and 28/28.25 for the leap and non-leap years respectively.

Many goods and services are affected by daily patterns of activity, which entail higher costs for producers, namely through the need of higher inventories, equipment and staff on certain days of the week.

For example, there is evidence that consumers buy more gasoline on certain days of the week, namely on Thursdays, Fridays, Saturdays and holidays, which results in line-ups and shortages at the pumps. In order to cope with the problem, gasoline retailers raise their price on those days to promote sales on other days. Furthermore, the elasticity of demand for gasoline is low. In other words, to reduce consumption by a small percentage, prices must be raised by a disproportionate percentage, which infuriates some consumers. On the other hand, consumers can buy their gasoline on other days. The alternative for retailers is to acquire larger inventories, larger tanks, more pumps and larger fleets of tanker trucks, all of which imply higher costs and translate into much higher prices. In other words, there are savings associated with more uniform daily patterns; and costs, with scattered daily patterns.

A similar consumer behaviour prevails for the purchases of food, which probably results in more expensive prices, namely through higher inventories, larger refrigerators, more numerous cash registers and more staff, than otherwise necessary.

Scattered daily patterns have been surprisingly observed for variables like births and deaths. Indeed, births are more frequent on certain days of the week, namely on working days to avoid overtime pay. This results from the practice of cesarean delivery and especially birth inducement now widely applied to encourage births on working days. A time series decomposition of monthly data for Québec in the 1990s revealed that 35% more births took place on Thursdays. A similarly analysis of the Ontario data revealed the same phenomenon. In this particular case, an appropriately scattered daily pattern reduces costs.

Deaths also occur more often on certain days of the week. Car accidents, drowning, skiing and other sporting accidents tend to occur on weekend days and on holidays. According to the Canadian Workmen Compensation Board, industrial accidents tend to occur more often on Friday afternoons when security is more lax.

In principle, stock series pertaining to one day display a particular kind of trading-day variations. Among other things, inventories must anticipate the activity (flow) of the following day(s). For such stock series, the monthly trading-day factor coincides with the daily weight of the day.

2.7.2 Models for Trading-Day Variations

A frequently applied deterministic model for trading-day variations is that developed by Young (1965),

$$y_t = D_t + u_t\ , \quad t=1,,...,n, \tag{2.31a}$$

$$D_t = \Sigma_{j=1}^{7} \alpha_j N_{jt}\ , \tag{2.31b}$$

where $u_t \sim WN(0, \sigma_u^2)$, $\Sigma_{j=1}^{7} \alpha_j = 0$, α_j, $j=1,...,7$ denote the effects of the seven days of the week, Monday to Sunday, and N_{jt} is the number of times day j is present in month t. Hence, the length of the month is $N_t = \Sigma_{j=1}^{7} N_{jt}$, and the cumulative monthly effect is given by (2.31b). Adding and subtracting $\bar{\alpha} = (\Sigma_{j=1}^{7} \alpha_j)/7$ to Eq. (2.31b) yields

$$D_t = \bar{\alpha} N_t + \Sigma_{j=1}^{7} (\alpha_j - \bar{\alpha}) N_{jt}\ . \tag{2.32}$$

Hence, the cumulative effect is given by the length of the month plus the net effect due to the days of the week. Since $\Sigma_{j=1}^{7} (\alpha_j - \bar{\alpha}) = 0$, model (2.32) takes into account the effect of the days present five times in the month. Model (2.32) can then be written as

$$D_t = \bar{\alpha} N_t + \Sigma_{j=1}^{6} (\alpha_j - \bar{\alpha}) (N_{jt} - N_{7t}), \tag{2.33}$$

with the effect of Sunday being $\alpha_7 = -\Sigma_{j=1}^{6} \alpha_j$.

Deterministic models for trading-day variations assume that the daily activity coefficients are constant over the whole range of the series. Appendix E presents the classical Young model in terms of monthly sums of dummy variables specified on the underlying daily series.

Stochastic model for trading-day variations have been rarely proposed. Dagum et al. (1992) developed such a model as follows

$$y_t = D_t + u_t\ , \ u_t \sim WN(0, \sigma_u^2), \quad t=1,,...,n, \tag{2.34}$$

where $$D_t = \Sigma_{j=1}^{6} \delta_{jt} X_{jt}, \quad \delta_{7t} = -\Sigma_{j=1}^{6} \delta_{jt}\ ,$$

and $X_{jt} = (N_{jt} - N_{7t})$ and the $\delta_{jt} = (\alpha_{jt} - \bar{\alpha}_t)$s are the daily coefficients which change over time following a stochastic difference equation,

$$(1 - B)^k \boldsymbol{\delta}_t = \boldsymbol{\eta}_t , \qquad (2.35)$$

where $\boldsymbol{\delta}_t = [\delta_{1t} \ldots \delta_{6t}]'$ and $\boldsymbol{\eta}_t \sim WN(\boldsymbol{0}_{6\times 1}, \sigma^2 \boldsymbol{I}_6)$. For k=1, model (2.35) reduces to a random walk and yields volatile estimates, for k=2, the daily pattern is more stable.

2.7.3 Sunday Opening of Stores

In the 1990s, the provincial legislations of Canada authorized the Sunday opening of stores, which used to be prohibited. As a result, the daily trading pattern changed drastically. To cope with the abrupt change, a first attempt was to estimate the daily pattern over the ranges covered by the *regulated regime* starting in 1981 and the *de-regulated regime* of Sunday opening starting in May 1992 and up to Dec. 1993. The estimates were obtained with the regression model due to Young (1965) described in Appendix E. This method requires six parameters to provide seven daily weights. Table 2.2 presents the resulting daily weights for the Sales by Canadian Department Stores.

The importance of Sunday rose from the regulated to the de-regulated regime. However the weight of Sunday was still below average (100). To a large extent, the activity of Monday shifted to Sunday. The daily weights of the de-regulated regime were based on only 19 observations and therefore unreliable and subject to large revisions.

Table 2.2. Daily patterns estimated for the sales by Canadian Department Stores in the regulated regime and the de-regulated regime (Sunday opening)

	regulated regime (01/81-04/92)	de-regulated (05/92-12/93)
Monday	104.9	67.4
Tuesday	77.8	61.9
Wednesday	94.6	112.1
Thursday	138.8	146.7
Friday	107.3	84.9
Saturday	155.2	141.8
Sunday	21.4	85.2

A more recent study examined two series from the province of New-Brunswick: the Total Retail Trade Sales and the Sales by Department Stores. Another approach was proposed by Quenneville et al. (1999) by means of a trigonometric model for the trading-day variations, as described in Appendix E. In this approach, each one of the six trigonometric regressor describes the seven days of the week.[4] The non-significant regressors can be dropped, thus providing a more parsimonious model. Two sets of trigonometric functions were used, one over the whole series from 1981 to 1996 and one over the months of Sunday opening. The second set captured the *change* in the daily pattern during the de-regulated regime. The de-regulated regime includes the following 23 months, which are quite sporadic:

Nov. and Dec. for 1991,
Sept. to Dec. for 1992, 1993, 1994 and 1995,
Aug. to Dec. for 1996.

In the case of Total Retail Trade, three parameters were required for the first set of trigonometric functions and two for the second set; in the case of Department Stores, five on the first set and four for the second set. Table 2.3 displays the resulting daily weights and the proportion of activity on each day of the week.

Table 2.3. Daily patterns estimated for Total Retail Trade and Department Stores in New Brunswick obtained with a trigonometric specification

	Total Retail				Department Stores			
	regulated		de-regulated		regulated		de-regulated	
	% activ.	weight	%activ.	weight	%activ.	weight	%activ.	weight
Monday	10	69.8	7	47.5	16	112.4	7	48.9
Tuesday	15	103.5	17	118.5	12	83.9	14	95.3
Wednesday	14	96.1	17	119.2	13	88.6	8	56.9
Thursday	17	116.9	11	73.7	19	132.0	24	167.2
Friday	23	160.5	18	126.4	19	134.6	5	32.6
Saturday	14	99.3	22	156.8	15	107.5	31	220.1
Sunday	7	53.8	8	57.9	6	41.0	11	79.0

For Total Retail Trade, de-regulation increased mainly the relative importance of Saturday and to a lesser extent that of Tuesday, Wednesday and Sunday; and similarly for Department stores. In both cases, de-regulation increased the importance of Sunday less than that of some of the other days. Indeed, for Total Retail the importance of Sunday increased by only 1% .

The de-regulation changed both daily patterns in a erratic manner, e.g. increasing the importance of one day and decreasing that of the next day. The explanation of these results may lie in the assumptions of the model. The model implicitly assumes that consumers adopt the de-regulated daily pattern

[4] If all six trigonometric regressors are used in the trigonometric approach, the results coincide with those of the classical regression approach due to Young (1965).

as soon as stores open on Sundays and immediately revert to their previous behaviour as stores no longer open on Sundays. In other words, consumers do not have time to adapt to the de-regulated regime, because the de-regulated months are sporadic; and vice versa. Another explanation could be the small number of data points for the de-regulated regime.

The model also included a dummy variable to determine whether Sunday opening increased the overall sales regardless of the day. For Total Retail, the parameter was not significant. For Department Stores, it was significant and entailed an increment of 2.2%. This result implies that the rest of the Retail Trade sector experienced a decline. The interpretation could be that consumers assumed that large stores, e.g. Department Stores, would be more likely to open on Sunday in the sporadically de-regulated months.

2.8 The Irregular Component

The irregular component in any decomposition model represents variations related to unpredictable events of all kinds. Most irregular values have a stable pattern, but some extreme values or outliers may be present. Outliers can often be traced to identifiable causes, for example strikes, droughts, floods, data processing errors. Some outliers are the result of displacement of activity from one month to the other.

Fig. 2.4e displays the irregular component of Sales by Canadian Department Stores, which comprises extreme values, namely in 1994, 1998, 1999 and Jan 2000. Most of these outliers have to do with the closure of some department stores and the entry of a large department store in the Canadian market.

As illustrated by Fig. 2.4e, the values of the irregular component may be very informative, as they quantify the effect of events known to have happened.

Note that it is much easier to locate outliers in the graph of the irregular component of Fig. 2.4e than in the raw series of Fig. 2.4a. Indeed the presence of seasonality hides the irregular fluctuations.

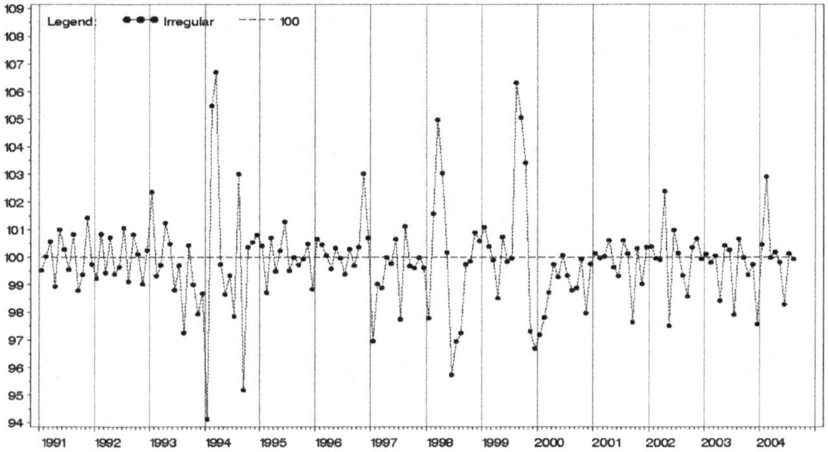

Fig. 2.4e. Estimate of the irregular component of Sales by Canadian Department Stores

2.8.1 Redistribution Outliers and Strikes

Some events can cause displacements of activity from one month to the next months, or vice versa. This phenomenon is referred to as redistribution outliers. We also deal with the strike effect under this headline. The outliers must be modelled and temporally removed from the series in order to reliably estimate the systematic components, namely the seasonal and trading-day components.

In January 1998, there was a severe Ice Storm which brought down major electrical transmission lines, causing black outs lasting from a few days to several weeks in populated parts of eastern Ontario and western Québec. The irregular component of New Motor Vehicles provided a measure of the effect, in the amount of 15% drop in sales. This drop was compensated in February.

The termination of a government grant program, e.g. for the purchase of a first house, can displace activity from a month to previous months. For example, in the 1990s the Province of Ontario had a grant program to encourage households to purchase a first house. At one point, the termination of the program was announced. Some buyers who considered purchasing a house decided to take advantage of the program. This caused a surge of activity in the months preceding the termination date and a short-fall in months following. Conversely, activity can be delayed, when a government program is expected to take effect at a given date.

Events like major snow storms and power black outs usually postpone activity to the next month, without much longer term effect.

2.8.2 Models for the Irregular Component and Outliers

The irregulars are most commonly assumed to follow a white noise process defined by

$$E(u_t) = 0,\ E(u_t^2) = \sigma_u^2 < \infty,\ E(u_t u_{t-k}) = 0\ \forall k \neq 0.$$

If σ_u^2 is assumed constant (homoscedastic condition), u_t is referred to as white noise in the "strict" sense. If σ_u^2 is finite but not constant (heteroscedastic condition), u_t is called white noise in the "weak" sense.

For inferential purposes, the irregular component is often assumed to be normally distributed and not correlated, which implies independence. Hence,

$$u_t \sim NID(0, \sigma_u^2).$$

There are different models proposed for the presence of outliers depending on how they impact the series under question. If the effect is transitory, the outlier is said to be additive; and if permanent, to be multiplicative.

Box and Tiao (1975) introduced the following intervention model to deal with different types of outliers,

$$y_t = \Sigma_{j=0}^{\infty} h_{t-j}\, x_{t-j} + \eta_t = \Sigma_{j=0}^{\infty} h_j B^j x_j + \eta_t = H(B) x_t + \eta_t,$$

where the observed series $\{y_t\}$ consists of an input series $\{x_t\}$ considered a deterministic function of time and a stationary process $\{\eta_t\}$ of zero mean and non-correlated with $\{x_t\}$. In such a case the mean of $\{y_t\}$ is given by the deterministic function $\Sigma_{j=0}^{\infty} h_{t-j}\, x_{t-j}$. The type of function assumed for $\{x_t\}$ and weights $\{h_j\}$ depend on the characteristic of the outlier or unusual event and its impact on the series.

If the outlier at time T is additive, in the senses that it will not permanently modifiy the level of $\{y_t\}$, that is $E(y_t)=0$, $\forall t \leq T$ and also $E(y_t) \to 0$ for $t \to \infty$, then an appropriate x_t is the impulse function defined by

$$x_t = P_t(T) = \begin{cases} 1, & t=T, \\ 0, & t \neq T. \end{cases}$$

If instead, the outlier at time T is multiplicative in the sense that it will modify the level of $\{y_t\}$, that is $E(y_t)=0$, $\forall t \leq T$ and also $E(y_t) \to c \neq 0$ for $t \to \infty$, then an appropriate x_t is the step function defined by

$$x_t = S_t(T) = \begin{cases} 1, & t \geq T, \\ 0, & t < T. \end{cases}$$

In fact, $S_t(T) - S_{t-1}(T) = (1-B)\, S_t(T) = P_t(T)$.

Once the deterministic function is chosen, the weights $\{h_t\}$ can follow different patterns dependent on the impact of the outlier. For example, if the outlier is additive and present at time $t=t_0$, an appropriate model is

$$H(B)x_t = \omega\, P_t(t_0) = \begin{cases} \omega, & t=t_0, \\ 0, & t \neq t_0, \end{cases}$$

where ω represents the effect of the outlier.

If the outlier is seasonal and additive starting at $t=t_0$, an appropriate model is

$$H(B)x_t = \omega\, P_t(t_0+ks) = \begin{cases} \omega, & t=t_0,\, t_0+s,\, t_0+2s, \ldots, \\ 0, & \textit{otherwise}. \end{cases}$$

For additive redistribution of outliers (displacement of activity), during period $t=t_0,\ t=t_0+1,...,\ t=t_0+k$ an appropriate model is

$$H(B)x_t = \omega_0 P_t(t_0) + \Sigma_{i=1}^{k} \omega_i P_t(t_0+i) = \begin{cases} \omega_0 = -\Sigma_{i=1}^{k} \omega_i,\ t=t_0, \\ \omega_1,\ t=t_0+1, \\ \vdots \\ \omega_k,\ t=t_0+k, \\ 0,\ otherwise, \end{cases}$$

where the weights $\boldsymbol{\omega} = [\omega_0\ \omega_1 \ldots \omega_0]$ measure the outlier effects during the $k+1$ periods $t_0, t_0+1,..., t_0+k$. The global effect is such that the sum of all the outlier effects cancel out.

For a sudden permanent change of level at time $t=t_0$, a step function can be used,

$$H(B)x_t = \omega\, S_t(t_0) = \begin{cases} \omega,\ t \geq t_0, \\ 0,\ otherwise, \end{cases}$$

where ω represent the level difference of the series before and after the outlier. Since $(1-B)\,S_t(T) = P_t(T)$, we can write

$$H(B)x_t = \frac{\omega}{(1-B)} P_t(t_0) = \begin{cases} \omega,\ t=t_0, \\ 0,\ otherwise, \end{cases}$$

where

$$\omega(1-B)^{-1} = \omega\,(1+B+B^2+...)\, P_t(t_0) = \Sigma_{j=0}^{\infty}\, \omega\, B^j P_t(t_0) = \Sigma_{j=0}^{\infty}\, \omega\, P_t(t_0+j).$$

For a sudden transitory change at time $t=t_0$, an appropriate model is

$$H(B)x_t = \frac{\omega}{(1-\delta B)} P_t(t_0) = \begin{cases} 0,\ t<t_0, \\ \omega,\ t=t_0, \\ \delta\omega,\ t=t_0+1, \\ \delta^2\omega,\ t=t_0+2, \\ \vdots \end{cases}$$

where ω denotes the initial effect and $0<\delta<1$ is the rate of decrease of the initial effect.

For a gradual permanent level change at time $t=t_0$, an appropriate model is

$$H(B)x_t = \frac{\omega}{(1-\delta B)} S_t(t_0) = \begin{cases} 0, & t<t_0, \\ \omega, & t=t_0, \\ (1+\delta)\omega, & t=t_0+1, \\ (1+\delta+\delta^2)\omega, & t=t_0+2, \\ \vdots & \\ \omega/(1-\delta), & t\to\infty, \end{cases}$$

where ω denotes the initial effect and the other terms described the cumulative effect, which converges to $\omega/(1-\delta)$. If $\delta=1$, a permanent slope change (a kink) occurs at time t_0 with no convergence.

3 The Cholette-Dagum Regression-Based Benchmarking Method – The Additive Model

3.1 Introduction

This chapter is important to the understanding of the following chapters, because several benchmarking and interpolation methods emerge as particular cases of the Cholette-Dagum regression-based method (1994), namely the benchmarking techniques of the Denton (1971) type and the interpolation methods of the Chow and Lin (1971 1976) type. The method is deemed regression-based because the benchmarking problem is cast as a classical regression model; this is particularly obvious from Eq. (3.18).

The benchmarking problem arises when time series data for the same target variable are measured at different time intervals with different levels of accuracy. Typically, one source of data is more frequent than the other: for example, monthly series on one hand and yearly series on the other, monthly and quarterly series, quarterly and yearly, daily and multi-weekly (e.g. 4- or 5-week). The more frequently observed series is usually less reliable than the less frequent one. For this reason, the latter is generally considered as benchmark. Discrepancies are usually observed between the yearly benchmarks (say) and the yearly sums of the more frequent series. Broadly speaking, benchmarking consists of combining the relative strengths of the two sources of data. More specifically, this has essentially consisted of adopting the movement of the more frequent series and the level of the benchmarks.

For convenience, the more frequently measured series will be referred to as "sub-annual", and the less frequent (the benchmarks) as "annual", unless it is relevant to be specific. Thus "sub-annual" data may be yearly, and the "annual" benchmarks may be quinquennial or decennial. The sub-annual and annual series are respectively denoted by

$$s_t,\ t=1,2,...,T, \tag{3.1a}$$
$$a_m,\ m=1,...,M, \tag{3.1b}$$

where $\{1,2,...,T\}$ refers to a set of *contiguous* months, quarters, days, etc.; and $\{1,2,...,M\}$ to a set of *not* necessarily of contiguous periods, e.g. there may not be a benchmark every "year". In some cases, benchmarks are available every second year, every fifth year, or even irregularly. In order words, the sub-annual series cannot contain missing values, whereas the annual series can. Furthermore, benchmarks may be sub-annual. Such benchmarks may be used to force the benchmarked series to start from a preset value, for example a monthly previously obtained benchmarked value considered as final; or to lead to a preset value, for example a monthly value pertaining to a redesigned survey (Helfand et al. 1977). Fig 10.1a illustrates the latter situation with the use of monthly benchmarks at the end of the series.

It is important to realize that in practice, yearly (*sic*) benchmarks become available several months after the current year is completed. As a result no yearly benchmark is ever available for the current year, hence there is a missing value. Nevertheless, as will be shown later, the most widely used benchmark methods make adjustments for the current year without benchmark, and thus ensure continuity between the estimates of the current year and those of the previous years with benchmarks. This entails an implicit forecast of the next benchmark(s).

Some practitioners of benchmarking explicitly forecast the unavailable benchmark, which embodies their subject matter expertise and may improve the benchmarking of the current sub-annual values. This approach also makes it necessary to also forecast the sub-annual series[1] for the sub-annual periods covered by the forecasted benchmark.

Benchmarking situations are very common in statistical agencies. In some cases, both the sub-annual and annual series originate from surveys or censuses; in others, none of the two series originate from surveys or only one does. The following examples illustrates some cases.

Until the end of the 1990s, Statistics Canada obtained the monthly Sales from the Monthly Wholesale and Retail Trade Survey and the yearly benchmarks from the Annual Census of Manufacturing. Similarly, Statistics Canada obtained the monthly values of Wages and Salaries from a survey and the yearly benchmarks from administrative data, namely the tax data anonymously supplied by Revenue Canada.

In Canada, the Census of the Population takes place every fifth year (1996, 2001, etc.). The yearly estimates of the Population are derived from the last quinquennial census value and from the yearly flows, adding to the stock of Population (Births and Immigration) and subtracting from the Population (Deaths and Emigration). These yearly flows are more accurate at the national level, but not so accurate at the provincial and sub-provincial

[1] This forecasting is usually carried out by ARIMA modelling.

levels because the estimates of the inter-provincial migrations are less reliable. When the quinquennial census data become available, demographers re-estimate the flows of the previous five years in the light of the newly available information. Conceptually, this is a form of benchmarking, although no formal benchmarking method is used.

In many situations none of the data originate from surveys. This is the case of the system of National Accounts of many countries. Indeed, many time series of the quarterly Accounts are interpolated from many sources to fit the concepts and definitions of the Accounts. The quarterly Accounts are also benchmarked to the yearly values provided by the Input-Output Accounts. The quarterly Accounts must also satisfy cross-sectional (contemporaneous) additivity constraints, which is a case of reconciliation. The Provincial Accounts are obtained by similar methods. Furthermore, they must cross-sectionally add up to the National Accounts. Benchmarking and reconciliation are critical processes in building the National Accounts (Bloem et al. 2001).

Benchmarking also occurs in the context of seasonal adjustment. Seasonally adjusting a monthly or quarterly time series causes discrepancies between the yearly sum of the raw (seasonally *un*-adjusted) series and the corresponding yearly sums of the seasonally adjusted series. Such seasonally adjusted series are then benchmarked to the yearly sums of the raw series. The benchmarking is usually performed within the seasonal adjustment method such as the X-11-ARIMA method (Dagum 1980 and 1988) and X-12 ARIMA method (Findley et al. 1998), using formal benchmarking methods.

Benchmarking requires that both the sub-annual and annual observations be dated with precision. This is achieved by means of a *starting date* and an *ending date*, which define a *reporting period*. The dates in Table 3.1 display the reporting periods for sequences of: (1) monthly data, (2) fiscal quarterly data, (3) fiscal yearly data, and (4) multi-weekly data. (The dates are displayed using the day/month/year format). The pair of dates on the left pertain to flow series, e.g. Sales; and the pair on the right, to stock data, e.g. Inventories, which are usually measured at the end of the reporting period. Note that the fiscal years and fiscal quarters illustrated are common to all banking institutions in Canada; and that the multi-weekly periods are typical to many of the large Retail and Wholesale Trade outlets.

The precise specification of the reporting periods is especially important for the benchmarks, because in some applications the reporting periods of the benchmarks change over time. At the micro level, the reporting periods may change for a given company. For example, a sequence of benchmarks may have quarterly reporting periods; the next sequence, monthly reporting periods; and the next sequence, multi-weekly reporting periods covering four or five weeks. These reporting periods contain the *timing* and the *duration* of the reported values, described next.

Table 3.1. Sample of reporting periods

Starting date	Ending date	Number days	Starting date	Ending date	Number days
Monthly flow data			Monthly stock data		
01/01/1998	31/01/1998	31	31/01/1998	31/01/1998	1
01/02/1998	28/02/1998	28	28/02/1998	28/02/1998	1
01/03/1998	31/03/1998	31	31/03/1998	31/01/1998	1
Fiscal quarter flow data			Fiscal quarter stock data		
01/11/1998	31/01/1999	92	31/01/1999	31/01/1999	1
01/02/1999	30/04/1999	89	30/04/1999	30/04/1999	1
01/05/1999	28/07/1999	89	28/07/1999	28/07/1999	1
Fiscal year flow data			Fiscal year stock data		
01/11/1997	31/10/1998	365	31/10/1998	31/10/1998	1
01/11/1998	31/10/1999	365	31/10/1999	31/10/1999	1
01/11/1999	31/10/2000	366	31/10/2000	31/10/2000	1
Multi-weekly flow data			Multi-weekly stock data		
07/01/1998	03/02/1998	28	03/02/1998	03/02/1998	1
04/02/1998	03/03/1998	28	03/03/1998	03/03/1998	1
04/03/1998	07/04/1998	35	07/04/1998	07/04/1998	1
08/04/1998	05/05/1998	28	05/05/1998	05/05/1998	1

In order to mathematically process the information in the reporting periods, it is useful to assign one integer value for each day of the 20-th and 21-st centuries in an ascending manner, e.g. 1 for Jan. 1 1900, 2 for Jan. 2 1900, etc.[2] This scheme thus stores the starting day, the starting month and the starting year in one single date variable; and similarly, for the ending day, the ending month and the ending year. The *timing* of the data can then be defined as the middle date of the reporting period, that is the starting date plus the ending date divided by 2; and the *duration*, as the number of days covered, that is the ending date minus starting date plus one day.

[2] This is how the spreadsheet program Excel® stores date variables. In a similar manner, program SAS® (Statistical Analysis System) assigns 0 to Jan. 1 1960, 1 to Jan. 2 1960, 2 to Jan. 3 1960, 366 to Jan 1 1961, and so forth.

The benchmarking problem arises because the annual sums of the sub-annual series are not equal to the corresponding annual values. In other words, there are (non-zero) *annual* (or temporal) *discrepancies*, d_m, between the annual benchmarks and the sub-annual values:

$$d_m = a_m - \Sigma_{t=t_{1m}}^{t_{Lm}} j_{mt}\, s_t \,,\; m=1,...,M, \tag{3.2a}$$

where t_{1m} and t_{Lm} are respectively first and last sub-annual periods, t, covered by the m-th benchmark, e.g. quarters 1 to 4, 5 to 8, and the j_{mt}'s are the coverage fractions here assumed equal to 1. Note that the quantity, $j_{m\bullet} = \Sigma_t\, j_{mt}$, is the number of sub-annual periods covered by the benchmark.

The discrepancies are more often expressed in terms of *proportional annual* (or temporal) *discrepancies*:

$$d_m^{(p)} = a_m \,/\, (\Sigma_{t=t_{1m}}^{t_{Lm}} j_{mt}\, s_t),\; m=1,...,M. \tag{3.2b}$$

which may be less, equal or greater than 1.0.

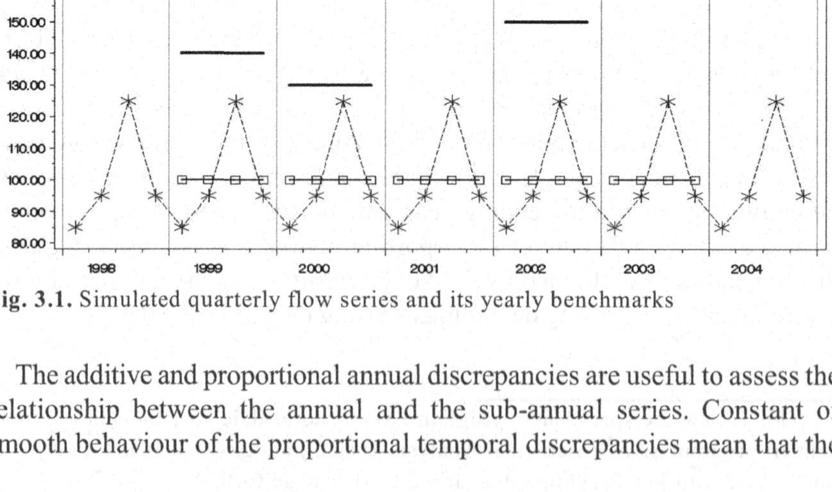

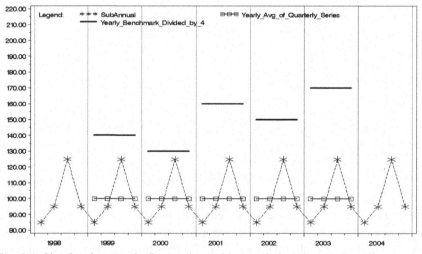

Fig. 3.1. Simulated quarterly flow series and its yearly benchmarks

The additive and proportional annual discrepancies are useful to assess the relationship between the annual and the sub-annual series. Constant or smooth behaviour of the proportional temporal discrepancies mean that the annual movement of the sub-annual series is consistent with that of the annual benchmarks. The erratic behaviour of discrepancies indicate mutually inconsistent pair of series.

The annual discrepancies (3.2a) are illustrated in Fig. 3.1 by the level difference between the yearly benchmarks and the corresponding yearly sums of the quarterly series. The yearly benchmarks and yearly sums are represented divided by 4, in order to give them a quarterly scale. The resulting values are repeated over each year in order to visually convey the quarters covered by each benchmark and by the yearly sums. Table 3.2 displays the benchmarks and the discrepancies. The simulated quarterly series repeats the values 85, 95, 125 and 95 from year to year, as shown by Fig. 3.1.

More generally, the benchmarks will be graphically represented divided by the number of sub-annual periods covered $j_{m\bullet}$ and repeated over these periods. This practice will be applied throughout the book.

Benchmarking usually consists of imposing the annual values onto the sub-annual values. In other words, the sub-annual series is modified so that the annual sums of the resulting sub-annual series be equal to the corresponding benchmark. That is

$$a_m - \Sigma_{t=t_{1m}}^{t_{Lm}} j_{mt}\, \hat{\theta}_t = 0, \; m=1,...,M, \tag{3.3}$$

where $\hat{\theta}_t$ is the benchmarked series.

Classical benchmarking assumes that the benchmarks are fully reliable and therefore *binding*, i.e. they satisfy (3.3). A broader definition of benchmarking recognizes the fact the benchmarks may be observed with error. Benchmarking then consists of optimally combining both the more frequent and less frequent data to increase their reliability. An improved set of annual values may then be obtained by taking the annual sums of the benchmarked series:

$$\hat{a}_m = \Sigma_{t=t_{1m}}^{t_{Lm}} j_{mt}\, \hat{\theta}_t, \; m=1,...,M. \tag{3.4}$$

Such benchmarks are called *non-binding*, because the benchmarked series not necessarily satisfy the constraints (3.3).

3.2 Simple Benchmarking Methods

This section reviews the simplest benchmarking methods, namely the prorating method and the Denton method, which are widely known. Both methods impose the annual benchmarks onto the sub-annual. Fig. 3.2a illustrates the prorating method using the data displayed in Fig. 3.1. The yearly benchmarks appear in Table 3.2; and the quarterly series repeat the values 85, 95, 125 and 95 from year to year. Prorating merely consists of multiplying the sub-annual values by the corresponding annual proportional annual discrepancies of (3.2b), displayed in Table 3.2a. For example the quarterly values of 1999 are multiplied by 1.4; those of 2000 by 1.3; and so forth. For year 2004 without benchmark, the quarterly values are multiplied by the closest proportional discrepancy of 2003 equal to 1.7; and similarly for 1998. This method is easily programmed in spreadsheets programs.

Table 3.2. Yearly benchmarks used for the simulated quarterly series

Starting date	Ending date	Benchmark value	Benchmark variance	Yearly Sums	Discrepancies Addit.	Discrepancies Prop.
01/01/99	31/12/99	560.0	0	400.0	160.0	1.40
01/01/00	31/12/00	520.0	0	400.0	120.0	1.30
01/01/01	31/12/01	640.0	0	400.0	240.0	1.60
01/01/02	31/12/02	600.0	0	400.0	200.0	1.50
01/01/03	31/12/03	680.0	0	400.0	280.0	1.70

Fig. 3.2b displays the proportional yearly discrepancies (3.2b) and the proportional corrections $\hat{\theta}_t / s_t$ produced by the prorating method (where $\hat{\theta}_t$ stands for the benchmarked series). For the years with benchmarks, the corrections coincide with the proportional discrepancies repeated over each year. The corrections are constant within each year. As a result, the prorating method exactly preserves the proportional movement within each year, i.e. $\hat{\theta}_t / s_t - \hat{\theta}_{t-1} / s_{t-1} = 0$; but produces potentially large and spurious discontinuities between the last quarter of a year and the first quarter of the next year, as illustrated in Fig. 3.2b for years 1999 to 2003. The prorating method thus assumes that socio-economic variables grow abruptly precisely between years. This assumption is most dubious: growth is much more likely to occur gradually within and between the years.

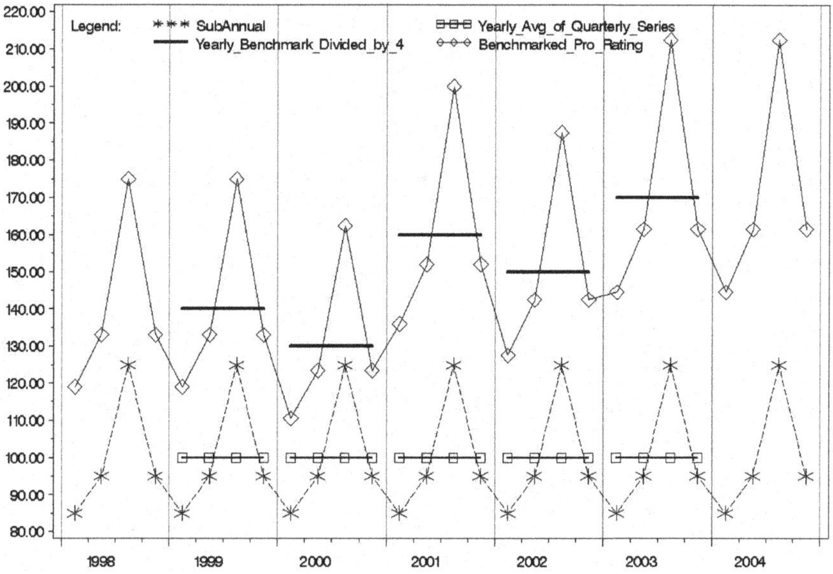

Fig. 3.2a. Simulated quarterly flow series benchmarked to yearly values, using the prorating method

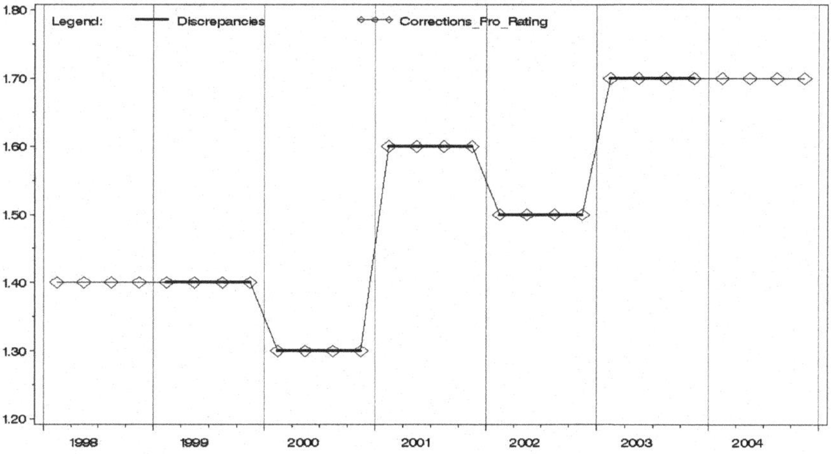

Fig. 3.2b. Proportional yearly discrepancies and proportional quarterly corrections corresponding to the prorating method

The widely used modified Denton[3] method (1971) also imposes the benchmarks onto the sub-annual series and solves the problem of large inter-annual discontinuities of the prorating method. Fig. 3.3a displays the

[3] The original Denton method assumed $\theta_0 = s_0$, which could introduce a spurious transient movement in the benchmarked series, as discussed in Chapter 6.

benchmarked series obtained by the proportional first difference Denton method and by the prorating method. The Denton method gradually distributes the discontinuities over all sub-annual periods.

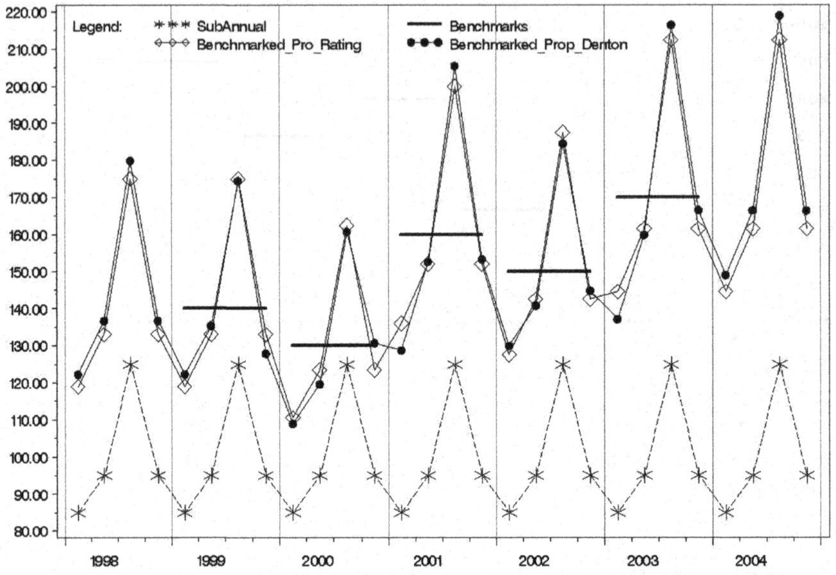

Fig. 3.3a. Simulated quarterly flow series benchmarked to yearly values, using the prorating method and the modified proportional first difference Denton method

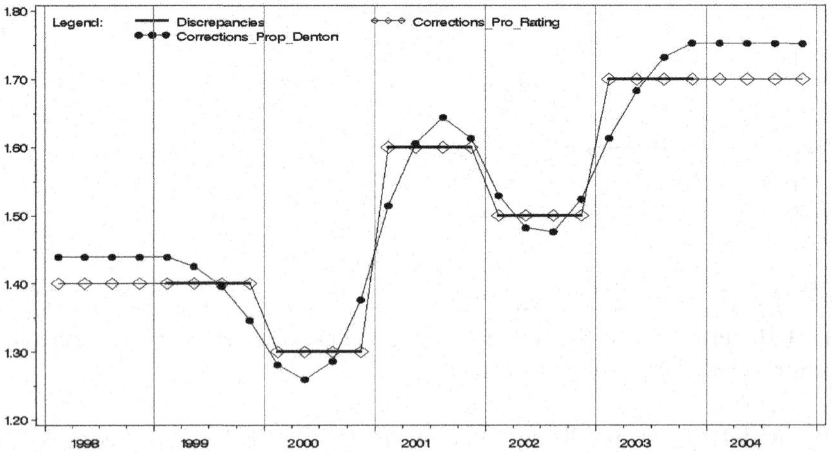

Fig. 3.3b. Proportional yearly discrepancies and proportional quarterly corrections corresponding to the prorating and the Denton methods

This gradual distribution becomes more obvious in Fig. 3.3b, which displays the proportional corrections $\hat{\theta}_t / s_t$ of the Denton and the prorating methods. The flatter and more gradual the corrections, the more movement preservation is achieved. In particular, the proportional first difference variant illustrated maximizes the preservation of the proportional sub-annual movement of the original series, by minimizing the change in proportional movement $\Sigma_{t=2}^{T} (\theta_t / s_t - \theta_{t-1} / s_{t-1})^2$ within and between the years.

The justification of movement preservation is that the original sub-annual movement is the only one available. Note that the Denton corrections display more amplitude than those of prorating. As a result, the Denton benchmarked series has more amplitude, although this is not obvious in Fig. 3.3a. The variants of the Denton method will be fully examined in Chapter 6.

Note that the graphical examination of the series and of the corrections is very useful to detect outliers and anomalies.

We now introduce the regression-based benchmarking model developed by Cholette and Dagum (1994), widely used by Statistics Canada and other statistical agencies. In some respects, the model generalizes the Denton approach to benchmarking.

3.3 The Additive Benchmarking Model

The Cholette-Dagum regression-based additive model consists of the following two equations:

$$\begin{aligned} s_t &= \Sigma_{h=1}^{H} r_{th} \beta_h + \theta_t + e_t \\ &= \Sigma_{h=1}^{H} r_{th} \beta_h + \theta_t + \sigma_t^{\lambda} e_t^{\dagger}, \quad t=1,...,T, \end{aligned} \qquad (3.5a)$$

where $E(e_t) = 0$, $E(e_t \, e_{t-\ell}) = \sigma_t^{\lambda} \sigma_{t-\ell}^{\lambda} \omega_\ell$ and

$$a_m = \Sigma_{t=t_{1m}}^{t_{Lm}} j_{mt} \theta_t + \varepsilon_m, \quad m=1,...,M, \qquad (3.5b)$$

where $E(\varepsilon_m) = 0$, $E(\varepsilon_m^2) = \sigma_{\varepsilon_m}^2$, $E(\varepsilon_m e_t) = 0$.

Eq. (3.5a) shows that the observed sub-annual series s_t is corrupted by deterministic time effects $\Sigma_{h=1}^{H} r_{th} \beta_h$ and autocorrelated error e_t. The latter can be written as $\sigma_t^{\lambda} e_t^{\dagger}$ where $e_t^{\dagger}$ is a standardized error of mean zero and unit variance. In the equation, θ_t stands for the true values which satisfy the annual constraints (3.5b). Eq. (3.5a) basically sets the sub-annual movement of the series under consideration. Eq. (3.5b), on the other hand, sets the level and accounts for the duration and timing of the benchmarks as explained below.

In the case of benchmarking, the deterministic regressors $r_{th}, h=1,...,H$ consist typically of a constant (i.e. H=1 and $r_{tH}=-1$), which captures the average level difference between the annual and the sub-annual data. This quantity is essentially the average of the discrepancies, d_m, given by (3.2a). In a survey context, the resulting estimated regression parameter, $\hat{\beta}_1$, is called "bias". This occurs when the sub-annual data is deemed "biased" with respect to the annual benchmarks due to under-coverage of the survey.

In some cases, a second regressor is used to capture a deterministic trend in the discrepancies, d_m. The next section discusses such a regressor.

The regressors r_{th} may be absent, in which case H=0.

In some cases, one can set H equal to 1, the regressor equal to $-s_t$, and s_t equal to 0. Eq. (3.5a) then becomes

$$0 = -s_t \beta_1 + \theta_t + e_t, \tag{3.6}$$

where the coefficient β_1 now denotes a multiplicative "bias". For example, if β_1 is equal to 1.10, the benchmarked series is on average 10% higher than the original series s_t. The resulting model is multiplicative with an additive error similar to Mian and Laniel (1993). In order to be valid, the error e_t must be independent of the regressor s_t, which implies the standard deviations cannot be derived from the coefficients of variations of the observed sub-annual series.[4] However, this is not the case for the Mian and Laniel model because the bias multiplies θ_t instead of s_t: $s_t = b \times \theta_t + e_t$. Chapter 5 on multiplicative benchmarking comments on their method.

[4] Ordinary and Generalized Least Square estimation require the assumption of independence between the regressors and the error.

The extended Gauss-Markov theorem of Appendix A is needed to justify Eq. (3.6), because the regressand is equal to zero.

The model consisting of (3.6) and (3.5b) is a particular case of the Chow and Lin (1971) interpolation method discussed in Chapter 7; the "related series" is the sub-annual series itself. If the error follows an autoregressive model of order 1 with parameter close to 1 and constant variance, the model approximates the interpolation model of Ginsburgh (1973), except the author pre-estimates parameter β_1 by regressing the benchmarks on the annual sums of the related series with ordinary least squares.

The error e_t in (3.5a) plays a critical role in benchmarking. The error model chosen governs the manner in which the annual discrepancies are distributed over the sub-annual observations. The error in (3.5a) is homoscedastic when λ=0; and heteroscedastic with standard deviation σ_t^λ otherwise. Parameter λ can take values 0, ½ and 1. However, if the standard deviations are constant, the process is homoscedastic regardless of the value of λ. Under independently distributed errors, the error is proportional to σ_t if λ=1/2; and proportional to σ_t^2 if λ=1.

The autocorrelations ω_ℓ of the standardized error $e_t^\dagger$ are usually those of a stationary and invertible auto-regressive moving average (**ARMA**) process (Box and Jenkins, 1976). The ARMA process is often a first or a second order autoregressive (AR) process.

For an AR(1), the autocorrelations are given by $\omega_\ell = \varphi^{|\ell|}$, $\ell = 0,...,T-1$, where $|\varphi|<1$ is the autoregressive parameter. The standardized error $e_t^\dagger = e_t / \sigma_t^\lambda$ then follows the model

$$e_t^\dagger = \varphi\, e_{t-1}^\dagger + v_t\,, \tag{3.7}$$

where the v_t's are the innovations assumed to be independent and identically distributed (i.i.d.).

For an AR(2), the autocorrelations are given by $\omega_0 = 1$, $\omega_1 = \varphi_1 / (1-\varphi_2)$, $\omega_\ell = \varphi_1\, \omega_{\ell-1} + \varphi_2\, \omega_{\ell-2}$, $\ell = 2,...,T-1$, where $\varphi_1 + \varphi_2 < 1$, $\varphi_2 - \varphi_1 < 1$ and $|\varphi_2| < 1$ to ensure stationarity. The standardized error $e_t^\dagger = e_t / \sigma_t^\lambda$ follows the model

$$e_t^\dagger = \varphi_1\, e_{t-1}^\dagger + \varphi_2\, e_{t-2}^\dagger + v_t\,, \tag{3.8}$$

where the v_t's are i.i.d. These models are desirable for stock series.

The choice of an ARMA model and the corresponding covariance matrix for the sub-annual errors is discussed in Chapter 4.

Eq. (3.5b) shows that the benchmarks a_m are equal to the temporal sums of the true sub-annual values θ_t, over their reporting periods plus a non-autocorrelated error ε_m with heteroscedastic variance.

The errors e_t and ε_m are assumed to be mutually uncorrelated because the sub-annual and the annual data come from two different sources (e.g. a survey and administrative records).

If a benchmark a_m is not subject to error ($\sigma^2_{\varepsilon_m} = 0$), it is fully reliable and binding; and, non-binding otherwise. Non-binding benchmarks are not benchmark measurements in a strict sense, but simply less frequent measurements of the target variable. Under non-binding benchmarks, benchmarking combines the sub-annual and the annual observations in an optimal manner. Under binding benchmarks, benchmarking adjust the sub-annual observations to the annual benchmarks, as in the classical cases of benchmarking.

The first and the last sub-annual periods covered by benchmark a_m are respectively denoted by t_{1m} and t_{2L}. The *coverage fractions* are denoted by j_{mt}: for a given sub-annual period t, j_{mt} is the proportion of the days of period t included in the reporting period of the m-th benchmark. For example in the case of a monthly sub-annual series and a benchmark a_m covering from June 21 2002 to June 30 2003, the coverage fractions j_{mt} are 10/30 for June 2002 and 1.0 for July 2002, August 2002 up to June 2003. The coverage fractions are the explanatory variables of the *relative duration* and *relative timing* of the benchmarks with respect to the sub-annual series.

More formally, let $\boldsymbol{A}_m$ be the set of dates (defined by their numerical values as explained in section 3.1) covered by the m-th benchmark and $\boldsymbol{B}_t$ the set of days covered by the month t partially or totally covered by the benchmark. The coverage fraction j_{mt} is then equal to the ratio of the number of dates in the intersection of $\boldsymbol{A}_m$ and $\boldsymbol{B}_t$ (i.e. $\boldsymbol{A}_m \cap \boldsymbol{B}_t$) to the number of dates in $\boldsymbol{B}_t$.

More precise coverage fractions can be obtained by replacing the dates in each set $\boldsymbol{A}_m \cap \boldsymbol{B}_t$ and $\boldsymbol{B}_t$ by the corresponding daily weights and by setting j_{mt}

equal to the ratio of the sum of the daily weights in set $\boldsymbol{A}_m \cap \boldsymbol{B}_t$ to the sum of those in set $\boldsymbol{B}_t$.

In the case of stock series, the duration of the sub-annual and annual periods is often one day; one week, in the case of the Canadian Labour Force series. The formalisation just described would work for benchmarks covering calendar periods. For example, period t would contain the last day of each month and the reporting periods of the benchmarks would contain the last day of each year.

In the case of benchmarks covering fiscal periods, this formalization does not work if some fiscal periods do not end on the last day of a month. In such cases, the coverage fractions reflect the temporal distance between the previous and following sub-annual ending dates. Thus for a benchmark pertaining to the April 10 1996 and two neighbouring sub-annual periods t and t+1 corresponding to the March 31 and April 30 1996, the fractions j_{mt} and $j_{m,t+1}$ applied to θ_t and θ_{t+1} are 20/30 and 10/30. Indeed the distance between the two sub-annual periods t and t+1 is 30 days, the benchmark is 10 days away from t (March 31) and 20 days way from t+1. In order to reflect the relative timing of the benchmark with respect to the two sub-annual periods, j_{mt} assigns more weight to t, namely 20/30; and less weight to t+1, namely 10/30. This kind of situation happens for fiscal data, typically at the micro economic level, at the level of businesses for example.

This scheme is equivalent to fitting a straight line between θ_t and θ_{t+1} and to reading the benchmark on the line at the particular date specified by the reference period. More precisely, this is equivalent to fitting the θ_t's in such a way that the observed benchmark lies on the straight line.

For a benchmark pertaining to a date earlier than the ending date of first sub-annual period t=1 or later than the ending of the last sub-annual period t=T, j_{mt} equal to 1.

The sums of the coverage fractions of benchmark m over t, i.e. $j_{m\bullet} = \Sigma_t \, j_{mt}$, is the number of sub-annual periods covered by the m-th benchmark, e.g. 12 months and 10/30 in the last example. If the daily pattern of activity is known, it can be used to produce more precise coverage fractions.

Coverage fractions in the neighbourhood of 0.50 (in the range 0.30 to 0.70 say) tend to produce large spurious fluctuations in the benchmarked series, if the benchmark covers less than two sub-annual periods. These fluctuations cannot be explained by the sub-annual nor the annual data. The problem is not so acute when the benchmarks cover two or more sub-annual periods ($j_{m\bullet} \geq 2$). However sub-annual benchmarks are not problematic, because their coverage fraction is equal to 1.

In most real applications, the number of benchmarks, M, is much smaller than the number of sub-annual observation, T. In some cases however, M is greater than T. This happens when benchmarking a monthly series to multi-weekly data covering four weeks. Such examples also lead to the problem of coverage fractions in the neighbourhood of 0.50 just mentioned. Some of the benchmarks must then be specified as non-binding; otherwise, the benchmarked monthly series is over-determined.

Section 9.5 of Chapter 9 on the calendarization of multi-weekly data discusses this problem in more detail. For the time being, we assume coverage fractions equal to 1.

3.4 A Conservative Specification of Deterministic Trends

In the case of benchmarking, the deterministic regressor is typically a constant, which captures the average level difference between the annual and the sub-annual data. This quantity is essentially a weigted average of the discrepancies, d_m, given by (3.2a). In a survey context, the resulting estimated regression parameter, $\hat{\beta}_1$, is often called "bias". This occurs when the sub-annual data is deemed "biased" with respect to the annual benchmarks due to under-coverage of the survey.

In some cases, a second regressor is needed to capture a deterministic trend in the discrepancies, d_m. However this trend must be conservative, i.e. to display a mild slope at both ends of the series. We now describe a procedure to obtain a deterministic trend regressor by means of constrained cubic splines.

Polynomial functions of time of degree higher than one are well known to display explosive behaviour at the ends of time series and to undergo heavy revisions when adding new data. Spline techniques can be used to make them level-off at chosen moments of time called knots. Note that the movement of *any* regressor x_t multiplied by a positive (say) regression coefficient β yields the same movement reduced or amplified depending on the value of the coefficient. In particular, if the regressor considered levels off at time t, $x_t\,\beta$ also levels off at t.

Let the deterministic trend c_t follow a polynomial function of time of degree $N \geq 3$, so that

$$c_t = \Sigma_{n=0}^{N} (t-\bar{t})^n \, b_n, \; t=1,...,T, \tag{3.9}$$

subject to $M<N$ constraints $c_{t_m^0} - c_{t_m^0-1} = 0$, $m=1,...,M$ at knots t_m^0, and where $\bar{t} = (T+1)/2$. In this case N=3 and M=2. The constraints can be written as

$$\Sigma_{n=0}^{3} (t_m^0 - \bar{t})^n \, b_n - \Sigma_{n=0}^{3} (t_m^0 - 1 - \bar{t})^n \, b_n = 0, \; m=1,2. \tag{3.10}$$

These constraints can be used to solve the last two parameters in terms of the first two parameters. Let vector $\boldsymbol{b}_B$ contain the last two parameters; and $\boldsymbol{b}_A$, the first two parameters. The trend can be written in matrix form as

$$\boldsymbol{c} = \boldsymbol{X}_A \boldsymbol{b}_A + \boldsymbol{X}_B \boldsymbol{b}_B, \tag{3.11}$$

where the columns of $\boldsymbol{X}_A$ contains the first two powers of time, $(t-\bar{t})^n$, n=0,1; and where $\boldsymbol{X}_B$ contains the last two powers of time, $(t-\bar{t})^n$, n=2,3. The constraints (3.10) can be written in matrix form as

$$\boldsymbol{Z}_A \boldsymbol{b}_A + \boldsymbol{Z}_B \boldsymbol{b}_B = 0, \tag{3.12}$$

where $\boldsymbol{Z}_A = [z_{m,n}^A]$ of dimension 2×2 contains the elements $z_{m,n}^A = (t_m^0 - \bar{t})^n - (t_m^0 - 1 - \bar{t})^n$, $m=1,2$, $n=0,1$; and where $\boldsymbol{Z}_B$ of same dimension contain the elements $z_{m,n}^B = (t_m^0 - \bar{t})^n - (t_m^0 - 1 - \bar{t})^n$, $m=1,2$, $n=2,3$. The constraints (3.12) imply $\boldsymbol{b}_B = -\boldsymbol{Z}_B^{-1} \boldsymbol{Z}_A \boldsymbol{b}_A$. Substituting $\boldsymbol{b}_B$ into (3.11) and rearranging yields

$$\boldsymbol{c} = [\boldsymbol{X}_A - \boldsymbol{X}_B \boldsymbol{Z}_B^{-1} \boldsymbol{Z}_A] \, \boldsymbol{b}_A = \boldsymbol{X} \boldsymbol{b}_A. \tag{3.13}$$

where $\boldsymbol{X}$ contains the two regressors levelling at the two knots.

For example, with T=15, t_1^0=3 and t_2^0=13, the resulting regressor is

$$\mathbf{X} = \begin{bmatrix} 1.00000 & -3.38255 \\ 1.00000 & -3.82550 \\ 1.00000 & -3.82550 \\ 1.00000 & -3.46309 \\ 1.00000 & -2.81879 \\ 1.00000 & -1.97315 \\ 1.00000 & -1.00671 \\ 1.00000 & 0.00000 \\ 1.00000 & 0.96644 \\ 1.00000 & 1.81208 \\ 1.00000 & 2.45638 \\ 1.00000 & 2.81879 \\ 1.00000 & 2.81879 \\ 1.00000 & 2.37584 \\ 1.00000 & 1.40940 \end{bmatrix}, \quad \tilde{\boldsymbol{X}} = \begin{bmatrix} 1.00000 & -3.82550 \\ 1.00000 & -3.82550 \\ 1.00000 & -3.82550 \\ 1.00000 & -3.46309 \\ 1.00000 & -2.81879 \\ 1.00000 & -1.97315 \\ 1.00000 & -1.00671 \\ 1.00000 & 0.00000 \\ 1.00000 & 0.96644 \\ 1.00000 & 1.81208 \\ 1.00000 & 2.45638 \\ 1.00000 & 2.81879 \\ 1.00000 & 2.81879 \\ 1.00000 & 2.81879 \\ 1.00000 & 2.81879 \end{bmatrix}. \qquad (3.14)$$

Fig. 3.4 displays the trend regressor, i.e. the second column of matrix $\boldsymbol{X}$ of Eq. (3.14) (circles); and a modified regressor (dots) obtained from $\boldsymbol{X}$. Its values are in the second column of matrix $\tilde{\boldsymbol{X}}$ of Eq. (3.14) The modified regressor ensures that the trend stays constant before the first knot t_1^0=2 and after the second knot t_2^0=13. The first t_1^0-2 rows (if any) are set equal to row t_1^0-1; and the last $T-t_2^0$ rows, equal to row t_2^0 of $\boldsymbol{X}$.

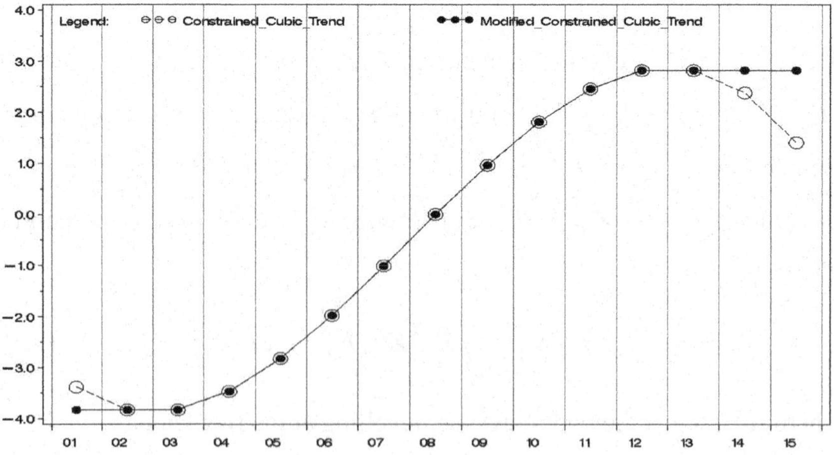

Fig. 3.4. Constrained cubic regressor

Note that depending on the location of the knots, this technique can cause the polynomial to level-off inside or outside the range $[1,...,T]$ (e.g. at $t_2^0=T+3$). In the context of benchmarking, we would recommend the knots be located either at $t_1^0=2$ and $t_2^0=T$, or at the first period covered by the first benchmark and at the period following the last period covered by the last benchmark. The second option provides more conservative extrapolations for the current periods without benchmark.

The actual regressor $\boldsymbol{R}$ proposed is $-\tilde{\boldsymbol{X}}$. This sign reversal facilitates interpretation. Namely a positive regression coefficient for the constant regressor (-1) means the benchmarked series is higher than the original series; and a negative coefficient, lower than the original.

3.5 Flow, Stock and Index Series

The annual measurement Eq. (3.5b) can accommodate flow, stock, and index series. As explained in Chapter 2, the annual values of flows (e.g. sales) correspond to the annual sums of the sub-annual values; the annual values of stocks series (e.g. inventories) pertain to one single sub-annual value, usually the last value of the "year".

For the purpose of benchmarking and interpolation, *index series* are such that their annual benchmarks pertain to the annual averages of a sub-annual series. Thus, the annual values of index series correspond to the average of the sub-annual values, whether the underlying variable is a flow or a stock. For example, the yearly measures of unemployment are usually considered to be the yearly averages of the monthly unemployment (stock) series. For index variables, we multiply each benchmark in Eq. (3.5b) by the number of periods covered, $j_{m\bullet}$; and its variance, by the square of $j_{m\bullet}$.

For flow series, the temporal sums in (3.5b) usually contain more than one term, i.e. $t_{1m} \le t_{Lm}$; the equality holds only in the case of sub-annual benchmarks. For stock series, the temporal sums in (3.5b) usually contain only one term, i.e. $t_{1m} = t_{Lm}$. However if the benchmarks falls between two sub-annual periods, there would be two terms associated with two coverage fractions between 0 and 1 exclusively, as discussed in conjunction to Eq. (3.5b).

3.6 Matrix Representation of the Model

Equations (3.5a) and (3.5b) can be written in matrix algebra as

$$\boldsymbol{s} = \boldsymbol{R}\boldsymbol{\beta} + \boldsymbol{\theta} + \boldsymbol{e},\ E(\boldsymbol{e}) = 0,\ E(\boldsymbol{e}\,\boldsymbol{e}') = \boldsymbol{V}_e, \tag{3.15a}$$

$$\boldsymbol{a} = \boldsymbol{J}\,\boldsymbol{\theta} + \boldsymbol{\varepsilon},\ E(\boldsymbol{\varepsilon}) = 0,\ E(\boldsymbol{\varepsilon}\,\boldsymbol{\varepsilon}') = \boldsymbol{V}_\varepsilon,\ E(\boldsymbol{e}\,\boldsymbol{\varepsilon}') = 0, \tag{3.15b}$$

where $\boldsymbol{s} = [s_1, ..., s_T]'$, $\boldsymbol{\theta} = [\theta_1, ..., \theta_T]'$, $\boldsymbol{a} = [a_1, ..., a_M]'$, and similarly for the other vectors and matrices.

Matrix $\boldsymbol{R}$ of dimension T by H contains the H regressors of Eq. (3.5a) and $\boldsymbol{\beta} = [\beta_1, ..., \beta_H]$ the H regression parameters.

As discussed in Chapter 4, the covariance matrix $\boldsymbol{V}_e$ of the sub-annual series considered may originate from a statistical model applied upstream of benchmarking, for example signal extraction. If such a covariance matrix is available it should be used in benchmarking to maintain the dependence structure of the sub-annual series as much as possible.

In most cases however, the covariance matrix must be built from an autocorrelation function, and from assumed *constant* standard deviations or from standard deviations based on *constant* CVs ($\sigma_t = c \times s_t$) as discussed in Chapter 4. The resulting covariance matrix has the following structure:

$$\boldsymbol{V}_e = \boldsymbol{\Xi}^\lambda \boldsymbol{\Omega}\, \boldsymbol{\Xi}^\lambda, \tag{3.16}$$

where $\boldsymbol{\Xi}$ is a diagonal matrix of standard deviations, $\sigma_1, ..., \sigma_T$, and $\boldsymbol{\Omega}$ contains the autocorrelations usually corresponding to an ARMA error model (McLeod 1975).[5] For an AR(1) error model with parameter $|\varphi|<1$, the elements of $\boldsymbol{\Omega}$ are given by $\omega_{ij} = \varphi^{|i-j|}$. For an AR(2) error model, the elements ω_{ij} are given in terms of lags $\ell = |i-j| : \omega(\ell) = 1$ for $\ell = 0$, $\omega(\ell) = \varphi_1/(1-\varphi_2)$, for $\ell = 1$, $\omega(\ell) = \varphi_1\, \omega_{\ell-1} + \varphi_2\, \omega_{\ell-2}$, for $\ell = 2, ..., T-1$, where $\varphi_1 + \varphi_2 < 1$, $\varphi_2 - \varphi_1 < 1$ and $|\varphi_2| < 1$ to ensure stationarity. Parameter λ can be equal to 0, ½ or 1. With $\lambda=0$, the errors are homoscedastic, and may be heteroscedastic otherwise.

[5] To each ARMA model satisfying the conditions of stationarity and invertibility corresponds a unique autocorrelation matrix.

3.6.1 Temporal Sum Operators

Matrix $\boldsymbol{J}$ is a temporal sum operator of dimension M by T, containing the coverage fractions:

$$\boldsymbol{J} = \begin{bmatrix} j_{11} & j_{12} & \cdots & j_{1T} \\ \vdots & \vdots & \ddots & \vdots \\ j_{M1} & j_{M2} & \cdots & j_{MT} \end{bmatrix}, \tag{3.17}$$

where $j_{mt} = 0$ for $t < t_{1m}$ or $t > t_{Lm}$, and j_{mt} is as defined in Eq. (3.5b) for $t_{1m} \leq t \leq t_{Lm}$. Each coverage fraction j_{mt} is placed in row m and column t of $\boldsymbol{J}$.

For example, for a quarterly sub-annual flow series starting in Jan. 1 1999 and ending in Sept. 30 2002, and calendar year benchmarks starting in Jan 1 1999 and ending in Dec 31 2001, matrix $\boldsymbol{J}$ is

$$\boldsymbol{J} = \begin{bmatrix} 1 & 1 & 1 & 1 & 0 & 0 & 0 & 0 & 0 & 0 & 0 & 0 & 0 & 0 & 0 \\ 0 & 0 & 0 & 0 & 1 & 1 & 1 & 1 & 0 & 0 & 0 & 0 & 0 & 0 & 0 \\ 0 & 0 & 0 & 0 & 0 & 0 & 0 & 0 & 1 & 1 & 1 & 1 & 0 & 0 & 0 \end{bmatrix}.$$

The first row of matrix $\boldsymbol{J}$ corresponds to the first benchmark; the second row to the second ; and so forth. The first column of $\boldsymbol{J}$ pertains to the first quarter of 1999; the second column, to the second quarter of 1999; ...; the fifth column, to the first quarter of 2000; and so forth.

If instead of a flow we have a stock series where the observations are measured on the last day of the fourth quarter, the $\boldsymbol{J}$ matrix is

$$\boldsymbol{J} = \begin{bmatrix} 0 & 0 & 0 & 1 & 0 & 0 & 0 & 0 & 0 & 0 & 0 & 0 & 0 & 0 & 0 \\ 0 & 0 & 0 & 0 & 0 & 0 & 0 & 1 & 0 & 0 & 0 & 0 & 0 & 0 & 0 \\ 0 & 0 & 0 & 0 & 0 & 0 & 0 & 0 & 0 & 0 & 0 & 1 & 0 & 0 & 0 \end{bmatrix}.$$

For a monthly sub-annual flow series starting in Jan. 1 1999 and ending in Sept. 30 2002, and calendar year benchmarks starting in Jan 1 1999 and ending in Dec 31 2001, matrix $\boldsymbol{J}$ is

$$\boldsymbol{J} = \begin{bmatrix} \boldsymbol{\mathit{1}} & \boldsymbol{\mathit{0}} & \boldsymbol{\mathit{0}} & 0 & 0 & 0 & 0 & 0 & 0 & 0 & 0 & 0 \\ \boldsymbol{\mathit{0}} & \boldsymbol{\mathit{1}} & \boldsymbol{\mathit{0}} & 0 & 0 & 0 & 0 & 0 & 0 & 0 & 0 & 0 \\ \boldsymbol{\mathit{0}} & \boldsymbol{\mathit{0}} & \boldsymbol{\mathit{1}} & 0 & 0 & 0 & 0 & 0 & 0 & 0 & 0 & 0 \end{bmatrix},$$

where $\boldsymbol{1}$ is a 1 by 12 row vector of ones, $\boldsymbol{0}$ is a 1 by 12 row vector of zeroes, and 0 is a scalar. For the corresponding stock series pertaining to the last day of December, the 11 first elements of vector $\boldsymbol{1}$ would be zeroes, and the 12-th element would be one.

Matrix $\boldsymbol{J}$ of dimension $M \times T$ can accommodate benchmarks which are irregularly spaced and have different durations. For example, some time periods t may be covered by quarterly benchmarks; some t, by yearly benchmarks; some t, by a quarterly and a yearly benchmarks (*sic*); and some t, not covered at all.

When all the benchmarks are regularly spaced (e.g. all yearly) and have the same duration and when all sub-annual periods t are covered by one benchmark, T is a multiple of M. Matrix $\boldsymbol{J}$ can then be expressed by means of Kronecker products:[6]

$$\boldsymbol{J} = \boldsymbol{I}_M \otimes \boldsymbol{1}.$$

For example, if the sub-annual series is quarterly with three yearly benchmarks (M=3), the Kronecker product yields

$$\boldsymbol{J} = \begin{bmatrix} \boldsymbol{1} & \boldsymbol{0} & \boldsymbol{0} \\ \boldsymbol{0} & \boldsymbol{1} & \boldsymbol{0} \\ \boldsymbol{0} & \boldsymbol{0} & \boldsymbol{1} \end{bmatrix} = \begin{bmatrix} 1 & 1 & 1 & 1 & 0 & 0 & 0 & 0 & 0 & 0 & 0 & 0 \\ 0 & 0 & 0 & 0 & 1 & 1 & 1 & 1 & 0 & 0 & 0 & 0 \\ 0 & 0 & 0 & 0 & 0 & 0 & 0 & 0 & 1 & 1 & 1 & 1 \end{bmatrix},$$

where $\boldsymbol{1}$ is a 1×4 row vector of ones, and $\boldsymbol{0}$ is a 1×4 row vector of zero. However, this situation seldom happens in practice, mainly because the current year is typically without benchmark. Nevertheless, this notation is convenient for theoretical analyses.

3.6.2 Solution of the Model

Model (3.15) can be written in matrix notation as

$$\begin{bmatrix} \boldsymbol{s} \\ \boldsymbol{a} \end{bmatrix} = \begin{bmatrix} \boldsymbol{R} & \boldsymbol{I}_T \\ \boldsymbol{0} & \boldsymbol{J} \end{bmatrix} \begin{bmatrix} \boldsymbol{\beta} \\ \boldsymbol{\theta} \end{bmatrix} + \begin{bmatrix} \boldsymbol{e} \\ \boldsymbol{\varepsilon} \end{bmatrix}; \qquad (3.18a)$$

[6] The elements of Kronecker product $\boldsymbol{A} \otimes \boldsymbol{B}$, of matrices $\boldsymbol{A}$ and $\boldsymbol{B}$ respectively of dimension $K \times L$ and $M \times N$, are given by $a_{ij}\, \boldsymbol{B}$, where a_{ij} are the elements of $\boldsymbol{A}$. The resulting product has dimension $KM \times LN$.

or simply as

$$y = X\alpha + u, \quad E(u) = 0, \; E(u\,u') = V_u = block(V_e, V_\varepsilon), \tag{3.18b}$$

where $y = \begin{bmatrix} s \\ a \end{bmatrix}$, $X = \begin{bmatrix} R & I_T \\ 0 & J \end{bmatrix}$, $\alpha = \begin{bmatrix} \beta \\ \theta \end{bmatrix}$, $u = \begin{bmatrix} e \\ \varepsilon \end{bmatrix}$, $V_u = \begin{bmatrix} V_e & 0 \\ 0 & V_\varepsilon \end{bmatrix}$ and I_T is an identity matrix of dimension $T{\times}T$.

Model (3.18b) has the form of a standard regression model, with generalized least square solution:

$$\hat{\alpha} = (X' V_u^{-1} X)^{-1}\, X' V_u^{-1} y \;\; \equiv \begin{bmatrix} \hat{\beta} \\ \hat{\theta} \end{bmatrix}, \tag{3.19a}$$

$$var[\hat{\alpha}] = (X' V_u^{-1} X)^{-1} \;\; \equiv \begin{bmatrix} var[\hat{\beta}] & cov[\hat{\beta}\,\hat{\theta}] \\ cov[\hat{\theta}\,\hat{\beta}] & var[\hat{\theta}] \end{bmatrix}. \tag{3.19b}$$

The H first elements of the vector $\hat{\alpha}=[\hat{\beta}'\,\hat{\theta}']'$ contain the estimates of $\hat{\beta}_1,\ldots,\hat{\beta}_H$; the last T elements, the estimates of $\hat{\theta}_1,\ldots,\hat{\theta}_T$. Similarly the first H rows and columns of $var[\hat{\alpha}]$ contains the covariance matrix of vector $\hat{\beta}$; and the last T rows and columns of $var[\hat{\alpha}]$, the covariance matrix of vector $\hat{\theta}$.

The solution (3.19) is obtained only if V_u is invertible, which requires that both V_e and V_ε be invertible. The implication of the latter is that all the diagonal elements of V_ε must be greater than zero, i.e. all the benchmarks must be non-binding.

In Appendix B we provide a proof for an alternative solution to Eq. (3.19), which allows binding benchmarks ($V_\varepsilon = 0$) as follows:

$$\hat{\beta} = -(R' J' V_d^{-1} J R)^{-1}\, R' J' V_d^{-1}\, [a - J s], \tag{3.20a}$$

$$var[\hat{\beta}] = (R' J' V_d^{-1} J R)^{-1}, \tag{3.20b}$$

$$\hat{\boldsymbol{\theta}} = \boldsymbol{s}^{\dagger} + \boldsymbol{V}_e \boldsymbol{J}' \boldsymbol{V}_d^{-1} [\boldsymbol{a} - \boldsymbol{J}\boldsymbol{s}^{\dagger}], \tag{3.20c}$$

$$var[\hat{\boldsymbol{\theta}}] = \boldsymbol{V}_e - \boldsymbol{V}_e \boldsymbol{J}' \boldsymbol{V}_d^{-1} \boldsymbol{J} \boldsymbol{V}_e + \boldsymbol{W}\, var[\hat{\boldsymbol{\beta}}]\, \boldsymbol{W}', \tag{3.20d}$$

where $\boldsymbol{s}^{\dagger} = \boldsymbol{s} - \boldsymbol{R}\hat{\boldsymbol{\beta}}$, $\boldsymbol{W} = \boldsymbol{R} - \boldsymbol{V}_e \boldsymbol{J}' \boldsymbol{V}_d^{-1} \boldsymbol{J}\boldsymbol{R}$ and $\boldsymbol{V}_d = [\boldsymbol{J}\boldsymbol{V}_e\boldsymbol{J}' + \boldsymbol{V}_\varepsilon]$ and where the invertibility of $\boldsymbol{V}_d$ now requires only the non-singularity of $\boldsymbol{J}\boldsymbol{V}_e\boldsymbol{J}'$. Note that $\boldsymbol{V}_d$ is the covariance matrix of the annual discrepancies $\boldsymbol{d} = [\boldsymbol{a} - \boldsymbol{J}\boldsymbol{s}]$. This solution is widely applied in contexts other than benchmarking.

In the absence of $\boldsymbol{R}\boldsymbol{\beta}$, the solution (3.20) simplifies to:

$$\hat{\boldsymbol{\theta}} = \boldsymbol{s} + \boldsymbol{V}_e \boldsymbol{J}' \boldsymbol{V}_d^{-1} [\boldsymbol{a} - \boldsymbol{J}\boldsymbol{s}], \tag{3.21a}$$

$$var[\hat{\boldsymbol{\theta}}] = \boldsymbol{V}_e - \boldsymbol{V}_e \boldsymbol{J}' \boldsymbol{V}_d^{-1} \boldsymbol{J} \boldsymbol{V}_e. \tag{3.21b}$$

Appendix C provides analytical formulae to directly calculate some of the matrix products in (3.20) and (3.21), with no need to generate the matrices to be multiplied. These formulae are very useful for programming.

The residuals associated with the sub-annual and the annual equations are respectively

$$\hat{\boldsymbol{e}} = (\boldsymbol{s} - \boldsymbol{R}\hat{\boldsymbol{\beta}}) - \hat{\boldsymbol{\theta}} \equiv \boldsymbol{s}^{\dagger} - \hat{\boldsymbol{\theta}} \equiv -\boldsymbol{V}_e \boldsymbol{J}' \boldsymbol{V}_d^{-1} [\boldsymbol{a} - \boldsymbol{J}\boldsymbol{s}^{\dagger}], \tag{3.22a}$$

$$\hat{\boldsymbol{\varepsilon}} = \boldsymbol{a} - \boldsymbol{J}\hat{\boldsymbol{\theta}}. \tag{3.22b}$$

Sometimes the variances of $\boldsymbol{e}$ and $\boldsymbol{\varepsilon}$ are known, hence there is no need to estimate the variances of the residuals. For most applications, it is necessary to estimate the residuals, calculate their variances and refit the model to obtain valid confidence intervals for the estimates. Of course, this is not necessary if the method is used in a numerical manner.

In a similar situation (group heteroscedasticity),[7] the practice is to take the average of the original variances and the average of the squared residuals, respectively:

$$\bar{\sigma}^{2\,(0)} = \Sigma_{t=1}^{T} \sigma_t^2 / T, \quad \bar{\sigma}^{2\,(1)} = \Sigma_{t=1}^{T} \hat{e}_t^2 / T. \tag{3.23}$$

[7] Each variance corresponds to the variance of a group.

The original standard deviations of (3.5a) are then multiplied by the ratio $\bar{\sigma}^{(1)}/\bar{\sigma}^{(0)}$, and the model is re-estimated on the basis of $\hat{\boldsymbol{V}}_e = \hat{\boldsymbol{\Xi}}\boldsymbol{\Omega}\hat{\boldsymbol{\Xi}}$.

We do not re-estimate the variances of the benchmarks, which are usually null or much smaller than the sub-annual variances. Under binding benchmarks ($\boldsymbol{V}_\varepsilon = 0$), refitting the model does *not* change the benchmarked series $\hat{\boldsymbol{\theta}}$ nor the parameters $\hat{\boldsymbol{\beta}}$, but does change their variances. It is then sufficient to multiply the covariance matrices $var[\hat{\boldsymbol{\beta}}]$ and $var[\hat{\boldsymbol{\theta}}]$ by $(\bar{\sigma}^{(1)}/\bar{\sigma}^{(0)})^2$.

The variance of any linear combination of the adjusted series and its variance is given by

$$\hat{\boldsymbol{\theta}}_L = \boldsymbol{L}\hat{\boldsymbol{\theta}}, \tag{3.24a}$$

$$var[\hat{\boldsymbol{\theta}}_L] = \boldsymbol{L}\, var[\hat{\boldsymbol{\theta}}]\boldsymbol{L}'. \tag{3.24b}$$

In particular, matrix $\boldsymbol{L}$ can be a temporal sum operator, taking sums over calendar years, calendar quarters, etc.

Guerrero (1990) and Guerrero and Peña (2000) provide a test to determine whether the original sub-annual series and the benchmarks are mutually compatible. In terms of our notation, the test consists of comparing the statistic

$$G = (\boldsymbol{a} - \boldsymbol{J}\boldsymbol{s}^\dagger)'\,(\boldsymbol{J}\hat{\boldsymbol{V}}_e\boldsymbol{J}')^{-1}\,(\boldsymbol{a} - \boldsymbol{J}\boldsymbol{s}^\dagger),$$

to a chi-square distribution with $M\!-\!H\!-\!L$ degrees of freedom, where $\boldsymbol{s}^\dagger = \boldsymbol{s} - \boldsymbol{R}\hat{\boldsymbol{\beta}}$ and L is the number of parameters in the ARMA models. If the statistic G is less than the tabulated value $\chi^2_n(\alpha)$, the sub-annual series and the benchmarks are compatible at significance level α. The test assumes benchmarks ($\boldsymbol{V}_\varepsilon = \boldsymbol{0}$) and the normal distribution of the errors.

3.7 Other Properties of the Regression-Based Benchmarking Method

The Cholette-Dagum additive benchmarking model contains very few degrees of freedom: $T+H$ parameters must be estimated from $T+M$ observations. This means that the benchmarked values will be very close to the bias-corrected series, $\boldsymbol{s}^{\dagger} = \boldsymbol{s} - \boldsymbol{R}\hat{\boldsymbol{\beta}}$. This proximity is in fact required to preserve the sub-annual movement of the sub-annual series. In other words, as already said before, the annual benchmarks are mainly useful to correct the level of the original sub-annual series.

Furthermore, Eq. (3.21b), i.e.

$$var[\hat{\boldsymbol{\theta}}] = \boldsymbol{V}_e - \boldsymbol{V}_e \boldsymbol{J}' \boldsymbol{V}_d^{-1} \boldsymbol{J} \boldsymbol{V}_e,$$

shows that, in the absence of a deterministic time regressor (H=0), the values of the covariance matrix of the benchmarked series, $var[\hat{\boldsymbol{\theta}}]$, are smaller than the original covariance matrix $\boldsymbol{V}_e$, because the term subtracted from $\boldsymbol{V}_e$, $\boldsymbol{V}_e \boldsymbol{J}' \boldsymbol{V}_d^{-1} \boldsymbol{J} \boldsymbol{V}_e$, is positive definite. In fact, the benchmark constraints whether binding or not can be interpreted as smoothing priors and will always reduce the variance of the adjusted series.

On the other hand, Eq. (3.20d), i.e.

$$var[\hat{\boldsymbol{\theta}}] = \boldsymbol{V}_e - \boldsymbol{V}_e \boldsymbol{J}' \boldsymbol{V}_d^{-1} \boldsymbol{J} \boldsymbol{V}_e + \boldsymbol{W} var[\hat{\boldsymbol{\beta}}] \boldsymbol{W}',$$

shows that in the presence of a time regressor (H >0) the resulting covariance $var[\hat{\boldsymbol{\theta}}]$ may be larger, equal or smaller than $\boldsymbol{V}_e$. This will depend on the value of $var[\hat{\boldsymbol{\beta}}]$, more precisely on whether $\boldsymbol{V}_e \boldsymbol{J}' \boldsymbol{V}_d^{-1} \boldsymbol{J} \boldsymbol{V}_e$ is larger, equal or smaller than $\boldsymbol{W} var[\hat{\boldsymbol{\beta}}] \boldsymbol{W}'$.

The values shown in Tables 3.3a and 3.3b illustrate the above observations, using the simulated quarterly series of Fig. 3.1. The regression-based additive model applied includes a constant regressor $\boldsymbol{R}$ equal to -1, a correlation matrix $\boldsymbol{\Omega}$ corresponding to an AR(1) error model $e_t = 0.729\, e_{t-1} + v_t$ with $\boldsymbol{\Xi} = \boldsymbol{I}_T$ and $\lambda = 1$. Table 3.3a displays the original quarterly series, s, the resulting quarterly series corrected for "bias", $\boldsymbol{s}^{\dagger} = \boldsymbol{s} - \boldsymbol{R}\hat{\boldsymbol{\beta}}$, the corresponding benchmarked series $\hat{\boldsymbol{\theta}}$, depicted in Fig. 3.5a. Table 3.3b displays the contributions to the variance of the benchmarked series $var[\hat{\boldsymbol{\theta}}]$, namely the *diagonal elements* of matrix $\boldsymbol{V}_e$, of matrix product $\boldsymbol{V}_e \boldsymbol{J}' \boldsymbol{V}_d^{-1} \boldsymbol{J} \boldsymbol{V}_e$ and of matrix product $\boldsymbol{W} var[\hat{\boldsymbol{\beta}}] \boldsymbol{W}'$.

As shown in Table 3.3b, the variance of the benchmarked series, $var[\hat{\boldsymbol{\theta}}]$, is smaller than the original variance $\boldsymbol{V}_e$, except for the last two quarters of year 2004 and the first two quarters of year 1998; these two years have no benchmark. For those two quarters, the positive contribution to the variance $\boldsymbol{W} var[\hat{\boldsymbol{\beta}}] \boldsymbol{W}'$ dominates the negative contribution of $\boldsymbol{V}_e \boldsymbol{J}' \boldsymbol{V}_d^{-1} \boldsymbol{J} \boldsymbol{V}_e$.

Table 3.3a. Simulated quarterly series benchmarked using the Cholette-Dagum additive regression-based model

Qu.	Year	t	s	$s^{\dagger}$	$\hat{\theta}$
I	1998	1	85.0	136.8	134.8
II	1998	2	95.0	146.8	144.1
III	1998	3	125.0	176.8	173.1
IV	1998	4	95.0	146.8	141.8
I	1999	5	85.0	136.8	129.9
II	1999	6	95.0	146.8	137.4
III	1999	7	125.0	176.8	163.8
IV	1999	8	95.0	146.8	128.9
I	2000	9	85.0	136.8	112.2
II	2000	10	95.0	146.8	120.6
III	2000	11	125.0	176.8	154.1
IV	2000	12	95.0	146.8	133.1
I	2001	13	85.0	136.8	138.3
II	2001	14	95.0	146.8	156.7
III	2001	15	125.0	176.8	189.1
IV	2001	16	95.0	146.8	155.8
I	2002	17	85.0	136.8	136.4
II	2002	18	95.0	146.8	142.2
III	2002	19	125.0	176.8	172.9
IV	2002	20	95.0	146.8	148.4
I	2003	21	85.0	136.8	149.4
II	2003	22	95.0	146.8	165.8
III	2003	23	125.0	176.8	198.1
IV	2003	24	95.0	146.8	166.7
I	2004	25	85.0	136.8	151.3
II	2004	26	95.0	146.8	157.3
III	2004	27	125.0	176.8	184.5
IV	2004	28	95.0	146.8	152.4

A final note on the properties of the variance of benchmarked series is in order. If the M benchmarks are binding (variance equal to 0), the rank of the covariance matrix $var[\hat{\boldsymbol{\theta}}]$ is T-M. Table 3.2 displays the yearly benchmarks used for the simulated quarterly series of Table 3.3a. Each yearly block of $var[\hat{\boldsymbol{\theta}}]$, consisting of rows 1 to 4 and columns 1 to 4, rows 5 to 8 and columns 5 to 8, etc., has rank 3 instead of 4. However, for year 1998 and 2004 without benchmarks, the yearly block has full rank. If M^* of the M benchmarks are binding the rank of $var[\hat{\boldsymbol{\theta}}]$ is T-M^*. These facts are relevant for the reconciliation of benchmarked series, as discussed in Chapters 11 to 14.

Fig. 3.5a illustrates the benchmarking process of the simulated quarterly series of Table 3.3. The regression-based methods perform benchmarking in two stages. The first stage corrects for "bias", by raising the quarterly series to the average level of the benchmarks, $\boldsymbol{s}^{\dagger} = \boldsymbol{s} - \boldsymbol{R}\hat{\boldsymbol{\beta}}$; if $\boldsymbol{R}$ is a constant (as in the example), this step exactly preserves the quarter-to-quarter movement.

Table 3.3b. Contributions to the variance of the benchmarked series shown in Table 3.3a

t	V_e	$V_e J' V_d^{-1} J V_e$	$W\,var[\hat{\boldsymbol{\beta}}]W'$	$var[\hat{\boldsymbol{\theta}}]$
1	1.0000	-0.0515	0.1498	1.0983
2	1.0000	-0.0969	0.1166	1.0197
3	1.0000	-0.1824	0.0779	0.8955
4	1.0000	-0.3432	0.0374	0.6942
5	1.0000	-0.6458	0.0058	0.3599
6	1.0000	-0.8053	0.0000	0.1947
7	1.0000	-0.7930	0.0013	0.2083
8	1.0000	-0.6843	0.0018	0.3175
9	1.0000	-0.6985	0.0003	0.3017
10	1.0000	-0.8054	0.0000	0.1946
11	1.0000	-0.8048	0.0001	0.1953
12	1.0000	-0.7003	0.0001	0.2997
13	1.0000	-0.7010	0.0000	0.2990
14	1.0000	-0.8054	0.0000	0.1946
15	1.0000	-0.8054	0.0000	0.1946
16	1.0000	-0.7010	0.0000	0.2990
17	1.0000	-0.7003	0.0001	0.2997
18	1.0000	-0.8048	0.0001	0.1953
19	1.0000	-0.8054	0.0000	0.1946
20	1.0000	-0.6985	0.0003	0.3017
21	1.0000	-0.6843	0.0018	0.3175
22	1.0000	-0.7930	0.0013	0.2083
23	1.0000	-0.8053	0.0000	0.1947
24	1.0000	-0.6458	0.0058	0.3599
25	1.0000	-0.3432	0.0374	0.6942
26	1.0000	-0.1824	0.0779	0.8955
27	1.0000	-0.0969	0.1166	1.0197
28	1.0000	-0.0515	0.1498	1.0983

The second stage performs a residual adjustment around $\boldsymbol{s}^{\dagger}$ to satisfy the binding benchmark constraints. The final benchmarked series converges to the bias-corrected series $\boldsymbol{s}^{\dagger}$ at the end (year 2004). This is a desirable feature, because the bias, $-\boldsymbol{R}\hat{\boldsymbol{\beta}}$, measures the average historical annual discrepancy between the benchmarks and the annual sums of the sub-annual series. This feature thus provides an acceptable implicit forecast of the next benchmark.

This should also minimize revisions to the preliminary benchmarked series when the next benchmark actually becomes available. The convergence is due to the AR(1) model assumed for the error e_t. In the absence of benchmarks (after 2003), the residual corrections for 2004 follow the deterministic recursion $e_t = 0.729\, e_{t-1}$ because there are no new innovations.

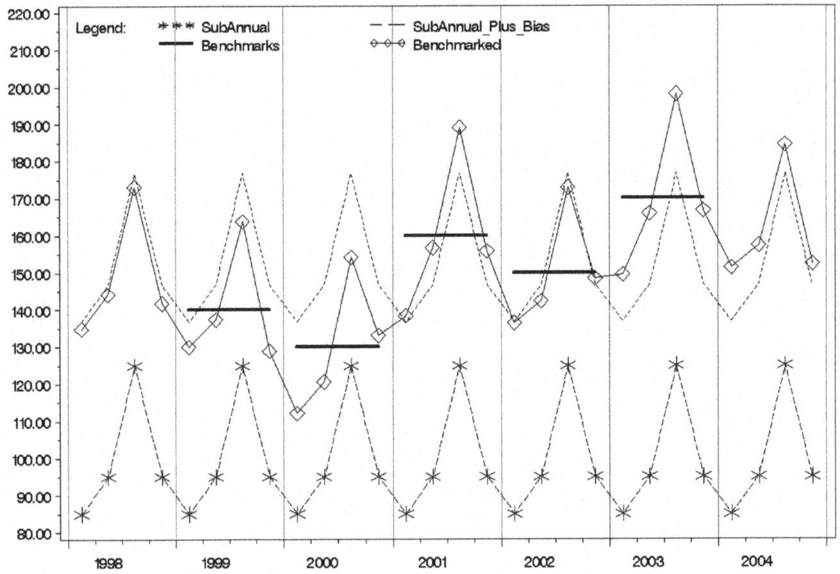

Fig. 3.5a. Simulated quarterly flow series benchmarked to yearly values, using the additive Cholette-Dagum method under a constant bias and an AR(1) error model with φ=0.729

Fig. 3.5b illustrates the benchmarking process of the simulated quarterly series using the same additive regression-based model, except that the bias follows a spline trend levelling-off at the first sub-annual period covered by the first benchmark and at the last sub-annual period covered by the last benchmark, in the first quarter of 1999 and the last quarter of 2003 respectively.

Fig. 3.5c displays the spline bias and the residual corrections around it, accompanied with the constant bias (used in Fig. 3.5a) and its corresponding residual corrections. In both cases, the residual corrections converge to their respective bias at both ends, which causes the current benchmarked series to converge to the respective series corrected for bias of Fig. 3.5b and Fig. 3.5b. Note that the total corrections are different only at the beginning and at the end of the series. In other words, the shape of the bias has little effect on the central years 2000 to 2002. Given the trend displayed by the yearly discrepancies, the spline bias is likely to lead to smaller revisions than the constant one.

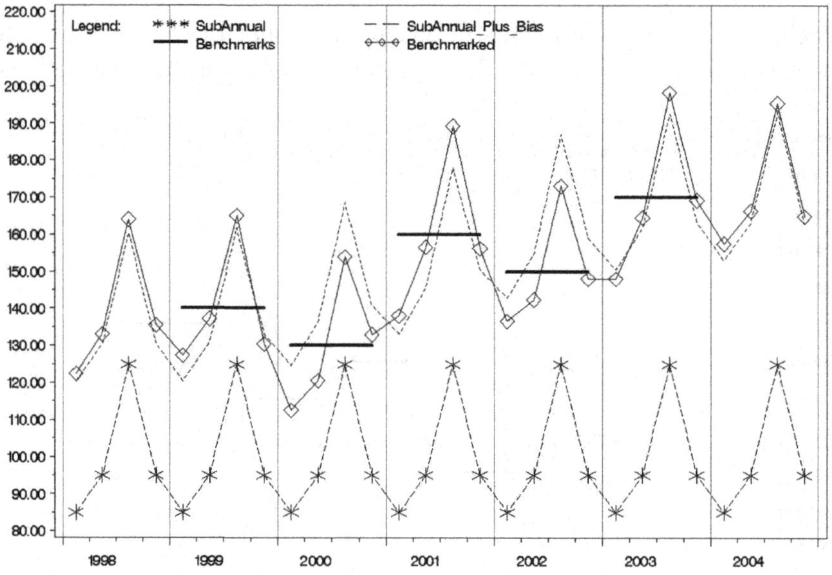

Fig. 3.5b. Simulated quarterly flow series benchmarked to yearly values, using the additive Cholette-Dagum method under a spline bias and an AR(1) error model with φ=0.729

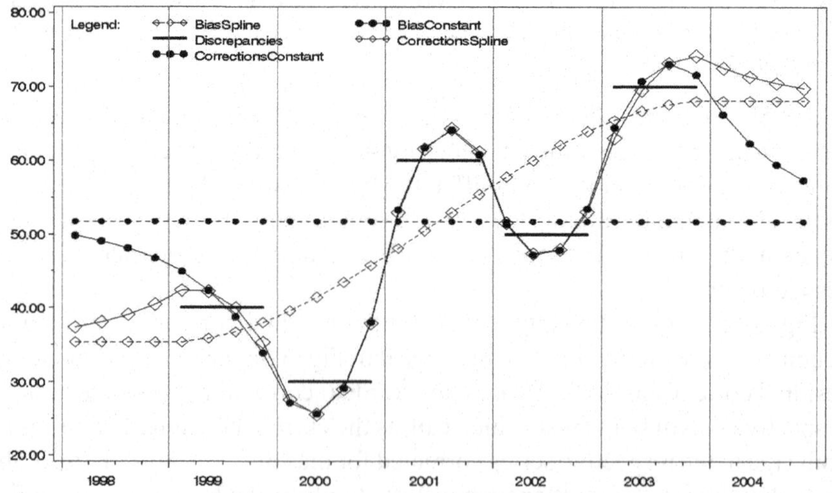

Fig. 3.5c. Yearly discrepancies and quarterly corrections under constant and spline bias

3.8 Proportional Benchmarking with the Regression-Based Model

Many time series display seasonal patterns with wide amplitude. This is the case for the Canadian Total Retail Trade series exhibited in Fig. 3.7a. Additive benchmarking assumes that all observations are equally accountable for the annual discrepancies regardless of their relative values. This assumption may not be appropriate in some cases and can produce negative benchmarked values for positively defined variables. Proportional benchmarking, on the other hand, assumes that seasonally high values are more accountable for the annual discrepancies than the seasonally low ones and rarely produces negative benchmarked values.

In order to achieve proportional benchmarking, the sub-annual series considered is multiplied by a re-scaling factor b given by,

$$b = (\Sigma_{m=1}^{M} a_m) \,/\, (\Sigma_{m=1}^{M} \Sigma_{t=t_{1m}}^{t_{Lm}} j_{mt}\, s_t). \tag{3.26}$$

This re-scaled series $s_t^{\dagger} = b \times s_t$ brings the sub-annual series to the average level of the benchmarks and exactly preserves the proportional movement and the growth rates of the original series: $s_t^{\dagger} / s_t - s_{t-1}^{\dagger} / s_{t-1} = 0$, t=2,...,T. Indeed $s_t^{\dagger} / s_t = s_{t-1}^{\dagger} / s_{t-1}$ implies $s_t^{\dagger} / s_{t-1}^{\dagger} - 1 = s_t / s_{t-1} - 1$. Note that depending of its value, the re-scaling factor shrinks or amplifies the seasonal pattern without ever producing negative values for $s_t^{\dagger}$, unless s_t contains negative values.

The re-scaled series $s_t^{\dagger} = b \times s_t$ is then benchmarked using the regression-based additive benchmarking model with no regressor $\boldsymbol{R}$. The covariance matrix used in the benchmarking is $V_e = \boldsymbol{\Xi}^{\lambda} \boldsymbol{\Omega} \boldsymbol{\Xi}^{\lambda}$, where λ is set to 1. Matrix $\boldsymbol{\Xi}$ is set to $diag(c \times \boldsymbol{s})$ where c is a constant coefficient of variation equal to 0.01 (say). The autocorrelation matrix $\boldsymbol{\Omega}$ is that of a regular first order autoregressive process. If the autoregressive parameter φ is close to 1, the benchmarked series closely approximate the proportional first difference Denton method illustrated in Fig. 3.3. If φ is not so close to 1, e.g. 0.8, we are still dealing with proportional benchmarking with a milder form of movement preservation as discussed in Chapter 4.

Note that with a constant regressor (H=1 and $\boldsymbol{R}$=−1), the series would be benchmarked under a mixed model: the re-scaling would be additive and the residual adjustment would be proportional.

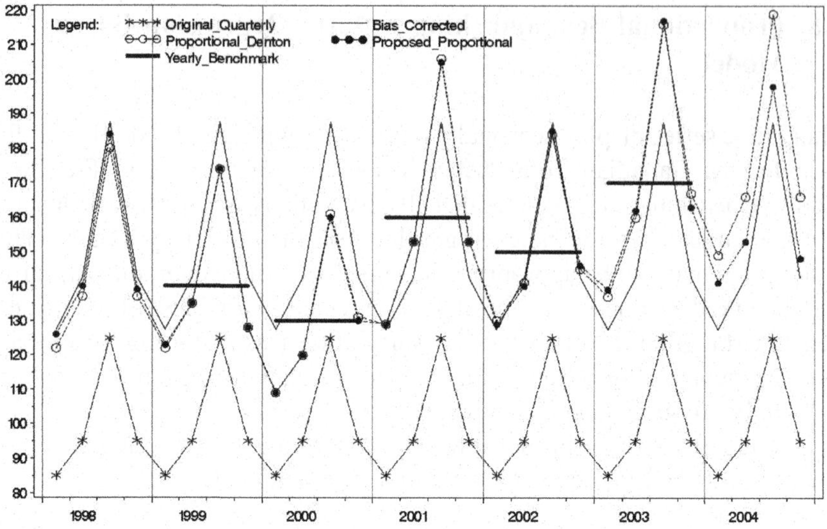

Fig. 3.6a. Simulated quarterly series benchmarked using the proportional Denton and the proposed proportional method with an AR(1) error model with φ=0.729

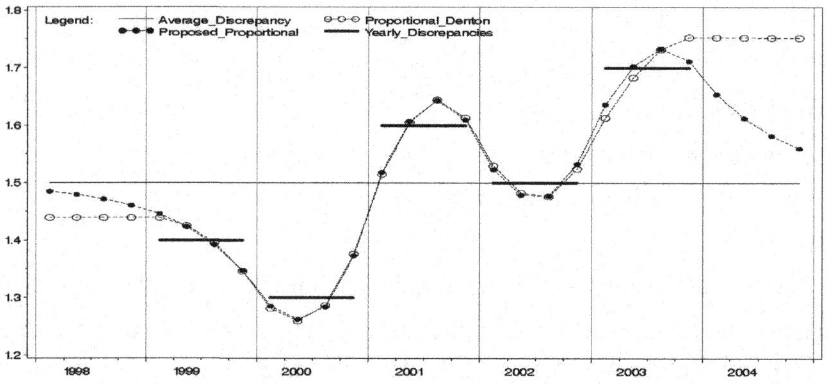

Fig. 3.6b. Yearly discrepancies and quarterly corrections under constant and spline bias

To users looking for a pragmatic benchmarking method routinely applicable to large numbers of seasonal time series, we recommend the proportional benchmarking method described in this section with first order autoregessive errors. The autoregressive parameter should be positive, with value in the neighbourhood of 0.90 for monthly (sub-annual) series, 0.90^3 for quarterly series, and $0.90^{1/30.4375}$ for daily series, 30.4375 being the average number of days per month. As described in Chapter 9, the benchmarking of daily series occurs mostly in the calendarization of data with short reporting periods (e.g. 4 and 5 weeks) to monthly values.

Fig 3.6a exhibits the simulated benchmarked quarterly series obtained from the proportional benchmarking model just proposed and the series obtained from the modified first difference Denton method. The former benchmarked series converges to the series corrected for "bias", whereas the latter does not. This is more obvious in Fig. 3.6b, where the proportional corrections under the proposed model converge to the bias.

The covariance matrix of the proportionally benchmarked series is approximated by Eq. (3.21). This covariance ignores the error associated with the multiplicative bias (3.26).

3.9 A Real Data Example: the Canadian Total Retail Trade Series

The Canadian monthly Retail Trade series considered originates from the Monthly Wholesale and Retail Trade Survey; and the yearly benchmarks, from the Annual (yearly) Census of Manufacturing.

Fig. 3.7a displays the original and the benchmarked Total Retail Trade Series, obtained with by the C-D additive benchmarking model. The model includes a constant regressor $\boldsymbol{R}$ equal to -1 to capture the "bias", a correlation matrix $\boldsymbol{\Omega}$ corresponding to an homoscedastic AR(1) error model $e_t = 0.90\, e_{t-1} + v_t$, with $\boldsymbol{\Xi} = \boldsymbol{I}_T$ and $\lambda = 1$. The benchmarks are specified as binding ($\boldsymbol{V}_\varepsilon = \boldsymbol{0}$).

Fig. 3.7b displays the yearly discrepancies, represented by their yearly average, and the *total corrections* $\hat{\theta}_t - s_t$ made to the original monthly data to obtain the benchmarked series. The total corrections consist of

(a) adding the bias correction, equal to 1,223,740 (with t-value exceeding 27), and
(b) adding a *residual correction* ($\hat{e}_t$) to the bias to satisfy the benchmarking constraints.

The bias correction of (a) exactly preserve the month-to-month movement, i.e. $s_t^{\dagger} - s_{t-1}^{\dagger} = s_t - s_{t-1}$; and the residual correction of (b) largely preserves the movement because $\hat{e}_t$ is much smaller than the bias and follows the AR(1) error model $e_t = 0.90\, e_{t-1} + v_t$.

The total corrections thus entirely distribute the yearly discrepancies over the original sub-annual series. In Fig. 3.7b, the area under each yearly discrepancy (steps) is equal to the area under the total corrections for the months covered by the benchmark. Put differently, the area between the corrections and the discrepancies cancels out for the months covered.

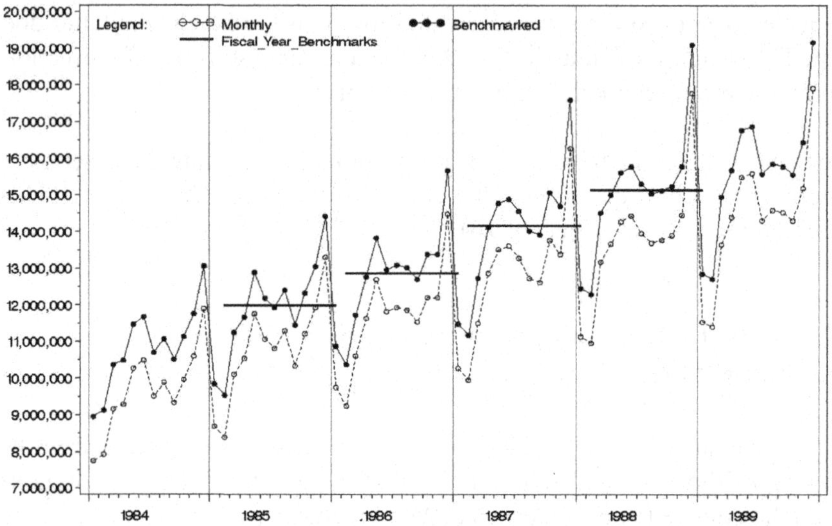

Fig. 3.7a. The Canadian Total Retail Trade series benchmarked to yearly values, using the additive Cholette-Dagum method under an AR(1) error model with φ=0.90

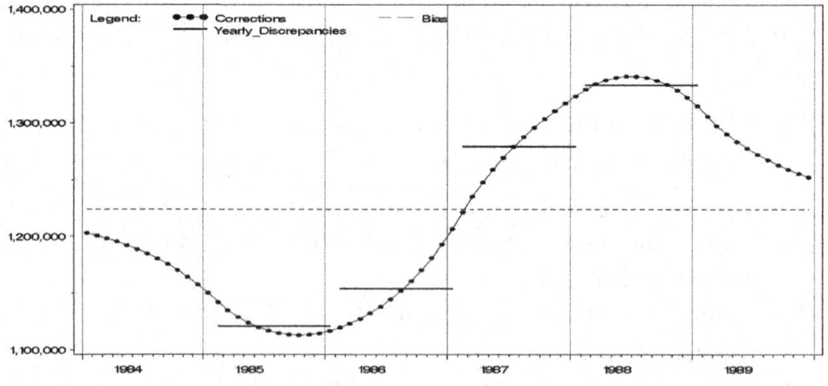

Fig. 3.7b. Yearly discrepancies and corresponding monthly corrections

The residual corrections converge to the bias, which embodies the historical average yearly discrepancy between the benchmarks and the corresponding yearly sums of the original monthly series. This convergence minimizes the revisions of the benchmarked series due to the eventual incorporation of the next benchmark. Indeed, this mechanism provides an implicit forecast of the next benchmark, equal to the yearly sum of the monthly series (what ever its value turns out to be), plus the bias estimate (1,223,740) based on the history of the series.

Note that the model provides corrections for all months whether they are covered by benchmarks or not.

Furthermore, it should be noted that the benchmarks cover fiscal years ranging from Feb. 1 to Jan. 31 of the following year, because this fiscal year best represents the situation. Calendar year estimates can then be obtained by taking the temporal sums of the benchmarked series over *all* the calendar years 1984 to 1989, i.e. by using Eq. (3.24).

The fiscal year benchmarks covering from Feb. 1 to Jan. 31 of the following year and the calendar year values appear in Table 3.4 with their standard deviations and their percent CVs, i.e. their standard deviation divided by the calendarized value and multiplied by 100%. The values have larger CVs for the calendar years less covered by the benchmarks, namely 1984 and 1989 at the beginning and the end of the series. The value of 1984 has the largest CV for lack of coverage by any benchmark.

Table 3.4. Calendarized yearly values obtained as a by-product of benchmarking

Starting date	Ending date	Fiscal value	Year	Calendarized value	Standard dev.	CV%
			1984	130,290,281	776,864	0.60
01/02/85	31/01/86	143,965,400	1985	142,935,112	86,867	0.06
01/02/86	31/01/87	154,377,100	1986	153,769,139	72,772	0.05
01/02/87	31/01/88	169,944,600	1987	168,965,381	72,350	0.04
01/02/88	31/01/89	181,594,000	1988	181,199,171	80,233	0.04
			1989	188,206,569	718,078	0.38

Coefficients of variations were available for the monthly series; they hovered around 0.8% and 0.9% with some exceptions. We assumed constant standard deviations, because constant target CVs call for a multiplicative benchmarking model discussed in the Chapter 5.

Different and more sophisticated adjustments of the Canadian Total Retail Trade Series will be presented in the coming chapters and contrasted with the present treatment.

4 Covariance Matrices for Benchmarking and Reconciliation Methods

4.1 Introduction

The covariance matrix V_e to be used for the sub-annual observations in benchmarking and reconciliation methods can originate from a statistical model previously applied to the series considered, such as in signal extraction (e.g. Hillmer and Trabelsi 1987, Durbin and Quenneville 1997, Chen et al. 1997) discussed in Chapter 8.[1] If available, it is desirable to use that covariance matrix as such for benchmarking or reconciliation adjustments, because it increases the efficiency of the estimator and preserves the statistical properties of the series.

In most cases however, the sub-annual covariance matrix has to be built from known or assumed standard deviations of the sub-annual series and from a known or assumed autocorrelation function, as follows:

$$V_e = \boldsymbol{\Xi}^{\lambda} \boldsymbol{\Omega}\, \boldsymbol{\Xi}^{\lambda}, \tag{4.1}$$

where $\boldsymbol{\Xi}$ is a diagonal matrix containing the standard deviations, matrix $\boldsymbol{\Omega}$ typically contains the autocorrelations of an AutoRegressive Moving Average (ARMA) process, and λ takes the value 0, 1/2 or 1. To each ARMA model satisfying the conditions of stationarity and invertibility corresponds a unique autocorrelation matrix.

In the case of series originating from a survey, covariance matrices should ideally be produced as a by-product of the survey process. In practice however, surveys provide only the variances, typically in the form of coefficients of variation $c_t = \sigma_t / s_t$. These can be converted into standard deviations and substituted in matrix $\boldsymbol{\Xi}$ in conjunction with an ARMA model embodied in matrix $\boldsymbol{\Omega}$ of Eq. (4.1). However, observations with larger standard deviations are thus specified as less reliable and therefore more modifiable. In some cases, this may have perverse effects for outliers values

[1] These matrices must be positive definite; if not, wild adjusted values and numerical problems occur.

with large standard deviations. As illustrated in Fig. 4.4b, these values may become even more extreme as a result of benchmarking or reconciliation. It is then preferable to assume constant target standard deviations or target coefficients of variations (CV) of the survey; or to use the average of the standard deviations or CVs.

In the total absence of any information regarding the covariance matrix of Eq. (4.1), the situation is no different from fitting any regression model to data without any available statistical information about the data. In the case of benchmarking, the model is chosen a priori, fitted to the data, and the result is evaluated.

In the context of the Cholette-Dagum regression-based benchmarking model discussed in Chapter 3, we need to specify the deterministic regressor $\boldsymbol{R}$ and the model followed by the sub-annual error $\boldsymbol{e}$, i.e. the content of $\boldsymbol{V}_e$ of Eq. (4.1). The specification of both the regressor and the model for the sub-annual error plays a critical role in distributing the annual discrepancies between the benchmarks and the corresponding temporal sums of the sub-annual values defined by,

$$d_m = a_m - \Sigma_{t=t_{1m}}^{t_{Lm}} j_{mt}\, s_t,\ m=1,\ldots,M. \tag{4.2a}$$

The presence of non-zero discrepancies makes obvious the presence of error in the sub-annual data. Indeed, the annual benchmarks brings new information, which is absent from the sub-annual series. The sub-annual series must then be adjusted in the light of this new information, and consequently the annual discrepancies must be distributed accordingly.

A constant regressor - or possibly a mild trend regressor - generally captures the part of the sub-annual error due to under-coverage and non-response, which mis-represents the level of the series. After this level correction, *residual* annual discrepancies are left, i.e.

$$d_m^{(r)} = a_m - \Sigma_{t=t_{1m}}^{t_{Lm}} j_{mt}\, (s_t - \Sigma_{h=1}^{H} r_{th}\, \hat{\beta}_h). \tag{4.2b}$$

Assuming a gradual deterioration of the sub-annual data, as is usually the case, the residual discrepancies should be distributed smoothly. For flow series, a smooth distribution of the residual discrepancies is generally achieved by AR(1) models (AutoRegressive model of order 1) with parameter values between 0.70 and 0.90. For stock series, this is better achieved by AR(2) error models which can be factorized as the product of two identical AR(1) models. This implies a positive first order parameter greater than 1 and a negative second order parameter such that the sum of the two smaller than one. To our knowledge, AR(2) error models have not been proposed for benchmarking. This will be discussed in section 4.2 and 4.3.

The behaviour of these error models will largely preserve the movement of the sub-annual series, which is the only one available. It should be emphasized that, if the series originates from a survey, the error model selected and embodied in V_e should not reflect the behaviour of the survey error (e.g. due to sample rotation). Benchmarking does *not* aim at correcting survey data for the survey error. Indeed, sample rotation is widely known to produce a more reliable sub-annual movement at the expense of the level. The level in turn is corrected by benchmarking. The resulting series is then more reliable in terms of movement and level. Furthermore, the ARMA modelling of the survey error on the basis of the sample rotation designs has not been pursued further because of the modelling difficulty and the fact that the sample design changes occasionally through time.

The next section establishes the relationship between AR(1) and AR(2) error models and the well known principle of sub-annual movement preservation.

4.2 Minimization of an Objective Function

Most of the mainstream benchmarking methods currently used were derived from the minimization of an objective function, or criterion, which preserves the sub-annual movement as much as possible. This section will show that the choice of the sub-annual covariance matrix V_e in generalized least square regression, also entails the minimization of an objective function.

The objective function of the modified additive first difference Denton[2] method is

$$\begin{aligned} &\min \{ \Sigma_{t=2}^{T} [(s_t - \theta_t) - (s_{t-1} - \theta_{t-1})]^2 \} \\ &\equiv \min \{ \Sigma_{t=2}^{T} [(\theta_t - \theta_{t-1}) - (s_t - s_{t-1})]^2 \} . \end{aligned} \tag{4.3a}$$

The first expression of the objective function specifies that the error, $e_t = s_t - \theta_t$, should change as little as possible from period $t-1$ to period t

---

[2] The original Denton method minimized the first correction, which assumed $\theta_0 = s_0$ and could introduce a transient spurious movement in the benchmarked series, as discussed in Chapter 6.

(subject to the benchmarking constraints). The second expression shows that the latter is equivalent to keeping the movement of the benchmarked series, $\theta_t - \theta_{t-1}$, as close as possible to that of the original series, $s_t - s_{t-1}$.

The objective function of the modified proportional first difference Denton method is

$$min \{ \Sigma_{t=2}^{T} [(s_t - \theta_t)/s_t - (s_{t-1} - \theta_{t-1})/s_{t-1}]^2 \}$$
$$\equiv min \{ \Sigma_{t=2}^{T} [\theta_t/s_t - \theta_{t-1}/s_{t-1}]^2 \}. \quad (4.3b)$$

The first expression of the objective function specifies that the ratio of the error $e_t = s_t - \theta_t$ to the observed series s_t changes as little as possible from period $t-1$ to period t (subject to the benchmarking constraints). The second expression states that this is equivalent to keeping the ratio of the benchmarked to the original series as constant as possible from period to period. This is the meaning of proportional movement preservation.

The objective function of the growth rate preservation method is very similar to (4.3a), albeit in the logs:

$$min \{ \Sigma_{t=2}^{T} [(ln\, s_t - ln\, \theta_t) - (ln\, s_{t-1} - ln\, \theta_{t-1})]^2 \}$$
$$\equiv min \{ \Sigma_{t=2}^{T} [(ln\, \theta_t - ln\, \theta_{t-1}) - (ln\, s_t - ln\, s_{t-1})]^2 \}$$
$$\equiv min \{ \Sigma_{t=2}^{T} [ln(\theta_t/\theta_{t-1}) - ln(s_t/s_{t-1})]^2 \}. \quad (4.3c)$$

The last expression of the objective function specifies that the growth rates of the benchmarked series θ_t/θ_{t-1} are as close as possible to that of those of the original series s_t/s_{t-1}.

The key to movement preservation is to keep the corrections $\hat{\theta}_t - s_t$ or $\hat{\theta}_t/s_t$ as constant as possible, given the benchmarking constraints. This can be verified graphically. However, when comparing two or more series, it is usually more convenient to examine the root mean square of z, $(z/(T-1))^{1/2}$, where z is the sum of squares of the objective functions (4.3).

The choice of the sub-annual covariance matrix V_e in regression analysis, also entails the minimization of an objective function. In fact textbooks show that both ordinary and generalized least square regression are based on the minimization of an objective function. This is also the case for the Cholette-Dagum regression-based benchmarking model with covariance matrix $V_e = \boldsymbol{\Xi}^{\lambda} \boldsymbol{\Omega} \boldsymbol{\Xi}^{\lambda}$.

Let $\boldsymbol{\Omega}^{-1} = \boldsymbol{P}'\boldsymbol{P}$, where $\boldsymbol{P}'\boldsymbol{P}$ is the Choleski factorization of a symmetric matrix. Matrix $\boldsymbol{P}$ is a lower triangular matrix; and $\boldsymbol{P}'$, upper triangular . The objective function is

$$min\ \boldsymbol{e}'\boldsymbol{V}_e^{-1}\boldsymbol{e} \rightarrow min\ \boldsymbol{e}'\boldsymbol{\Xi}^{-\lambda}\boldsymbol{\Omega}^{-1}\boldsymbol{\Xi}^{-\lambda}\boldsymbol{e} \tag{4.4a}$$

$$\rightarrow min\ (\boldsymbol{e}'\boldsymbol{\Xi}^{-\lambda})\boldsymbol{P}'\boldsymbol{P}(\boldsymbol{\Xi}^{-\lambda}\boldsymbol{e}) \tag{4.4b}$$

$$\rightarrow min\ \boldsymbol{e}^{\dagger\prime}\boldsymbol{P}'\boldsymbol{P}\boldsymbol{e}^{\dagger} \rightarrow min\ \boldsymbol{v}'\boldsymbol{v}, \tag{4.4c}$$

where in our case $\boldsymbol{e} = \boldsymbol{s} - \boldsymbol{R}\boldsymbol{\beta} - \boldsymbol{\theta}$, $\boldsymbol{\Xi}$ is a diagonal matrix with elements equal to $\sigma_1,...,\sigma_T$, vector $\boldsymbol{e}^{\dagger}$ is the standardized error $\boldsymbol{\Xi}^{-\lambda}\boldsymbol{e}$, and vector $\boldsymbol{v}$ contains the innovations $v_1,...,v_T$ of an ARMA process. Equation (4.4) shows that minimizing $\boldsymbol{e}'\boldsymbol{V}_e^{-1}\boldsymbol{e}$ amounts to minimizing the sum of squares of the innovations ($\boldsymbol{v}'\boldsymbol{v}$) driving the ARMA process. It should be noted that the innovations v_t originate from the annual discrepancies; indeed without the discrepancies there are no innovations.

In the case of an AR(1) model, the standardized error $e_t^{\dagger} = e_t/\sigma_t^{\lambda}$ follows the model:

$$e_t^{\dagger} = \varphi\, e_{t-1}^{\dagger} + v_t\,, \tag{4.5}$$

where $0 \le \varphi < 1$. The elements ω_{ij} of the autocorrelation matrix $\boldsymbol{\Omega}$ are given by $\varphi^{|i-j|}$:

$$\boldsymbol{\Omega} = \begin{bmatrix} 1 & \varphi & \varphi^2 & \cdots & \varphi^{T-1} \\ \varphi & 1 & \varphi & \cdots & \varphi^{T-2} \\ \varphi^2 & \varphi & 1 & \cdots & \varphi^{T-3} \\ \vdots & \vdots & \vdots & \ddots & \vdots \\ \varphi^{T-1} & \varphi^{T-2} & \varphi^{T-3} & \cdots & 1 \end{bmatrix}. \tag{4.6}$$

The elements of $\boldsymbol{\Omega}^{-1}$, $\boldsymbol{P}$ and $\boldsymbol{P}\boldsymbol{e}^{\dagger} = \boldsymbol{v}$ are given by

$$\boldsymbol{\Omega}^{-1} = 1/(1-\varphi^2) \begin{bmatrix} 1 & -\varphi & 0 & 0 & \cdots & 0 & 0 & 0 \\ -\varphi & (1+\varphi^2) & -\varphi & 0 & \cdots & 0 & 0 & 0 \\ \vdots & \vdots & \vdots & \vdots & \ddots & \vdots & \vdots & \vdots \\ 0 & 0 & 0 & 0 & \cdots & -\varphi & (1+\varphi^2) & -\varphi \\ 0 & 0 & 0 & 0 & \cdots & 0 & -\varphi & 1 \end{bmatrix}, \tag{4.7a}$$

$$P = 1/\sqrt{1-\varphi^2} \begin{bmatrix} \sqrt{1-\varphi^2} & 0 & 0 & \dots & 0 & 0 & 0 \\ -\varphi & 1 & 0 & \dots & 0 & 0 & 0 \\ \vdots & \vdots & \vdots & \ddots & \vdots & \vdots & \vdots \\ 0 & 0 & \dots & 0 & -\varphi & 1 & 0 \\ 0 & 0 & 0 & \dots & 0 & -\varphi & 1 \end{bmatrix}, \tag{4.7b}$$

$$P\,\Xi^{-\lambda}\,e = P\,e^{\dagger} = 1/\sqrt{1-\varphi^2} \begin{bmatrix} \sqrt{1-\varphi^2} \times e_1/\sigma_1^{\lambda} \\ -\varphi\, e_1/\sigma_1^{\lambda} + e_2/\sigma_2^{\lambda} \\ \vdots \\ -\varphi\, e_{T-1}/\sigma_{T-1}^{\lambda} + e_T/\sigma_T^{\lambda} \end{bmatrix} = v\,. \tag{4.7c}$$

Given Eqs. (4.7), the objective function (4.4c) can be written in scalar algebra as

$$min\ \{(e_1/\sigma_1^{\lambda})^2 + \Sigma_{t=2}^{T}\,(e_t/\sigma_t^{\lambda} - \varphi e_{t-1}/\sigma_{t-1}^{\lambda})^2/\sigma_v^2\} = min\ \{\Sigma_{t=1}^{T}\, v_t^2\}, \tag{4.8a}$$

where σ_v^2 is the variance of the innovations equal to $(1-\varphi^2)$, given that the variance of the standardized errors e_t/σ_t^{λ} is 1. Minimizing (4.8a) is equivalent to minimizing

$$\sigma_v^2 \times min\ \{(e_1/\sigma_1^{\lambda})^2\,(1-\varphi^2) + \Sigma_{t=2}^{T}\,(e_t/\sigma_t^{\lambda} - \varphi e_{t-1}/\sigma_{t-1}^{\lambda})^2\}. \tag{4.8b}$$

In the context of the C-D regression-based benchmarking method, the error e_t is equal to $s_t - \theta_t - \Sigma_{h=1}^{H}\, r_{th}\,\beta_h$. The minimization consists of *estimating* the values of θ_t, t=1,...,T, and of β_h, h=1,...,H, for which the minimum of (4.8b) is reached.

Setting λ equal to 0, the objective function (4.8b) becomes

$$\sigma_v^2 \times min\ \{e_1^2\,(1-\varphi^2) + \Sigma_{t=2}^{T}\,(e_t - \varphi e_{t-1})^2\} = min\ \{\Sigma_{t=1}^{T}\, v_t^2\}. \tag{4.9}$$

Note that if H=0, the error e_t is equal to $(s_t - \theta_t)$. Furthermore, if φ is close to 1, the first term inside the braces tends to 0 and Eq. (4.9) tends to the objective function of the modified first difference Denton (4.3a). (The leading multiplicative factor does not influence the minimization.)

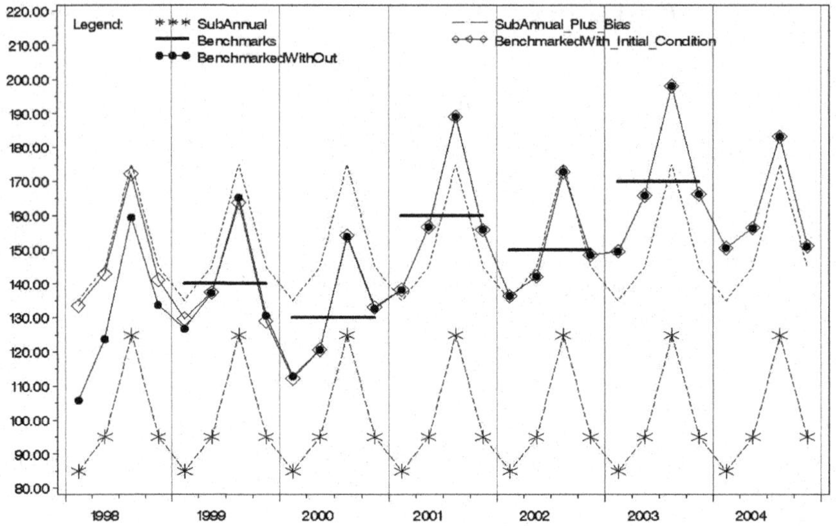

Fig. 4.1a. Simulated quarterly flow series benchmarked to yearly values, using an AR(1) error model with φ=0.729, with and without initial condition

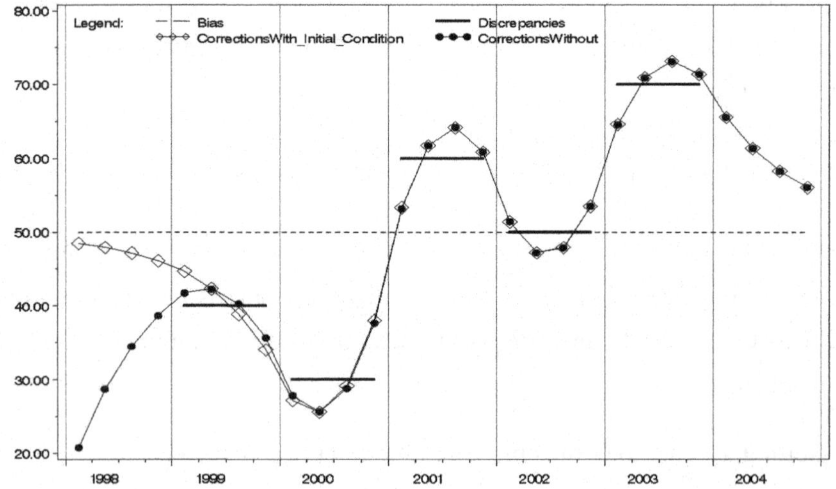

Fig. 4.1b. Yearly discrepancies and corresponding quarterly corrections for series in Fig. 4.1a

We now give an explanation for the need of the first term in (4.9), $e_1^2\,(1-\varphi^2)$ and the corresponding term in (4.8b). The presence of this *initial condition*[3] makes $\boldsymbol{\Omega}^{-1}$ invertible and symmetric with respect to both diagonals.[4] This double symmetry causes the AR(1) model to behave in the same manner *both* at the beginning and the end of the sub-annual series.

Fig. 4.1a shows the results obtained by benchmarking the simulated quarterly series of Fig. 3.1, with and without the initial condition in (4.9). The regression-based additive benchmarking model applied includes a constant regressor (r_t=−1), a correlation matrix $\boldsymbol{\Omega}$ corresponding to an AR(1) error model $e_t = 0.729\, e_{t-1} + v_t$, with and without the initial condition. Both benchmarked series are very similar except at the beginning of the series. This difference appears more clearly in Fig. 4.1b, which displays the annual discrepancies $d_m = a_m - \Sigma_{t=t_{1m}}^{t_{2m}} s_t$, the corrections $s_t - \hat{\theta}_t$, the estimated bias $\hat{\beta}_1$ and the residuals $\hat{e}_t = s_t + \hat{\beta}_1 - \hat{\theta}_t$.

At the end of the series (year 2004), both sets of residuals converge to zero according to the rule $\hat{e}_t = 0.729\, \hat{e}_{t-1}$ for t=25,...,28, because there are no benchmarks after 2003 (t=24). As a result, the benchmarked series converges to the original sub-annual series corrected for bias $s_t^{\dagger} = s_t + \hat{\beta}_1$.

At the beginning of the series (1998), the residuals obtained *with* the initial condition behave in the same manner as at the end. On the contrary, the residuals obtained *without* the initial condition diverge exponentially from zero, according to the model $\hat{e}_t = 0.729^{-1}\, \hat{e}_{t+1} \approx 1.372\, \hat{e}_{t+1}$ for t=4,...,1, because there are no benchmarks before 1999 (t=5). This fluctuations in spurious and not supported by data. Consequently, the corresponding benchmarked series of Fig. 4.1a diverges from the original sub-annual series corrected for bias $s_t^{\dagger} = s_t + \hat{\beta}_1$ for t=4,...,1.

The conclusion is that the initial condition of autoregressive models and more generally of ARMA model are desirable.

The next section explains how objective functions such as (4.9), with initial condition, entails different variants of movement preservation depending on the AR model and the parameter values chosen.

---

[3] This initial condition is not to be confused with the initial condition in the original Denton method, namely $\theta_0 = s_0$ or $e_0 = 0$.

[4] By omitting this term, i.e. omitting the first row of (4.7b) and (4.7c), the first entry of $\boldsymbol{\Omega}^{-1}$ is $\omega_{11} = \varphi^2$ instead of 1.

4.3 Weak versus Strong Movement Preservation

In the context of additive benchmarking, strong movement preservation is represented by the objective function (4.3a). On the other hand, weak movement preservation is represented by objective functions such as (4.9), which tend to repeat a fraction φ of the error from period to period. Setting $\varphi = 0.90$ largely preserves the movement of the sub-annual series in view of (4.3a), especially when used with a constant regressor ($r_t = -1$) which entirely preserves the movement: $s_t^{\dagger} = s_t + \beta_1 \Rightarrow s_t^{\dagger} - s_{t-1}^{\dagger} \equiv s_t - s_{t-1}$.

More movement preservation can be achieved by setting φ closer to 1, e.g. $\varphi = 0.999$. But this will be at the expense of a much slower convergence to the sub-annual series corrected for bias, $s_t^{\dagger} = s_t - \Sigma_{h=1}^{H} r_{th} \hat{\beta}_h$, at both ends.

In our experience with a large number of real and simulated series, $\varphi = 0.90$ gives an excellent movement preservation for monthly series. Furthermore, this AR(1) model provides a reasonable implicit forecast of the next annual benchmark. This forecast is equal to the annual sum of the sub-annual series (whatever it turns out to be) plus the historical "bias" ($-\Sigma_{h=1}^{H} r_{th} \hat{\beta}_h$). This implicit forecast also minimizes the revisions of the benchmarked series resulting from the incorporation of the next benchmark when it becomes available.

This is illustrated by the corrections of the additively benchmarked Canadian Total Retail Trade series displayed in Fig. 3.7a and Fig. 3.7b. The value $\varphi = 0.90$ is adequate for monthly series. For quarterly series, however, the value $\varphi = 0.90$ provides too slow a convergence at the ends. A more adequate value for quarterly series is $\varphi = 0.90^3 = 0.729$, which provides as much speed of convergence in one quarter, as the value $\varphi = 0.90$ for monthly series over three months. Fig. 3.5 illustrates the benchmarked simulated quarterly series obtained with $\varphi = 0.729$ and the corresponding corrections.

Fig. 4.2a displays the simulated quarterly series benchmarked under the same model, with constant regressor ($r_t = -1$), $\boldsymbol{\Xi} = \boldsymbol{I}_T$ and a correlation matrix $\boldsymbol{\Omega}$ corresponding to three AR(1) error models with values $\varphi = 0.999$, $\varphi = 0.80$ and $\varphi = 0.40$ respectively. The benchmarked series with smaller parameter values converge faster to the sub-annual series corrected for bias $s_t^{\dagger} = s_t + 50$. The benchmarked series with φ=0.40, converges too fast; that with φ=0.999, too slowly. These observations are more obvious in Fig. 4.2b which displays the corresponding annual discrepancies and the corrections $\hat{\theta}_t - s_t$. The convergence to the bias equal to 50 is obvious. This convergence in turn causes the convergence of the benchmarked series to the original sub-annual series corrected for bias in Fig. 4.2a.[5]

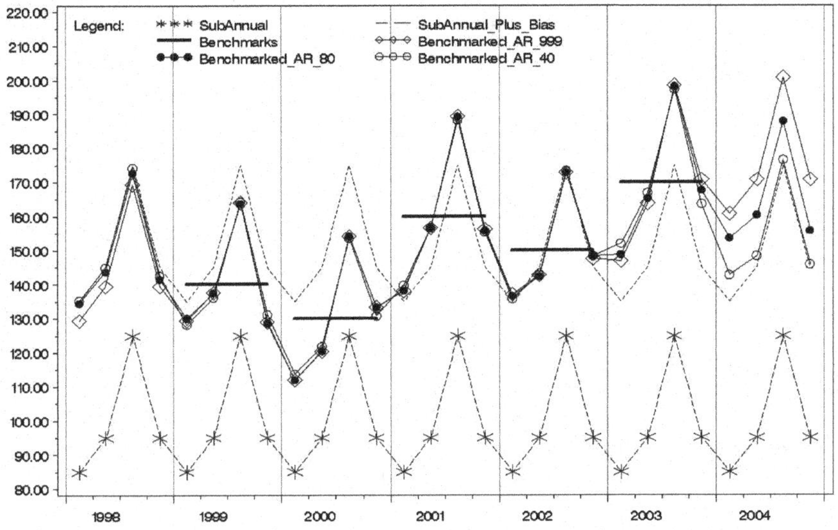

Fig. 4.2a. Simulated quarterly flow series benchmarked to yearly values, using AR(1) error models with parameters φ=0.999, φ=0.80 and φ=0.40

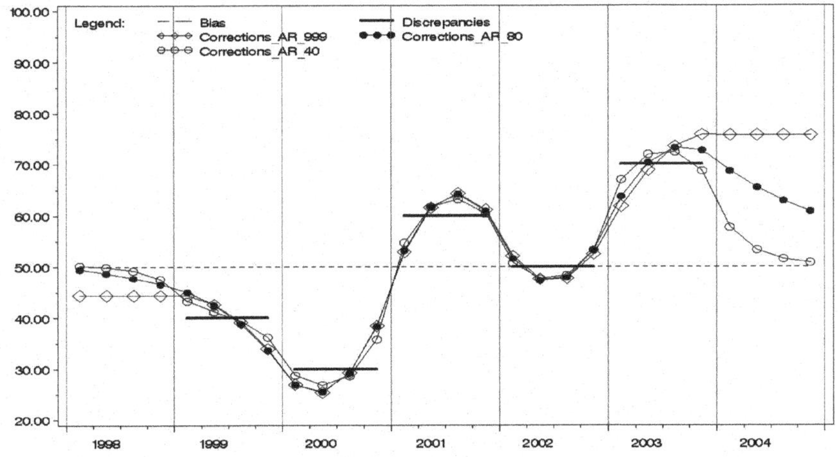

Fig. 4.2b. Yearly discrepancies and corresponding quarterly corrections for series in Fig. 4.2a

[5] The estimates of the bias vary somewhat with the AR(1) model selected. In order to simplify the graphs, we display the "true" value (50) used in generating the data of the simulated quarterly series.

Strong movement preservation, approximated with $\varphi = 0.999$ (Bournay and Laroque, 1979), assumes that the next annual discrepancy will be higher than the last one, if the last annual discrepancy is larger than the previous one; and, conversely. Moreover, the implicit forecast of the next benchmark is based on the last two discrepancies, which is highly questionable. These observations also apply to the first difference variants of the Denton method (with $\varphi = 1$) illustrated in Fig 3.3b.

In summary, weaker movement preservation offers the best compromise between strong movement preservation and optimality of the implicit forecast of the next benchmark. In the context of the Cholette-Dagum regression-based benchmarking method, experience with many series favours $\varphi = 0.90$ for monthly series and $\varphi = 0.90^3$ for quarterly series. With these respective values, quarterly errors converge to zero (at the ends) over one quarter as fast as monthly errors over three months. The idea is to homogenize the speed of convergence across series of different periodicities. These observations are also valid for proportional movement ($\theta_t / s_t - \theta_{t-1} / s_{t-1}$) preservation and for growth rate preservation ($\theta_t / \theta_{t-1} - s_t / s_{t-1}$).

The estimated errors are represented in Fig. 4.2b, by the distance between the corrections and the constant. The innovations $\hat{v}_t = \hat{e}_t - 0.80\,\hat{e}_{t-1}$ for instance are not mutually independent since they are trending. The innovations are negative from 1998 to the middle of 2000, from the middle of 2001 to the middle of 2002, and beyond the middle of 2003; and, positive in the remaining dates. locations. time periods. Choosing a more complex ARMA model for the error would eliminate some of the trending of the innovations, but would also produce spurious movements, not supported by the yearly data. The goal is not to obtain random innovations, but to distribute the annual discrepancies in a rational manner supported by the data. In order to produce random innovations in Fig. 4.2b, the residuals would basically have to behave stepwise like the repeated residual annual discrepancies around the fitted constant.

Fig. 4.3a displays the behaviour of the same three models for the same simulated quarterly series (of Fig. 4.2a), treated as a stock series with benchmarks in the fourth quarter of years 1999 to 2003. Again the models include a constant regressor ($r_t = -1$), $\boldsymbol{\Xi} = \boldsymbol{I}_T$ and a correlation matrix $\boldsymbol{\Omega}$ corresponding to three homoscedastic AR(1) models with values $\varphi = 0.999$, $\varphi = 0.80$ and $\varphi = 0.40$ respectively. The presence of seasonality in Fig. 4.3a obscures the results as is often the case.

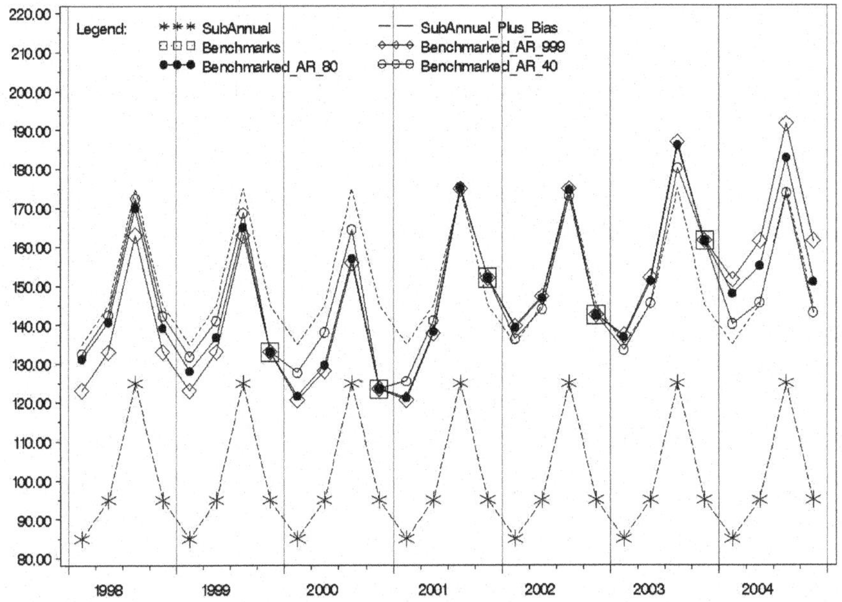

Fig. 4.3a. Simulated quarterly stock series benchmarked to yearly values, using AR(1) error models with parameters φ=0.999, φ=0.80 and φ=0.40

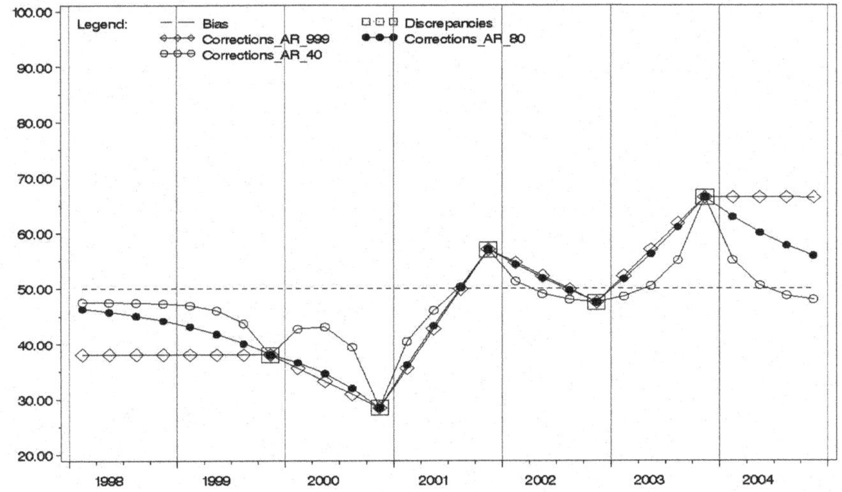

Fig. 4.3b. Yearly discrepancies and corresponding quarterly corrections for series in Fig. 4.3a

On the other hand, the absence of seasonality in Fig 4.3b reveals problems with the corrections $\hat{\theta}_t - s_t$, namely the presence of kinks at the dates of the benchmarks and a spurious movement reversal between the last quarters of

1999 and 2000. The AR(1) with φ=0.999 model produces no reversals, but kinks and lack of converge at the ends.

These kinks are unlikely and un-supported by the data. Indeed it is hard to support that economic activity alters its course so abruptly and precisely at the dates of the benchmarks. A smooth curve passing through the annual discrepancies makes more economic sense. As for the movement reversal, it is spurious and un-supported by the data.

Kinks - and movement reversals especially - alter the movement of the sub-annual series, thus violating the movement preservation principle. After seasonal adjustment,[6] the seasonally adjusted benchmarked series and the corresponding trend-cycle estimate will be seriously distorted, particularly for short-term trend-cycle analysis. This example also illustrates the usefulness of examining the graph of the discrepancies and the corrections $\hat{\theta}_t - s_t$ to assess the adequacy of the AR error model chosen and its parameter values.

In some applications, there is often no sub-annual series, just benchmarks. In such cases, the annual discrepancies coincide with the benchmarks because the sub-annual series is 0. This situation can be conveyed by Fig. 4.3b, where the annual discrepancies are now the benchmarks; and the corrections, the benchmarked series. In fact the benchmarked series have become an interpolated series, since the sub-annual series does not exist. The kinks and movement reversals displayed then become more obvious and clearly unacceptable; so is their effect on the trend-cycle component.

Such kinks and movement reversals can also occur for index and flow series, for example when benchmarks are available every second year; more generally when sub-annual periods are not covered by benchmarks and are embedded between benchmarks as illustrated below in Fig. 4.6. The common denominator in all these cases is that the benchmarks are *not temporally contiguous*. Benchmarks are temporally contiguous when the starting date of each benchmark is equal to the ending date of the previous benchmark plus one day.

In these cases, AR(1) error models do not preserve movement, whereas AR(2) error models with negative second order parameter provide better adjustments.

For an autoregressive processes of order 2, the elements ω_{ij} of the auto-correlation matrix $\boldsymbol{\Omega}$ are given in terms of lags $\ell = |i-j|: \omega(\ell)=1$ for $\ell=0$,

[6] Most statistical agencies publish sub-annual time series with the corresponding seasonally adjusted series and often with trend-cycle estimates as well. Benchmarking takes places upstream of seasonal adjustment.

$\omega(\ell) = \varphi_1 / (1-\varphi_2)$, for $\ell=1$, $\omega(\ell) = \varphi_1 \omega_{\ell-1} + \varphi_2 \omega_{\ell-2}$, for $\ell = 2,\ldots,T-1$, where $\varphi_1 + \varphi_2 < 1$, $\varphi_2 - \varphi_1 < 1$ and $|\varphi_2| < 1$ to ensure stationarity.

Given (4.4), the objective function of an AR(2) model is

$$\begin{aligned} &min\,\{\, (e_1/\sigma_1^{\lambda})^2 + (e_2/\sigma_2^{\lambda} - \omega(1)\times e_1/\sigma_1^{\lambda})^2/(1-\omega^2(1)) \\ &\Sigma_{t=3}^{T} (e_t/\sigma_t^{\lambda} - \varphi_1 e_{t-1}/\sigma_{t-1}^{\lambda} - \varphi_2 e_{t-2}/\sigma_{t-2}^{\lambda})^2/\sigma_v^2 \,\} \;=\; min\,\{\Sigma_{t=1}^{T} v_t^2\}, \end{aligned} \tag{4.10a}$$

which is equivalent to minimizing

$$\begin{aligned} &\sigma_v^2 \times min\,\{\, (e_1/\sigma_1^{\lambda})^2/\sigma_v^2 + [e_2/\sigma_2^{\lambda} - \omega(1)\times e_1/\sigma_1^{\lambda}]^2/(1-\omega^2(1))\,\sigma_v^2 \\ &\Sigma_{t=3}^{T} (e_t/\sigma_t^{\lambda} - \varphi_1 e_{t-1}/\sigma_{t-1}^{\lambda} - \varphi_2 e_{t-2}/\sigma_{t-2}^{\lambda})^2 \,\} \;=\; min\,\{\Sigma_{t=1}^{T} v_t^2\}, \end{aligned} \tag{4.10b}$$

where $\omega(1) = \varphi_1/(1-\varphi_2)$ and $\sigma_v^2 = [(1-\varphi_2)^2 - \varphi_1^2]/[(1-\varphi_2)/(1+\varphi_2)]$. The objective function now has two initial conditions, namely the two first terms of Eqs. (4.10). These terms play the same role as the initial condition of AR(1) processes, namely ensuring similar behaviour at the beginning and the end of the series.

Setting λ equal to 0, the objective function corresponding to an AR(2) model is:

$$\begin{aligned} &\sigma_v^2 \times min\,\{\, e_1^2/\sigma_v^2 + (e_2 - \omega(1)\times e_1)^2/(1-\omega^2(1))\,\sigma_v^2 \\ &\Sigma_{t=3}^{T} (e_t - \varphi_1 e_{t-1} - \varphi_2 e_{t-2})^2 \,\} \;=\; min\,\{\Sigma_{t=1}^{T} v_t^2\}. \end{aligned} \tag{4.11}$$

We consider two AR(2) error models, one with parameters $\varphi_1 = 1.80$ and $\varphi_2 = -0.81$, and the other with parameters $\varphi_1 = 1.40$ and $\varphi_2 = -0.49$. These two AR(2) models can be factorized as $(1-0.90B)\;(1-0.90B)\,e_t = v_t$ and $(1-0.70B)\;(1-0.70B)\,e_t = v_t$ respectively, where B is the backshift operator such that $B^k e_t \equiv e_{t-k}$. In other words, these AR(2) error models are the products of two AR(1) models. The first model can be written as

$$\begin{aligned} &(1-0.90B)(1-0.90B)\,e_t = v_t \;\Rightarrow (1-2\times 0.90B + 0.90^2 B^2)\,e_t = v_t \\ &\Rightarrow e_t - 2\times 0.90\,e_{t-1} + 0.90^2 e_{t-2} = v_t \;\Rightarrow e_t = 1.80\,e_{t-1} - 0.81\,e_{t-2} + v_t; \end{aligned}$$

and, similarly for the second model.

These AR(2) error models satisfy the stationarity conditions and converge to zero at both ends of the series because the sum of φ_1 and φ_2 is lower than 1. This convergence of e_t ensures that the benchmarked series converges to the original sub-annual series (or the sub-annual series corrected for "bias" when H>0).

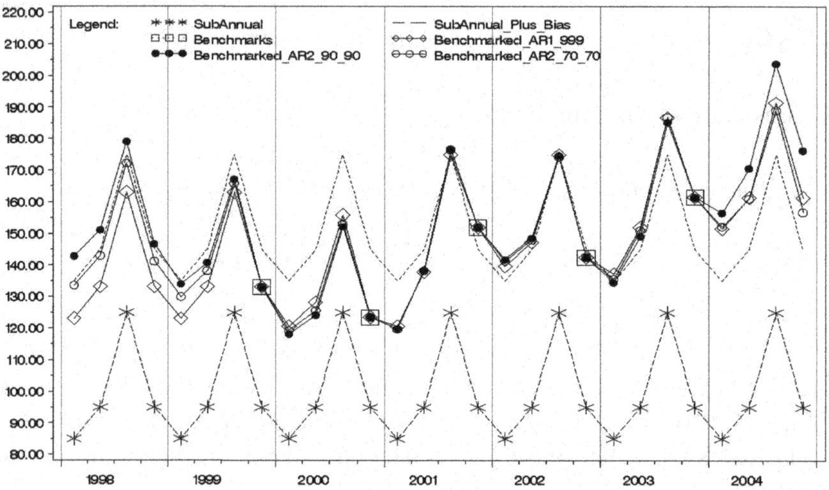

Fig. 4.4a. Simulated quarterly stock series benchmarked to yearly values, using AR(2) error models with parameters φ_1=1.80, φ_2 =-0.81 (AR2_90_90) and φ_1=1.40, φ_2=-0.49 (AR2_70_70)

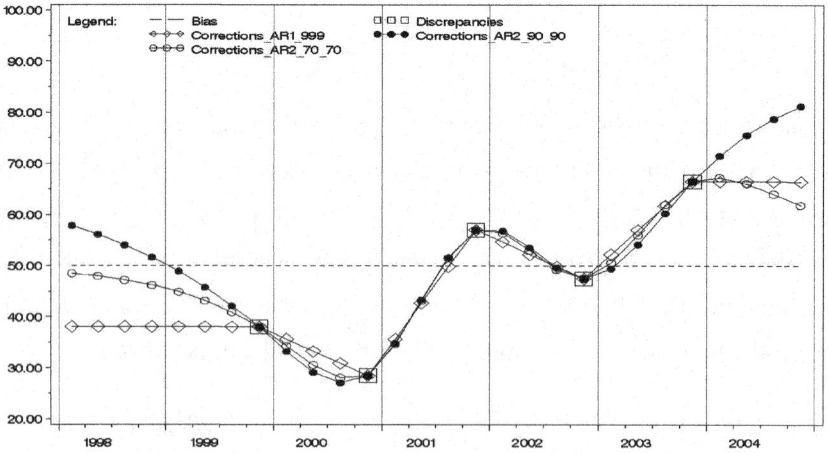

Fig. 4.4b. Yearly discrepancies and corresponding quarterly corrections for series in Fig. 4.4a

Fig. 4.4 illustrates the behaviour of these two AR(2) error models (labelled AR2_90_90 and AR2_70_70) for our simulated quarterly series, treated as a stock. Again the presence of seasonality in Fig. 4.4a obscures the analysis of the results. The absence of seasonality in Fig. 4.4b reveals that both AR(2) models eliminate the movement reversals observed in Fig. 4.3b. The AR(2) model with φ_1=1.40 and φ_2=-0.49 (curve AR2_70_70) provides an excellent fit at both ends, where the corrections converge to the bias $s_t^\dagger = s_t + 50$, which cause the benchmarked series to converge to the series corrected for bias in Fig. 4.4a. In principle, the AR(2) with φ_1=1.80 and φ_2=-0.81 (AR2_90_90) also converges but would require more time to do so; the model could be more appropriate for monthly series.

Fig. 4.5 illustrates the behaviour of the same two AR(2) models for our simulated quarterly series, treated as a flow. Fig. 4.5b reveals that AR(2) models could be used in the benchmarking of flow series, especially in cases of sub-annual benchmarks. The AR(2) with φ_1=1.40 and φ_2=-0.49 (curve AR2_70_70) produces a fit very close to the AR(1) also depicted (curve AR1_80). The AR(2) with φ_1=1.80 and φ_2=-0.81 (AR2_900) does not converge fast enough to the quarterly series. For the sake of the parsimony principle we would recommend the AR(1) model.

Fig. 4.6a and 4.6b illustrates that in the case of flow series with sub-annual periods without benchmarks (quarters of 2002 in the figures), AR(2) error models would be preferable. Indeed the corrections corresponding to the AR(1) model (curve AR1_80) displays a movement reversal in 2002 unsupported by data. The movement would be more obvious with lower autoregressive parameters.

Note that an AR(2) error model with parameters values $\varphi_1 = 2\times0.999$ and $\varphi_2 = -0.999^2$ approximates the modified Denton second difference model objective function

$$min\,\{\, \Sigma_{t=3}^{T}\, (e_t - 2e_{t-1} + e_{t-2})^2 \,\},$$

which produces straight line behaviour at the ends.

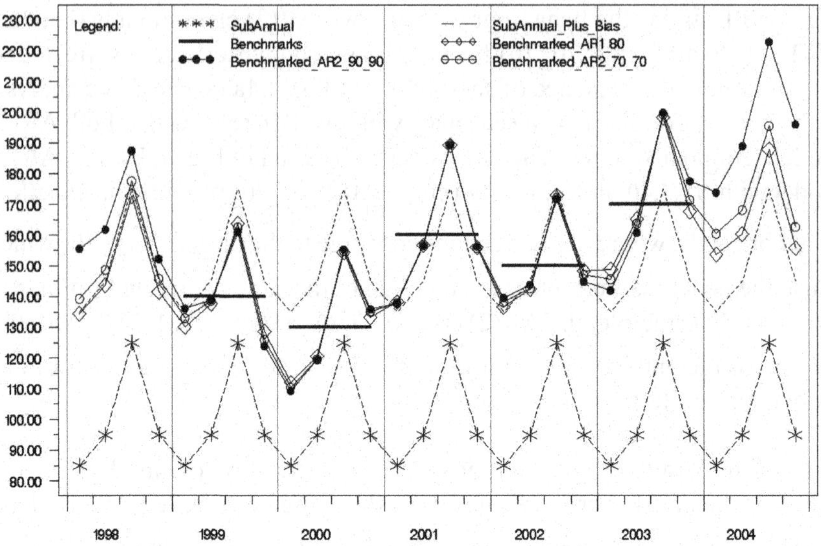

Fig. 4.5a. Simulated quarterly flow series benchmarked to yearly values, using AR(2) error models with parameters φ_1=1.80, φ_2=-0.81 (AR2_90_90) and φ_1=1.40, φ_2=-0.49 (AR2_70_70)

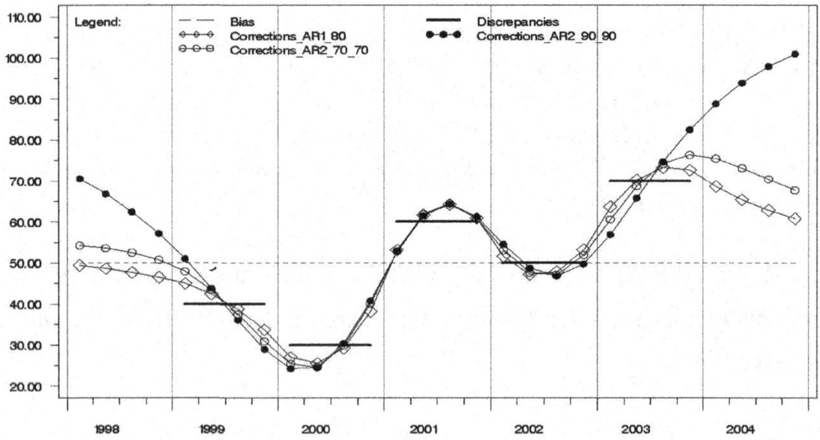

Fig. 4.5b. Yearly discrepancies and corresponding quarterly corrections for series in Fig. 4.5a

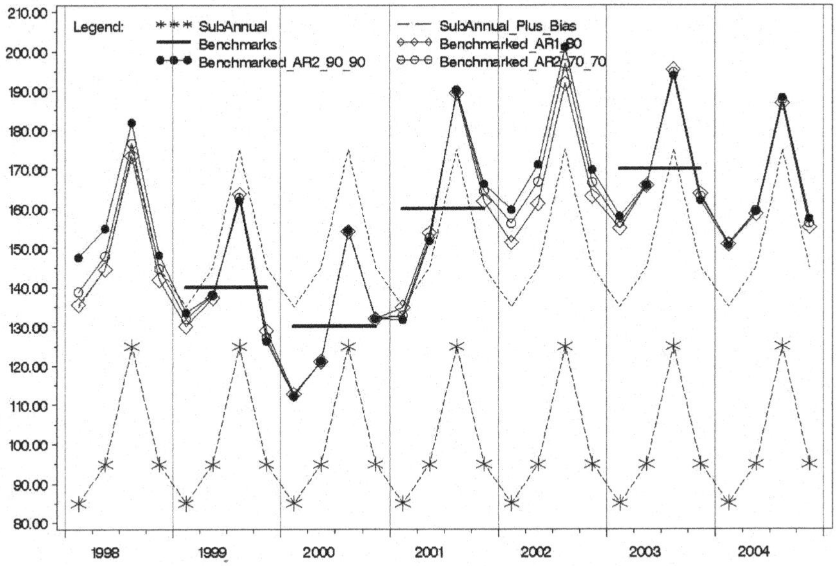

Fig. 4.6a. Simulated quarterly flow series benchmarked to yearly values, using an AR(1) error model with $\varphi = 0.80$ and AR(2) models with parameters $\varphi_1 = 1.80$, $\varphi_2 = -0.81$ (AR2_90_90) and $\varphi_1 = 1.40$, $\varphi_2 = -0.49$ (AR2_70_70)

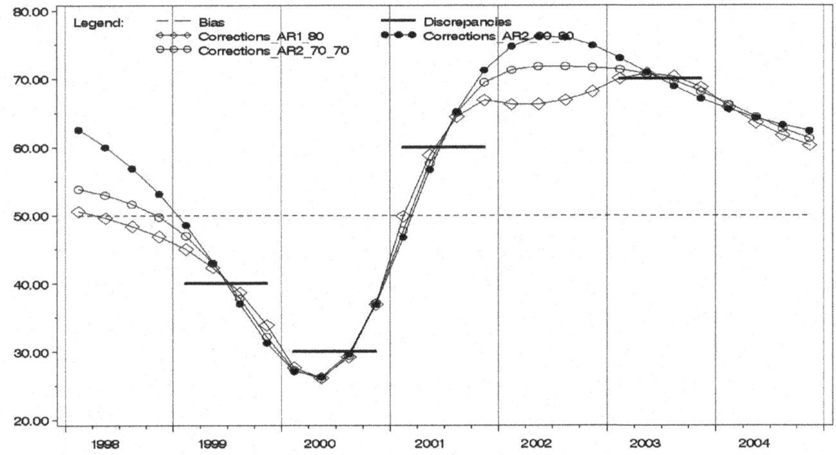

Fig. 4.6b. Yearly discrepancies and corresponding quarterly corrections for series in Fig. 4.6a

4.4 Weak versus Strong Proportional Movement Preservation

In many cases, the sub-annual series is strongly affected by seasonality, and the assumption of homoscedastic additive errors does not reflect the presence of a sample autocorrelation at seasonal lags. In such cases, the assumption of a proportional error is more appropriate.

In the proportional error model, the standard deviations are defined as the product of constant CVs and the sub-annual observations: $\sigma_t = c \times s_t$. Assuming an AR(1) model and setting H equal to 0 and λ equal to 1 in (4.4), the objective function (4.8b) becomes

$$((1-\varphi^2)/c) \times min \{ (e_1/s_1)^2/(1-\varphi^2) + \Sigma_{t=2}^{T} (e_t/s_t - \varphi\, e_{t-1}/s_{t-1})^2 \}. \quad (4.12)$$

Note that with H=0, the error e_t is equal to $(s_t - \theta_t)$. Furthermore, if φ is close to 1, the first term inside the braces tends to 0 and Eq. (4.12) tends to the objective function of the modified proportional first difference Denton (4.3b). (The leading multiplicative factor does not influence the minimization.)

With a moderate value of the autoregressive parameter, e.g. φ=0.729, the standardized error, $(s_t - \theta_t)/s_t$ tends to largely repeat from period t-1 to period t. This behaviour largely preserves the proportional movement of the sub-annual series and inherits its seasonal pattern.

Fig. 4.7 displays two benchmarked series produced by the additive regression-based benchmarking model, with no regressor $\boldsymbol{R}$ (H=0), and an AR(1) model with parameter $\varphi = 0.729$. The additively benchmarked series is obtained with $\lambda = 0$; and the proportional with $\lambda = 1$. The proportionally benchmarked series displays larger seasonal amplitude. Indeed, proportional benchmarking modifies the larger sub-annual values more than the smaller ones, which is not the case for additive benchmarking. The basic assumption in the proportional approach is that larger values (the seasonally larger ones in this case) account for more of the annual discrepancies than the smaller ones. This assumption is widely accepted by practitioners in statistical agencies. The assumption also tends to avoid negative benchmarked values. Note that both benchmarked series converge to the original sub-annual series at the ends.

In the case of stock series, the assumption of proportional error following an AR(1) models cause the same problems encountered with additive AR(1) models, namely spurious movements un-supported by data, as observed in the previous section. The solution is proportional AR(2) error models.

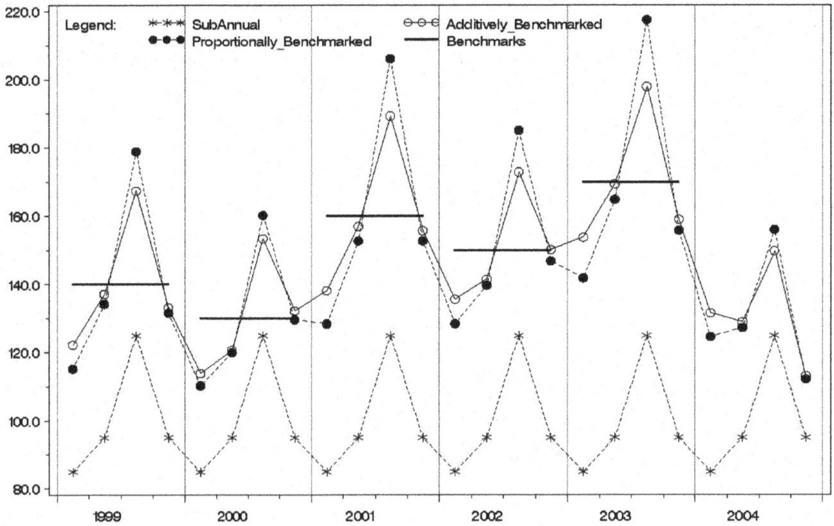

Fig. 4.7. Simulated quarterly flow series proportionally and additively benchmarked to yearly values, using an AR(1) error model with parameter φ=0.729

4.5 Minimizing the Size of the Corrections

Selecting a diagonal autocorrelation matrix $\boldsymbol{\Omega} = \boldsymbol{I}$ results into the following objective function

$$min \; \{ \Sigma_{t=1}^{T} \, (e_t / \sigma_t^{\lambda})^2 \}, \tag{4.14}$$

which minimizes the size of the weighted error. With λ=0, the error is homoscedastic.

If the coefficients of variation are constant or assumed constant, the standard deviations are given by $\sigma_t = c \, s_t$. Setting λ=1/2 leads to an objective function which minimizes the proportional size of the error with respect to the original sub-annual values:

$$(1/c) \times min \, \{ \Sigma_{t=1}^{T} \, e_t^2 / s_t \}. \tag{4.15}$$

In the context of the regression-based benchmarking method, this approximates prorating (except for the years without benchmark).

Prorating consists of multiplying the sub-annual observations by the annual proportional discrepancy (2.2b). For example if the ratio of the benchmark to the corresponding sub-annual sum is 1.10, the sub-annual values covered by the benchmark are multiplied by 1.10.

Note that choosing $\lambda=1$, leads to the minimization of the proportional size of the error with respect to the square of sub-annual values: $(1/c)\times$ $min\,\{\Sigma_{t=1}^{T}\, e_t^2\,/\,s_t^2\}$.

Objective function (4.15) is often used in reconciliation as discussed in Chapter 11. When restoring the additivity of a table of scalars or of time series, it is often desired that the corrections be proportional to the values of the original table. Under (4.15), iterative proportional fitting ("raking") approximates reconciliation, if the marginal totals are specified as exogeneous, i.e. pre-determined.

4.6 Other ARMA Error Models and Movement Preservation

This section illustrates the impact of other ARMA processes, namely moving averages processes (MA), autoregressive moving average processes (ARMA) and seasonal autoregressive processes (SAR), on the distribution of the annual discrepancies and the movement preservation principle.

As shown in section 4.3, it is more useful to examine the corrections resulting from an ARMA process than the corresponding benchmarked series. Hence we present only the graphs of the corrections $\hat{\theta}_t - s_t$. As mentioned earlier, in cases of interpolation (i.e. $s_t=0$) these corrections coincide with the interpolated series.

Fig.4.8 exhibits the behaviour of two MA(1) error models $e_t = v_t - \psi\, v_{t-1}$, for flow series: one with moving average parameter $\psi = 0.80$ and the other with $\psi = 0.40$. In both cases, the MA corrections tend to flatten within each year and therefore produce jumps between the years, which makes them unacceptable for flow series.

Fig. 4.9 exhibits the behaviour of the same two MA(1) error models for the corresponding stock series. Again both models produce abrupt and spurious movements unsupported by the data.

Fig. 4.10 and 4.11 display the correction of two MA(1) error models with negative parameters for flow and stock series respectively. In both figures, the corrections exhibit kinks and large movement reversals which are not substantiated by the data. The movements reversal hugely alter the movement of the series compared to the AR(1) error model also displayed in the figure. MA(1) error models with negative coefficients are therefore highly inadequate for benchmarking.

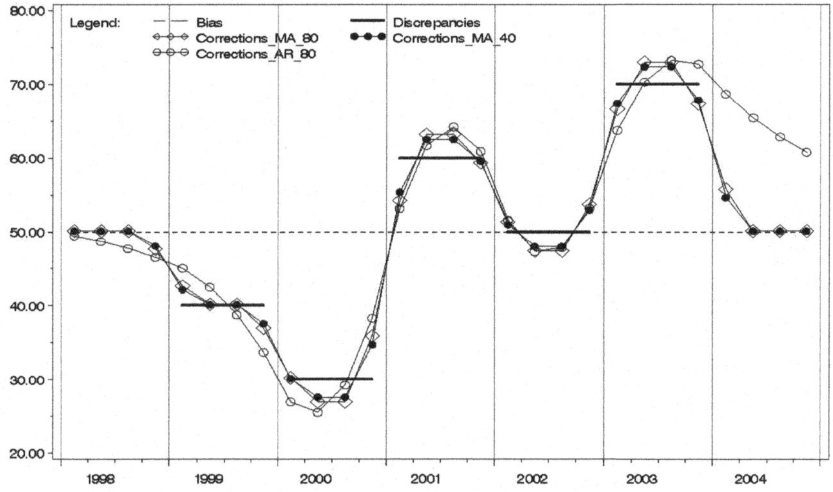

Fig. 4.8. Quarterly corrections obtained for flows, using MA(1) error models with parameters ψ =0.80 and ψ =0.40

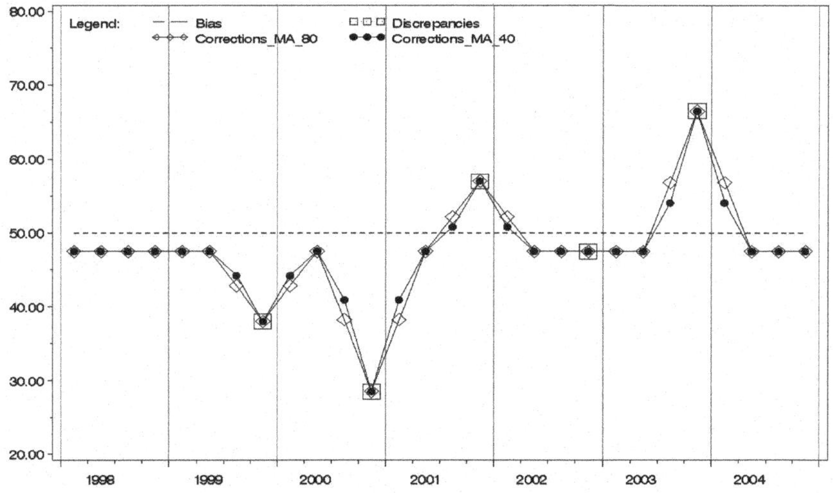

Fig. 4.9. Quarterly corrections obtained for stocks, using MA(1) error models with parameters ψ =0.80 and ψ =0.40

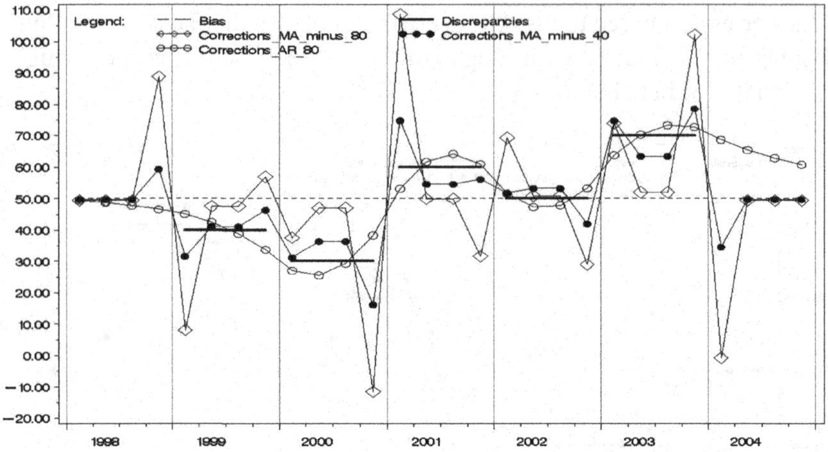

Fig. 4.10. Quarterly corrections obtained for flows, using MA(1) error models with parameters $\psi = -0.80$ and $\psi = -0.40$

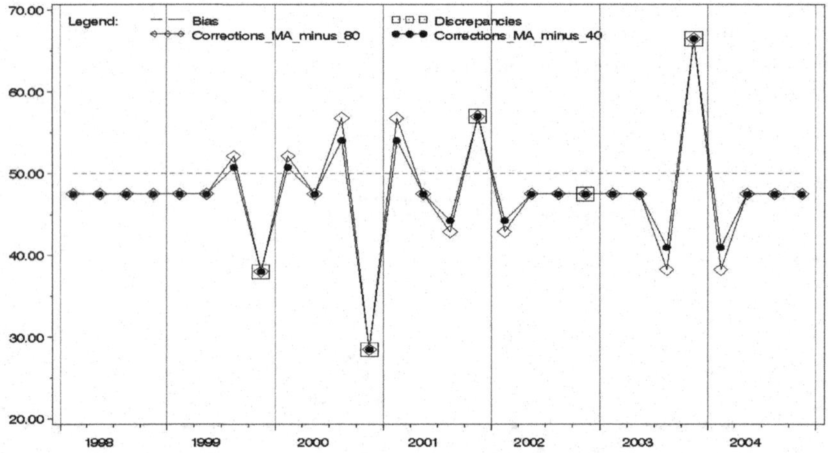

Fig. 4.11. Quarterly corrections obtained for stocks, using MA(1) error models with parameters $\psi = -0.80$ and $\psi = -0.40$

Fig. 4.12 and 4.13 depicts the behaviour of an ARMA(1,1), a first order autoregressive moving average model, $e_t = \varphi\, e_{t-1} + v_t \;\; - \psi\, v_{t-1}$, with $\varphi = \mathbf{0.80}$ and $\psi = \mathbf{0.70}$ for flow and stock series respectively. In the case of flows, the ARMA(1,1) corrections (curve ARMA_80_70) tend to flatten within each year, compared to the AR(1) corrections with first $\varphi = \mathbf{0.80}$ also displayed, and to suddenly change level between the years. In the case of stocks, the ARMA(1,1) corrections display movement reversals un-supported by the data.

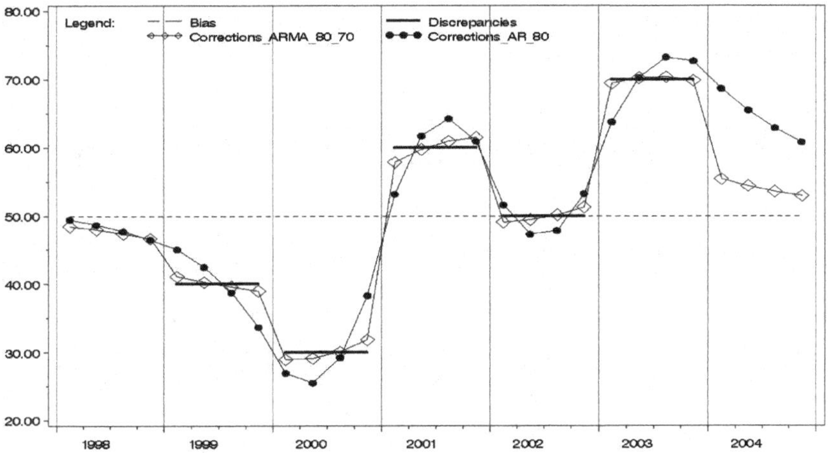

Fig. 4.12. Quarterly corrections obtained for flows, using an ARMA(1,1) error model with parameters φ=0.80 and ψ=0.70

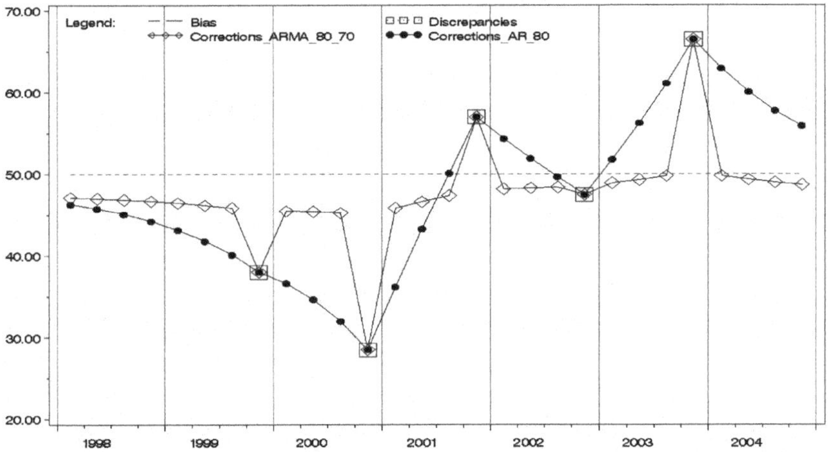

Fig. 4.13. Quarterly corrections obtained for stocks, using an ARMA(1,1) error model with parameters φ=0.80 and ψ=0.70

It should be noted that ARMA(1,1) error models, with positive parameters with $\varphi > \psi$, reduce the impact of the innovation of time $t-1$ by an amount give by the value of ψ times the innovation. The innovation of the fourth quarter of 2003 in Fig. 4.13 is incorporated in the model in that quarter through term v_t $(e_t = \varphi\, e_{t-1} + v_t - \psi\, v_{t-1})$; and largely cancelled out in the first quarter of 2004 by the moving average parameter through term $-\psi\, v_{t-1}$ $(e_{t+1} = \varphi\, e_t + v_{t+1} - \psi\, v_t)$. Without the moving average parameter, the whole innovation is built into the future of the model, as illustrated by curve AR_80 in the quarters of 2004.

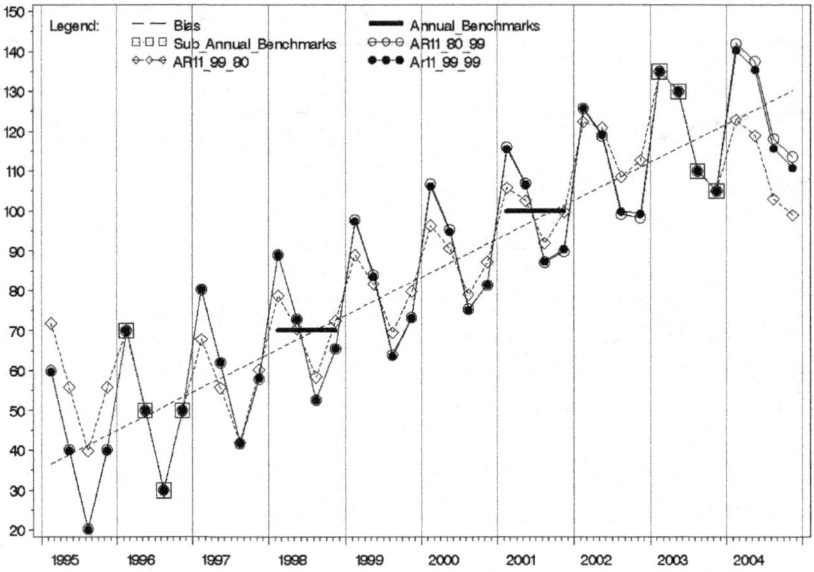

Fig. 4.14. Behaviour of seasonal AR(1)(1) error models with parameters φ=0.80 and Φ =0.99, φ=0.99 and Φ =0.80, and φ=0.99 and Φ=0.99

In this new example of Fig. 4.14, we deal with the interpolation of a seasonal series with seasonality borrowed from quarterly benchmarks and with level governed by annual benchmarks. The example contains two yearly benchmarks for 1998 and 2001 and four quarterly (sub-annual) benchmarks for 1996 and four quarterly benchmarks 2003. The seasonal pattern of the quarterly benchmarks in 1996 differs from that in 2003. The original quarterly series is equal to zero and the deterministic regressors are a constant and time, resulting in a linear "bias". The goal of benchmarking in this case is to produce quarterly interpolations consistent with the quarterly and the yearly benchmarks. This is achieved by using a seasonal autoregressive SAR(1) model, $(1-\varphi B)(1-\Phi B^4)\,e_t = v_t$ with three sets of parameters: $\varphi = 0.80$ and $\Phi = 0.99$, $\varphi = 0.99$ and $\Phi = 0.80$ and $\varphi = 0.99$ and $\Phi = 0.99$.[7] The interpolations obtained display a smooth transition from the seasonal pattern prevailing in 1996 to that in 2003. However, the interpolations with set of parameters $\varphi = 0.99$ and $\Phi = 0.80$ (curve AR11_99_80) exhibit a shrinking seasonal amplitude for the years without

[7] This model is thus equivalent to $e_t = \varphi e_{t-1} + \Phi e_{t-4} - \varphi \Phi e_{t-5} + v_t$.

quarterly benchmarks. This is due to the seasonal autoregressive parameter equal to 0.80 instead of 0.99 (close to 1.0) for the other two models.

Seasonal ARMA processes can only be used in conjunction with sub-annual benchmarks which determine the realization of the seasonal pattern based on data. Without sub-annual benchmarks, the resulting interpolations are most likely to display a spurious seasonal pattern. The sub-annual benchmarks need not be clustered in only a few years as in the example. However they must be supplied at regular intervals for *each* quarter (not necessarily in the same year), in order to avoid spurious seasonality.

4.7 Guidelines on the Selection of Sub-Annual Error Models

This sections provides guidelines concerning the choice of sub-annual error models. These guidelines are based on results obtained from a large number of applications with simulated and real data, which confirmed the preference for autoregressive error models. The AR models and corresponding parameter values produced satisfactory convergence to zero at both ends. Hence, the benchmarked series converged to the original sub-annual series corrected for bias $s_t^\dagger = s_t - \Sigma_{h=1}^{H} r_{th} \hat{\beta}_h$, which is a desirable property.

Table 4.1. Guidelines regarding sub-annual ARMA error models for benchmarking

	Error model	Parameter values
For temporally contiguous benchmarks		
Monthly	$(1-\varphi B)\, e_t = v_t$	$0.70 \le \varphi \le 0.90$
Quarterly	same	$0.70^3 \le \varphi \le 0.90^3$
For non temporally contiguous benchmarks		
Monthly	$(1-\varphi B)(1-\varphi B)\, e_t = v_t$	$0.70 \le \varphi \le 0.80$
Quarterly	same	$0.70^3 \le \varphi \le 0.80^3$

Table 4.1 displays the AR error models for benchmarking with a range of parameter values. The choice of the appropriate model depends on whether the benchmarks are temporally contiguous or not. Benchmarks are contiguous when the starting date of each benchmark is equal to the ending date of the previous benchmark plus one day; and, non-contiguous otherwise. Thus, benchmarks of stock series are not contiguous in Fig. 4.3a and 4.4a. Non-contiguous benchmarks can also occur for index and flow series as illustrated in Fig. 4.6, where the quarters of year 2002 are not covered by benchmarks and are embedded between benchmarks.

Furthermore, the choice of value for parameter φ depends on the periodicity of the sub-annual data. The purpose is to obtain a comparable convergence between quarterly and monthly series after a year. The degree of convergence is given by the autocorrelation function. For example the autocorrelation function of an AR(1) model with parameter φ=0.80 is equal to 0.07 at lag $\ell=12$. This implies that 93% of the convergence is achieved after 12 months of the last year without benchmark. The AR(2) model in Table 4.1 $(1-\varphi B)(1-\varphi B)\, e_t = v_t$ can be written as $e_t = \varphi_1\, e_{t-1} - \varphi_2\, e_{t-2} + v_t$ where $\varphi_1 = 2\varphi$ and $\varphi_2 = -\varphi^2$. For AR(1) and AR(2) error models, the autocorrelations are respectively given by

$$\omega_\ell = \varphi^{|\ell|},\ \ell = 0, \ldots, T-1, \tag{4.16}$$

and

$$\omega_0 = 1,\ \omega_1 = \varphi_1 / (1 - \varphi_2),$$
$$\omega_\ell = \varphi_1\, \omega_{\ell-1} + \varphi_2\, \omega_{\ell-2},\ \ell = 2, \ldots, T-1, \tag{4.17}$$

where ℓ is the lag, $|\varphi|<1$, and $\varphi_1 + \varphi_2 < 1$, $\varphi_2 - \varphi_1 < 1$, $|\varphi_2| < 1$.

4.8 The Covariance Matrix of the Benchmarks

The covariance matrix of the annual benchmarks V_ε is usually set equal to 0; which specifies them as binding. One reason is that the benchmarks are predetermined. For example, the yearly benchmarks used by the quarterly National Accounts originate from a more general accounting framework, namely the yearly Input-Output Accounts[8]. In other words, the quarterly Accounts must be consistent with the yearly Accounts.

Another reason for binding benchmarks (V_ε=0) is that they originate from yearly censuses, which are not subject to survey error. This assumption may be legitimate if all respondents to the census have the same fiscal year. This is the case in some industries. For example, the Canadian Federal and Provincial governments have a common fiscal year ranging from April 1 to March 31 of the following year; banks have a common fiscal year ranging from Nov. 1 to Oct. 31 of the following year; school boards and academic institution, from Sept. 1 to Aug. 31 of the following year. Taking the cross-sectional sums of fiscally homogeneous fiscal year data produces a tabulated

[8] The National Accounts cover only the final goods and services produced in the economy. They are a sub-set of the Input-Output Accounts, which cover both intermediate and final goods and services.

benchmarks which may be used as such in benchmarking, provided the coverage fractions, j_{mt}, are specified accordingly. Calendar year values may then be obtained as a by-product of benchmarking, by merely taking the yearly sums of the benchmarked series.

In many cases however, the fiscal years of the respondents are not homogeneous. For example, the Census of Manufacturing accepts the fiscal year data of the respondents as valid. As a result, the data may contain up to 12 different fiscal years, covering from Jan. 1 to Dec. 31 of the same year, Feb. 1 to Jan. 31 of the following year, and so forth. This problem is dealt with in the following manner:

(1) The reported values are arbitrarily classified in year y. For example, any reported value with ending date in year y is classified in year y.

(2) The yearly benchmark of year y is then obtained by taking the cross-sectional sum of the reported values classified in year y.

The resulting tabulated benchmark may thus implicitly cover up to 23 consecutive months: from Feb. 1 of year y-1 to Dec. 31 of year y. As a result, the timing of the benchmarks is late and their duration is too long. The benchmarks therefore under-estimate the calendar year values in periods of growth and over-estimate in periods of decline, because their timing is late. In the case of turning points, the benchmarks also underestimate the peaks and over-estimates the troughs of the business-cycle, because their duration is too long, 23 months instead of 12. As discussed in Chapter 9 on calendarization, this treatment of the fiscal year data raises doubts about their reliability and the wisdom of specifying them as binding.

Note that the same situation prevails with quarterly benchmarks covering any three consecutive months. The resulting tabulated quarterly values cover up to five months. However, in the case of homogeneous fiscal quarters, pertaining to the banking sector for instance, the fiscal quarter benchmarks may be used as binding in benchmarking, provided the coverage fractions are correctly specified. Chapter 9 provides more details on fiscal data.

The CVs of the benchmarks originating from reported data covering various fiscal period could be specified by subject matter expert. For example, if the expert deems that the benchmarks in a specific industry have a confidence interval of plus or minus 2 percent, this implies CVs equal to 1 percent, under the Normal distribution. Such CVs can be translated into variances, by multiplying the CVs and the benchmarks and squaring. The covariance is usually specified as a diagonal matrix with diagonal elements equal to the variances.

5 The Cholette-Dagum Regression-Based Benchmarking Method - The Multiplicative Model

5.1 Introduction

The additive regression-based benchmarking model discussed in Chapter 3 is appropriate when the variances of the sub-annual error are independent of the level of the series. However, for most time series, this condition is not present, and additive benchmarking is sub-optimal. When the error variances are proportional to the level of the series, a multiplicative model is more appropriate. In such cases, all the components of the benchmarking model and the resulting benchmarked estimates tend to have constant coefficients of variation (CV)s, $\hat{\sigma}_t / \hat{\theta}_t \approx c$, as intended by the survey design of the original series.

The seasonal component of most time series display a seasonal amplitude which increases and shrinks with the level of the series. This is illustrated in Fig. 5.3a for the Canadian Total Retail Trade series. Additively benchmarking such series can produce negative benchmarked values. Since benchmarking basically alters the level of series, it is desirable to use proportional (Chapter 3, section 3.8) or multiplicative benchmarking. Both methods tend to preserve growth rates, which are widely used to compare the behaviour of various socio-economic indicators. The basic assumption in the proportional and multiplicative benchmarking is that seasonally larger values account for more of the annual discrepancies than the smaller ones. This assumption is widely accepted by practitioners in statistical agencies. The assumption also reduces the possibility of negative benchmarked values.

What distinguishes the multiplicative benchmarking model from proportional model is the ability of the former to include a multiplicative deterministic regression part $\boldsymbol{R\beta}$, whereas the proportional model can only contain an additive regression part.

Another important reason for the wide application of the multiplicative benchmarking model is that the proportional discrepancies are standardized, thus facilitating comparisons among series.

In this chapter, we focus on regression-based benchmarking, but several results also apply to interpolation, temporal dis-aggregation and signal extraction, discussed in Chapters 7 and 8 respectively.

5.2 The Multiplicative Benchmarking Model

Like the regression-based additive model discussed in Chapter 3, the multiplicative version consists of two equations. The first equation is now multiplicative:

$$s_t = b_t \times \theta_t \times e_t \;=\; b_t \times \theta_t \times (1 + e_t^*)$$
$$= b_t \times \theta_t \times (1 + c_t e_t^\dagger), \qquad (5.1a)$$

where $b_t = (\Pi_{h=1}^{H} \beta_h^{r_{th}})$ is the deterministic regression part. The error e_t is centred on 1 and equal to $(1 + e_t^*) = (1 + c_t e_t^\dagger)$, where e_t^* is a deviation from 1 which is close to 0. The standardized error denoted as $e_t^\dagger$ has mean 0 and unit variance, and c_t is the standard deviation of e_t^*, with $E(e_t^*) = 0$, $E(e_t^* \; e_{t-\ell}^*) = c_t c_{t-\ell} \; \omega_\ell$, where ω_ℓ is the autocorrelation of the standardized error.

Applying the logarithmic transformation to Eq (5.1a) yields

$$ln s_t = ln b_t + ln\theta_t + ln(1 + e_t^*) \;=\; ln b_t + ln\theta_t + ln(1 + c_t e_t^\dagger). \qquad (5.1b)$$

Assuming that c_t is positive and small (as is often the case), we can make use of the approximation $ln\, x \approx x - 1$ to approximate $ln(1 + c_t e_t^\dagger)$ by $c_t e_t^\dagger$. Hence

$$ln s_t \approx ln b_t + ln\theta_t + \; c_t e_t^\dagger, \; t=1,...,T. \qquad (5.1c)$$

Taking the logarithmic transformation of $b_t = (\Pi_{h=1}^{H} \beta_h^{r_{th}})$ yields $ln b_t =$ $\Sigma_{h=1}^{H} ln\beta_h^{r_{th}} = \Sigma_{h=1}^{H} r_{th} \, ln\beta_h$. Substituting in (5.1c) yields

$$s_t^* = \Sigma_{h=1}^{H} r_{th} \; \beta_h^* + \theta_t^* + c_t e_t^\dagger,$$
$$s_t^* = \Sigma_{h=1}^{H} r_{th} \; \beta_h^* + \theta_t^* + e_t^*, \; t=1,...,T, \qquad (5.2a)$$

where the asterisk denotes the logarithmic transformation was applied.

The second equation remains as in the additive regression-based model:

$$a_m = \Sigma_{t=t_{1m}}^{t_{Lm}} j_{mt} \, \theta_t + \varepsilon_m, \; m=1,..., M, \tag{5.2b}$$

where $E(\varepsilon_m) = 0$, $E(\varepsilon_m^2) = \sigma_{\varepsilon_m}^2$.

The multiplicative model requires that both the sub-annual and the annual observations be positive; the model avoids negative adjusted values.

Equation (5.2a) shows that the (logarithm of the) observed sub-annual series s_t^* is contaminated by autocorrelated error e_t^* and the possible presence of deterministic time effects $\Sigma_{h=1}^{H} r_{th} \, \beta_h^*$. In the equation, θ_t^* stands for the true values satisfying the annual constraints. The errors e_t^* can be expressed as the product of the coefficients of variation c_t and a standardized error $e_t^{\dagger}$. As a result, the coefficients of variation used or assumed in the multiplicative model play the role of the standard deviations in the additive model.

Eq. (5.2a) basically sets the sub-annual *growth rate* movement of the series under consideration.

The deterministic regressor $\boldsymbol{R}$ is typically a constant (i.e. H=1 and r_{Ht}=−1), which captures the average proportional level difference between the annual and the sub-annual data. This quantity is a weighted average of the proportional annual discrepancies,

$$d_m^{(p)} = a_m \, / \, (\Sigma_{t=t_{1m}}^{t_{Lm}} j_{mt} \, s_t), \; m=1,...,M.$$

In a survey context, the estimated regression parameter $\hat{\beta}_1 = exp(\hat{\beta}_1^*)$ is interpreted as a multiplicative "bias".

In some cases, a second regressor is used to capture a deterministic trend in the proportional discrepancies, $d_m^{(p)}$, as discussed in section 3.4, or the regressor $\boldsymbol{R}$ may be absent, in which case H=0.

The error e_t^* in (5.1a) plays a critical role in benchmarking. The error model chosen governs the manner in which the annual discrepancies are distributed over the sub-annual observations. The error manifests itself in the presence of benchmarks and the corresponding annual discrepancies.

The error e_t^* is homoscedastic when λ=0; and heteroscedastic with standard deviation c_t^{λ}, otherwise. Parameter λ can take values 0, 1/2 and 1. Note that if the CVs are constant, the logarithmic error is homoscedastic regardless of the value of λ. The autocorrelations ω_ℓ of the standardized error $e_t^\dagger$ are usually those of an autoregressive process of order 1 or 2, i.e. an AR(1) model or an AR(2) model as discussed in Chapter 4.

Eq. (5.2b) shows that the benchmarks a_m are equal to the temporal sums of the true sub-annual values θ_t, over their reporting periods plus a non-autocorrelated error ε_m with heteroscedastic variance. The first and the last sub-annual periods covered by benchmark a_m are respectively denoted by t_{1m} and t_{Lm}. The *coverage fractions* are denoted by j_{mt}: for a given sub-annual period t, j_{mt} is the proportion of the days of period t included in the reporting period of the m-th benchmark. Various types of coverage fractions are extensively discussed in section 3.1 and 3.3. Finally it should be noted that Eq. (5.2b) accommodates flow, stock and index series, as discussed in section 3.5.

The errors e_t and ε_m are assumed to be mutually uncorrelated since the sub-annual and the annual data come from two different sources (e.g. a survey and administrative records).

The annual equation (5.2b) remains additive, because the benchmarks measure the annual sums of the true sub-annual values. Indeed, taking the logs of a_m and θ_t in (5.2b) would specify the benchmarked values to annually multiply to the benchmarks. The resulting benchmarked series $\hat{\theta}_t$ would take a level drastically different from that of the original sub-annual series.

However, in the simple case where all the coverage fractions j_{mt} are equal to 1 and each benchmark refers to a single time period (e.g. stock series), and Eq. (5.2b) may be replaced by

$$a_m^* = \Sigma_{t=t_{1m}}^{t_{1m}} j_{mt}\, \theta_t^* + \varepsilon_m^* \equiv \theta_{t_{1m}}^* + \varepsilon_m^* \,, \qquad (5.3)$$

where the summations are now trivial. The model is then completely linear in the parameters $\boldsymbol{\beta}^*$ and $\boldsymbol{\theta}^*$ and may be solved using the linear solution

(3.20) and (3.21) of the additive model of Chapter 4. The resulting covariance matrices $var[\hat{\boldsymbol{\beta}}^*]$ and $var[\hat{\boldsymbol{\theta}}^*]$ may be converted into the original scale using

$$\hat{\theta}_t = exp(\hat{\theta}_t^*), \tag{5.4a}$$

$$v_{ti} = \{exp(v_{ti}^*) - 1\}\, \hat{\theta}_t\, \hat{\theta}_i. \tag{5.4b}$$

Note that the coefficients of variation of $\hat{\theta}_t$ (*sic*) are then given by the square root of the diagonal elements of $var[\hat{\boldsymbol{\theta}}^*]$.

For all cases different from that just described regarding the coverage fractions, Eq. (5.1b) is non-linear in the parameters $\boldsymbol{\theta}^*$.

5.3 Matrix Representation

Eq. (5.2a) and (5.2b) can be written in matrix algebra as:

$$\boldsymbol{s}^* = \boldsymbol{R}\, \boldsymbol{\beta}^* + \boldsymbol{\theta}^* + \boldsymbol{e}^*,\ E(\boldsymbol{e}^*) = 0,\ E(\boldsymbol{e}^*\boldsymbol{e}^{*\prime}) = \boldsymbol{V}_{e^*}, \tag{5.5a}$$

$$\boldsymbol{a} = \boldsymbol{J}\, exp(\boldsymbol{\theta}^*) + \boldsymbol{\varepsilon},\ E(\boldsymbol{\varepsilon}) = 0,\ E(\boldsymbol{\varepsilon}\boldsymbol{\varepsilon}') = \boldsymbol{V}_{\varepsilon},\ E(\boldsymbol{e}^*\boldsymbol{\varepsilon}') = 0. \tag{5.5b}$$

As discussed in Chapter 4, the covariance matrix $\boldsymbol{V}_{e^*}$ of the log transformed sub-annual series may originate from a statistical model applied upstream of benchmarking, for example signal extraction. Whenever such a covariance matrix is available it should be used directly in benchmarking, in order to maintain the dependence structure of the sub-annual series as much as possible.

In most cases, however, the covariance matrix $\boldsymbol{V}_{e^*}$ has to be built from known or assumed standard deviations and an autocorrelation function, as follows:

$$\boldsymbol{V}_{e^*} = \boldsymbol{\Xi}^{\lambda}\, \boldsymbol{\Omega}\, \boldsymbol{\Xi}^{\lambda}, \tag{5.5c}$$

where λ equal to 0, 1/2 or 1, and where $\boldsymbol{\Xi}$ is a diagonal matrix containing the constant CVs, $c_1,\dots,c_T$ equal to c, and $\boldsymbol{\Omega}$ contains the autocorrelations corresponding to an ARMA process. For an AR(1) error model with parameter $|\varphi|<1$, the elements ω_{ij} of $\boldsymbol{\Omega}$ are given by $\varphi^{|i-j|}$. For an AR(2) error model, the elements ω_{ij} of the autocorrelation matrix $\boldsymbol{\Omega}$ are given in terms of lags $\ell = |i-j|$ such that $\omega(\ell)=1$ for $\ell=0$, $\omega(\ell) = \varphi_1/(1-\varphi_2)$, for $\ell=1$, and $\omega(\ell) = \varphi_1\,\omega_{\ell-1} + \varphi_2\,\omega_{\ell-2}$, for $\ell = 2,\ldots,T-1$, where $\varphi_1 + \varphi_2 < 1$, $\varphi_2 - \varphi_1 < 1$ and $|\varphi_2| < 1$ to ensure stationarity.

Matrix $\boldsymbol{J}$ is a temporal sum operator of dimension M by T, containing the coverage fractions:

$$\boldsymbol{J} = \begin{bmatrix} j_{11} & j_{12} & \cdots & j_{1T} \\ \vdots & \vdots & \ddots & \vdots \\ j_{M1} & j_{M2} & \cdots & j_{MT} \end{bmatrix},$$

where $j_{mt} = 0$ for $t_{Lm} < t < t_{1m}$, and j_{mt} is as defined in Eq. (3.5b) for $t_{1m} \le t \le t_{Lm}$. The coverage fractions j_{mt} are placed in row m and column t. More details are given in section 3.6.1.

5.4 Non-Linear Estimation of the Multiplicative Model

This section provides a non-linear algorithm to estimate the multiplicative benchmarking model, based on linearized regression.

The multiplicative model (5.5) can be written as a single equation,

$$\boldsymbol{y} = f(\boldsymbol{a}^*) + \boldsymbol{u}, \quad E(\boldsymbol{u}) = 0,\; E(\boldsymbol{u}\,\boldsymbol{u}') = \boldsymbol{V}_u\,, \tag{5.6a}$$

where
$$\boldsymbol{y} = \begin{bmatrix} \boldsymbol{s}^* \\ \boldsymbol{a} \end{bmatrix}, \; \boldsymbol{a}^* = \begin{bmatrix} \boldsymbol{\beta}^* \\ \boldsymbol{\theta}^* \end{bmatrix}, \; f(\boldsymbol{a}^*) = \begin{bmatrix} \boldsymbol{R}\boldsymbol{\beta}^* + \boldsymbol{\theta}^* \\ \boldsymbol{J}\boldsymbol{\theta} \end{bmatrix},$$

$$\boldsymbol{u} = \begin{bmatrix} \boldsymbol{e}^* \\ \boldsymbol{\varepsilon} \end{bmatrix}, \; \boldsymbol{V}_u = \begin{bmatrix} \boldsymbol{V}_e & \boldsymbol{0} \\ \boldsymbol{0} & \boldsymbol{V}_\varepsilon \end{bmatrix}, \; \boldsymbol{\theta} = exp(\boldsymbol{\theta}^*). \tag{5.6b}$$

Model (5.6) can be linearly approximated by a Taylor expansion of $f(\boldsymbol{a}^*)$ at appropriate initial values of the parameters, $\boldsymbol{a}_0^* = [\boldsymbol{\beta}_0^{*\prime}\; \boldsymbol{\theta}_0^{*\prime}]'$ provided below Eq. (5.10c):

$$f(\boldsymbol{a}^*) \approx \boldsymbol{f}_0 + \boldsymbol{X}_0 (\boldsymbol{a}^* - \boldsymbol{a}_0^*) \,, \tag{5.7a}$$

where

$$\boldsymbol{a}_0^* = \begin{bmatrix} \boldsymbol{\beta}_0^* \\ \boldsymbol{\theta}_0^* \end{bmatrix}, \; \boldsymbol{f}_0 = f(\boldsymbol{a}_0^*) = \begin{bmatrix} \boldsymbol{R}\boldsymbol{\beta}_0^* + \boldsymbol{\theta}_0^* \\ \boldsymbol{J}\boldsymbol{\theta}_0 \end{bmatrix}, \; \boldsymbol{X}_0 = \left[\frac{\partial f(\boldsymbol{a}^*)}{\partial \boldsymbol{a}^*} \right]_{\boldsymbol{a}^* = \boldsymbol{a}_0^*} = \begin{bmatrix} \boldsymbol{R} & \boldsymbol{I}_T \\ \boldsymbol{0} & \boldsymbol{J}_0 \end{bmatrix},$$

$$\boldsymbol{J}_0 = \boldsymbol{J}\, diag(\boldsymbol{\theta}_0) = \begin{bmatrix} j_{11}\theta_{0,1} & j_{12}\theta_{0,2} & \cdots & j_{1T}\theta_{0,T} \\ \vdots & \vdots & \ddots & \vdots \\ j_{M1}\theta_{0,1} & j_{M2}\theta_{0,2} & \cdots & j_{MT}\theta_{0,T} \end{bmatrix}, \tag{5.7b}$$

where $\boldsymbol{X}_0$ denotes the derivative of the $f(\boldsymbol{a}^*)$ at the initial values $\boldsymbol{a}_0^*$ and $\boldsymbol{\theta}_0 = exp(\boldsymbol{\theta}_0^*)$.

The linearized regression model is then $\boldsymbol{y} \approx \boldsymbol{f}_0 + \boldsymbol{X}_0(\boldsymbol{a}^* - \boldsymbol{a}_0^*) + \boldsymbol{u}$, or

$$\boldsymbol{y}_0 \approx \boldsymbol{X}_0 \boldsymbol{a}^* + \boldsymbol{u}, \tag{5.8a}$$

where
$$\boldsymbol{y}_0 = \boldsymbol{y} - \boldsymbol{f}_0 + \boldsymbol{X}_0 \boldsymbol{a}_0^* = \begin{bmatrix} \boldsymbol{s}^* \\ \boldsymbol{a} - \boldsymbol{J}\boldsymbol{\theta}_0 + \boldsymbol{J}_0 \boldsymbol{\theta}_0^* \end{bmatrix} = \begin{bmatrix} \boldsymbol{s}^* \\ \boldsymbol{a}_0 \end{bmatrix}. \tag{5.8b}$$

Applying generalized least squares to (5.8a) yields a revised estimate of $\boldsymbol{a}^*$

$$\hat{\boldsymbol{a}}^* = (\boldsymbol{X}_0' \boldsymbol{V}_u^{-1} \boldsymbol{X}_0)^{-1} \boldsymbol{X}_0' \boldsymbol{V}_u^{-1} \boldsymbol{y}_0 . \tag{5.9}$$

Letting the revised estimates $\hat{\boldsymbol{a}}^*$ play the role of new initial values (i.e. setting $\boldsymbol{a}_0^*$ to $\hat{\boldsymbol{a}}^*$), vectors $\boldsymbol{a}_0^*$ and $\boldsymbol{y}_0$ and matrix $\boldsymbol{X}_0$ are updated, and $\boldsymbol{a}^*$ is re-estimated using (5.9). This process can be carried out until the absolute change in $\hat{\boldsymbol{a}}^*$ is smaller than the convergence parameter, e.g. 0.00001. However, this solution does not allow binding benchmarks.

Noting that matrices $\boldsymbol{X}_0$ and $\boldsymbol{V}_u$ have the same form (and to a large extent the same content) as in the additive benchmarking model of Chapter 3, the alternative solution (3.20) which allows for binding benchmarks is applicable to the multiplicative model. In other words, result (3.20) holds *mutatis mutandis* for each iteration of (5.9), which leads to the following iterative solution:

$$\hat{\boldsymbol{\beta}}^* = -(\boldsymbol{R}' \boldsymbol{J}_0' \boldsymbol{V}_d^{-1} \boldsymbol{J}_0 \boldsymbol{R})^{-1} \boldsymbol{R}' \boldsymbol{J}_0' \boldsymbol{V}_d^{-1} [\boldsymbol{a}_0 - \boldsymbol{J}_0 \boldsymbol{s}^*], \quad (5.10a)$$

$$var[\hat{\boldsymbol{\beta}}^*] = (\boldsymbol{R}' \boldsymbol{J}_0' \boldsymbol{V}_d^{-1} \boldsymbol{J}_0 \boldsymbol{R})^{-1}, \quad (5.10b)$$

$$\hat{\boldsymbol{\theta}}^* = \boldsymbol{s}^\dagger + \boldsymbol{V}_{e*} \boldsymbol{J}_0' \boldsymbol{V}_d^{-1} [\boldsymbol{a}_0 - \boldsymbol{J}_0 \boldsymbol{s}^\dagger], \quad (5.10c)$$

where $\boldsymbol{s}^\dagger = \boldsymbol{s}^* - \boldsymbol{R}\hat{\boldsymbol{\beta}}^*$, $\boldsymbol{V}_d = [\boldsymbol{J}_0 \boldsymbol{V}_{e*} \boldsymbol{J}_0' + \boldsymbol{V}_\varepsilon]$, $\boldsymbol{J}_0 = \boldsymbol{J}\, diag(\boldsymbol{\theta}_0)$,
$\boldsymbol{a}_0 = \boldsymbol{a} - \boldsymbol{J}\boldsymbol{\theta}_0 + \boldsymbol{J}_0 \boldsymbol{\theta}_0^*$ and $\boldsymbol{\theta}_0 = exp(\boldsymbol{\theta}_0^*)$.

The initial values $\boldsymbol{\theta}_0^*$ needed to obtain $\boldsymbol{\theta}_0$, $\boldsymbol{J}_0$, $\boldsymbol{a}_0$ and $\boldsymbol{V}_d$ come from the previous iteration. For the first iteration, these values are obtained from a linear variant of model (5.5), with the second equation replaced by $\tilde{\boldsymbol{a}} = \tilde{\boldsymbol{J}}\, \boldsymbol{\theta}^* + \boldsymbol{\varepsilon}$, where $\tilde{\boldsymbol{a}}$ and $\tilde{\boldsymbol{J}}$ are respectively equal to $\boldsymbol{a}$ and $\boldsymbol{J}$ divided by the number of sub-annual periods covered $j_{m,\bullet}$, i.e. $\tilde{a}_m = ln(a_m / j_{m,\bullet})$ and $\tilde{j}_{mt} = j_{mt} / j_{m,\bullet}$. This model is solved using (3.14) or (3.15) *mutatis mutandis*. The justification for these initial values is the proximity of the arithmetic average to the geometric average for positive data.

After convergence, the covariance matrix of the benchmarked series may be obtained from the following formula

$$var[\hat{\boldsymbol{\theta}}^*] = \boldsymbol{V}_{e*} - \boldsymbol{V}_{e*} \boldsymbol{J}_0' \boldsymbol{V}_d^{-1} \boldsymbol{J}_0 \boldsymbol{V}_{e*} + \boldsymbol{W} var[\hat{\boldsymbol{\beta}}^*]\, \boldsymbol{W}', \quad (5.10d)$$

where $\boldsymbol{W} = \boldsymbol{R} - \boldsymbol{V}_{e*} \boldsymbol{J}_0' \boldsymbol{V}_d^{-1} \boldsymbol{J}_0 \boldsymbol{R}$. The square roots of the diagonal elements of $var[\hat{\boldsymbol{\theta}}^*]$ provide the estimated coefficients of variation of $\hat{\boldsymbol{\theta}} = exp(\hat{\boldsymbol{\theta}}^*)$.

In the absence of $\boldsymbol{R}\boldsymbol{\beta}$, the solution (5.10a) and (5.10b) respectively simplify to:

$$\hat{\boldsymbol{\theta}}^* = \boldsymbol{s} + \boldsymbol{V}_{e*} \boldsymbol{J}_0' \boldsymbol{V}_d^{-1} [\boldsymbol{a}_0 - \boldsymbol{J}_0 \boldsymbol{s}], \quad (5.11a)$$

$$var[\hat{\boldsymbol{\theta}}^*] = \boldsymbol{V}_{e*} - \boldsymbol{V}_{e*} \boldsymbol{J}_0' \boldsymbol{V}_d^{-1} \boldsymbol{J}_0 \boldsymbol{V}_{e*}, \quad (5.11b)$$

where $\boldsymbol{V}_d = [\boldsymbol{J}_0 \boldsymbol{V}_{e*} \boldsymbol{J}_0' + \boldsymbol{V}_\varepsilon]$, $\boldsymbol{J}_0 = \boldsymbol{J}\, diag(\boldsymbol{\theta}_0)$,

$$\boldsymbol{a}_0 = \boldsymbol{a} - \boldsymbol{J}\boldsymbol{\theta}_0 + \boldsymbol{J}_0\boldsymbol{\theta}_0^* \text{ and } \boldsymbol{\theta}_0 = exp(\boldsymbol{\theta}_0^*).$$

In most situations, this non-linear estimation procedure requires less than five iterations, with a convergence criterion equal to 0.0000001. This rapid convergence is due to the fact that the problem is not highly non-linear. Indeed, by trial and error it is possible to find an alternative set of benchmarks $\boldsymbol{a}^{\ddagger}$, which satisfies both $ln\, \boldsymbol{a}^{\ddagger} = \boldsymbol{J}\boldsymbol{\theta}^*$ and $\boldsymbol{a} = \boldsymbol{J}\boldsymbol{\theta}$. These alternative benchmarks would allow us to use the linear solution (3.20) and (3.21) applied to the logarithmically transformed data.

In principle, the multiplicative model does not produce negative benchmarked values. However, there are situations leading to negative benchmarked values in proportional benchmarking (e.g. the modified proportional Denton) which may cause numerical problems in multiplicative benchmarking, namely singularity of the matrix $\boldsymbol{V}_d$ in (5.10) and (5.11) and non-convergence of the iterative solution.

These situations occur when the benchmarks dramatically change from one year to the next, while the sub-annual series changed very little in comparison[1]; or, when the benchmarks change little, while the annual sums of the sub-annual series change dramatically.

Such numerical problems indicate that the benchmarked series cannot be properly fitted under both the benchmarking constraints and the non-negativity constraints implicit in the logarithmic transformation of the multiplicative model.

The residuals associated with the sub-annual and annual equations are obtained using the following formulae

$$\hat{\boldsymbol{e}}^* = \hat{\boldsymbol{\theta}}^* - (\boldsymbol{s}^* - \boldsymbol{R}\hat{\boldsymbol{\beta}}^*) \equiv \hat{\boldsymbol{\theta}}^* - \boldsymbol{s}^{\dagger *} = \boldsymbol{V}_{e*}\boldsymbol{J}_0^{\prime}\boldsymbol{V}_d^{-1}[\boldsymbol{a}_0 - \boldsymbol{J}_0\boldsymbol{s}^{\dagger *}], \quad (5.12a)$$

$$\hat{\boldsymbol{\varepsilon}} = \boldsymbol{a} - \boldsymbol{J}\hat{\boldsymbol{\theta}}. \quad (5.12b)$$

When the variances of $\boldsymbol{e}^*$ and $\boldsymbol{\varepsilon}$ are known, there is no need to estimate the residuals. However, for most applications, it is necessary to estimate the residuals, calculate their variances and refit the model to obtain valid confidence intervals for the estimates.

---

[1] Drastic fluctutations may occur in the manufacturing of some large items, like aircrafts, ships, nuclear plants.

In a similar situation (group heteroscedasticity)[2], the practice is to take the average of the original logarithmic variances c_t^2 and the average of the squared residuals, respectively:

$$\bar{c}^{2\,(0)} = \Sigma_{t=1}^{T} c_t^2 / T, \qquad \bar{c}^{2\,(1)} = \Sigma_{t=1}^{T} \hat{e}_t^{*\,2} / T. \tag{5.13}$$

The original CVs of Eq. (5.1a) are then multiplied by the ratio $\bar{c}^{(1)} / \bar{c}^{(0)}$, and the model is re-estimated on the basis of $\hat{\boldsymbol{V}}_{e^*} = \hat{\boldsymbol{\Xi}} \boldsymbol{\Omega} \hat{\boldsymbol{\Xi}}$.

We do not re-estimate the variances of the benchmarks, which are usually null or much smaller than the variances of the sub-annual series. Under binding benchmarks ($\varepsilon = 0$ and $V_\varepsilon = 0$), refitting the model does *not* change the benchmarked series $\hat{\boldsymbol{\theta}}^*$ nor the parameters $\hat{\boldsymbol{\beta}}^*$, but does change their variances. It is then sufficient to multiply the covariance matrices $var[\hat{\boldsymbol{\beta}}^*]$ and $var[\hat{\boldsymbol{\theta}}^*]$ by $(\bar{c}^{(1)} / \bar{c}^{(0)})^2$.

Durbin and Quenneville (1997) cast the regression-based multiplicative benchmarking model in a state space representation. Their approach will be discussed in Chapter 8 in the context of signal extraction and benchmarking.

Mian and Laniel (1993) presented a mixed model in which the bias is multiplicative and the error is additive: $s_t = b \times \theta_t + e_t$. Their results are almost identical to those of the multiplicative model, because as shown in section 5.2 an additive error can be expressed multiplicatively and vice-versa.

Appendix C provides formulae for some of the matrix products in (5.10) and (5.11). These formulae would be most useful for those intending to program multiplicative benchmarking.

5.5 Other Properties of the Regression-Based Multiplicative Benchmarking Model

The Cholette-Dagum multiplicative benchmarking model contains very few degrees of freedom: *T+H* parameters must be estimated from *T+M* observations. This means that the benchmarked values will be very close to the bias-corrected series, $\boldsymbol{s}^\dagger = \boldsymbol{s} / exp(\boldsymbol{R}\hat{\boldsymbol{\beta}}^*)$. This proximity is in fact required to preserve the growth rates of the sub-annual series. As discussed

[2] Each squared CV correspond to the logarithmic variance of a group.

before, most of the improvement introduced by benchmarking to the sub-annual series lies in correcting the level.

Table 5.1a. Simulated quarterly series benchmarked using the Cholette-Dagum regression-based multiplicative model

Starting date	Ending dates	t	s	$s^{\dagger}$	$\hat{\theta}$
01/01/98	31/03/98	1	85.0	128.3	126.7
01/04/98	30/06/98	2	95.0	143.4	141.0
01/07/98	30/09/98	3	125.0	188.6	184.4
01/10/98	31/12/98	4	95.0	143.4	139.0
01/01/99	31/03/99	5	85.0	128.3	123.0
01/04/99	30/06/99	6	95.0	143.4	135.2
01/07/99	30/09/99	7	125.0	188.6	173.9
01/10/99	31/12/99	8	95.0	143.4	128.0
01/01/00	31/03/00	9	85.0	128.3	109.6
01/04/00	30/06/00	10	95.0	143.4	120.4
01/07/00	30/09/00	11	125.0	188.6	160.5
01/10/00	31/12/00	12	95.0	143.4	129.5
01/01/01	31/03/01	13	85.0	128.3	128.0
01/04/01	30/06/01	14	95.0	143.4	152.5
01/07/01	30/09/01	15	125.0	188.6	206.2
01/10/01	31/12/01	16	95.0	143.4	153.4
01/01/02	31/03/02	17	85.0	128.3	129.6
01/04/02	30/06/02	18	95.0	143.4	140.6
01/07/02	30/09/02	19	125.0	188.6	184.6
01/10/02	31/12/02	20	95.0	143.4	145.1
01/01/03	31/03/03	21	85.0	128.3	138.5
01/04/03	30/06/03	22	95.0	143.4	161.6
01/07/03	30/09/03	23	125.0	188.6	217.0
01/10/03	31/12/03	24	95.0	143.4	162.9
01/01/04	31/03/04	25	85.0	128.3	140.8
01/04/04	30/06/04	26	95.0	143.4	153.4
01/07/04	30/09/04	27	125.0	188.6	198.2
01/10/04	31/12/04	28	95.0	143.4	148.6

Furthermore, Eq. (5.11b) shows that, in the absence of a deterministic time regressor (H=0), the values of covariance matrix of the benchmarked series $var[\hat{\boldsymbol{\theta}}^*]$ are smaller than the original covariance matrix, $\boldsymbol{V}_{e^*}$, because the term subtracted from $\boldsymbol{V}_{e^*}$, i.e. $\boldsymbol{V}_{e^*} \boldsymbol{J}_0' \boldsymbol{V}_d^{-1} \boldsymbol{J}_0 \boldsymbol{V}_{e^*}$, is positive definite. In fact, the benchmarking constraints whether binding or not can be interpreted as smoothing priors and will always reduce the variance of the adjusted series.

Table 5.1b. Contributions to the variance of the benchmarked series shown in Table 5.1a

t	V_{e^*}	$-V_{e^*}J_0^{\prime}V_d^{-1}J_0V_{e^*}$	$W\,var[\hat{\beta}^*]W^{\prime}$	$var[\hat{\theta}^*]$
1	1.0000	-0.0485	0.1531	1.1046
2	1.0000	-0.0913	0.1204	1.0292
3	1.0000	-0.1717	0.0820	0.9103
4	1.0000	-0.3231	0.0412	0.7181
5	1.0000	-0.6080	0.0079	0.3999
6	1.0000	-0.7962	0.0001	0.2039
7	1.0000	-0.8286	0.0017	0.1730
8	1.0000	-0.6798	0.0014	0.3216
9	1.0000	-0.6615	0.0001	0.3386
10	1.0000	-0.7853	0.0000	0.2147
11	1.0000	-0.8433	0.0000	0.1567
12	1.0000	-0.7092	0.0001	0.2909
13	1.0000	-0.6616	0.0001	0.3384
14	1.0000	-0.7886	0.0000	0.2114
15	1.0000	-0.8514	0.0000	0.1486
16	1.0000	-0.7079	0.0000	0.2921
17	1.0000	-0.6711	0.0001	0.3290
18	1.0000	-0.7891	0.0001	0.2110
19	1.0000	-0.8420	0.0000	0.1580
20	1.0000	-0.7030	0.0002	0.2971
21	1.0000	-0.6509	0.0009	0.3500
22	1.0000	-0.7780	0.0011	0.2231
23	1.0000	-0.8491	0.0001	0.1510
24	1.0000	-0.6602	0.0051	0.3449
25	1.0000	-0.3509	0.0363	0.6854
26	1.0000	-0.1865	0.0768	0.8904
27	1.0000	-0.0991	0.1159	1.0168
28	1.0000	-0.0527	0.1493	1.0967

On the other hand, Eq. (5.10d) shows that in the presence of a deterministic regressor (H>0) the resulting covariance $var[\hat{\theta}^*]$ may be larger, equal or smaller than V_{e^*}. This will depend on the value of $var[\hat{\beta}^*]$, more precisely on whether $V_{e^*}\,J_0^{\prime}\,V_d^{-1}\,J_0\,V_{e^*}$ is greater, equal or smaller than $W\,var[\hat{\beta}^*]\,W^{\prime}$.

The values shown in Tables 5.1a and 5.1b illustrate the above observations, using the simulated quarterly series of Fig. 2.1. The regression-based multiplicative model includes a constant regressor $\boldsymbol{R}$ equal to -1, a correlation matrix $\boldsymbol{\Omega}$ corresponding to an AR(1) error model $e_t^* = 0.729\, e_{t-1}^* + v_t$ with $\boldsymbol{\Xi} = \boldsymbol{I}_T$ and $\lambda = 1$. Table 5.1a displays the original quarterly series, $\boldsymbol{s}$, the resulting quarterly series corrected for "bias", $\boldsymbol{s}^\dagger = \boldsymbol{s} \times exp(\hat{\beta}_1^*) = \boldsymbol{s} \times \hat{\beta}_1$, the corresponding benchmarked series $\hat{\boldsymbol{\theta}}$, depicted in Fig. 5.1a.

Table 5.1b exhibits the contributions to the variance of the benchmarked series $var[\hat{\boldsymbol{\theta}}^*]$, namely the *diagonal elements* of matrix $\boldsymbol{V}_{e^*}$ of matrix product $\boldsymbol{V}_{e^*} \boldsymbol{J}_0' \boldsymbol{V}_d^{-1} \boldsymbol{J}_0 \boldsymbol{V}_{e^*}$ and of matrix product $\boldsymbol{W}\, var[\hat{\boldsymbol{\beta}}]\, \boldsymbol{W}'$.

As shown, the variance of the benchmarked series, $var[\hat{\boldsymbol{\theta}}^*]$, is smaller than the original variance $\boldsymbol{V}_{e^*}$, except for the last two quarters of year 2004 and the first two quarters of year 1998; these two years have no benchmark. For these four quarters, the positive contribution to the variance $\boldsymbol{W}\, var[\hat{\boldsymbol{\beta}}^*]\, \boldsymbol{W}'$ dominates the negative contribution of $\boldsymbol{V}_{e^*} \boldsymbol{J}_0' \boldsymbol{V}_d^{-1} \boldsymbol{J}_0 \boldsymbol{V}_{e^*}$.

A final note on the properties of the variance of benchmarked series is of order. If the M benchmarks are binding (variance equal to 0), the rank of the covariance matrix $var[\hat{\boldsymbol{\theta}}^*]$ is *T-M*. Table 3.2b of Chapter 3 displays the yearly benchmarks used for the simulated quarterly series of Table 5.1a. Each yearly block of $var[\hat{\boldsymbol{\theta}}^*]$, consisting of rows 1 to 4 and columns 1 to 4, rows 5 to 8 and columns 5 to 8, etc., has rank 3 instead of 4. However, for 1998 and 2004 without benchmarks, the yearly block has full rank. If M^* of the M benchmarks are binding the rank of $var[\hat{\boldsymbol{\theta}}^*]$ is *T-M**. These facts are relevant for the reconciliation of benchmarked series, as discussed in Chapters 11 to 14.

Fig. 5.1a exhibits the series of Table 5.1a. The multiplicative regression-based method performs benchmarking in two stages. The first stage corrects for "bias", by placing the quarterly series at the level of the benchmarks, $\boldsymbol{s}^\dagger = \boldsymbol{s} \times \hat{\beta}_1$, which exactly preserves the quarter-to-quarter growth rates. The second stage performs a residual adjustment around $\boldsymbol{s}^\dagger$ to satisfy the binding benchmarks. The final benchmarked series converges to the bias-corrected series $\boldsymbol{s}^\dagger$ at both ends (years 1998 and 2004). This is a desirable feature, because the multiplicative bias, $\hat{\beta}_1$ measures the average historical annual proportional discrepancy between the benchmarks and the annual sums of the sub-annual series. This feature thus provides an implicit forecast of the next benchmark, which will minimize revisions to the preliminary benchmarked series when the next benchmark actually becomes available. The convergence is due to the AR(1) error model selected. In the absence of benchmarks after 2003, the errors follow the recursion $e_t^* = 0.729\, e_{t-1}^*$ in 2004.

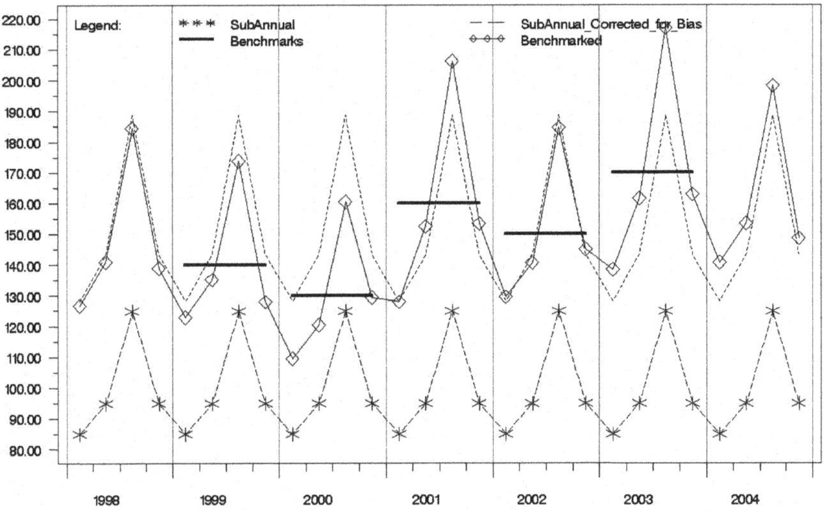

Fig. 5.1a. Simulated quarterly flow series benchmarked to yearly values, using the multiplicative Cholette-Dagum method under constant bias and AR(1) error model with parameter φ=0.729

Fig. 5.1b illustrates the benchmarking process of the simulated quarterly series using the same multiplicative regression-based model, except that the bias follows a spline trend levelling-off at the first and last sub-annual periods covered by the first and last benchmarks (in the first quarter of 1999 and the last quarter of 2003 respectively).

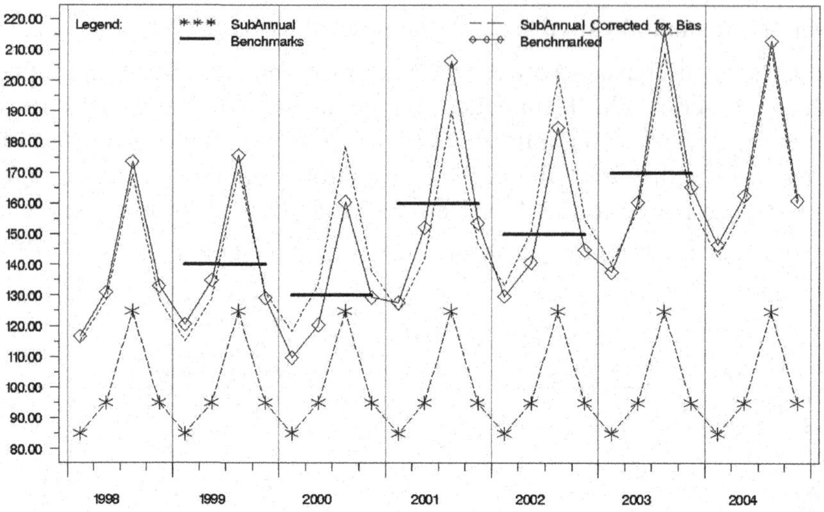

Fig. 5.1b. Simulated quarterly flow series benchmarked to yearly values, using the multiplicative Cholette-Dagum method under a spline bias and an AR(1) error model with φ=0.729

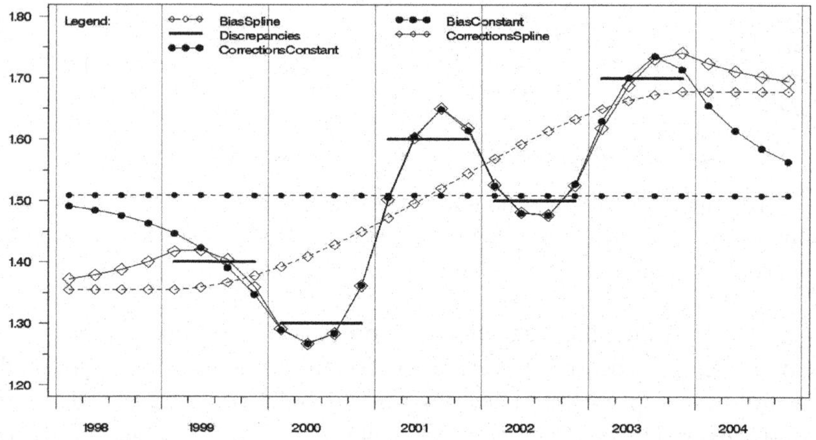

Fig. 5.1c. Yearly discrepancies and quarterly corrections under constant and spline bias

Fig. 5.1c displays both spline and constant bias and the residual corrections around them. In both cases, the latter converge to their respective bias at both ends, which causes the current benchmarked series to converge to the respective series corrected for bias. Note that the total corrections are different only at the beginning and at the end of the series. In other words, the shape of the bias has little effect on the central years 2000 to 2001. Given the trend displayed by the yearly discrepancies, the spline bias is likely to lead to smaller revisions than the constant one.

Fig. 5.2 displays the multiplicatively benchmarked series of Fig. 5.1a with the corresponding additively benchmarked series, obtained by the regression-based model under the same options. The multiplicatively benchmarked series displays a wider seasonal amplitude. Multiplicative benchmarking assumes that lower observations account for less, and larger observations for more, of the annual discrepancies.

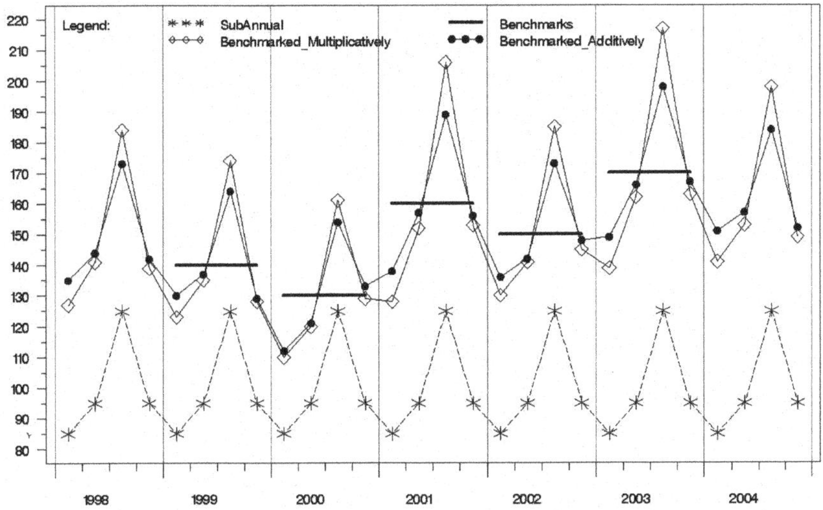

Fig. 5.2. Simulated quarterly series benchmarked, using the multiplicative and the additive variants of the Cholette-Dagum method, under an AR(1) error model with parameter φ=0.729

5.6 A Real Data Example: The Canadian Total Retail Trade Series

The Canadian monthly Retail Trade series originates from the Monthly Wholesale and Retail Trade Survey; and the yearly benchmarks, from the Annual (yearly) Census of Manufacturing.

Fig. 5.2a displays the original and the benchmarked Total Retail Trade Series obtained from the multiplicative benchmarking model. This model includes a constant regressor ***R*** to capture the bias and an AR(1) model for the logarithmic error: $e_t^* = 0.90\, e_{t-1}^* + v_t$; the error is homoscedastic with constant coefficients of variation ($\boldsymbol{\Xi} = \boldsymbol{I}_T$ and $\lambda = 1$). The benchmarks are specified as binding ($\boldsymbol{V}_\varepsilon = \boldsymbol{0}$) for simplicity.

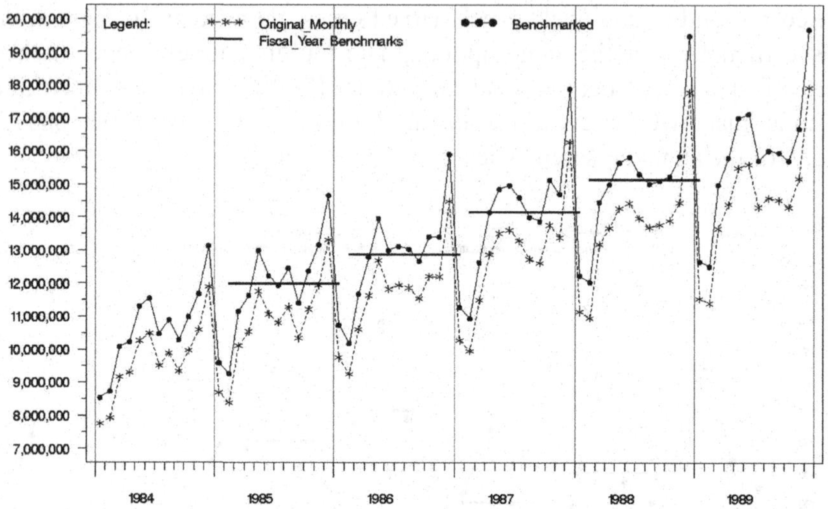

Fig. 5.3a. The Canadian Total Retail Trade series benchmarked to yearly values, using the multiplicative Cholette-Dagum method under an AR(1) error model with parameter φ=0.90

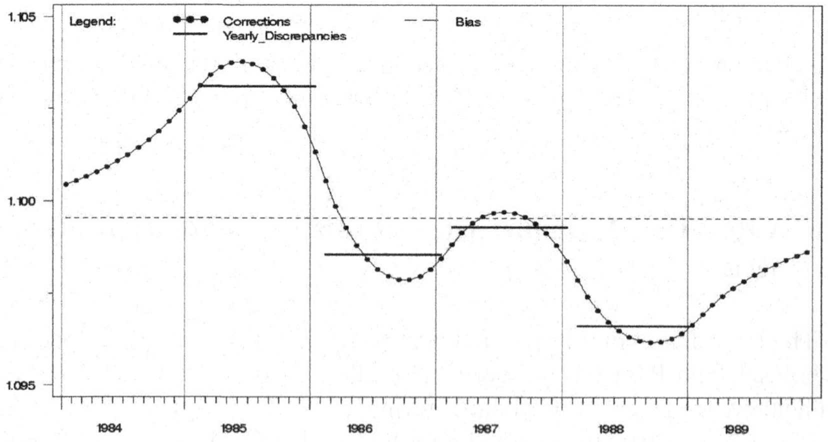

Fig. 5.3b. Proportional yearly discrepancies and corresponding multiplicative monthly corrections

Fig. 5.3b shows the proportional yearly discrepancies and the *proportional corrections* $\hat{\theta}_t / s_t$ made to the original monthly series to obtain the benchmarked series. The latter is obtained by

(a) multiplying the original series by the bias estimate equal to $exp(0.0949) = 1.0995$ to obtain the series corrected for bias $s_t^\dagger = s_t \times 1.0995$

(b) and multiplying $s_t^\dagger$ by *residual corrections* ($\hat{e}_t^*$) around the bias to satisfy the benchmarks.

The logarithmic bias estimate $\hat{\beta}_1^*$ was equal to 0.0949 with a significant *t*-value exceeding 82. The resulting bias correction exactly preserves the month-to-month growth rate movement $s_t^\dagger / s_{t-1}^\dagger = s_t / s_{t-1}$. The residual correction of (b) largely preserves the growth rates because $\hat{e}_t^*$ is much smaller than the bias and follows the AR(1) model $e_t^* = 0.90\, e_{t-1}^* + v_t$

These total proportional corrections distribute the proportional yearly discrepancies over the original monthly series.

The residual corrections converge to the bias at both ends. The bias captures the historical average proportional discrepancy between the benchmarks and the corresponding yearly sums of the original monthly series. This convergence will minimize the revision of the benchmarked when the next benchmark becomes available. Indeed, this mechanism provides an implicit forecast of the next benchmark, equal to the yearly sum of the monthly series (whatever its value turns out to be) times the bias estimate (1.0995), based on the history of the series.

Note that the proportional annual discrepancies of the Canadian Total Retail series of Fig. 5.3b become slightly smaller between 1985 and 1988; whereas the additive annual discrepancies of Fig. 3.7b become larger.

Fig. 5.4 exhibits the Canadian Total Retail Trade series additively and multiplicatively benchmarked using the regression-based. The additive model assumes constant standard deviations;[3] and the multiplicative, constant CVs. The remaining options are the same, namely constant bias and the same AR(1) error model with φ=0.90 for the standardized sub-annual error. The multiplicatively benchmarked series displays more seasonal amplitude than the additively benchmarked; this is particularly obvious in 1988. The main reason is that the multiplicative constant bias correction equal to 1.0995

[3] Under constant CVs, the additive model would be a mixed model, with an additive constant bias and proportional errors.

amplifies the seasonal and other variations in the series. (A bias value lower than 1 would reduce the amplitude.) The additive constant bias on the other hand does not alter the amplitude. The effect of the bias varies with the level of the series in the multiplicative model, whereas it remains constant in the additive model.

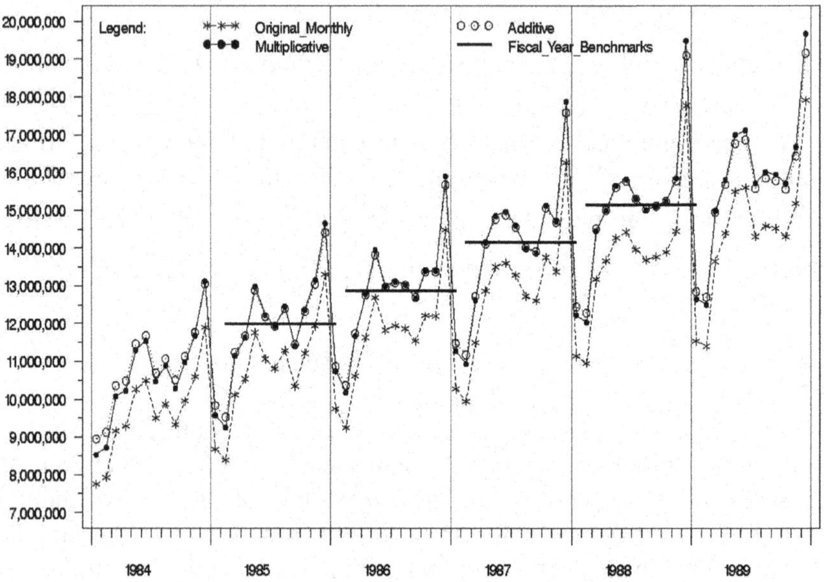

Fig. 5.4. The Canadian Total Retail Trade series benchmarked to yearly values, using the multiplicative and the additive variants of the Cholette-Dagum method under an AR(1) error model with φ=0.90

The multiplicative model fits the data better than the additive model. The average and the variance of the residuals are equal to -979 and $79{,}274^2$ for the additive model; and, 2,981 and $27{,}406^2$ for the multiplicative model after converting the residual into the original units of measurements. Furthermore, the CVs of the benchmarked series displayed in Table 5.2 (for years 1988 and 1989) are smaller for the multiplicative model than for the additive model. This goodness of fit criterion should always be used to choose between the additive and multiplicative models of the regression-based benchmarking method.

Fig. 5.5a displays the CVs available for the original monthly Canadian Total Retail Trade series. These actual CVs do change through time because of operational problems encountered in certain months, namely in July 1987 and the end of 1989. Fig. 5.5b displays the proportinal corrections $\hat{\theta}_t / s_t$

obtained using the actual CVs and constant CVs, under constant multiplicative bias and the same AR(1) with parameter φ=0.90. The transient movements in the corrections cause by the changing CVs distorts the movements of the corresponding benchmarked series. These transient movements do not occur for the corrections associated with constant CVs. (We do not depict the two benchmarked series, as they look quite identical.)

Table 5.2. The Canadian Total Retail Trade Series benchmarked using the additive and multiplicative models of the regression-base method (1988 and 1989)

Year	Month	Original	Additively Benchmarked	CV	Multiplicatively Benchmarked	CV
1988	1	11,134,013	12,457,401	0.37	12,229,264	0.12
1988	2	10,959,374	12,288,829	0.38	12,031,896	0.12
1988	3	13,177,788	14,512,002	0.31	14,461,525	0.12
1988	4	13,666,311	15,004,029	0.29	14,992,496	0.11
1988	5	14,267,530	15,607,536	0.26	15,647,682	0.11
1988	6	14,432,944	15,774,048	0.24	15,825,687	0.10
1988	7	13,960,825	15,301,848	0.23	15,305,672	0.09
1988	8	13,691,315	15,031,078	0.23	15,008,779	0.09
1988	9	13,773,109	15,110,419	0.24	15,097,847	0.09
1988	10	13,900,743	15,234,379	0.25	15,237,997	0.10
1988	11	14,453,461	15,782,162	0.27	15,845,028	0.11
1988	12	17,772,990	19,095,440	0.25	19,486,745	0.12
1989	1	11,537,416	12,852,229	0.42	12,652,645	0.14
1989	2	11,402,084	12,707,790	0.48	12,507,509	0.16
1989	3	13,652,705	14,950,214	0.44	14,979,859	0.17
1989	4	14,392,411	15,682,543	0.44	15,794,823	0.18
1989	5	15,487,026	16,770,519	0.43	16,999,346	0.19
1989	6	15,595,082	16,872,600	0.45	17,120,897	0.20
1989	7	14,305,348	15,577,488	0.50	15,707,407	0.20
1989	8	14,584,459	15,851,759	0.50	16,016,104	0.21
1989	9	14,521,628	15,784,572	0.51	15,949,104	0.21
1989	10	14,297,343	15,556,366	0.53	15,704,544	0.21
1989	11	15,182,860	16,438,355	0.50	16,678,910	0.22
1989	12	17,909,814	19,162,134	0.44	19,676,363	0.22

Higher CVs specify the corresponding observations to be more alterable (than those with lower CVs) in the benchmarking process. However, higher CVs do not necessarily imply that the observations will be altered more. In the example of Fig. 5.5b, the alteration of July 1987 is mild despite the much larger CV.

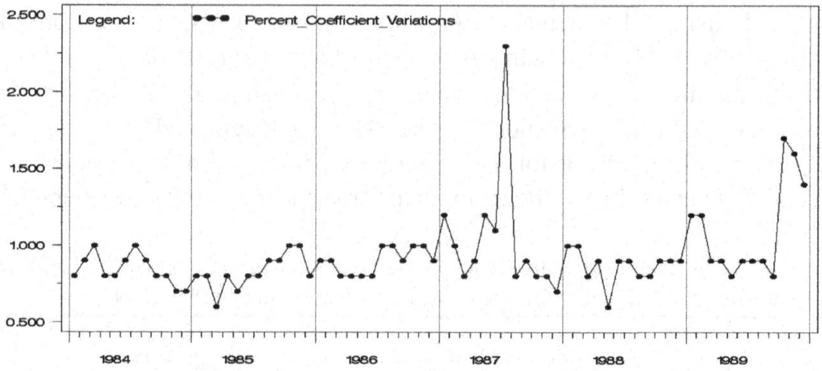

Fig. 5.5a. Coefficients of variations of the Canadian Total Retail Trade Series

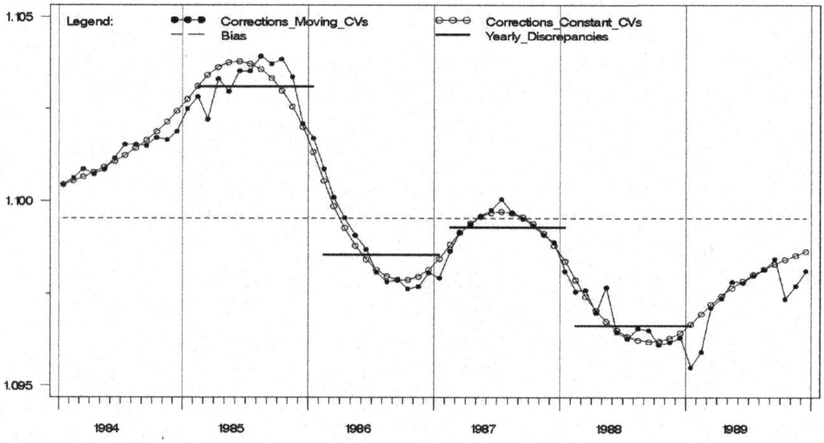

Fig. 5.5b. Multiplicative corrections obtained for the Canadian Total Retail Trade Series, using an AR(1) error model with φ=0.90 under moving and constant coefficients of variations

The use of varying CVs can cause unwanted effects. For example, a large positive outlier with large CVs will often become even larger when the corresponding annual discrepancy is positive. In the context of the regression-based benchmarking method, it is then advisable to use constant CVs given by the target CV of the survey or by an average of the available CVs. Signal extraction examined in Chapter 8 offers a solution to these unwanted effects, and in this case using the actual CVs is recommended.

6 The Denton Method and its Variants

6.1 Introduction

This chapter discusses the classical Denton (1971) method for benchmarking and interpolation. It presents both the original additive and proportional variants with first and second order differencing. The initial conditions in the original variants entail some shortcomings. The modified Denton variants presented correct those shortcomings. We show that the modified variants can be approximated by the additive regression-based benchmarking method of Chapter 3.

Denton developed benchmarking methods based on the principle of *movement preservation* which are still widely applied. According to this principle, the benchmarked series θ_t should reproduce the movement in the original series s_t. The justification for this is that the sub-annual movement of original series is the only one available. Denton proposed a number of definitions of movement preservation, each corresponding to a variant of his method.

(1) The additive first difference variant keeps additive corrections $(\theta_t - s_t)$ as constant as possible under the benchmarking constraints, as a result θ_t tends to be parallel to s_t. This is achieved by minimizing the sum of squares of the period-to-period first differences of the original series in the benchmarked series: $\Delta\,(\theta_t - s_t)$.

(2) The proportional first difference variant keeps the ratio of the benchmarked series to the original as constant as possible under the benchmarking constraints, as a result θ_t tends to have the same growth rate as s_t. This is achieved by minimizing the sum of squares of $\Delta(\theta_t / s_t)$.

(3) The additive second difference variant keeps the corrections $(\theta_t - s_t)$ as linear as possible, by minimizing the sum of squares of $\Delta^2(\theta_t - s_t)$.

(4) The proportional second difference variant keeps the ratio of the benchmarked series to the original as linear as possible, by minimizing the sum of squares of $\Delta^2(\theta_t/s_t)$.

The *original* Denton method has two major short-comings:
(1) The method introduces a transient movement at the beginning of the series, which defeats the stated principle of movement preservation and
(2) entails an implicit forecast of the next discrepancy at the end of the series, on the basis of the last two discrepancies only.

The first limitation was solved by Helfand et al. (1977) in their multiplicative first difference variant. Cholette (1979) solved the problem in the original Denton variants. This resulted in a modified variant for each variant of the Denton method.

The second limitation can be solved by the regression-based benchmarking model with a constant parameter and autoregressive errors, which perform a residual adjustment to satisfy the constraints. The constant captures the average annual discrepancy and thus provides a forecast of the next discrepancy based on historical behaviour.

6.2 The Original and Modified Additive First Difference Variants of the Denton Method

The underlying model of the Denton benchmarking method is the following

$$s_t = \theta_t + e_t \, , \tag{6.1a}$$

$$a_m = \Sigma_{t=t_{1,m}}^{t_{L,m}} j_{m,t} \, \theta_t \, , \tag{6.1b}$$

where s_t and θ_t are respectively the observed sub-annual series and the benchmarked series.[1] The behaviour of the error $e_t = \theta_t - s_t$ is specified by an objective function.

The objective function of the *original* additive first difference Denton method is

$$\begin{aligned} &min \{ (\theta_1 - s_1)^2 + \Sigma_{t=2}^{T} [(\theta_t - s_t) - (\theta_{t-1} - s_{t-1})]^2 \} \\ &\equiv min \{ (\theta_1 - s_1)^2 + \Sigma_{t=2}^{T} [(\theta_t - \theta_{t-1}) - (s_t - s_{t-1})]^2 \}, \end{aligned} \tag{6.2}$$

[1] We use the notation of the previous chapters instead of that in Denton (1971).

because Denton imposes the constraint $\theta_0 = s_0$. This *Denton initial condition* forces the benchmarked series to equal the original series at time zero and results in the minimization of the first correction $(\theta_1 - s_1)$.

Note that without the Denton initial condition (first term), Eq. (6.2) specifies that the difference between the benchmarked and the original series $(\theta_t - s_t)$ must be as constant as possible through time. The *modified* first difference variant precisely omits this first term to solve the short-coming (1) mentioned in the previous section.

In both the original and modified variants, an objective function is minimized subject to the benchmarking constraints (6.1b).

The objective function can be written in matrix algebra as

$$f(\boldsymbol{\theta},\boldsymbol{\gamma}) = (\boldsymbol{\theta} - \boldsymbol{s})' \boldsymbol{D}' \boldsymbol{D} (\boldsymbol{\theta} - \boldsymbol{s}) - 2\boldsymbol{\gamma}'(\boldsymbol{a} - \boldsymbol{J}\boldsymbol{\theta})$$

$$= \boldsymbol{\theta}' \boldsymbol{D}' \boldsymbol{D} \boldsymbol{\theta} - 2\boldsymbol{\theta}' \boldsymbol{D}' \boldsymbol{D} \boldsymbol{s} + \boldsymbol{s}' \boldsymbol{D}' \boldsymbol{D} \boldsymbol{s} - 2\boldsymbol{\gamma}' \boldsymbol{a} + 2\boldsymbol{\gamma}' \boldsymbol{J} \boldsymbol{\theta}\,, \tag{6.3}$$

where $\boldsymbol{a}$ stands for the benchmarks, $\boldsymbol{s}$ and $\boldsymbol{\theta}$ respectively denote the observed original series and benchmarked series, and $\boldsymbol{\gamma}$ contains the Lagrange multipliers associated with the linear constraints $\boldsymbol{a} - \boldsymbol{J}\boldsymbol{\theta} = \boldsymbol{0}$. Matrix $\boldsymbol{J}$ is the temporal sum operator discussed in section 3.6.1. For flow series, matrix $\boldsymbol{J}$ is equal to Kronecker product $\boldsymbol{J} = \boldsymbol{I}_M \otimes \boldsymbol{1}$, where $\boldsymbol{1}$ is unit row vector of order 4 or 12 for quarterly and monthly series respectively. For stock series, the unit vector $\boldsymbol{1}$ contains zeroes in the first three or eleven columns for quarterly and monthly series respectively. Matrix $\boldsymbol{D}$ is a regular (non-seasonal) difference operator.

All variants of the Denton method, original and modified, depend on the specification of the difference operator chosen in objective function (6.3).

We now consider two first order difference operators:

$$\underset{T \times T}{\boldsymbol{D}_0^1} = \begin{bmatrix} 1 & 0 & 0 & \cdots \\ -1 & 1 & 0 & \cdots \\ 0 & -1 & 1 & \cdots \\ \vdots & \vdots & \vdots & \ddots \end{bmatrix}, \quad \underset{(T-1) \times T}{\boldsymbol{D}^1} = \begin{bmatrix} -1 & 1 & 0 & 0 & \cdots \\ 0 & -1 & 1 & 0 & \cdots \\ 0 & 0 & -1 & 1 & \cdots \\ \vdots & \vdots & \vdots & \vdots & \ddots \end{bmatrix}. \tag{6.4}$$

The original first difference variant with the Denton initial condition ($\theta_0 = s_0$) is obtained when $\boldsymbol{D} = \boldsymbol{D}_0^1$; and the modified first difference variant without the initial condition, when $\boldsymbol{D} = \boldsymbol{D}^1$ of Eq. (6.4).

The necessary conditions for optimization require that the derivative of the objective function (6.3) with respect to the parameters be equal to zero:

$$\partial(f(\boldsymbol{\theta},\boldsymbol{\gamma}))/\partial\boldsymbol{\theta} = 2\boldsymbol{D}'\boldsymbol{D}\boldsymbol{\theta} - 2\boldsymbol{D}'\boldsymbol{D}\boldsymbol{s} + 2\boldsymbol{J}'\boldsymbol{\gamma} = \boldsymbol{0} \tag{6.5a}$$

$$\partial(f(\boldsymbol{\theta},\boldsymbol{\gamma}))/\partial\boldsymbol{\gamma} = 2\boldsymbol{J}\boldsymbol{\theta} - 2\boldsymbol{a} = \boldsymbol{0}. \tag{6.5b}$$

The sufficient condition for a minimum is that the matrix of the second order derivatives $\boldsymbol{A} = \partial(f(\boldsymbol{\theta},\boldsymbol{\gamma}))/\partial\boldsymbol{\theta}^2$ be positive definite. Matrix $\boldsymbol{A} = \boldsymbol{D}'\boldsymbol{D}$ is indeed positive definite because it is of the form $\boldsymbol{A} = \boldsymbol{B}'\boldsymbol{B}$.

Eqs. (6.5) can be written as

$$\boldsymbol{D}'\boldsymbol{D}\boldsymbol{\theta} + \boldsymbol{J}'\boldsymbol{\gamma} = \boldsymbol{D}'\boldsymbol{D}\boldsymbol{s}\,, \tag{6.6a}$$

$$\boldsymbol{J}\boldsymbol{\theta} = \boldsymbol{a} \equiv \boldsymbol{J}\boldsymbol{s} + (\boldsymbol{a} - \boldsymbol{J}\boldsymbol{s}), \tag{6.6b}$$

where $\boldsymbol{a}$ is replaced by $\boldsymbol{J}\boldsymbol{s} + (\boldsymbol{a} - \boldsymbol{J}\boldsymbol{s})$ which provides a more convenient result. Eqs. (6.6) may be expressed as

$$\begin{bmatrix} \boldsymbol{D}'\boldsymbol{D} & \boldsymbol{J}' \\ \boldsymbol{J} & \boldsymbol{0} \end{bmatrix} \begin{bmatrix} \boldsymbol{\theta} \\ \boldsymbol{\gamma} \end{bmatrix} = \begin{bmatrix} \boldsymbol{D}'\boldsymbol{D} & \boldsymbol{0}_{T\times M} \\ \boldsymbol{J} & \boldsymbol{I}_M \end{bmatrix} \begin{bmatrix} \boldsymbol{s} \\ (\boldsymbol{a} - \boldsymbol{J}\boldsymbol{s}) \end{bmatrix}. \tag{6.7}$$

The solution (6.8a) is provided by Denton (1971 Eq. (2.2)) with a different notation:[2]

$$\begin{bmatrix} \hat{\boldsymbol{\theta}} \\ \hat{\boldsymbol{\gamma}} \end{bmatrix} = \begin{bmatrix} \boldsymbol{D}'\boldsymbol{D} & \boldsymbol{J}' \\ \boldsymbol{J} & \boldsymbol{0} \end{bmatrix}^{-1} \begin{bmatrix} \boldsymbol{D}'\boldsymbol{D} & \boldsymbol{0}_{T\times M} \\ \boldsymbol{J} & \boldsymbol{I}_M \end{bmatrix} \begin{bmatrix} \boldsymbol{s} \\ (\boldsymbol{a} - \boldsymbol{J}\boldsymbol{s}) \end{bmatrix} \tag{6.8a}$$

$$\Rightarrow \begin{bmatrix} \hat{\boldsymbol{\theta}} \\ \hat{\boldsymbol{\gamma}} \end{bmatrix} = \boldsymbol{W} \begin{bmatrix} \boldsymbol{s} \\ (\boldsymbol{a} - \boldsymbol{J}\boldsymbol{s}) \end{bmatrix} = \begin{bmatrix} \boldsymbol{I}_T & \boldsymbol{W}_\theta \\ \boldsymbol{0}_{M\times T} & \boldsymbol{W}_\gamma \end{bmatrix} \begin{bmatrix} \boldsymbol{s} \\ (\boldsymbol{a} - \boldsymbol{J}\boldsymbol{s}) \end{bmatrix}, \tag{6.8b}$$

[2] In Denton's paper z corresponds to $\boldsymbol{s}$; $\boldsymbol{x}$ corresponds to $\boldsymbol{\theta}$; $\boldsymbol{B}'$, to $\boldsymbol{J}$; $\boldsymbol{r}$, to $(\boldsymbol{a} - \boldsymbol{J}\boldsymbol{s})$; and $\boldsymbol{A}$, to $\boldsymbol{D}'\boldsymbol{D}$.

where the partition consisting of the first T columns and rows of $\boldsymbol{W}$ contains the identity matrix, and the partition consisting of the first T columns and rows $T+1$ to $T+M$ contains only 0s. Eq. (6.8b) thus imply

$$\hat{\boldsymbol{\theta}} = \boldsymbol{s} + \boldsymbol{W}_{\theta}\,(\boldsymbol{a} - \boldsymbol{J}\boldsymbol{s}). \tag{6.8c}$$

In the case of the Denton original variant, the matrix inversion in (6.8a) may be performed by parts. Indeed with $\boldsymbol{D}=\boldsymbol{D}_0^1$ of Eq. (6.4), partition $\boldsymbol{D}'\boldsymbol{D}$ is invertible. Furthermore the elements of the inverse of $(\boldsymbol{D}_0^{1\prime}\,\boldsymbol{D}_0^1)$ are conveniently known to be $min(i,j)$, i=1,...,T; j=1,...,T,

$$(\boldsymbol{D}_0^{1\prime}\,\boldsymbol{D}_0^1)^{-1} \equiv (\boldsymbol{D}_0^1)^{-1}\,(\boldsymbol{D}_0^{1\prime})^{-1} = \begin{bmatrix} 1 & 1 & 1 & \dots & 1 \\ 1 & 2 & 2 & \dots & 2 \\ \vdots & \vdots & \vdots & \ddots & \vdots \\ 1 & 2 & 3 & \dots & T \end{bmatrix}, \tag{6.9a}$$

$$\text{where } (\boldsymbol{D}_0^1)^{-1} = \begin{bmatrix} 1 & 0 & 0 & \dots & 0 \\ 1 & 1 & 0 & \dots & 0 \\ \vdots & \vdots & \vdots & \ddots & \vdots \\ 1 & 1 & 1 & \dots & 1 \end{bmatrix}. \tag{6.9b}$$

The inversion by part in (6.8a) leads to the following alternative solution:

$$\hat{\boldsymbol{\theta}} = \boldsymbol{s} + (\boldsymbol{D}'\boldsymbol{D})^{-1}\boldsymbol{J}'\,(\boldsymbol{J}(\boldsymbol{D}'\boldsymbol{D})^{-1}\boldsymbol{J}')^{-1}\,(\boldsymbol{a} - \boldsymbol{J}\boldsymbol{s}), \tag{6.10}$$

also provided by Denton (*Ibid* in text below Eq. (2.2)). This solution requires the inversion of a much smaller matrix of dimension $M\times M$, instead of a $(T+M)\times(T+M)$ in solution (6.8a). Indeed, the inversion of $\boldsymbol{D}'\boldsymbol{D}$ (with $\boldsymbol{D}=\boldsymbol{D}_0^1$) in Eq. (6.10) is avoided since the inverse is known from (6.9a). On the other hand, the presence of the Denton initial condition associated with $\boldsymbol{D}_0^1$ introduces a transient effect which jeopardizes the principle of movement preservation.

In the 1960s and early 1970s, the inversion of the relatively large $T\times T$ matrix $\boldsymbol{D}'\boldsymbol{D}$ posed a problem, not to mention the inversion of the $(T+M)\times(T+M)$ matrix in Eq. (6.8a); nowadays such inversions no longer pose a problem.

The use of the exact first difference operator $\boldsymbol{D}^1$ of Eq. (6.4) requires solution (6.8a), which involves the inversion of a $(T{+}M){\times}(T{+}M)$ matrix; but avoids the Denton initial condition $\theta_0 = s_0$, i.e. the minimization of $(\theta_1 - s_1)^2$.

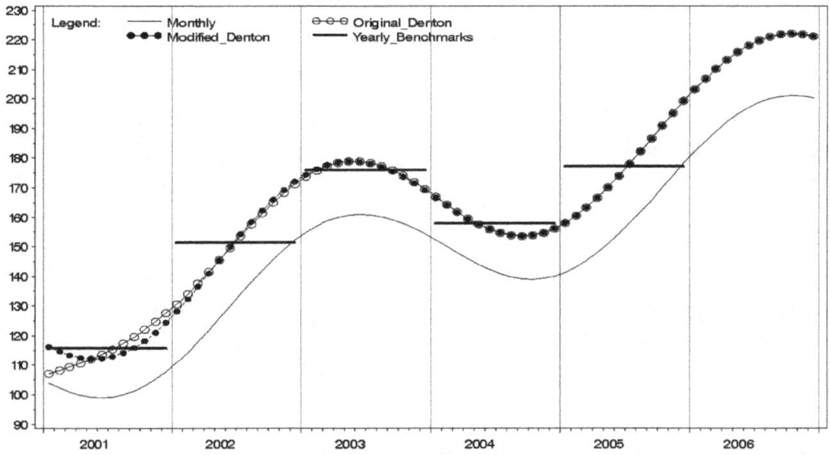

Fig. 6.1a. Monthly benchmarked series obtained with the *original* first difference Denton variant and the corresponding *modified* variant

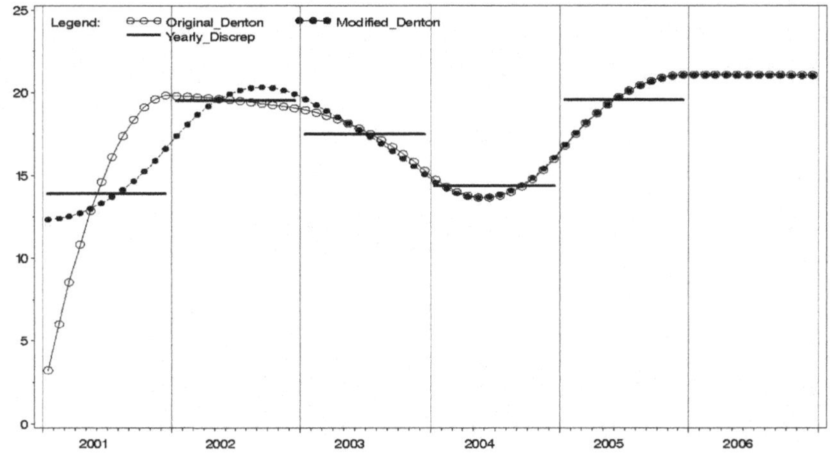

Fig. 6.1b. Additive corrections to the observed sub-annual series obtained with the *original* first difference Denton variant and the corresponding *modified* variant

In the Denton (1971) article, the undesirable effect of the initial condition on the movement was concealed by the large seasonal component (with 100% amplitude) of the series exemplified. In order to illustrate the transient movement at the beginning of the series, we selected a non-seasonal series depicted in Fig. 6.1a. It is now apparent that the benchmarked series obtained with the original Denton variant increases at the beginning of 2001 while in fact the un-benchmarked series declines.

On the other hand, the modified variant without the Denton initial condition reproduces the decline, and thus ensures maximum movement preservation. Indeed, by omitting the first term $(\theta_1 - s_1)^2$ from Eq. (6.2) the modified variant (based on Eq. (6.8a) with $\boldsymbol{D} = \boldsymbol{D}^1$) truly maximizes the movement preservation to the extent permitted by the benchmarking constraints (6.1b).

The spurious movement caused by the Denton initial condition $(\theta_0 = s_0)$ is most visible in the corrections $(\theta_t - s_t)$ displayed in Fig. 6.1b. The movement distortion with the original variant is more acute in the first year (2001) and becomes negligible in the third year (2003).

On the other hand, the corrections corresponding to the modified method are as constant (flat) as possible and thus maximizes movement preservation.

6.2.1 Preserving Continuity with Previous Benchmarked Values

It has been argued in the literature that the Denton initial condition is desirable, because it ensures continuity with the previously obtained benchmarked values. This erroneous interpretation is still commonly held among some practitioners of benchmarking and interpolation. To ensure continuity with the previously benchmarked value, the best way is to specify a monthly (sub-annual) benchmark equal to a selected estimate of the previously obtained benchmarked series (Dec. 2000 in the example depicted in Fig. 6.2a). The corresponding un-benchmarked value is also required to calculate the monthly discrepancy. In such cases, matrix $\boldsymbol{J}$ becomes

$$\underset{6\times 61}{\boldsymbol{J}} = \begin{bmatrix} 1 & \boldsymbol{0}_{1\times12} & \boldsymbol{0}_{1\times12} & \boldsymbol{0}_{1\times12} & \boldsymbol{0}_{1\times12} & \boldsymbol{0}_{1\times12} \\ 0 & \boldsymbol{1}_{1\times12} & \boldsymbol{0}_{1\times12} & \boldsymbol{0}_{1\times12} & \boldsymbol{0}_{1\times12} & \boldsymbol{0}_{1\times12} \\ 0 & \boldsymbol{0}_{1\times12} & \boldsymbol{1}_{1\times12} & \boldsymbol{0}_{1\times12} & \boldsymbol{0}_{1\times12} & \boldsymbol{0}_{1\times12} \\ \vdots & \vdots & \vdots & \vdots & \vdots & \vdots \end{bmatrix},$$

instead of

$$\underset{5\times 60}{\boldsymbol{J}} = \begin{bmatrix} \boldsymbol{1}_{1\times12} & \boldsymbol{0}_{1\times12} & \boldsymbol{0}_{1\times12} & \boldsymbol{0}_{1\times12} & \boldsymbol{0}_{1\times12} \\ \boldsymbol{0}_{1\times12} & \boldsymbol{1}_{1\times12} & \boldsymbol{0}_{1\times12} & \boldsymbol{0}_{1\times12} & \boldsymbol{0}_{1\times12} \\ \vdots & \vdots & \vdots & \vdots & \vdots \end{bmatrix} = \boldsymbol{I}_5 \otimes \boldsymbol{1}_{1\times12}$$

without the monthly benchmark of 2000.

Fig. 6.2a shows that the original variant with the Denton initial condition causes an abrupt decline between the previously obtained benchmarked value of Dec. 2000 (square) and the Jan. 2001 value (1st circle). The modified variant causes the benchmarked series to start from the previously obtained benchmarked value.

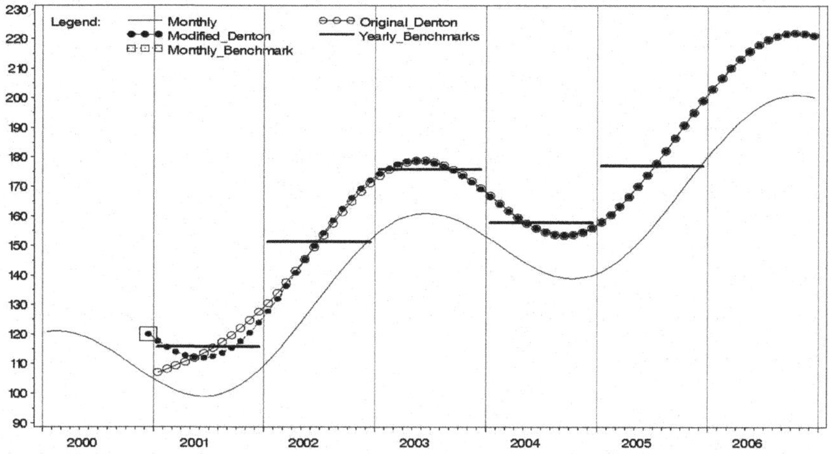

Fig. 6.2a. Monthly benchmarked series obtained from the *original* first difference Denton variant and from the corresponding *modified* variant with a monthly benchmark in Dec. 2000

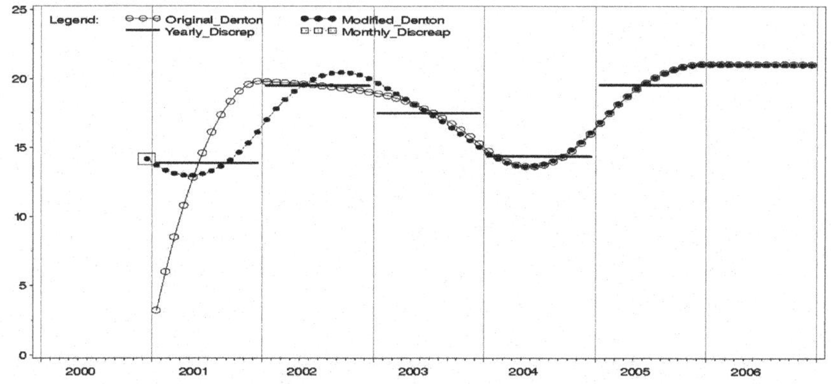

Fig. 6.2b. Additive corrections to the observed sub-annual series obtained from the *original* first difference Denton variant and the corresponding modified variant with a monthly benchmark in Dec. 2000

The spurious movement with the original variant is most noticeable in the corrections displayed in Fig 6.2b. Furthermore, if a benchmark becomes available every year and benchmarking is performed every year, the transient movement recurs every year. In other words, the yearly benchmarking process may induce a spurious seasonal movement into the series, in regime of more or less constant discrepancies.

Despite its unwanted properties, the first difference operator, $\boldsymbol{D}_0^1$ from Eq. (6.4), is also used in temporal distribution and interpolation of the Chow-Lin type (1971) and related techniques. The operator $\boldsymbol{D}_0^1$ introduces the same kind of spurious movement in the estimates as in benchmarking and also biases the estimation of the regression parameters (of the "related series") towards zero, as discussed at the end of section 7.6.

Denton also proposed another first difference operator similar to $\boldsymbol{D}_0^1$ of Eq. (6.4), where the Denton initial condition applies to the last observation instead of the first:

$$\underset{T\times T}{\boldsymbol{D}_0^1} = \begin{bmatrix} -1 & 1 & 0 & \dots & 0 & 0 \\ \vdots & \vdots & \vdots & \ddots & 0 & 0 \\ 0 & 0 & 0 & \dots & -1 & 1 \\ 0 & 0 & 0 & \dots & 0 & 1 \end{bmatrix}.$$

This operator results in minimizing $(\theta_T - s_T)$ instead of $(\theta_1 - s_1)$ in the objective function (6.2). This is even more questionable since the spurious movement affects the latest current observations which are the most important for decision making.

6.2.2 Approximation of the Original and Modified Denton Variants by the Additive Regression-Based Model

The original variants of the Denton method (with the Denton initial condition) are a particular case of the additive Cholette-Dagum regression-based model of Chapter 3, repeated here for convenience:

$$\boldsymbol{s} = \boldsymbol{R\beta} + \boldsymbol{\theta} + \boldsymbol{e},\ E(\boldsymbol{e}) = 0,\ E(\boldsymbol{e}\,\boldsymbol{e}') = \boldsymbol{V}_e \qquad (6.11a)$$

$$\boldsymbol{a} = \boldsymbol{J}\,\boldsymbol{\theta} + \boldsymbol{\varepsilon},\ E(\boldsymbol{\varepsilon}) = 0,\ E(\boldsymbol{\varepsilon}\,\boldsymbol{\varepsilon}') = \boldsymbol{V}_\varepsilon,\ E(\boldsymbol{e}\,\boldsymbol{\varepsilon}') = 0. \qquad (6.11b)$$

If $\boldsymbol{R\beta}$ is absent, the solution is

$$\hat{\boldsymbol{\theta}} = \boldsymbol{s} + \boldsymbol{V}_e\,\boldsymbol{J}'\,\boldsymbol{V}_d^{-1}\,[\boldsymbol{a} - \boldsymbol{J}\boldsymbol{s}], \qquad (6.11c)$$

where $\boldsymbol{V}_d = [\boldsymbol{J}\boldsymbol{V}_e\boldsymbol{J}' + \boldsymbol{V}_\varepsilon]$. The solution (6.10) of the original additive first difference variant of the Denton method is obtained, by setting $\boldsymbol{V}_\varepsilon$ equal to $\boldsymbol{0}$ and replacing $\boldsymbol{V}_e$ by $(\boldsymbol{D}_0^{1\prime}\boldsymbol{D}_0^1)^{-1}$ where the difference operator $\boldsymbol{D}_0^1$ is defined in Eq. (6.4). The original additive second difference variant is obtained, by replacing $\boldsymbol{V}_e$ by $(\boldsymbol{D}_0^{2\prime}\boldsymbol{D}_0^2)^{-1}$ where $\boldsymbol{D}_0^2 = \boldsymbol{D}_0^1\boldsymbol{D}_0^1$.

Similarly, the modified (without the Denton initial condition) first difference variants can be approximated by the additive regression-based benchmarking model (6.11). This is achieved by omitting $\boldsymbol{R\beta}$, setting $\boldsymbol{V}_\varepsilon$ equal to $\boldsymbol{0}$ and setting the covariance matrix $\boldsymbol{V}_e$ equal to that of an autoregressive model $e_t = \varphi e_{t-1} + v_t$ with positive parameter slightly smaller than 1, e.g. 0.999 (Bournay and Laroque 1979). The resulting covariance matrix is

$$\boldsymbol{V}_e = (\boldsymbol{D}_\varphi^{1\prime}\boldsymbol{D}_\varphi^1)^{-1} = \boldsymbol{D}_\varphi^{-1}(\boldsymbol{D}_\varphi^{-1})' = \begin{bmatrix} 1 & \varphi & \varphi^2 & \dots & \varphi^{T-1} \\ \varphi & 1 & \varphi & \dots & \varphi^{T-2} \\ \vdots & \vdots & \vdots & \ddots & \vdots \\ \varphi^{T-1} & \varphi^{T-2} & \varphi^{T-3} & \dots & 1 \end{bmatrix}, \tag{6.12}$$

where $\boldsymbol{D}_\varphi^1$ is now a $T \times T$ quasi-difference operator:

$$\boldsymbol{D}_\varphi^1 = \begin{bmatrix} \sqrt{1-\varphi^2} & 0 & 0 & \dots \\ -\varphi & 1 & 0 & \dots \\ 0 & -\varphi & 1 & \dots \\ \vdots & \vdots & \vdots & \ddots \end{bmatrix} / \sqrt{1-\varphi^2}, \tag{6.13a}$$

with

$$\boldsymbol{D}_\varphi^{-1} = \begin{bmatrix} \sqrt{1-\varphi^2} & 0 & 0 & \dots \\ \varphi\sqrt{1-\varphi^2} & 1 & 0 & \dots \\ \varphi^2\sqrt{1-\varphi^2} & \varphi & 1 & \dots \\ \vdots & \vdots & \vdots & \ddots \end{bmatrix} / \sqrt{1-\varphi^2}. \tag{6.13b}$$

In order to obtain a closer approximation of the modified first difference variant, it is desirable to first bring the level of the original series to that of the benchmarks by adding the following constant

$$b = [\Sigma_{m=1}^{M} (a_m - \Sigma_{t=t_{1m}}^{t_{Lm}} j_{mt} \, s_t)] / [\Sigma_{m=1}^{M} \Sigma_{t=t_{1m}}^{t_{Lm}} j_{m,t}]. \tag{6.14}$$

This constant is sometimes referred to as the "bias" of the sub-annual series with respect to the level of the benchmarks. The solution (6.11c) applies except that $\boldsymbol{s}$ is replaced by $\boldsymbol{s}^{\dagger} = \boldsymbol{s} + b$. The solution coincides with (6.10) with $(\boldsymbol{D}_0^{1\prime} \boldsymbol{D}_0^1)^{-1}$ replaced by $\boldsymbol{V}_e = (\boldsymbol{D}_\varphi^{1\prime} \boldsymbol{D}_\varphi^1)^{-1}$.

6.2.3 Preferred Variant of Movement Preservation

As discussed in Chapter 4 sections 4.3 and 4.4, it is better to use a value of φ not so close to 1, e.g. 0.90 for monthly series and 0.90^3 for quarterly series. The resulting variant is preferable to both the original and the modified first difference variants of the Denton variants. Indeed the constant given by Eq. (6.14) provides an implicit forecast of the next annual discrepancy, which is based on the historical average of the annual discrepancies. Such a constant can alternatively be estimated by setting $\boldsymbol{R}$ equal to -1 in model (6.11), in which case solution (3.20) applies. The constant can be viewed as the expected value of the next discrepancy given the past ones.

Note that $\boldsymbol{s}^{\dagger} = \boldsymbol{s} + b$ exactly preserves the month-to-month movement for the whole temporal range of the series: i.e. $(s_t^{\dagger} - s_{t-1}^{\dagger}) \equiv (s_t - s_{t-1})$. The residual adjustment required to satisfy the benchmarking constraints is performed by the autoregressive error model $e_t = \varphi e_{t-1} + v_t$. With φ equal to 0.90, the AR(1) model causes the benchmarked series to converge to the original series corrected for "bias" $s_t^{\dagger} = s_t + b$ at both ends, as exemplified by Fig. 6.3a.

This behaviour is most obvious in Fig 6.3b, where the corrections converge to the average value of the discrepancies (the "bias") at both ends. The original Denton variant on the other hand assumes that the next annual discrepancy will be in the neighbourhood of the last one. Generally speaking the Denton method assumes that the next discrepancy will be larger or comparable to the last one, if the last two discrepancies increase; and smaller or comparable, if they decrease. The former case is illustrated at the end of the series of Fig. 6.3b; and the latter, at the beginning.

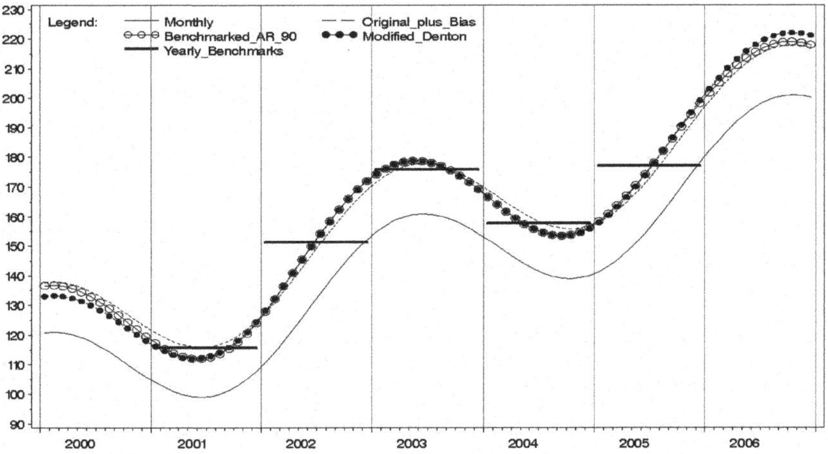

Fig. 6.3a. Monthly benchmarked series obtained with the modified first difference Denton variant and with the additive regression-based model with autoregressive errors (φ=0.90)

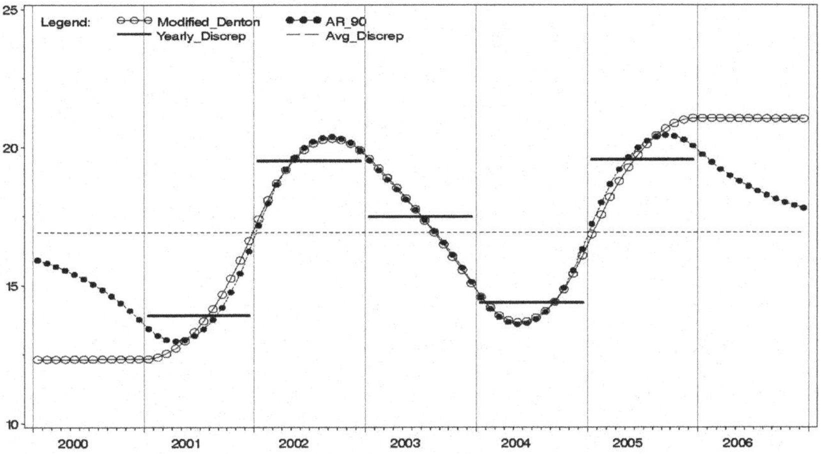

Fig. 6.3b. Additive corrections to the observed sub-annual series obtained from the modified first difference Denton variant and from the additive regression-based model with autoregressive errors (φ=0.90)

6.3 The Proportional First Difference Variants of the Denton Method

Many time series display large seasonal patterns. Additive benchmarking tends to correct the sub-annual values equally regardless of their relative values. This can produce negative benchmarked values for non negative variables (e.g. Sales). Proportional benchmarking, on the other hand, corrects seasonally higher values more than seasonally low ones and is less likely to produce negative benchmarked values. Furthermore, proportional benchmarking can be seen as an approximation of the preservation of period-to-period growth rate of the original series in the benchmarked series.

The objective function of the original proportional first difference Denton method is fact

$$\begin{aligned} &\min \{ [(\theta_1 - s_1)/s_1]^2 + \Sigma_{t=2}^{T} [(\theta_t - s_t)/s_t - (\theta_{t-1} - s_{t-1})/s_{t-1}]^2 \} \\ \equiv\; &\min \{ (\theta_1/s_1 - 1)^2 + \Sigma_{t=2}^{T} (\theta_t/s_t - \theta_{t-1}/s_{t-1})^2 \} . \end{aligned} \quad (6.15)$$

because of the assumption $\theta_0 = s_0$. As a result the first proportional correction $(\theta_1 - s_1)/s_1$ is minimized creating a spurious movement in the benchmarked series similar to that observed for the additive first difference. Note that without the Denton initial condition (first term), Eq. (6.15a) specifies that the difference between the benchmarked and the original series $(\theta_t - s_t)/s_t$ must be as constant as possible through time. The *modified* first proportional difference variant precisely omits this first term to solve the problem.

In both the original and modified proportional variants, the objective function (6.3) is minimized. The *original* variant is achieved by replacing the difference operator $\boldsymbol{D}$ by $\boldsymbol{D}_0^1\,\boldsymbol{\Xi}$, where $\boldsymbol{\Xi}$ is a diagonal matrix containing the values of s_t. Preferably, matrix $\boldsymbol{\Xi}$ should contain the relative values $s_t/\bar{s}$ of s_t, in order to standardize the data and avoid potential numerical problems. By setting $\boldsymbol{D} = \boldsymbol{D}_0^1\,\boldsymbol{\Xi}$ in Eq. (6.3) and (6.5) to (6.9) are still applicable and yield the same solution (6.10) repeated here:

$$\hat{\boldsymbol{\theta}} = \boldsymbol{s} + (\boldsymbol{D}'\boldsymbol{D})^{-1}\boldsymbol{J}'(\boldsymbol{J}(\boldsymbol{D}'\boldsymbol{D})^{-1}\boldsymbol{J}')^{-1}(\boldsymbol{a} - \boldsymbol{J}\boldsymbol{s}) . \quad (6.16)$$

Like in the additive first difference variant, the presence of the Denton initial condition with the first difference operator, $\boldsymbol{D}_0^1$ from Eq. (6.4), introduces a transient effect which jeopardizes the principle of movement preservation.

The *modified* proportional first difference variant is achieved by replacing the difference operator $\boldsymbol{D}$ by $\boldsymbol{D}^1\,\boldsymbol{\Xi}$ in Eq. (6.3). The use of the exact first difference operator $\boldsymbol{D}^1$ of Eq. (6.4) requires solution (6.8) but avoids the Denton initial condition $\theta_0 = s_0$, i.e. the minimization of $(\theta_1 - s_1)/s_1$.

Fig. 6.4a depicts the artificial series used by Denton (1971) to exemplify his method. As mentioned before, the original quarterly series is mainly seasonal, repeating the values 50, 100, 150 and 100 over five year (we added a sixth year). The benchmarks are 500, 400, 300, 400 and 500 for years 2001 to 2005 respectively; there is no benchmark for year 2006.

The benchmarked series obtained with the original proportional first difference Denton variant displays a spurious transient movement at the beginning, caused by the initial condition. However, the movement is concealed in Fig 6.4 by the very large seasonal amplitude of the series.

The spurious movement in 2001 becomes apparent in the proportional corrections $\hat{\theta}_t / s_t$ exhibited in Fig 6.4b. On the contrary, the corrections corresponding to the modified variant are as flat as permitted by the discrepancies, thus maximizing the preservation of the proportional movement.

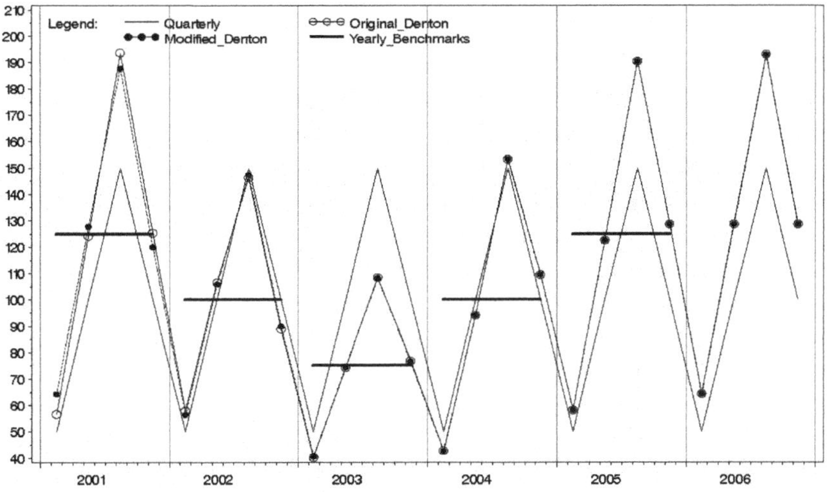

Fig. 6.4a. Quarterly benchmarked series obtained with the *original* proportional first difference Denton variant and the corresponding *modified* variant

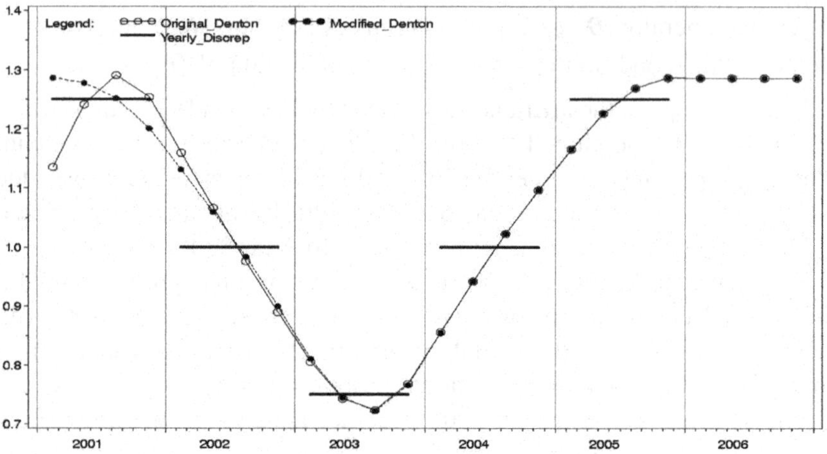

Fig. 6.4b. Proportional corrections to the observed sub-annual series obtained with the *original* proportional first difference Denton variant and the corresponding *modified* variant

The initial constraint ($\theta_0 = s_0$) of the original Denton variant in Fig. 6.4b actually defeats the purpose of maintaining historical continuity with previously benchmarked values. The way to ensure continuity series is to specify a sub-annual benchmark, as described in section 6.2.1.

An alternative to ensure continuity is to run the whole series each time a new benchmark is available, using the modified variant. This entails applying the benchmarking method to the original and the new data: i.e. to the whole unbenchmarked series and to all the benchmarks. Note that adding one more year of benchmark and sub-annual data at the end modifies mainly the last and the second last years of the previously available benchmarked series; the other years basically remain unaltered.

Whether or not to publish these modified values is a matter of revision policy. Revision policies are used for the publication of seasonally adjusted values. For example, each time an observation is added to the series only the previous estimate is usually revised; when a complete year is available, only the last three years of the series are revised.

6.3.1 Approximation of the Original and Modified Proportional Variants by the Additive Regression-Based Model

The modified proportional first difference variant (without the Denton initial condition) may be approximated with the additive regression-based benchmarking method of Chapter 3, reproduced by Eq. (6.11). This is achieved by omitting $\boldsymbol{R\beta}$, by setting $\boldsymbol{V}_\varepsilon$ equal to $\boldsymbol{0}$ and by setting the covariance matrix V_e equal to that of an AR(1) error model $e_t/s_t = \varphi e_{t-1}/s_{t-1} + v_t$ with positive parameter slightly smaller than 1, e.g. 0.999. The covariance matrix then takes the value

$$V_e = (\Xi D_\varphi^{1\prime} D_\varphi^1 \Xi)^{-1} = \begin{bmatrix} \sigma_1^2 & \sigma_1\sigma_2\varphi & \sigma_1\sigma_3\varphi^2 & \cdots \\ \sigma_2\sigma_1\varphi & \sigma_2^2 & \sigma_2\sigma_3\varphi & \cdots \\ \sigma_3\sigma_1\varphi^2 & \sigma_3\sigma_2\varphi & \sigma_3^2 & \cdots \\ \vdots & \vdots & \vdots & \ddots \end{bmatrix} 1/(1-\varphi^2), \qquad (6.17)$$

where $\sigma_t = s_t/\bar{s}$. Like in the additive variants, the factor $1/(1-\varphi^2)$ may be dropped (if $V_\varepsilon = \mathbf{0}$).

A closer approximation of the modified proportional first difference variant is achieved by first raising the original series to the level of the benchmarks with the following constant factor:

$$b = (\Sigma_{m=1}^{M} a_m) \;/\; (\Sigma_{m=1}^{M} \Sigma_{t=t_{1m}}^{t_{2m}} j_{mt}\, s_t). \qquad (6.18)$$

The solution (6.16) still applies except that s is replaced by $s^\dagger = s \times b$. The solution coincides with (6.16) with $(D_0^{1\prime} D_0^1)^{-1}$ replaced by V_e of (6.17). Note that $s^\dagger$ exactly preserves the growth rates of s: $s_t^\dagger/s_t - s_{t-1}^\dagger/s_{t-1} = 0$, t=2,...,T. Indeed $s_t^\dagger/s_t = s_{t-1}^\dagger/s_{t-1}$ implies $s_t^\dagger/s_{t-1}^\dagger - 1 = s_t/s_{t-1} - 1$. Also note that depending of its (positive) value, the re-scaling factor shrinks or amplifies the seasonal pattern without ever producing negative values for $s_t^\dagger$, unless s_t contains negative values.

6.3.2 Preferred Variant of Proportional Movement Preservation

As discussed in section 6.2, a better alternative to the modified proportional first difference Denton variant and to its approximation is to apply the additive regression-based model (6.11) as just described except with an autoregressive parameter not so close to 1, e.g. 0.90 for monthly series and 0.90^3 for quarterly series. Fig. 6.5a displays the benchmarked series obtained from this model (with φ = 0.729) from the modified proportional first difference Denton variant. The former series converges at both ends to the original series corrected for the anticipated annual proportional discrepancy (the "bias"), i.e. $s^{\dagger} = s \times b$. The wide seasonal amplitude of the series makes it hard to see the convergence in Fig. 6.5a.

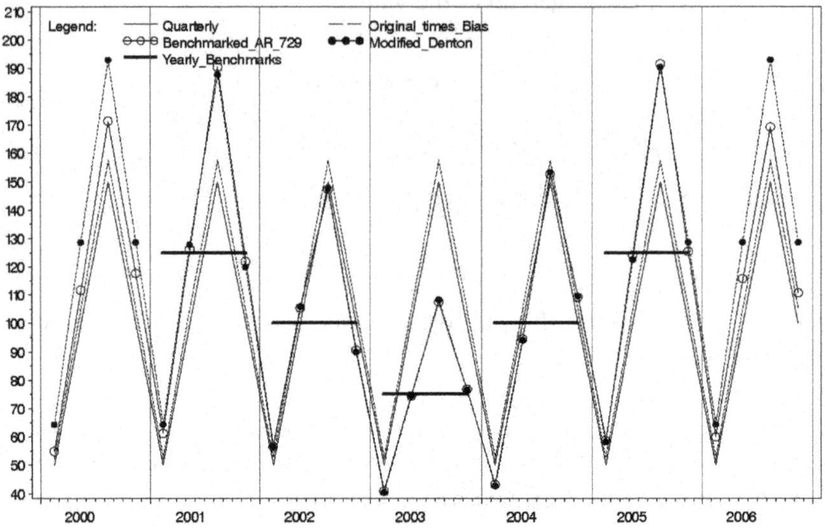

Fig. 6.5a. Quarterly benchmarked series obtained from the modified proportional first difference Denton variant and the additive regression-based benchmarking model with proportional autoregressive errors φ=0.729

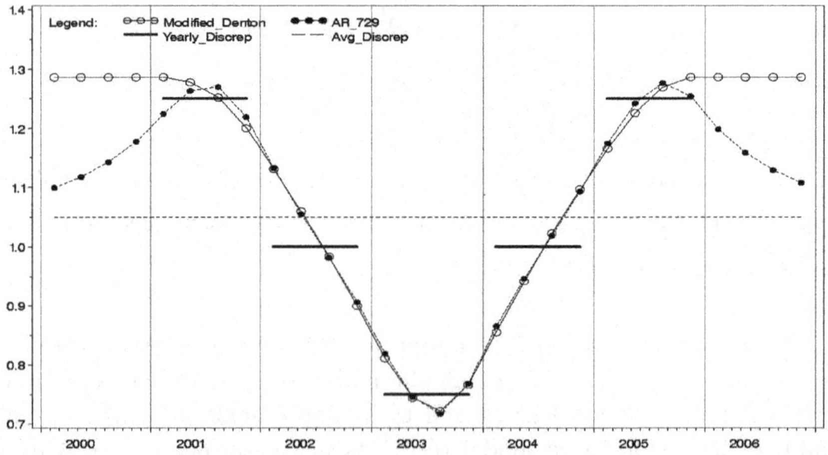

Fig. 6.5b. Porportional corrections to the observed sub-annual series obtained from the modified proportional first difference Denton variant and the additive regression-based benchmarking model with proportional autoregressive errors (φ=0.729)

On the other hand, the proportional corrections $\hat{\theta}_t / s_t$ exhibited in Fig. 6.5b display the convergence clearly. The corrections associated with autoregressive errors converge to the average proportional discrepancy, which means the benchmarked series converges to $\boldsymbol{s}^\dagger = \boldsymbol{s} \times \boldsymbol{b}$, whereas the corrections associated with the modified Denton do not converge.

Fig. 6.5a shows that for seasonal series, illustrations displaying the original and benchmarked series are not very useful compared to those displaying proportional corrections. Proportional corrections remove the seasonal component which conceal other variations, i.e. they focus on the change brought about by benchmarking.

The proportional benchmarking method with autoregressive errors would be most appropriate for users requiring a routinely applicable procedure to systems of time series. On the basis of large empirical studies, we found that appropriate values of the autoregressive parameter in the neighbourhood of 0.90 for monthly (sub-annual) series yield satisfactory results. If $\varphi = 0.90$ is selected in the monthly error models, the corresponding values for quarterly series should be 0.90^3. As a result, the convergence achieved in three months in a monthly error model is the same as that achieved in one quarter in the quarterly model.

Similarly if $\varphi = 0.90$ is used in monthly error model, the corresponding value for the daily model should be $\varphi = 0.90^{1/30.4375}$, where 30.4375 is the average number of days in a month. The convergence achieved in 30 or 31 days is then comparable to that achieved in one month by the monthly model. The benchmarking of daily values occurs mostly in the calendarization of data with short reporting periods (e.g. 4 and 5 weeks), as discussed in Chapter 9. Section 3.8 provides more details on proportional benchmarking.

Table 6.1 displays the observed quarterly values and the benchmarked values obtained with the original and modified proportional first difference variants and with the corresponding second difference variants, for the example of Denton (1971). The benchmarks are 500, 400, 300, 400 and 500 for years 2001 to 2005 respectively; year 2006 is without benchmark. The differences between the original and the modified variants occur at the beginning of the series.

Table 6.1. Results obtained from the *original* proportional first and second difference Denton variants with initial conditions and the corresponding *modified* variants without initial conditions, for the Denton (1971) numerical example

		First difference variants				Second difference variants			
		Original		Modified		Original		Modified	
Date	(1)	(2)	(4)	(3)	(4)	(2)	(4)	(3)	(4)
2001/1	50.0	56.8	1.135	64.3	1.287	55.0	1.099	66.5	1.330
2001/2	100.0	124.2	1.242	127.8	1.278	121.9	1.219	128.5	1.285
2001/3	150.0	193.6	1.291	187.8	1.252	194.2	1.295	185.9	1.239
2001/4	100.0	125.4	1.254	120.0	1.200	129.0	1.290	119.1	1.191
2002/1	50.0	58.0	1.160	56.6	1.131	60.5	1.211	56.8	1.135
2002/2	100.0	106.7	1.067	106.0	1.060	109.2	1.092	106.7	1.067
2002/3	150.0	146.4	0.976	147.5	0.983	144.8	0.965	147.5	0.984
2002/4	100.0	89.0	0.890	90.0	0.900	85.4	0.854	89.0	0.890
2003/1	50.0	40.3	0.806	40.5	0.811	38.7	0.774	40.1	0.802
2003/2	100.0	74.3	0.743	74.4	0.744	73.3	0.733	74.2	0.742
2003/3	150.0	108.5	0.724	108.3	0.722	110.2	0.735	109.2	0.728
2003/4	100.0	76.9	0.769	76.7	0.767	77.8	0.778	76.5	0.765
2004/1	50.0	42.8	0.856	42.8	0.855	42.6	0.852	42.1	0.842
2004/2	100.0	94.2	0.942	94.1	0.941	93.9	0.939	93.5	0.935
2004/3	150.0	153.4	1.023	153.4	1.023	153.6	1.024	154.0	1.027
2004/4	100.0	109.6	1.096	109.7	1.097	109.8	1.098	110.4	1.104
2005/1	50.0	58.3	1.166	58.3	1.166	58.0	1.160	58.3	1.165
2005/2	100.0	122.6	1.226	122.6	1.226	121.3	1.213	121.6	1.216
2005/3	150.0	190.4	1.269	190.4	1.269	189.5	1.263	189.4	1.263
2005/4	100.0	128.7	1.287	128.7	1.287	131.2	1.312	130.7	1.307
2006/1	50.0	64.3	1.287	64.3	1.287	68.0	1.361	67.6	1.352
2006/2	100.0	128.7	1.287	128.7	1.287	140.9	1.409	139.7	1.397
2006/3	150.0	193.0	1.287	193.0	1.287	218.7	1.458	216.3	1.442
2006/4	100.0	128.7	1.287	128.7	1.287	150.7	1.507	148.7	1.487

(1) Observed quarterly series (3) Modified variants
(2) Original variants with the Denton initial condition (4) Proportional corrections

6.4 Other Variants of the Denton Method

For the sake of completeness, we also discuss rarely applied variants of the Denton method, namely the additive and the proportional second difference variants.

6.4.1 The Additive Second Difference Variants

Consider the following second order difference operators:

$$\underset{T\times T}{\boldsymbol{D}_0^2} = \begin{bmatrix} 1 & 0 & 0 & 0 & \cdots \\ -2 & 1 & 0 & 0 & \cdots \\ 1 & -2 & 1 & 0 & \cdots \\ 0 & 1 & -2 & 1 & \cdots \\ \vdots & \vdots & \vdots & \vdots & \ddots \end{bmatrix}, \quad \underset{(T-2)\times T}{\boldsymbol{D}^2} = \begin{bmatrix} 1 & -2 & 1 & 0 & 0 & \cdots \\ 0 & 1 & -2 & 1 & 0 & \cdots \\ 0 & 0 & 1 & -2 & 1 & \cdots \\ \vdots & \vdots & \vdots & \vdots & \vdots & \ddots \end{bmatrix}, \qquad (6.19)$$

where $\boldsymbol{D}_0^2 = \boldsymbol{D}_0^1 \boldsymbol{D}_0^1$ ($\boldsymbol{D}_0^1$ is defined in Eq. (6.4)) and $\boldsymbol{D}^2$ is the exact second order difference operator. Substituting $\boldsymbol{D} = \boldsymbol{D}_0^2$ in Eq. (6.10) provides the solution to the original second difference variant. The inverse of $\boldsymbol{D}_0^{2\prime}\boldsymbol{D}_0^2$ is given by

$$(\boldsymbol{D}_0^{2\prime}\boldsymbol{D}_0^2)^{-1} \equiv (\boldsymbol{D}_0^2)^{-1}(\boldsymbol{D}_0^{2\prime})^{-1}, \text{ where } (\boldsymbol{D}_0^2)^{-1} = \begin{bmatrix} 1 & 0 & 0 & \cdots & 0 \\ 2 & 1 & 0 & \cdots & 0 \\ 3 & 2 & 1 & \cdots & 0 \\ \vdots & \vdots & \vdots & \ddots & \vdots \end{bmatrix}.$$

Given $\boldsymbol{D}_0^2$ of Eq. (6.19), the objective function of the original second order difference variant is

$$\begin{aligned} min\,\{ & (\theta_1 - s_1)^2 + [(\theta_2 - s_2) - 2(\theta_1 - s_1)]^2 + \\ & \Sigma_{t=3}^{T} [(\theta_t - s_t) - 2(\theta_{t-1} - s_{t-1}) + (\theta_{t-2} - s_{t-2})]^2 \}, \end{aligned} \qquad (6.20)$$

where the first two quadratic terms are initial conditions similar to that in the original first difference variants.

Substituting $\boldsymbol{D}$ equal to $\boldsymbol{D}^2$ in Eq. (6.8a) provides the modified second order difference variant without the Denton initial conditions. The corresponding objective function is given by Eq. (6.20) without the first two terms.

Fig. 6.6a displays the monthly series of Fig. 6.1a benchmarked using the original second difference variant of the Denton method and using the modified second difference variant. The original variant creates a spurious movement at the beginning of the series similar to that observed for the original first difference variant exhibited in Fig. 6.1a.

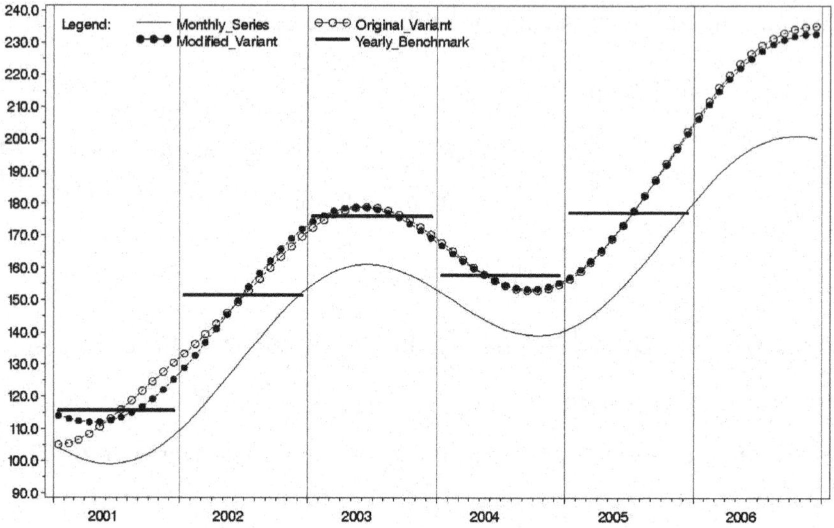

Fig. 6.6a. Monthly benchmarked series obtained with the *original* second difference Denton variant and the corresponding *modified* variant

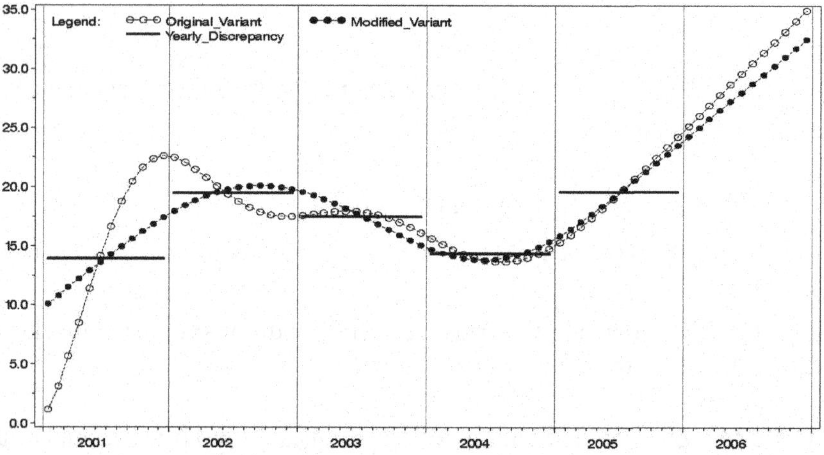

Fig. 6.6b. Additive corrections to the observed sub-annual series obtained with the *original* second difference variant and the corresponding *modified* variant

The corrections corresponding to the modified and original second difference variant provide a clearer picture. The spurious movement at the beginning of the series is more noticeable. At the end of the series, both second difference variants implicitly assume that the next yearly discrepancy will be much larger, which is not supported by the historical yearly discrepancies.

Fig 6.7a displays the benchmarked series obtained from the modified first difference and second difference variants of the Denton method. The second difference variant produces substantially larger values than the first difference variant for 2006.

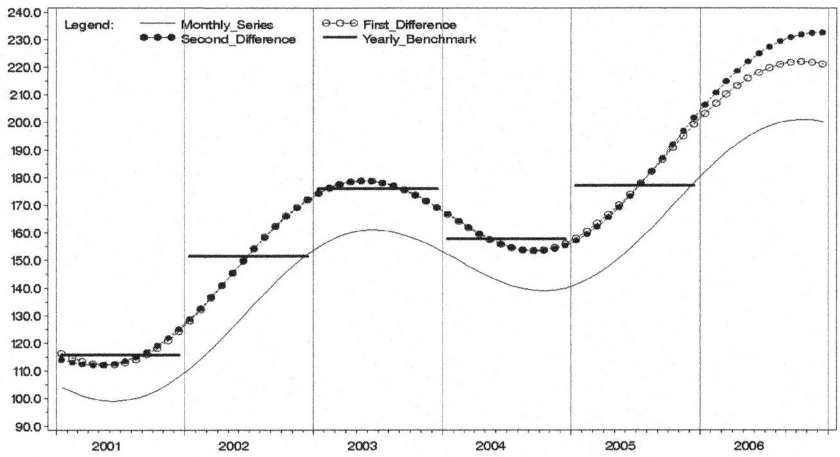

Fig. 6.7a. Monthly benchmarked series obtained with the modified first and second difference Denton variants

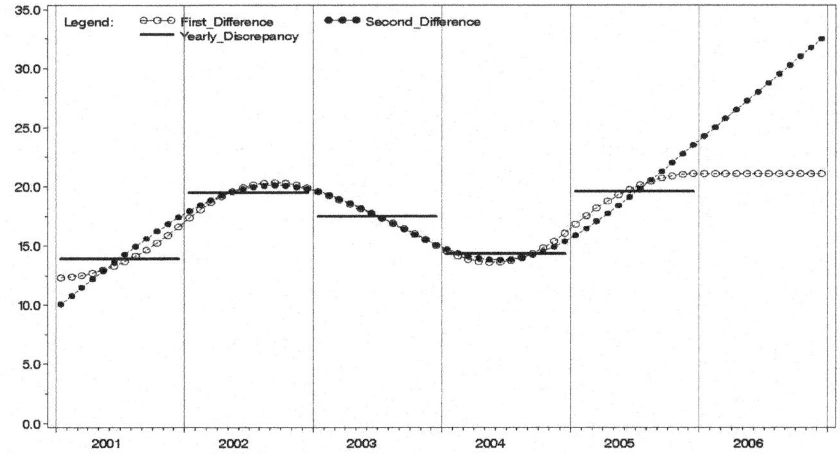

Fig. 6.7b. Additive corrections to the observed sub-annual series obtained with the modified first and second difference Denton variants

Again the corrections depicted in Fig. 6.7b provide a clearer picture. At both ends, the second difference corrections follow a straight line and induce volatility in the current year without benchmark, compared to the first difference corrections. Compared to the first difference variant, the second difference variant is likely to entail substantial revisions in the preliminary benchmarked series on the availability of the 2006 benchmark.

6.4.2 The Proportional Second Different Variants

The original proportional second difference variant of the Denton method is obtained by setting the difference operator $\boldsymbol{D}$ equal to $\boldsymbol{D}_0^2\,\boldsymbol{\Xi}$, where $\boldsymbol{D}_0^2 = \boldsymbol{D}_0^1\,\boldsymbol{D}_0^1$. The operator $\boldsymbol{D}_0^2$ is given by Eq. (6.19) and $\boldsymbol{\Xi}$ is a diagonal matrix containing the values of s_t or preferably the relative values of $s_t/\bar{s}$ to standardize the data. The solution is provided by (6.10).

Given $\boldsymbol{D}_0^2$ from Eq. (6.19), the objective function of the original proportional second difference variant is

$$\min\{((\theta_1-s_1)/s_1)^2+[(\theta_2-s_2)/s_2-2(\theta_1-s_1)/s_1]^2+ \\ \Sigma_{t=3}^{T}[(\theta_t-s_t)/s_t-2(\theta_{t-1}-s_{t-1})/s_{t-1}+(\theta_{t-2}-s_{t-2})/s_{t-2}]^2\}, \qquad (6.21)$$

where the two first quadratic terms are the detrimental initial conditions. In the modified proportional second difference variant, these two terms are omitted. This modified variant is obtained by setting the difference operator $\boldsymbol{D}$ equal to $\boldsymbol{D}^2\,\boldsymbol{\Xi}$, where $\boldsymbol{D}^2$ is given is Eq. (6.19). The solution is provided by Eq. (6.8a).

Fig. 6.8a exhibits the quarterly series exemplified by Denton (1971). The benchmarked series displayed were obtained with the modified proportional first difference and second difference variants. The series are quite similar except for the current year 2006 without benchmark where the second difference variant is substantially larger.

The reason can be seen more clearly in the proportional corrections $\hat{\theta}_t/s_t$ displayed in Fig. 6.8b. Under the second difference method the corrections are specified to behave linearly, which is the case for year 2006 with no benchmark to constrain the corrections in that year. As a result the preliminary benchmarked series of 2006 is likely to be subject to heavy revision under the second difference variant, given the past behaviour of the yearly discrepancies.

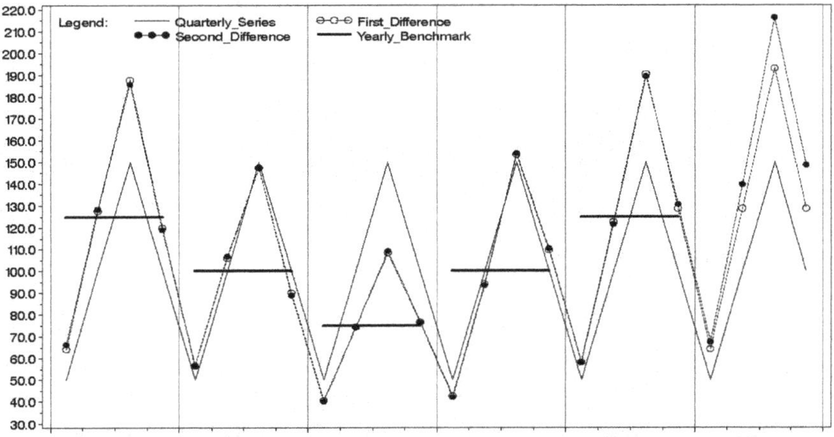

Fig. 6.8a. Quarterly benchmarked series obtained with the modified proportional first and second difference Denton variants

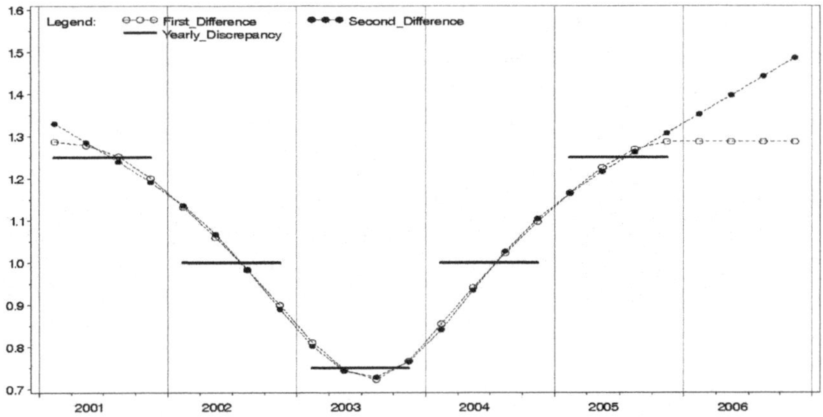

Fig. 6.8b. Proportional corrections to the observed sub-annual series obtained with the modified proportional first and second difference Denton variants

7 Temporal Distribution, Interpolation and Extrapolation

7.1 Introduction

Low frequency data (e.g. quinquennial, yearly) are usually detailed and precise but not very timely. High frequency data (e.g. monthly, quarterly) on the other hand are less detailed and precise but more timely. Indeed producing high frequency data at the same level of detail and precision typically requires more resources and imposes a heavier response burden to businesses and individuals.

At the international level, the problem is compounded by the fact that different countries produce data with different timing and duration (different frequencies) which complicates comparisons. In some cases, different countries produce at the same duration (frequency) but with different timings, for example every fifth year but not in the *same* years. Moreover, in some cases, the data are irregularly spaced in time or display gaps.

Current environmental and socio-economic analyses require an uninterrupted history of frequent and recent data about the variables of interest. Studies pertaining to long term phenomena, e.g. green house gazes and population growth, need a very long history of data. On the contrary, studies pertaining to fast growing leading edge industries, like bio-engineering, nano-technologies require no more than two decades of data.

Researchers and academics often face situations of data with different durations or timings and with temporal gaps. This complicates or impedes the development of a quarterly models (say) providing frequent and timely predictions. Benchmarking, interpolation and temporal disaggregation (distribution) address these issues by estimating high frequency values from low frequency data and available higher frequency data; and by providing frequent extrapolations where needed.

The Interpolation and temporal distribution problems are related to that of benchmarking. Indeed, it is possible to produce quarterly interpolations and temporal distributions from low frequency data specified as binding benchmarks. For example, benchmarking a quarterly *indicator series*, e.g. a seasonal pattern or proxy variable, to yearly data generates quarterly estimates. This practice is very widely used in statistical agencies. As a result, many of the concepts and notation used in benchmarking also apply to interpolation and temporal distribution.

We therefore continue to refer to the high frequency data as "sub-annual" values; and, to the low frequency data, as "annual" values, unless it is appropriate to specify their exact frequency, e.g. yearly, quinquennial. Note that the "sub-annual" data may be yearly, and the "annual" benchmarks may be quinquennial or decennial. Where applicable, the sub-annual (high frequency) indicator series is denoted by

$$s_t,\ t=1,2,...,T, \tag{7.1a}$$

where $\{1,2,...,T\}$ refers to a set of *contiguous* months, quarters, days. In the absence of an indicator, the index $t=1,...,T$ denotes the time periods for which interpolations, temporal distributions and extrapolations are needed.

The "annual" (low frequency) observations are denoted by

$$a_m,\ m=1,...,M, \tag{7.1b}$$

as in previous chapters on benchmarking. In fact, in interpolation and temporal distribution, the "annual" low frequency data are typically specified as binding benchmarks. The set $\{1,2,...,M\}$ may refer to a set of *non-contiguous* reporting periods, for example there may not be a yearly benchmark every "year".

Like in benchmarking, the first and the last sub-annual periods covered by a_m are respectively denoted by t_{1m} and t_{Lm}. The *coverage fraction* j_{mt} for a given sub-annual period *t*, is the proportion of the days of period *t* covered by the m-th annual value, more precisely the proportion of the days included in the *reporting periods* of the data point. As explained in Chapter 3, the coverage fractions are the explanatory variables of the *relative duration* and *relative timing* of the low frequency data (benchmarks) with respect to the high frequency (sub-annual) data. For the moment, we assume all coverage fractions are equal to 0 or 1.

The problem of interpolation usually arises in the context of stock series, where the yearly value should typically be equal to that of the fourth quarter or twelfth month. The purpose is to obtain estimates for the other quarters or months. This is a problem of producing missing values within the temporal range of the data.

The problem of temporal distribution, also referred to as temporal disaggregation, is usually associated with flow series, where low frequency annual data correspond to the annual sums or averages of the corresponding higher frequency data. Temporal distribution is a standard practice in the system of National Accounts, because of the excessive cost of collecting frequent data. High frequency data is therefore obtained from temporal distribution of the annual data. This process usually involves indicators which are available on a high frequency basis and deemed to behave like the target variable.

Extrapolation refers to the generation of values outside the temporal range of the data, whether the series under consideration is a flow or a stock. Note that the extrapolation process may be backward or forward, i.e. toward the past or the future.

Despite the conceptual differences between interpolation, temporal distribution and extrapolation, they are considered similar from the viewpoint of regression-based estimation. In this regard, Chow and Lin (1971 p 374) themselves wrote: " *... our methods treats interpolation and distribution identically.* [...] *This distinction is unjustified from the view point of the theory of estimation ...*".

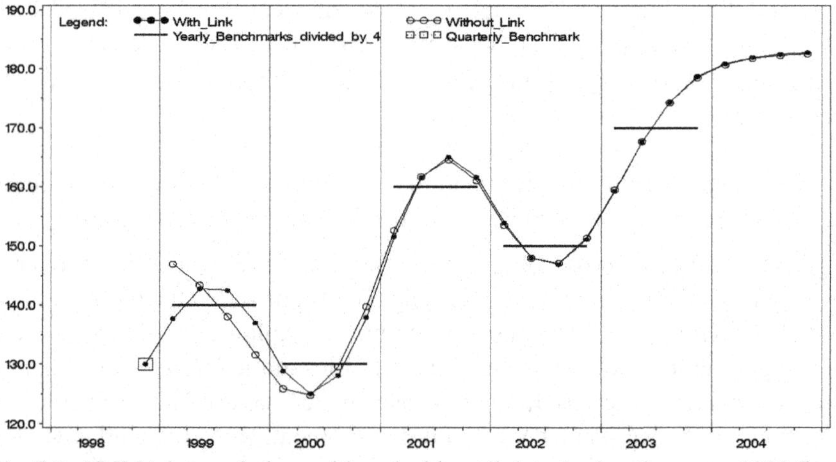

Fig. 7.1. ARIMA interpolations with and without link to the fourth quarter 1998 final value

Our experience with temporal distribution suggests that the distinction between interpolation and distribution breaks down in practice. Indeed monthly values are often specified as benchmarks in conjunction with yearly values for monthly flow series; and similarly with quarterly flow series. In fact, interpolation can also occur for flow variables. For example, immigration to a province may be measured every fifth year, and the years and perhaps the quarters need to be interpolated.

Fig. 7.1 displays two interpolated quarterly flow series, one without quarterly benchmark (circles); and, the other with a quarterly value at the beginning treated as binding benchmark, to force the estimated series (dots) to start from a historical "final" value. The series without quarterly *link benchmark* produces a discontinuity with respect to the last previously obtained final value. Sometimes such link benchmarks are used to force the series to end at a specific value (e.g. Helfand et al. 1977). Furthermore, yearly benchmarks are generally not available for the current year, which implies that extrapolations are generated for the sake of timeliness of the series. Also, in some cases yearly benchmarks are not available every year but every second or fifth year. This entails interpolation in the narrow sense between the available data points and temporal distribution over the periods covered by the yearly data.

Several methods have been developed for interpolation and temporal distribution. We will limit the discussion to the methods more widely applied in statistical agencies The methods analysed can be grouped as (1) ad hoc, (2) the Cholette-Dagum regression-based methods, (3) the Chow-Lin method and its variants and (4) the ARIMA model-based methods. Methods (2) to (3) produce smooth values, whereas the ad hoc techniques generate kinks in the estimated series.

In recent years there has been several attempts to use structural time series models cast in state space representation for interpolations and temporal distributions (e.g. Harvey and Pierse 1984, Gudmundson 1999, Hotta and Vasconcellos 1999, Proietti 1999, Aadland 2000). These methods are not yet fully corroborated by empirical applications and will not be discussed.

7.2 Ad Hoc Interpolation and Distribution Methods

The following method of interpolation is widely used for stock series as a rule of thumb. Recall that for stock series, the benchmarks (i.e. the low frequency data) cover one sub-annual period, i.e. $t_{1m} = t_{Lm}$. The ad hoc method consists of linearly interpolating between each two neighbouring annual observations:

$$\hat{\theta}_t = a_m + (t - t_{Lm})\, b, \quad t_{Lm} \le t < t_{1,m+1}, \quad m=1,...,M-1, \tag{7.2a}$$

where $b = (a_{m+1} - a_m) / (t_{1,m+1} - t_{Lm})$ is the slope. At both ends of the series, the method merely repeats the first and the last annual value:

$$\hat{\theta}_t = a_1, \quad t<t_{1m}, m=1; \quad \hat{\theta}_t = a_M, \quad t \geq t_{2M}. \tag{7.2b}$$

The method is often applied for yearly Inventories and Population series.

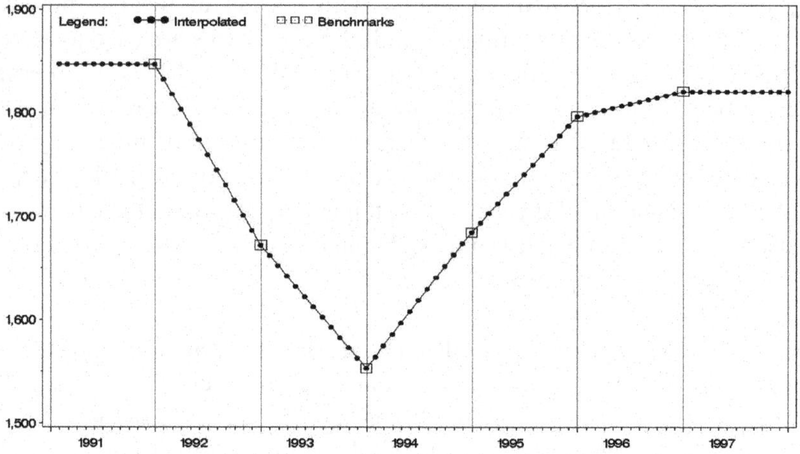

Fig. 7.2. Ad hoc interpolations for the Stock of Capital

Fig. 7.2 illustrates the ad hoc interpolations of the Stock of Capital for an industry. The benchmarks pertain to the last day of each year; and the interpolations, to the last day of each month. The method produces kinks in the series. In the absence of any knowledge on specific sub-annual movement, it is probably more reasonable to assume a gradual behaviour without kinks.

There is a corresponding method for flow series, often called the step-wise method. Recall that for flow series, the low frequency values treated as benchmarks cover at least one sub-annual periods, i.e. $t_{1m} \leq t_{Lm}$. The ad hoc method consists of averaging each available annual (less frequent) value over the sub-annual periods covered by a_m:

$$\hat{\theta}_t = a_m / (t_{Lm} - t_{1,m} + 1), \quad t_{1m} \leq t \leq t_{L,m}, \; m=1,...,M. \tag{7.3a}$$

If there are temporal gaps between benchmarks, the method linearly interpolates between the neighbouring interpolated values from Eq. (7.3a):

$$\hat{\theta}_t = \hat{\theta}_{t_{Lm}} + (t - t_{Lm})\, b, \quad t_{Lm} \leq t < t_{1,m+1}, \; m=1,...,M-1, \tag{7.3b}$$

where $b = (\hat{\theta}_{t_{1,m+1}} - \hat{\theta}_{t_{Lm}}) / (t_{1,m+1} - t_{Lm})$. At the ends, the method merely repeats the first and the last interpolated values from Eq. (7.3a) at the beginning and at the end:

$$\hat{\theta}_t = \hat{\theta}_{t_{1m}}, \quad t<t_{1m}, m=1; \; \hat{\theta}_t = \hat{\theta}_{t_{LM}}, \quad t \geq t_{LM} . \tag{7.3c}$$

In fact, the step-wise interpolation method (7.3) includes method (7.2) as a particular case. Both methods (7.2) and (7.3) are easy to implement in spreadsheet programs.

Fig. 7.3 displays the interpolations obtained by the step-wise method for an artificial flow series. The benchmarks pertain to fiscal years ranging from June to May of the following year. The method produces steps in the series. It is very doubtful that the level of activity changes precisely between the (fiscal) years.

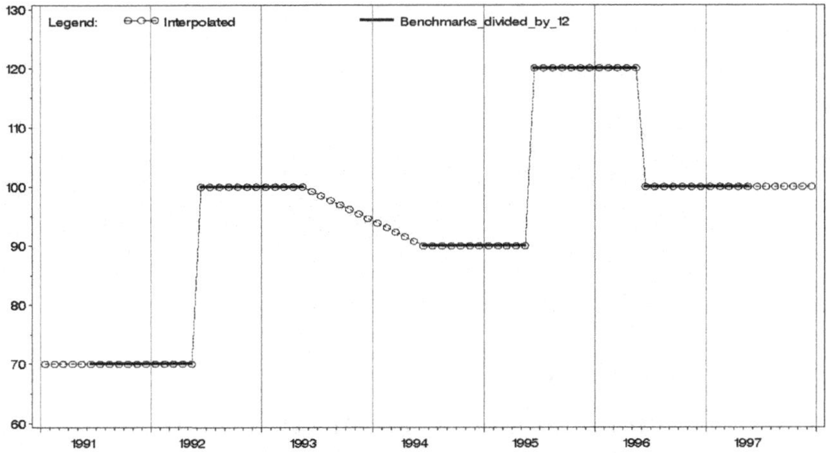

Fig. 7.3. Step-Wise interpolations for an artificial flow series

In some cases, statistical agencies use a *growth rate extrapolation method.* Yearly growth rates are developed from related series and subject matter expertise. The method consists of applying the growth rates to the same quarter of the previous year to obtain current *projections*:

$$s_t = g_t \times s_{t-4}, \quad t=5,...,T,$$

where the g_ts are the annual growth rates. The initial base year values, s_t, t=1,...,4, are developed from more detailed analysis and investigation.

The current projections are benchmarked as each yearly benchmark becomes available, and the future growth rates are applied to the previously obtained benchmarked values. The prorating benchmarking method would be recommended in this situation. Under prorating, the projections are multiplied by the proportional discrepancies: $a_m / (\Sigma_{t=t_{1m}}^{t_{Lm}} s_t)$.

Despite its ad hoc character, the method offers flexibility in building the growth rates from the best *currently available mix* of series and information, depending on the circumstances. Furthermore, the projection and prorating methods can be easily implemented in spreadsheet programs.

7.3 Interpolation and Temporal Distribution Based on Regression Methods

This section shows how regression-based models like those discussed for benchmarking can be used for interpolation and temporal distribution. It is appropriate to recall the additive Cholette-Dagum regression-based model of Chapter 3

$$\boldsymbol{s} = \boldsymbol{R}\boldsymbol{\beta} + \boldsymbol{\theta} + \boldsymbol{e}, \; E(\boldsymbol{e}) = 0, \; E(\boldsymbol{e}\,\boldsymbol{e}') = \boldsymbol{V}_e, \tag{7.4a}$$

$$\boldsymbol{a} = \boldsymbol{J}\,\boldsymbol{\theta} + \boldsymbol{\varepsilon}, \; E(\boldsymbol{\varepsilon}) = 0, \; E(\boldsymbol{\varepsilon}\,\boldsymbol{\varepsilon}') = \boldsymbol{V}_\varepsilon, \; E(\boldsymbol{e}\,\boldsymbol{\varepsilon}') = 0; \tag{7.4b}$$

and the corresponding multiplicative variant of Chapter 5

$$\boldsymbol{s}^* = \boldsymbol{R}\boldsymbol{\beta}^* + \boldsymbol{\theta}^* + \boldsymbol{e}^*, \; E(\boldsymbol{e}^*) = 0, \; E(\boldsymbol{e}^*\boldsymbol{e}^{*\prime}) = \boldsymbol{V}_{e^*}, \tag{7.5a}$$

$$\boldsymbol{a} = \boldsymbol{J}\,\boldsymbol{\theta} + \boldsymbol{\varepsilon}, \; E(\boldsymbol{\varepsilon}) = 0, \; E(\boldsymbol{\varepsilon}\,\boldsymbol{\varepsilon}') = \boldsymbol{V}_\varepsilon, \; E(\boldsymbol{e}^*\boldsymbol{\varepsilon}') = 0, \tag{7.5b}$$

where the asterisk denotes the logarithmic transformation was applied, and $\boldsymbol{\theta}$ is the series to be estimated. Note that Eq. (7.5a) implies $s_t = (\Pi_{h=1}^{H} \beta_h^{r_{th}}) \times \theta_t \times (1 + e_t^*)$. In both models, matrix $\boldsymbol{R}$ may contain related series as in Chow-Lin (1971) or functions of time as in Chow-Lin (1976). The solutions of models (7.4) and (7.5) are given by Eqs. (3.20) and (5.10) respectively.

The main distinction between benchmarking and temporal disaggregation (or interpolation) is the following. In benchmarking (in the strict sense), the annual measurements $\boldsymbol{a}$ series pertain to the target variable, and the series $\boldsymbol{s}$

to be benchmarked consists of sub-annual measurements of the *same* variable.

In temporal disaggregation (or interpolation), the series $\boldsymbol{s}$ is an *indicator* or a *proxy* for the target variable to be estimated sub-annually. The indicator selected should closely approximate the sub-annual behaviour of the target variable. The indicator may have an order of magnitude different from that of the low frequency data specified as benchmarks. For example the indicator may be in percentage; and the benchmarks in billions of dollars. Benchmarking and indicator is widely used practise in statistical agencies. Alternatively, no indicator is available; the indicator is then set to a constant (e.g. 0 nor 1), and matrix $\boldsymbol{R}$ contains related series as in the Chow and Lin (1976) method.

In benchmarking as well as temporal distribution, the component $\boldsymbol{R\beta}$ of (7.4) models the difference between the sub-annual data $\boldsymbol{s}$ and the series to be estimated $\boldsymbol{\theta}$. The latter must satisfy the level of $\boldsymbol{a}$ as specified by Eq. (7.4b).

The error model distributes the remaining difference between $\boldsymbol{s}^{\dagger} = \boldsymbol{s} - \boldsymbol{R\beta}$ and $\boldsymbol{\theta}$. This line of reasoning also applies to the multiplicative (logarithmic) model (7.5), *mutatis mutandis*. The error model is usually specified to follow an autoregressive model of order 1 or 2. The principle is to distribute the residual annual discrepancies in a manner supported by the data. For example, if the discrepancies hover around a constant, autoregressive models of order 1 or 2 can be appropriate. If enough benchmarks are sub-annual, a seasonal error model could be entertained.

Fig 7.4 illustrates three interpolated series pertaining to a stock variable measured on the last day of each year. All three series are obtained from the additive model (7.4), where $\boldsymbol{s}$ is set to $\boldsymbol{0}$ and the deterministic regressor $\boldsymbol{R}$ is set to -1. For the AR(1) interpolations with dots ("•"), the error follows a first order autoregressive error model: $e_t = 0.999\, e_{t-1} + v_t$. The resulting interpolations approximate the *ad hoc* interpolations of Fig. 7.2.

For the AR(1) interpolations with diamonds ("◇"), the error follows the model: $e_t = 0.80\, e_{t-1} + v_t$. The resulting interpolations display movement reversals between benchmarks and spikes at the benchmarks which are particularly obvious in the last quarters of 1995 and 2005. As discussed in Chapter 4, AR(1) models are not advisable when some of the benchmarks cover only one sub-annual observation, which typically occurs for stock series.

For the AR(2) interpolations, the error follows a second order autoregressive model: $e_t = 1.60\, e_{t-1} - 0.64\, e_{t-2} + v_t$. This model can be factorized as $(1-0.80B)(1-0.80B)e_t = v_t$, where B is the backshift operator such that $B^k e_t \equiv e_{t-k}$. In other words, the model is the product of two AR(1) models.

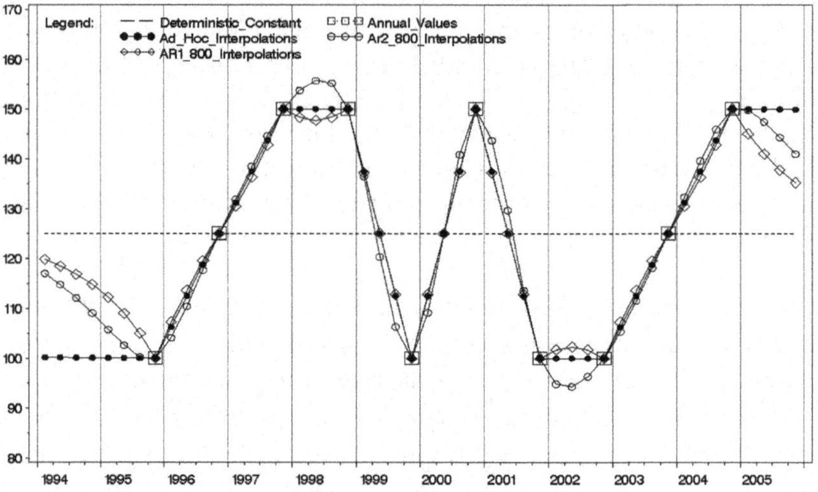

Fig. 7.4. Quarterly interpolations between yearly stock data obtained from the additive regression-based method with a constant indicator under three error models

The AR(2) error model produce an interpolated series with no kinks and with larger amplitude than the AR(1). This larger amplitude can be observed between the two yearly stock values of the fourth quarters of 1997 and 1998. It is indeed plausible that the AR(2) interpolated series should have its downward turning points between these two benchmarks. Furthermore, the much lower neighbouring 1996 and 1999 benchmarks do suggest that the turning-point is likely to occur above the 1997 and 1998 yearly values. The assumption behind this line of reasoning is that the series evolves gradually within each year. This would be a safer assumption in the absence of sub-annual data supporting a more complex behaviour.

The AR(2) model provides an excellent fit at both ends, since the interpolated series, more precisely the extrapolations, converge to the indicator series. This convergence is justified by the fact that the yearly stock data have historically hovered about the constant level represented by the deterministic part. The speed of convergence could be slowed down with by replacing 0.80 by 0.90 (say); and accelerated, by replacing 0.80 by 0.70.

The residuals $\hat{e}_t$ are represented by the distance between the interpolations and the constant. The resulting innovations $\hat{v}_t = \hat{e}_t - 0.999\hat{e}_{t-1}$ and $\hat{v}_t =$

$\hat{e}_t - 1.60\hat{e}_{t-1} + 0.64\hat{e}_{t-2}$ are positive during the periods of growth of the series, namely in 1996, 1997, 2000, 2003 and 2004; and negative during the periods of decline, namely in 1999 and 2001. Choosing a more complex ARMA model for the error would produce spurious interpolations, not supported by the yearly data. The goal is not to obtain random innovations, but to distribute the annual discrepancies in a reasonable manner supported by the data.

In many applications, the use of deterministic regressors is advisable. For example, a constant regressor can bring the indicator to the level of the low frequency data (benchmarks) as illustrated in Fig. 7.4. A linear trend or cubic spline trend, as discussed in section 3.4, can operate a level adjustment and a change in direction indicated by the benchmarks.

Fig. 7.5 displays such a case, where the fiscal year flow data pertaining to Interest Payments on Consumer Loans tend to evolve upwards. The temporal distributions are obtained by the additive model (7.4), where s is set to 0, the deterministic regressor $\boldsymbol{R}$ contains the spline cubic trend of section 3.4. The error model is an AR(1) with parameter $\varphi = 0.90$ and constant variance. The estimated constant is equal to 663.34 with t-value equal to 25.43; and the slope 6.57 with t-value equal to 3.53.

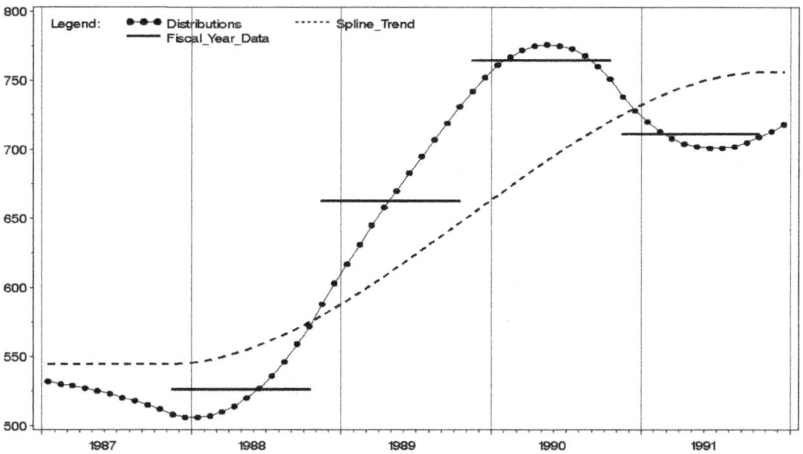

Fig. 7.5. Monthly distributions of fiscal year flow data pertaining to Interest Payments on Consumer Loans, obtained from the additive regression-based interpolation model with constant indicator and spline trend regressor

This application of the additive regression model (7.4) replicates the Chow and Lin (1976) method, where these authors proposed that the regressors can be functions of time, and not necessarily related socio-economic variables as in their previous paper (1971). In most cases, the use of the latter is more appropriate.

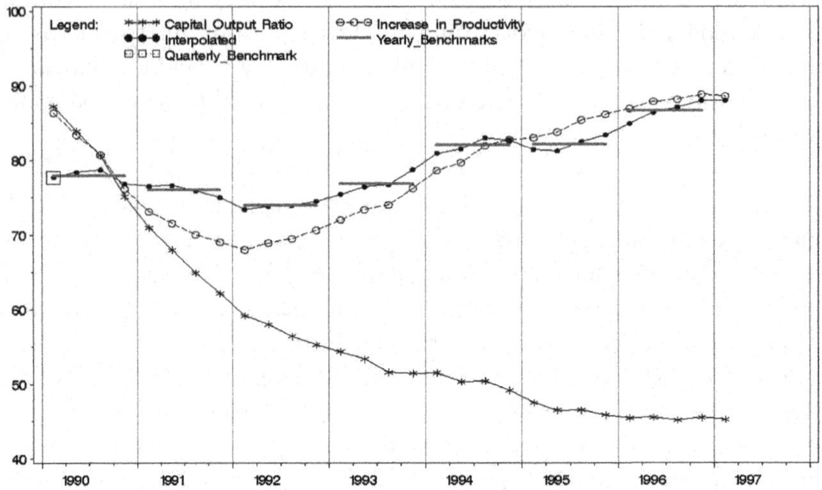

Fig. 7.6a. Interpolated quarterly values of Capacity Utilization using the Capital-Output Ratio as an indicator

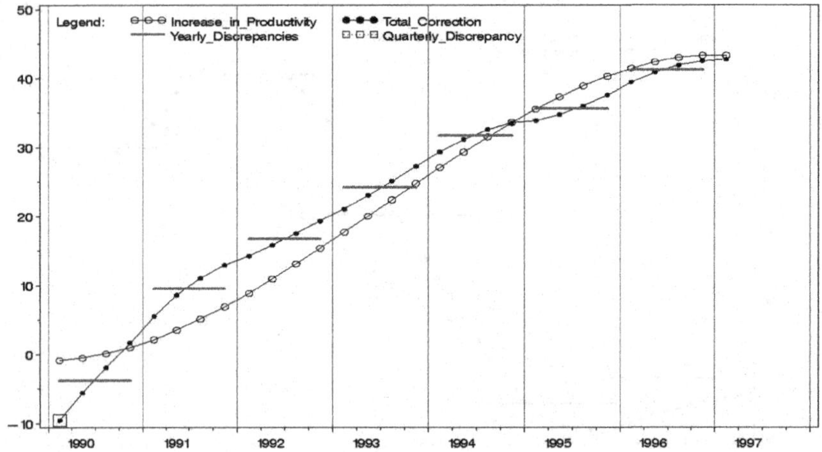

Fig. 7.6b. Additive corrections made to the Capital-Output Ratio, with cubic spline trend designed to correct for increases in productivity

The interpolation of Capacity Utilization Index provides a example involving an indicator series, namely the quarterly Capital-Output Ratio. This indicator is here re-scaled before hand by a multiplicative constant, for it to vary between 0 and 100%. The re-scaled Capital-Output Ratio is benchmarked to the yearly Capacity Utilization survey estimates, as displayed in Fig. 7.6a. The output capacity is assumed to be a gradual function of the Capital-Output Ratio: the more capital, the more output can be achieved.

With technical progress however, less and less capital is required in the long run to produce the same level of output. This is indeed the case for the Printing industry, due to the revolution of desktop publishing. As a result the Capital-Output ratio has been declining gradually and steadily, while the Capacity Utilization rate mainly increased from 1990 to 1997. At first glance, the Capital-Output ratio of the industry seems to be a poor indicator.

However, the solution is to include a trend in the interpolation model to account for the increased productivity of capital. Once corrected for productivity, the Capital-Output ratio becomes a good indicator. Fig. 7.6a displays the Capital-Output Ratio corrected for increase of productivity (circles). Fig 7.6b displays the spline trend.

The interpolation model is given by (7.4), where $\boldsymbol{R}$ contains the following regressors. The first column of $\boldsymbol{R}$ contains a constant equal to -1 to bring the indicator to the level of the benchmarks; and the second column, a cubic spline trend designed to level-off at both ends, namely in the first quarter of 1990 and 1997. The error model is an AR(1) with parameter φ=0.90^3: $e_t = 0.729\, e_{t-1} + v_t$. This model causes the extrapolations to converge to the Capital-Output Ratio corrected for productivity depicted in Fig. 7.6a.

The Capacity Utilization is an index series, and the yearly benchmarks pertain to the average of the year. The quarterly benchmark in the first quarter of 1990 forces the interpolated series to start at a specific point, which in our case is an already published interpolated value considered "final". This quarterly benchmark also avoids revisions to the published interpolated values before 1990 and ensures historical continuity with these values.

The model is applied in a seven year moving manner. This moving average application is required, because the spline trend (which captures gains in productivity) can only be valid over a restricted number of years, e.g. seven years.

A variant of the model could be to set $\boldsymbol{s}$ equal to zero and replace the constant in the first column of $\boldsymbol{R}$ by minus the Capital-Output ratio. The change of sign results into a positive value for β_1 which is easier to interpret: e.g. $\beta_1 = 1.10$ means the benchmarked series is 10% higher than the indicator.

Maximum output capacity is the ratio of the actual output divided by the capacity utilization of the year or the quarter. For example, if goods for

$5,000 millions are produced in the second quarter of 2004 with a capacity utilization equal to 80% (given by the benchmarked series), the potential output capacity of that quarter is $5,000/0.80 = $6,250 under the assumption of constant returns to scale.

The calculation also applies to the yearly values, provided by the benchmarks. If goods for $18,000 millions are produced in 2004 and the benchmark is equal to 85% , the potential output capacity is approximately equal to $18,000/0.85 = $21,176. In order to obtain a better approximation, it is necessary to calculate the potential output capacity for each quarter of 2004 and take the yearly sum over the four quarters.

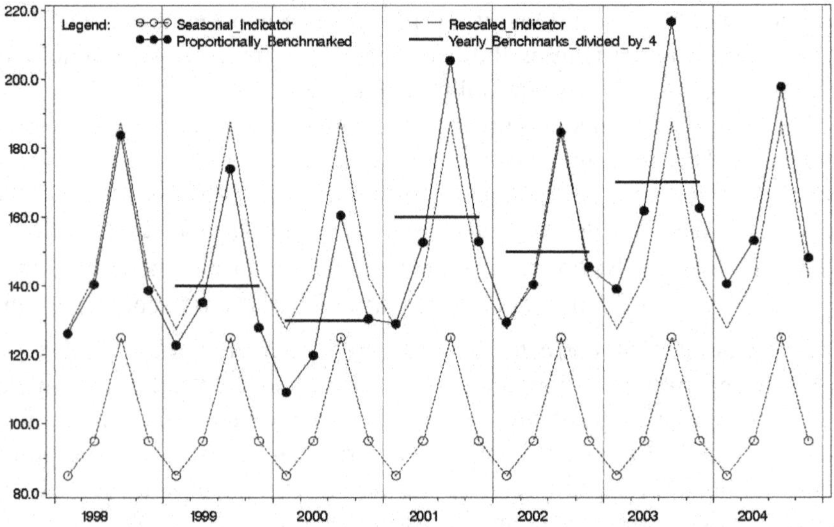

Fig. 7.7. Temporal distributions obtained by proportionally benchmarking a seasonal pattern to yearly values

In many cases, subject matter experts generate interpolations and temporal distributions by proportionally benchmarking a (sub-annual) seasonal pattern to annual values of stock or flows series. Fig. 7.7 illustrates a quarterly seasonal pattern (in percent) and the corresponding temporal distributions obtained for a flow series. The seasonal pattern apportions the annual values to the quarterly periods.

Proportional benchmarking, described in section 3.8, consists of re-scaling the indicator by a factor equal to $(\Sigma_{m=1}^{M} a_m) / (\Sigma_{m=1}^{M} \Sigma_{t=t_{1m}}^{t_{Lm}} j_{mt}\, s_t)$. This brings the indicator to the average level of the low frequency data treated as binding benchmarks. The *re-scaled indicator* $s_t^{\dagger}$ exactly preserves the original proportional movement, $s_t^\dagger / s_t = s_{t-1}^\dagger / s_{t-1}$; and consequently, the original growth rates since $s_t^\dagger / s_t = s_{t-1}^\dagger / s_{t-1}$ implies $s_t^\dagger / s_{t-1}^\dagger = s_t / s_{t-1}$. The re-scaling amplifies and shrinks the seasonal pattern depending on the level of the benchmarks. Furthermore this re-scaling cannot produce negative values. Benchmarking then performs a residual proportional adjustment around $s_t^\dagger$ to satisfy the benchmarks. This is achieved by setting $\boldsymbol{s}$ equal to the re-scaled series $\boldsymbol{s}^\dagger$, by omitting the regressor $\boldsymbol{R}$ in model (7.4) and by specifing a first order proportional autoregressive model for the error: $e_t / s_t = \varphi\, e_{t-1} / s_{t-1} + v_t$ with φ close to 1.

In the case depicted, the re-scaling factor is equal to 1.5 and the error model is an AR(1) with parameter φ=0.729. In most real cases, the order of magnitude of the low frequency data (benchmarks) is very different from that of the indicator. In such cases it is impossible to usefully represent the benchmarks and the indicator in the same figure.

If $\boldsymbol{s}$ is set to zero and the regressor is set to minus the indicator ($-\boldsymbol{s}$) in the additive model, Eq. (7.4a) becomes $\boldsymbol{0} = -\boldsymbol{s}\,\beta_1 + \boldsymbol{\theta} + \boldsymbol{e}$. The coefficient β_1 now denotes a multiplicative "bias", which re-scales the "regressor" to the level of the benchmarks. The resulting model is mixed: multiplicative bias with an additive non-proportional error. Indeed, in order to be valid, the error $\boldsymbol{e}$ must be independent of the regressor $\boldsymbol{s}$, which implies that the standard deviations cannot be derived from the coefficients of variations of the sub-annual regressor[1]. As a result the model is partly multiplicative (or proportional) and partly additive. This variant of the additive regression-based model (7.4) replicates the results of the Chow and Lin (1971) interpolation method, with only one "related series".

Fig. 7.8 displays a case of monthly distribution between fiscal quarter data pertaining to Interest Payments on Mortgages in the banking sector. The multiplicative regression-based model of Eq (7.5) is applied. The original series $\boldsymbol{s}$ is set to 100, matrix $\boldsymbol{R}$ contains trend and seasonal regressors. The error model is an AR(1): $e_t^* = 0.90\, e_{t-1}^* + v_t$ with constant variance (where the asterisks indicate the logarithmic transformation).

The two trend regressors are those of a cubic spline, which level-off at both ends, as described in section 3.4. The estimated coefficients are 6.8887 and 0.0160 (in the logs) with *t*-values equal 271.6332 and 10.4856

---

[1] Ordinary and Generalized Least Square estimation require the assumption of independence between the regressors and the error.

respectively. The two seasonal regressors are trigonometric, $-sin(\lambda_1 q)$ and $-cos(\lambda_2 q)$ displayed in Table D.3 of Appendix D, where q denotes the month of the year. The estimated coefficients are respectively -0.0228 and 0.0099 (in the logs) with t-values equal to -2.8233 and 2.2860.

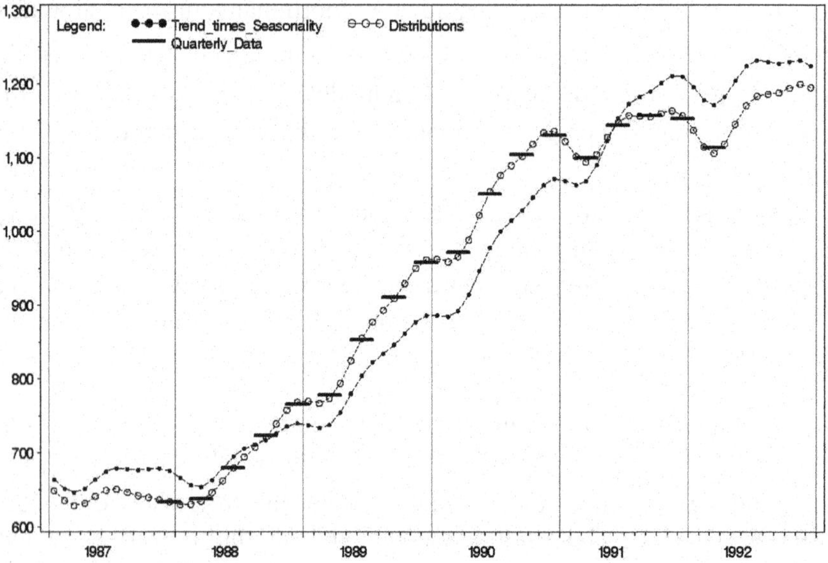

Fig. 7.8. Interpolating monthly values using a multiplicative variant of the Chow-Lin (1976) model, using trend and seasonal regressors

The seasonal regressors capture the seasonality present in the quarterly benchmarks. The estimated seasonality displays a mild seasonal trough in the first few months of every year. The resulting monthly seasonal pattern contains no sub-quarterly seasonality. The values of the seasonal pattern anticipate the relative values of the unavailable fiscal quarter values of 1992 and 1987.

The error model causes the extrapolations to converge to the regression part, i.e. the product of the trend and seasonality, and do incorporate the historical seasonal pattern. The values depicted are transformed back to the original scale. The multiplicative relation between the trend and the seasonal components causes an amplification of the seasonal pattern as the trend increases.

This application of the Cholette-Dagum regression-based multiplicative model (7.5) can be viewed as a multiplicative variant of the additive Chow-Lin (1976) method.

7.4 The Chow-Lin Regression-Based Method and Dynamic Extensions

A classical method frequently applied in the system of National Accounts for temporal distribution and interpolation is due to Chow and Lin (1971, 1976). The authors assume there is an available series $\boldsymbol{y}_{\bullet}$ ranging over n consecutive quarters (say) and a set of related indicator series $\boldsymbol{x}$ available monthly. The problem is to obtain the corresponding $3n$ monthly estimates in the temporal range of $\boldsymbol{y}_{\bullet}$. The underlying assumption is that the true monthly observations $\boldsymbol{y}$ satisfy a multiple regression model with p related series $\boldsymbol{x}_1, ..., \boldsymbol{x}_p$:

$$\boldsymbol{y} = \boldsymbol{X\beta} + \boldsymbol{u}, \quad E(\boldsymbol{u}) = \boldsymbol{0}, \; E(\boldsymbol{u}\,\boldsymbol{u}') = \boldsymbol{V}, \tag{7.6}$$

where $\boldsymbol{y}$ is a $3n\times 1$ vector of monthly non-observable data, $\boldsymbol{X}$ is a $3n\times p$ matrix of the related indicator series and $\boldsymbol{u}$ is a random error assumed to follow a autoregessive model of order 1 (AR(1)).

Chow and Lin do not discuss the nature of the related indicators. Presumably the indicators should be socio-economic variables deemed to behave like the target variable. In the absence of such variables, they suggest using functions of time.

The n-dimensional quarterly series $\boldsymbol{y}_{\bullet}$ must also satisfy (7.6), which implies

$$\boldsymbol{y}_{\bullet} = \boldsymbol{Cy} = \boldsymbol{CX\beta} + \boldsymbol{Cu} = \boldsymbol{X}_{\bullet}\boldsymbol{\beta} + \boldsymbol{u}_{\bullet}, \; E(\boldsymbol{u}_{\bullet}\boldsymbol{u}_{\bullet}') = \boldsymbol{CVC}' = \boldsymbol{V}_{\bullet}, \tag{7.7}$$

where $\boldsymbol{C} = \boldsymbol{I}_n \otimes \boldsymbol{c}$ with $\boldsymbol{c} = [\,1\;1\;1\,]$ for temporal distribution and $\boldsymbol{c} = [\,0\;0\;1\,]$ for interpolation and $\otimes$ denotes the Kronecker product introduced in section 3.6.1.

They introduce the $m\times 1$ vector $\boldsymbol{z}$ to include possible extrapolations outside the temporal range of the quarterly data $\boldsymbol{y}_{\bullet}$ and the corresponding regressors $\boldsymbol{X}_z$ are assumed available. Hence,

$$\boldsymbol{z} = \boldsymbol{X}_z\boldsymbol{\beta} + \boldsymbol{u}_z, \tag{7.8}$$

which coincides with (7.6) in the particular case of temporal distribution and interpolation.

The best linear un-biased estimator of $\hat{z}$ is

$$\hat{z} = A y_{\bullet} = X_z \hat{\beta} + V_{z\bullet} V_{\bullet}^{-1} \hat{u}_{\bullet} , \tag{7.9}$$

where $V_{z\bullet} = E(u_z u_{\bullet}^{\prime})$ and A is an $m \times n$ matrix such that $A X_{\bullet} - X_z = 0$. The latter constraint ensures the unbiasedness property of the estimator. Indeed $E(\hat{z} - z) = E[A(X_{\bullet}\beta + u_{\bullet}) - (X_z \beta + u_z)] = 0$ if $A X_{\bullet} - X_z = 0$.

By determining the value of matrix A, Chow and Lin obtain:

$$\hat{\beta} = (X_{\bullet}^{\prime} V_{\bullet}^{-1} X_{\bullet})^{-1} X_{\bullet}^{\prime} V_{\bullet}^{-1} y_{\bullet} , \tag{7.10a}$$

which is the generalized least square estimator of the regression coefficients using the n quarterly observations. Hence,

$$\hat{z} = A y_{\bullet} = X_z \hat{\beta} + V_{z\bullet} V_{\bullet}^{-1} \hat{u}_{\bullet} , \tag{7.10b}$$

$$var(\hat{z}) = (V_z - V_{z\bullet} V_{\bullet}^{-1} V_{\bullet z}) + (X_z - V_{z\bullet} V_{\bullet}^{-1} X_{\bullet}) (V_{z\bullet} V_{\bullet}^{-1} X_{\bullet})^{-1} (X_z^{\prime} - X_{\bullet}^{\prime} V_{\bullet}^{-1} V_{\bullet z}) , \tag{7.10c}$$

where $\hat{u}_{\bullet} = y_{\bullet} - X_{\bullet}\hat{\beta}$ and $V_{z\bullet} = V_z C^{\prime}$.

The estimation of z requires the covariance matrix V of the regression residuals.The authors estimate a first order autoregressive model to the error u from the observed residuals $\hat{u}_{\bullet} = y_{\bullet} - X_{\bullet}\hat{\beta}$.

It should be noted that the estimates $\hat{z}$ consists of two components. The first, $X_z \hat{\beta}$ results from applying the regression coefficients $\hat{\beta}$ to the related monthly indicators. The second component is an estimate of the residuals u_z by applying $V_{z\bullet} V_{\bullet}^{-1}$ to the estimated quarterly residuals $\hat{u}_{\bullet}$

This method is a particular case of the regression-based additive model (7.4), if s and V_{ε} are equal to zero. Indeed, the solution of model (7.4) given by Eqs. (3.20) coincides with the Chow-Lin solution. Their notation replaces matrices R, J, JR, V_e and V_d of Eqs. (3.20) by $-X$, C, $X_{\bullet}$, V and $V_{\bullet}$

respectively; and vectors $\boldsymbol{\theta}$, $\boldsymbol{a}$ and $\boldsymbol{e}$, by y, $y_{\bullet}$ and $\boldsymbol{u}$. The change of sign results from setting $\boldsymbol{R}$ equal to minus the regressors.

Chow and Lin derive their solution by means of the Minimum Variance Linear Unbiased Estimator approach. On the other hand, the solution of Cholette-Dagum model described in Chapter 3 is derived by means of generalized least squares applied to a single regression model containing both the sub-annual equation and the annual equation.

Chow and Lin (1971) propose a method to identify the covariance matrix $\boldsymbol{V}_e$ under the assumption that the errors follows an AR(1) model. The model can be estimated from matrix $\boldsymbol{V}_{\bullet}$ equal to $\boldsymbol{CVC}'$, if there is a sufficient number of observations.

In order to obviate this difficulty of identifying a reliable AR(1) model from $\boldsymbol{V}_{\bullet}$, some authors have attempted to extend the Chow-Lin method in a dynamic way.

Grégoir (1995), Santos Silva and Cardoso (2001) and more recently Di Fonzo (2003) have proposed a few dynamic variants of the Chow-Lin method. The dynamic extension include lagged target and/or related series. In a recent study, Di Fonzo (2003) applies the approach to the yearly sums of the quarterly U.S. Personal Consumption (1954-1983) and the quarterly U.S. Personal Disposable Income x_t, as the related series. He assesses the performance of the following four dynamic model extensions:

(1) $$(1-\varphi B)y_t = \alpha + \beta x_t + e_t\,,$$

(2) $$(1-\varphi B)y_t = \beta x_t + e_t\,,$$

(3) $$(1-\varphi B)\ln y_t = \alpha + \beta x_t + e_t\,,$$

(4) $$(1-\varphi B)\ln y_t = \beta x_t + e_t\,,$$

The models are compared to the Chow-Lin method characterized by

(5) $$y_t = \alpha + \beta x_t + e_t,\;\; (1-\varphi B)\,e_t = v_t\,,$$

with and without the constant α, and where $(1-\varphi B)\,e_t = v_t$ which implies $e_t = \varphi\, e_{t-1} + v_t$ and $e_t = (1-\varphi B)^{-1}\, v_t$.

Model (2) above can be written as

(2a)
$$y_t = (1-\varphi B)^{-1}\beta x_t + (1-\varphi B)^{-1}v_t$$
$$\Rightarrow\ y_t = (1+\varphi B+\varphi^2 B^2+...)\beta x_t + (1+\varphi B+\varphi^2 B^2+...)v_t$$
$$= x_t\beta + x_{t-1}\varphi\beta + x_{t-2}\varphi^2\beta + \ ...\ + v_t + \varphi v_{t-1} + \varphi^2 v_{t-2} + ...\ ,$$

and similarly for models (1), (3) and (4). The dependent variable y_t, Consumption, thus depends on a stream of Incomes, $x_t, x_{t-1}, x_{t-2}, ...$, and not just on the contemporaneous value x_t. The propensity to consume is now dynamic.

Model (4) yields preferable results over the other three models and the original Chow-Lin model (5), in terms of errors in levels and percent changes of the distributed series. The errors are measured with respect to the Personal Consumption also available quarterly.

The comparison also includes two static variants of the Chow-Lin method models where the errors are assumed to follow a random walk (Fernandez 1981) and a (1,1,0) model (Litterman 1983), with implicit initial condition $(u_0 = 0)$ in both cases, through the use of difference operator $\boldsymbol{D}_0^1$ of Eq. (6.4). Model (4) is still superior in terms of quarterly percent changes, but Litterman's model is superior in annual percent changes. However, the initial condition in Fernandez (1981) and Litterman (1983) tend to bias the regression coefficients towards zero and to introduce a spurious movement at the beginning of the series. The problem is illustrated by Fig. 6.2b.

Next we show how models (1) to (4) can be treated as particular cases of the Cholette-Dagum regression-base model of Eq. (7.4). Model (2) can be written in matrix algebra as

$$\boldsymbol{D}_\varphi \boldsymbol{y} = \boldsymbol{X\beta} + \boldsymbol{e} \Rightarrow \boldsymbol{D}_\varphi^{-1}\boldsymbol{D}_\varphi \boldsymbol{y} = \boldsymbol{D}_\varphi^{-1}\boldsymbol{X\beta} + \boldsymbol{D}_\varphi^{-1}\boldsymbol{v}$$
$$\Rightarrow \boldsymbol{y} = \boldsymbol{X}_\varphi\boldsymbol{\beta} + \boldsymbol{e},\ E(\boldsymbol{e}) = 0,\ E(\boldsymbol{e}\,\boldsymbol{e}') = \boldsymbol{V}_e, \tag{7.11}$$

where
$$\boldsymbol{D}_\varphi = 1/\sqrt{1-\varphi^2}\begin{bmatrix} \sqrt{1-\varphi^2} & 0 & 0 & \dots & 0 & 0 & 0 \\ -\varphi & 1 & 0 & \dots & 0 & 0 & 0 \\ \vdots & \vdots & \vdots & \ddots & \vdots & \vdots & \vdots \\ 0 & 0 & \dots & 0 & -\varphi & 1 & 0 \\ 0 & 0 & 0 & \dots & 0 & -\varphi & 1 \end{bmatrix},$$

$$\boldsymbol{D}_{\varphi}^{-1} = 1/\sqrt{1-\varphi^2}\begin{bmatrix} \sqrt{1-\varphi^2} & 0 & 0 & \dots & 0 & 0 \\ \varphi\sqrt{1-\varphi^2} & 1 & 0 & \dots & 0 & 0 \\ \varphi^2\sqrt{1-\varphi^2} & \varphi & 1 & \dots & 0 & 0 \\ \vdots & \vdots & \vdots & \ddots & \vdots & \vdots \\ \varphi^{T-1}\sqrt{1-\varphi^2} & \varphi^{T-2} & \varphi^{T-3} & \dots & \varphi & 1 \end{bmatrix},$$

$$\boldsymbol{V}_e = (\boldsymbol{D}_{\varphi}^{1\prime}\boldsymbol{D}_{\varphi}^{1})^{-1} = \boldsymbol{D}_{\varphi}^{-1}(\boldsymbol{D}_{\varphi}^{-1})' = \begin{bmatrix} 1 & \varphi & \varphi^2 & \dots & \varphi^{T-1} \\ \varphi & 1 & \varphi & \dots & \varphi^{T-2} \\ \vdots & \vdots & \vdots & \ddots & \vdots \\ \varphi^{T-1} & \varphi^{T-2} & \varphi^{T-3} & \dots & 1 \end{bmatrix}.$$

Rearranging Eq. (7.11) as $\boldsymbol{0} = -\boldsymbol{X}_{\varphi}\boldsymbol{\beta} + \boldsymbol{y} - \boldsymbol{e}$ has the same form as Eq. (7.4a) where $\boldsymbol{s}$ is equal to $\boldsymbol{0}$, $\boldsymbol{R}=-\boldsymbol{X}_{\varphi}$, $\boldsymbol{\theta}=\boldsymbol{y}$ and $\boldsymbol{e}$ as $-\boldsymbol{e}$. This also applies to model (1).

Proceeding similarly with models (3) to (4) leads to the multiplicative variant of the regression-based model (7.5a): $\boldsymbol{s}^* = \boldsymbol{R}\boldsymbol{\beta}^* + \boldsymbol{\theta}^* + \boldsymbol{e}^*$ where the asterisks denote the logarithmic transformation.

Note that the autoregressive parameter must be estimated, because it has an economic meaning in the dynamic model of Eqs. (2a) and Eq. (7.11). This estimation may be carried out by means of iterative generalized least squares over the range of acceptable values of the parameter.

7.5 ARIMA Interpolation, Temporal Distribution and Extrapolation

One should bear in mind that AutoRegressive Integrated Moving Average (ARIMA) models describe the *temporal behaviour* of time series, but do not explain the causality of the movements observed. As a result, they can extrapolate seasonality and existing trends, but cannot meaningfully predict turning points in the economy. The behaviour of socio-economic time series can be modelled because of physical, legal and institutional constraints. For example, the population of a country cannot double from one year to the next. In normal circumstances, the production capacity cannot change by 25% from year to year. This also applies to transportation, education, technology, climatic seasonality, etc. Employers and employees, producers and customers, are bound by mutual self-interest or contracts. As a result, neighbouring values of time series data tend to be highly correlated and therefore subject to descriptive modelling.

Most socio-economic time series display a non-stationary behaviour, in mean and variance. They typically show a trend with variance and autocovariance changing with time. To properly identify an ARIMA model, a series must be stationary in the first and second order moments. Stationarity in mean is usually achieved by differencing of a low order or by fitting and removing a polynomial or exponential function. In the first case, it is said that the series follows a difference stationary model; and in the second, a trend stationary model. Stationarity in the second order moments is obtained by means of data transformations, the most commonly applied being the logarithmic and the square root. Other adjustments can be identified by means of the Box-Cox transform (1964).

7.5.1 Trend Stationary Models

In the case of trend stationary models, the series is made stationary in mean by fitting a trend and a seasonal component if applicable. The example of Fig. 7.5 illustrates such a model, where the additive model (7.4) contained trend regressors (in matrix $\boldsymbol{R}$) to account for a trend in the fiscal year data.

The example of Fig. 7.8 also illustrates trend stationary models. The multiplicative model (7.5) contained regressors (in matrix $\boldsymbol{R}$) to account for trend and seasonality in the fiscal quarter data.

The example of Fig. 7.9 shows two yearly data points for 1998 and 2001 and four quarterly (sub-annual) data points of 1996 and four quarterly data points of 2003. These yearly and quarterly data are specified as binding benchmarks. The seasonal pattern of the quarterly data of 1996 differs from that of 2003. The goal is to produce quarterly interpolations consistent with both the quarterly and yearly data.

The three sets of interpolations displayed are obtained from the additive model (7.4). The indicator $\boldsymbol{s}$ is set to $\boldsymbol{0}$ and the regressor $\boldsymbol{R}$ to a linear trend in the three cases. The error e_t follows a seasonal autoregressive AR(1)(1) models,

$$(1-\varphi B)\,(1-\Phi B^4)\,e_t = v_t \;\Rightarrow\; e_t = \varphi e_{t-1} + \Phi e_{t-4} - \varphi\Phi\, e_{t-5} + v_t\,,$$

with three sets of parameters: $\varphi = 0.80$ and $\Phi = 0.99$, $\varphi = 0.99$ and $\Phi = 0.80$, and $\varphi = 0.99$ and $\Phi = 0.99$.

The three sets of interpolations display a smooth transition from the seasonal pattern prevailing in 1996 to that in 2003. However, the interpolations with set of parameters $\varphi = 0.99$ and $\Phi = 0.80$ (diamonds "◇")

exhibit a shrinking seasonal amplitude for the years without quarterly benchmarks. This is attributable to the seasonal autoregressive parameter equal to 0.80 instead of 0.99 for the other two error models.

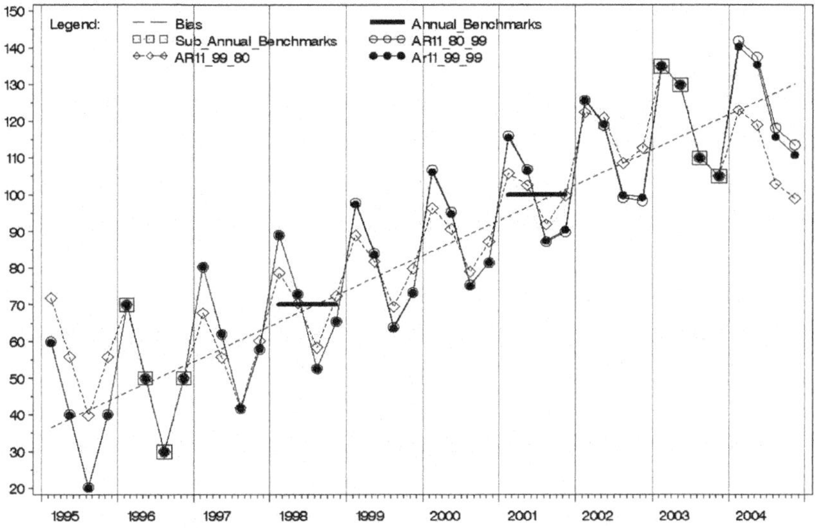

Fig. 7.9. Interpolating between annual and sub-annual benchmarks

Note that seasonal autoregressive (and other seasonal ARMA) processes can only be used in conjunction with sub-annual data which determine the realization of the seasonal pattern. Without sub-annual data, the resulting interpolations are most likely to display a spurious seasonal pattern. The sub-annual benchmarks need not be clustered in only a few years as in the example. However they must be supplied at regular intervals for *each* quarter (not necessarily in the same year).

7.5.2 Difference Stationary Models

The difference stationary model (Nelson and Plosser, 1982), applicable for interpolation and temporal distribution, consists of two equations:

$$\boldsymbol{a} = \boldsymbol{J}\boldsymbol{Z}\boldsymbol{\delta} + \boldsymbol{J}\boldsymbol{\eta} + \boldsymbol{\varepsilon}, \qquad E(\boldsymbol{\varepsilon}) = 0,\ E(\boldsymbol{\varepsilon}\boldsymbol{\varepsilon}') = \boldsymbol{V}_{\varepsilon}, \tag{7.12a}$$

$$\boldsymbol{D}\,\boldsymbol{\eta} = \boldsymbol{\upsilon}, \qquad E(\boldsymbol{\upsilon}) = 0,\ E(\boldsymbol{\upsilon}\boldsymbol{\upsilon}') = \boldsymbol{V}_{\upsilon}, \tag{7.12b}$$

where the $M\times 1$ vector $\boldsymbol{a}$ ($M \leq T$) contains annual and/or sub-annual treated as benchmarks. Matrix $\boldsymbol{J}$ of dimension $M\times T$ accounts for data which are irregularly spaced and have different durations. Some of the high frequency periods may be covered by quarterly data; some, by yearly data; and some not covered at all.

Eq. (7.12a) states that the temporal sums of the series $\boldsymbol{\theta} = \boldsymbol{Z\delta} + \boldsymbol{\eta}$ are equal to the potentially irregularly spaced data. Matrix $\boldsymbol{V}_{\varepsilon}$ is a diagonal matrix with positive elements tending to zero. Matrix $\boldsymbol{J}$ is a temporal sum operator fully described in section 3.6.1.

The $T\times1$ vector $\boldsymbol{\eta}$ denotes the stochastic part of series to be estimated, matrix $\boldsymbol{D}$ is a finite difference operator of dimension $N\times T$ ($N<T$), $\boldsymbol{V}_{\upsilon}$ is the autocovariance matrix of an AutoRegressive Moving Average (ARMA) process. Matrix $\boldsymbol{Z}$ of dimensions $T\times K$ contains intervention variables or related series.

Note that $\boldsymbol{Z}$ is implicitly subject to differencing. Indeed, if $M=T$, $\boldsymbol{J}=\boldsymbol{I}_T$ and letting $\boldsymbol{\varepsilon}\to\boldsymbol{0}$, Eq. (7.12a) may be written as $\boldsymbol{a} - \boldsymbol{Z\delta} = \boldsymbol{\eta}$. Substituting in Eq. (7.12b) yields $\boldsymbol{D}(\boldsymbol{a} - \boldsymbol{Z\delta}) = \boldsymbol{\upsilon}$ and $\boldsymbol{Da} = \boldsymbol{DZ\delta} + \boldsymbol{\upsilon}$.

In the context of state space modelling, Eq. (7.12a) can be viewed as the observation equation, relating the regressors to the data; and Eq. (7.12b) as the transition equation, which described the behaviour of some of the parameters though time. Eq. (7.12b) equation states that applying differences of a finite order to the stochastic part of the series makes it stationary in the mean and variance. The differenced series fluctuates about zero, according to an ARMA model.

In the cases of first order differences $(1-B)\,z_t \equiv z_t - z_{t-1}$ and of second order differences $(1-B)^2 z_t \equiv (1-2B+B^2)\,z_t \equiv z_t - 2z_{t-1} + z_{t-2}$, the difference operators are respectively

$$\underset{(T-1)\times T}{\boldsymbol{D}^1} = \begin{bmatrix} -1 & 1 & 0 & \cdots \\ 0 & -1 & 1 & \cdots \\ \vdots & \vdots & \vdots & \ddots \end{bmatrix}, \quad \underset{(T-2)\times T}{\boldsymbol{D}^2} = \begin{bmatrix} 1 & -2 & 1 & 0 & \cdots \\ 0 & 1 & -2 & 1 & \cdots \\ \vdots & \vdots & \vdots & \vdots & \ddots \end{bmatrix}, \qquad (7.13a)$$

In the case of seasonal quarterly differences of order 1 $(1-B^4)\,z_t \equiv z_t - z_{t-4}$, the operator is

$$\underset{(T-4)\times T}{\boldsymbol{D}_4^1} = \begin{bmatrix} -1 & 0 & 0 & 0 & 1 & 0 & 0 & \cdots \\ 0 & -1 & 0 & 0 & 0 & 1 & 0 & \cdots \\ \vdots & \vdots & \vdots & \vdots & \vdots & \vdots & \vdots & \ddots \end{bmatrix}. \qquad (7.13b)$$

In the case of combined regular and seasonal quarterly first differences $(1-B)(1-B^4)z_t \equiv (1-B-B^4+B^5)z_t \equiv z_t - z_{t-1} - z_{t-4} + z_{t-5}$, the operator is of dimension $(T-5)\times T$:

$$D = D^1 D_4^1 = \begin{bmatrix} 1 & -1 & 0 & 0 & -1 & 1 & 0 & \dots \\ 0 & 1 & -1 & 0 & 0 & -1 & 1 & \dots \\ \vdots & \vdots & \vdots & \vdots & \vdots & \vdots & \vdots & \ddots \end{bmatrix}. \tag{7.13c}$$

The use of seasonal differences requires a sufficient number of sub-annual benchmarks. For example if the series being interpolated is to be quarterly, at least four quarterly values are needed for different quarters. Otherwise, the fitted values will be spurious, i.e. purely artificial and unsupported by the data.

Note that at least one annual value is needed in the case of first differences (embodied in matrix $\boldsymbol{D}$); at least two, in the case of second differences. Otherwise the lack of degrees of freedom impedes the estimation of the model.

Model (7.12) can be written as a single equation

$$\begin{bmatrix} \boldsymbol{a} \\ \boldsymbol{0} \end{bmatrix} = \begin{bmatrix} \boldsymbol{JZ} & \boldsymbol{J} \\ \boldsymbol{0} & \boldsymbol{D} \end{bmatrix} \begin{bmatrix} \boldsymbol{\delta} \\ \boldsymbol{\eta} \end{bmatrix} + \begin{bmatrix} \boldsymbol{\varepsilon} \\ -\boldsymbol{\upsilon} \end{bmatrix}, \tag{7.14a}$$

or

$$\boldsymbol{y} = \boldsymbol{X\alpha} + \boldsymbol{u}, \quad E(\boldsymbol{uu}') = \boldsymbol{V}_u = \begin{bmatrix} \boldsymbol{V}_\varepsilon & \boldsymbol{0} \\ \boldsymbol{0} & \boldsymbol{V}_\upsilon \end{bmatrix}. \tag{7.14b}$$

The generalized least square solution of (7.14b) is

$$\hat{\boldsymbol{\alpha}} = (\boldsymbol{X}'\boldsymbol{V}_u^{-1}\boldsymbol{X})^{-1}\boldsymbol{X}'\boldsymbol{V}_u^{-1}\boldsymbol{y} \tag{7.15a}$$

$$var[\hat{\boldsymbol{\alpha}}] = (\boldsymbol{X}'\boldsymbol{V}_u^{-1}\boldsymbol{X})^{-1}. \tag{7.15b}$$

The estimator of the high time frequency values and their covariance matrix are given by

$$\hat{\boldsymbol{\theta}} = \boldsymbol{X}^\dagger \hat{\boldsymbol{\alpha}}, \tag{7.15c}$$

$$var[\hat{\boldsymbol{\theta}}] = \boldsymbol{X}^\dagger \, var[\hat{\boldsymbol{\alpha}}] \, \boldsymbol{X}^{\dagger\prime}, \tag{7.15d}$$

where $\boldsymbol{X}^\dagger = [\boldsymbol{Z} \;\; \boldsymbol{I}_T]$.

This solution requires that the low frequency (annual) values be non-binding. In other words, the diagonal elements of $\boldsymbol{V}_\varepsilon$ must be positive definite. However, it is possible to specify the benchmarks as quasi-binding, by setting the diagonal elements to a small number, e.g. 0.001; the resulting interpolations or temporal distributions satisfy the benchmarks in practice. If this is not satisfactory, the series can be benchmarked using formulae of Eq. (3.21), where $\boldsymbol{s}$ and $\boldsymbol{V}_e$ are respectively set equal to $\hat{\boldsymbol{\theta}}$ and $var[\hat{\boldsymbol{\theta}}]$ from Eq. (7.15).

In the absence of regressor $\boldsymbol{Z}$, matrix $\boldsymbol{X}$ of Eq. (7.14b) becomes $[\boldsymbol{J}'\ \boldsymbol{D}']'$ and $\boldsymbol{a} = \boldsymbol{\eta}$, and Eqs. (7.15a) and (7.15b) directly provides the estimates of the ARIMA interpolations, i.e. $\hat{\boldsymbol{\theta}} = \hat{\boldsymbol{a}} = \hat{\boldsymbol{\eta}}$.

Appendix A shows that the solution (7.15) is the Minimum Variance Linear Unbiased Estimator of the series.

When differencing is needed, the most straightforward way to obtain interpolated or distributed estimates which exactly satisfy the low frequency benchmarks is to minimize an objective function subject to constraints, as suggested by Boot et al. (1967) and by Cohen et al. (1971). The least square estimate of $\boldsymbol{\theta}$ minimizes the constrained objective function:

$$\begin{aligned} f(\boldsymbol{\theta},\boldsymbol{\gamma}) &= \boldsymbol{\theta}'\boldsymbol{D}'\boldsymbol{V}_\upsilon^{-1}\boldsymbol{D}\boldsymbol{\theta} - 2\boldsymbol{\gamma}'(\boldsymbol{a} - \boldsymbol{J}\boldsymbol{\theta}) \\ &= \boldsymbol{\theta}'\boldsymbol{D}'\boldsymbol{V}_\upsilon^{-1}\boldsymbol{D}\boldsymbol{\theta} - 2\boldsymbol{\gamma}'\boldsymbol{a} + 2\boldsymbol{\gamma}'\boldsymbol{J}\boldsymbol{\theta}\,, \end{aligned} \tag{7.16}$$

where $\boldsymbol{\gamma}$ denotes the Lagrange multipliers associated with the linear constraints $\boldsymbol{a} - \boldsymbol{J}\boldsymbol{\theta} = \boldsymbol{0}$. The necessary conditions for a minimum require that the derivative of the function with respect to the parameters be equal to zero:

$$\partial(f(\boldsymbol{\theta},\boldsymbol{\gamma}))/\partial\boldsymbol{\theta} = 2\boldsymbol{D}'\boldsymbol{V}_\upsilon^{-1}\boldsymbol{D}\boldsymbol{\theta} + 2\boldsymbol{J}'\boldsymbol{\gamma} = \boldsymbol{0} \tag{7.17a}$$

$$\partial(f(\boldsymbol{\theta},\boldsymbol{\gamma}))/\partial\boldsymbol{\gamma} = 2\boldsymbol{J}\boldsymbol{\theta} - 2\boldsymbol{a} = \boldsymbol{0}. \tag{7.17b}$$

These conditions may be expressed as

$$\begin{bmatrix} \boldsymbol{D}'\boldsymbol{V}_\upsilon^{-1}\boldsymbol{D} & \boldsymbol{J}' \\ \boldsymbol{J} & \boldsymbol{0} \end{bmatrix} \begin{bmatrix} \boldsymbol{\theta} \\ \boldsymbol{\gamma} \end{bmatrix} = \begin{bmatrix} \boldsymbol{0} \\ \boldsymbol{a} \end{bmatrix},$$

which can be solved linearly as:

$$\begin{bmatrix} \hat{\boldsymbol{\theta}} \\ \hat{\boldsymbol{\gamma}} \end{bmatrix} = \begin{bmatrix} \boldsymbol{D}'\boldsymbol{V}_{\upsilon}^{-1}\boldsymbol{D} & \boldsymbol{J}' \\ \boldsymbol{J} & \boldsymbol{0} \end{bmatrix}^{-1} \begin{bmatrix} \boldsymbol{0} \\ \boldsymbol{a} \end{bmatrix} = \begin{bmatrix} \boldsymbol{W}_{AA} & \boldsymbol{W}_{AB} \\ \boldsymbol{W}_{BA} & \boldsymbol{W}_{BB} \end{bmatrix} \begin{bmatrix} \boldsymbol{0} \\ \boldsymbol{a} \end{bmatrix}. \tag{7.18}$$

This in turn implies

$$\hat{\boldsymbol{\theta}} = \boldsymbol{W}_{AB}\,\boldsymbol{a}, \tag{7.19a}$$

because $\boldsymbol{W}_{AA}$ multiplies $\boldsymbol{0}$. The partition $\boldsymbol{W}_{AA}$ on the other hand is the covariance matrix of $\hat{\boldsymbol{\theta}}$.

$$var[\hat{\boldsymbol{\theta}}] = \boldsymbol{W}_{AA}. \tag{7.19b}$$

Indeed, solution (7.15) converges to solution (7.19), in terms of expected value and covariance matrix, as the diagonal matrix $\boldsymbol{V}_{\varepsilon}$ tend to $\boldsymbol{0}$.

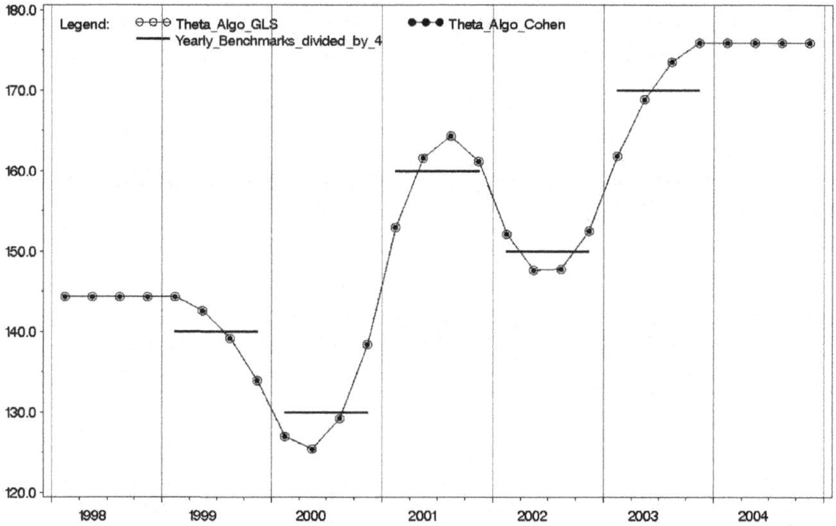

Fig. 7.10. Distributed quarterly values between yearly benchmarks under first differences, using solution (7.15) (circles) and (7.19) (dots)

Fig. 7.10 depicts the temporal distributions obtained under first differences using the generalized least square solution (7.15) and the Cohen et al. (1971) solution (7.19). The two curves are virtually identical. First differences cause the interpolations or distribution to be as flat (constant) as possible, under the constraint that they must satisfy the annual benchmarks. This is achieved at the beginning and end of the series because there are no benchmark to satisfy. Similarly, under second differences the extrapolations would behave linearly at the ends.

7.5.3 The Stram and Wei Approach

Other approaches have been proposed to generate ARIMA interpolations and distributions, namely that of Stram and Wei (1986) and Wei and Stram (1990). In the absence of sub-annual data pertaining to the target variable, Wei and Stram (1990) suggest that the sub-annual ("high frequency") ARIMA model should be derived from the observed annual ("lower frequency) error model. This entail using relationships linking the model of the annual sums $z_t = (1+B+B^2+B^{m-1})\,x_t$ to that of sub-annual series x_t to be interpolated.

If the "true" sub-annual values of x_t follow a (p, d, q) model, the annual sums z_t follow a (p^*, d^*, q^*). Symbols p and p^* denote the respective number of autoregressive parameters; d and d^*, the order of the differencing; and q and q^*, the number of moving average parameters; and m, is the number of seasons. The value of p^* is equal to p; that of d^*, to d; and that of q^*, to the integer part of $(p+d+1-(q-p-d-1)/m)$. For example if the model of x_t is a (1,1,1), that of z_t is a (1,1,3) with $m>2$.

Usually the model of z_t is less parsimonious than the model of x_t. This raises the issue of the availability of sufficient data to identify and properly estimate the annual model.

Rossana and Seater (1995) examined the temporal aggregation problem from an empirical point of view and concluded (p. 450):

> *"Temporal aggregation causes a severe loss of information about the time series processes driving many economic variables.* [...] *Temporal aggregation systematically alters the time series properties of the data so that even the supra-annual variation in the underlying data is totally lost. Moreover, the aggregated data have excessive long-term persistence."*

The temporal aggregation in question may apply to the underlying observed or un-observed monthly or quarterly series. Spectral analysis (e.g. Koopman 1974) is useful to better understand these authors conclusions. Fig. 7.11 displays the gain function of yearly sums of monthly series. The weights of yearly sums are divided by 12 to obtain gain functions in the range 0 to 100%. The yearly sum operator reduces the amplitude of shorter business cycles in comparison to the longer ones. It can be seen, for example, that the yearly sum operator preserves only 93.6% of the 60-months business cycles (first square at frequency 6/360), only 82.8% of the 36-month business cycles (second square at 10/360). As a result, part of these business cycle movements cannot be retrieved from yearly flow data, even if the theoretical monthly model is known.

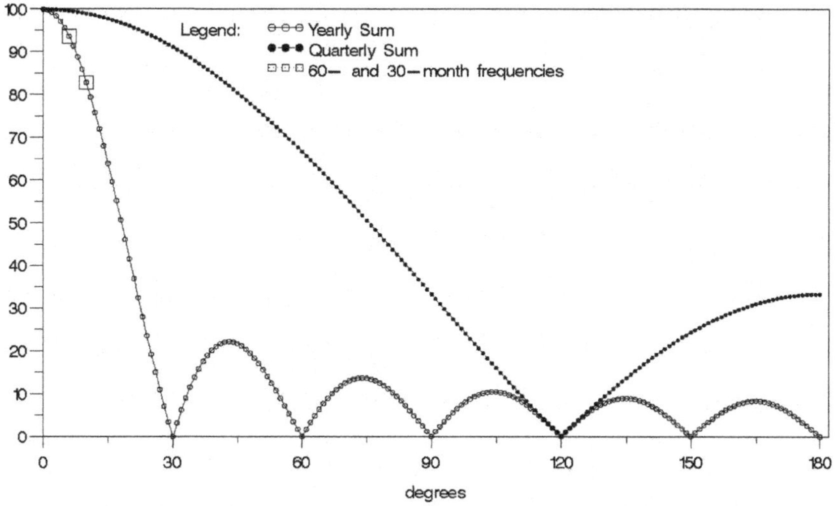

Fig. 7.11. Gain functions of the yearly sum and of the quarterly sum operators of monthly series

The yearly sum operator completely eliminate the seasonal cycles (frequencies 30, 60, 90, 120, 150 and 180 over 360). All the seasonal fluctuations are irretrievably lost.

The above authors add:

> *"Quarterly data do not seem to suffer badly from temporal aggregation distortion, nor are they subject to the construction problems affecting monthly data. They therefore may be the optimal data for econometric analysis."*

This is confirmed by the gain function of the quarterly sum operator in Fig. 7.11. The operator largely preserves the business-cycles, reduces the seasonal frequencies and completely eliminates the 3-month seasonal cycle (120/360). This implies monthly seasonality cannot be fully recovered from quarterly flow data.

7.6 Combining Sub-Annual and Annual Forecasts

Model (7.4) can also be used to optimally combine "sub-annual" forecasts with "annual" forecasts (see among others Cholette 1982, de Alba 1988, Guerrero 1989 and 1990, Pankratz 1989, Trabelsi and Hillmer 1989). Vector $\boldsymbol{s}$ of Eq. (7.4a) contains the sub-annual forecasts with covariance matrix $\boldsymbol{V}_e$; vector $\boldsymbol{s}$ may also contain historical values (with variance 0). Vector $\boldsymbol{a}$ of Eq. (7.31b) contains the annual forecasts with covariance matrix $\boldsymbol{V}_\varepsilon$. A regressor $\boldsymbol{R}$ may be required to capture a mild trend between the sub-annual and annual forecasts or trading-day variations. The approach can also be used to impose cross-sectional aggregation constraints to systems of series, as discussed in Chapter 11.

This variant of model (7.4) is useful in situations like the following.
(a) Monthly unemployment rate data are available up to December 2001 and hover around a yearly average of 7.0%. Twenty four months of forecasts are produced, using ARIMA modelling for example.
(b) Two yearly forecasts for the 2002 and 2003 generated from a yearly econometric model are available; these forecasts predict unemployment to be 9.5% and 8.0% respectively .
(c) In order to anticipate the benefits paid each month to the unemployed under that scenario, government needs monthly forecasts of the number of unemployed persons from January 2002 to December 2003, which take into account or conform to the two yearly forecasts.

It has been shown that monthly and quarterly ARIMA models are highly successful at forecasting seasonal patterns, but cannot meaningfully forecast turning points. Conversely, yearly econometric models can predict turning points, in a relatively successful manner, but can obviously not forecast seasonal patterns. Benchmarking the sub-annual ARIMA forecasts to the yearly forecasts combines the best features of the two kinds of predictions.

The approach of combining sub-annual and annual forecasts can also be used to assess the current sub-annual behaviour of socio-economic variables in regard to target values pertaining to future years. For example, consider the atmospheric carbon dioxide (CO_2) concentrations (ppmv) derived from in *situ* air samples collected at Mauna Loa Observatory (Hawaii) depicted in

Fig. 7.12.[2] In order to unclutter the figure, we took the quarterly averages of the monthly readings taken on the 15-th day of each month. Let us assume that governments agree to stabilize the CO_2 at the average yearly level of 2003 in 2012. The target value is also depicted in the figure, with the unconstrained and constrained ARIMA extrapolations.

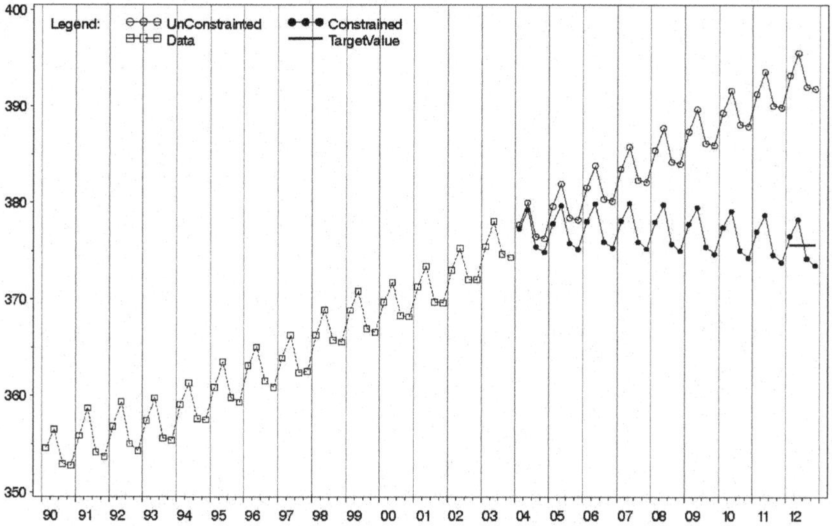

Fig. 7.12. Un-constrained and constrained ARIMA extrapolations of carbon dioxide concentrations at Hawaii

The ARIMA model was obtained in the following manner. A seasonal ARIMA model $(0,1,0)(0,1,1)_4$ was identified and fitted to the data from 1990 to 2003:

$$(1-B)(1-B^4)z_t = e_t,$$

$$e_t = (1-\psi B^4)v_t,\ E(v_t)=0,\ E(v_t^2)=\sigma_v^2\ . \qquad (7.20a)$$

The estimated values of the moving average parameter ψ is 0.774 with a t-value equal to 6.02. The variance of the innovations is $\sigma_v^2 = 0.1095$, $\sigma_e^2 = 0.1095\times(1+0.7447^2)=0.1702$. The Porte-manteau Ljung-Box (1978) test does not reject the null hypothesis of non-correlated residuals with a probability values of 0.4812 at lag 12.

The ARIMA model can be written as

[2] These data are publically available on the internet site http://cdiac.esd.ornl.gov /trends/co2/contents.htm.

$$z_t - z_{t-4} = z_{t-1} - z_{t-5} + e_t, \; t=1,...,92,$$
$$e_t = v_t - 0.7447 v_{t-5}, \quad E(v_t)=0, \; E(v_t^2) = 0.1095, \qquad (7.20b)$$

where z_t stands for the quarterly data and the error follows a seasonal moving average model. The model states that same-quarter changes tend to repeat for quarter to quarter; or put differently quarter-to-quarter changes tend to repeat from year to year $z_t - z_{t-1} = z_{t-4} - z_{t-5} + e_t$. In this case the error e_t is very small compared to the data.

In fact, ARIMA modelling of seasonal time series typically consists of identifying the correlations structure left in e_t, after regular and seasonal differencing. This means that we are modelling change in annual change.

The un-constrained extrapolations and their covariance matrix are produced by model (7.12) with solution (7.15) where the $\boldsymbol{Z\delta}$ is absent. As a result the model reduces to

$$\begin{bmatrix} \boldsymbol{a} \\ \boldsymbol{0} \end{bmatrix} = \begin{bmatrix} \boldsymbol{J} \\ \boldsymbol{D} \end{bmatrix} \boldsymbol{\theta} + \begin{bmatrix} \boldsymbol{\varepsilon} \\ -\boldsymbol{e} \end{bmatrix} \quad \text{or} \quad \boldsymbol{y} = \boldsymbol{X\theta} + \boldsymbol{u}, \; E(\boldsymbol{uu}') = \boldsymbol{V}_u = \begin{bmatrix} \boldsymbol{V}_\varepsilon & \boldsymbol{0} \\ \boldsymbol{0} & \boldsymbol{V}_e \end{bmatrix}. \qquad (7.21)$$

In model (7.21), vector $\boldsymbol{\theta}$ of dimension 92×1 (T=92) contains the interpolations and extrapolations to be estimated. The regular and monthly seasonal difference operator $\boldsymbol{D}$ of (7.13c) has dimension 87×92. The 87×87 covariance matrix is equal to $\boldsymbol{V}_e = \sigma_e^2 \boldsymbol{\Omega}$, where $\boldsymbol{\Omega}$ denotes the autocorrelation matrix of the seasonal moving average model $e_t = v_t - 0.7447 v_{t-5}$. Since the innovations v_t have variance equal to 0.1095, σ_e^2 is equal to 0.1095×(1+0.7447²) =0.1702.

The 56×1 vector $\boldsymbol{a}$ contains the quarterly data from 1990 to 2003 specified as quarterly benchmarks. As a result matrix $\boldsymbol{J}$ is as follows,

$$\boldsymbol{J}_{56\times 92} = \begin{bmatrix} \boldsymbol{I}_{56} & \boldsymbol{0}_{56\times 36} \end{bmatrix},$$

where the first 56 columns of $\boldsymbol{J}$ pertains to the 56 data points from the first quarter of 1990 to the fourth quarter of 2003 and the last 36 columns pertain to the un-constrained extrapolations from the first quarter of 2004 to the fourth quarter of 2012. Each row of $\boldsymbol{J}$ pertains to one benchmark. Matrix $\boldsymbol{V}_\varepsilon$ is diagonal with element equal to 0.001.

In the case of (7.21), the solution (7.15) reduces to $\hat{\boldsymbol{\theta}} = (\boldsymbol{X}'\boldsymbol{V}_u^{-1}\boldsymbol{X})^{-1}$ $\boldsymbol{X}'\boldsymbol{V}_u^{-1}\boldsymbol{y}$ with $var[\hat{\boldsymbol{\theta}}] = (\boldsymbol{X}'\boldsymbol{V}_u^{-1}\boldsymbol{X})^{-1}$, where $\boldsymbol{X}$ is defined in (7.21) . This solution yields the unconstrained forecasts displayed in Fig. 7.12 and their covariance matrix.

The constrained extrapolations displayed in Figs 7.12 and 7.13 are obtained by applying model (7.4) with the benchmarking solution (3.21), $\hat{\boldsymbol{\theta}} = \boldsymbol{s} + \boldsymbol{V}_e\boldsymbol{J}'\boldsymbol{V}_d^{-1}\ [\boldsymbol{a} - \boldsymbol{J}\boldsymbol{s}]$ and $var[\hat{\boldsymbol{\theta}}] = \boldsymbol{V}_e - \boldsymbol{V}_e\boldsymbol{J}'\boldsymbol{V}_d^{-1}\boldsymbol{J}\boldsymbol{V}_e$, where $\boldsymbol{V}_e$ is now equal to $var[\hat{\boldsymbol{\theta}}]$ of (7.15b), and $\boldsymbol{a}$ contains the 2012 target average value, and matrix $\boldsymbol{J} = \left[\boldsymbol{0}_{1\times 88}\ \ 0.25\boldsymbol{1}_{1\times 4}\right]$. The resulting covariance matrix provides the confidence intervals of the constrained extrapolations displayed in Fig 7.13.

It is possible to obtain the same constrained extrapolations directly from model (7.15) with solution (7.15), by also including the target benchmark value of 2012 in vector $\boldsymbol{a}$ of Eq. (7.12b). The resulting 57×1 vector $\boldsymbol{a}$ contained the 56 quarterly data points and the target value 2012 in the last (57-th) row. As a result matrix $\boldsymbol{J}$ is as follows

$$\boldsymbol{J} = \begin{bmatrix} \boldsymbol{I}_{56} & \boldsymbol{0}_{56\times 32} & \boldsymbol{0}_{56\times 4} \\ \boldsymbol{0}_{1\times 56} & \boldsymbol{0}_{1\times 32} & 0.25\boldsymbol{1}_{1\times 4} \end{bmatrix},$$

where the first 56 columns of $\boldsymbol{J}$ pertains to the 56 quarterly data points and the last four columns pertain to the four quarters of 2012. The last (57-th) row pertains to the target value of 2012. Again, the target value is specified as a sum benchmark. Matrix $\boldsymbol{V}_\varepsilon$ is diagonal with diagonal element equal to 0.001. Solution (7.15) yields the constrained forecasts and their covariance matrix.

Fig. 7.13 focuses on the extrapolations beyond 2003. The un-constrained and the constrained extrapolations are accompanied by their 95% confidence intervals, under the Normality assumption. The target value is clearly outside the confidence intervals of the un-constrained ARIMA extrapolations (light curves with circles "o"). This suggests that in the absence of drastic action on the part of governments the CO_2 concentration will substantially exceed (miss) the target 2012 stabilization value.

Note the effect of the 2012 benchmark on the confidence intervals of the constrained extrapolations. The confidence intervals (light curves with dots "•") widen from 2004 to 2008 and begin to shrink in 2009 as the 2012 benchmark approaches. The confidence intervals of the un-constrained interpolations on the other hand widen much faster, without ever shrinking.

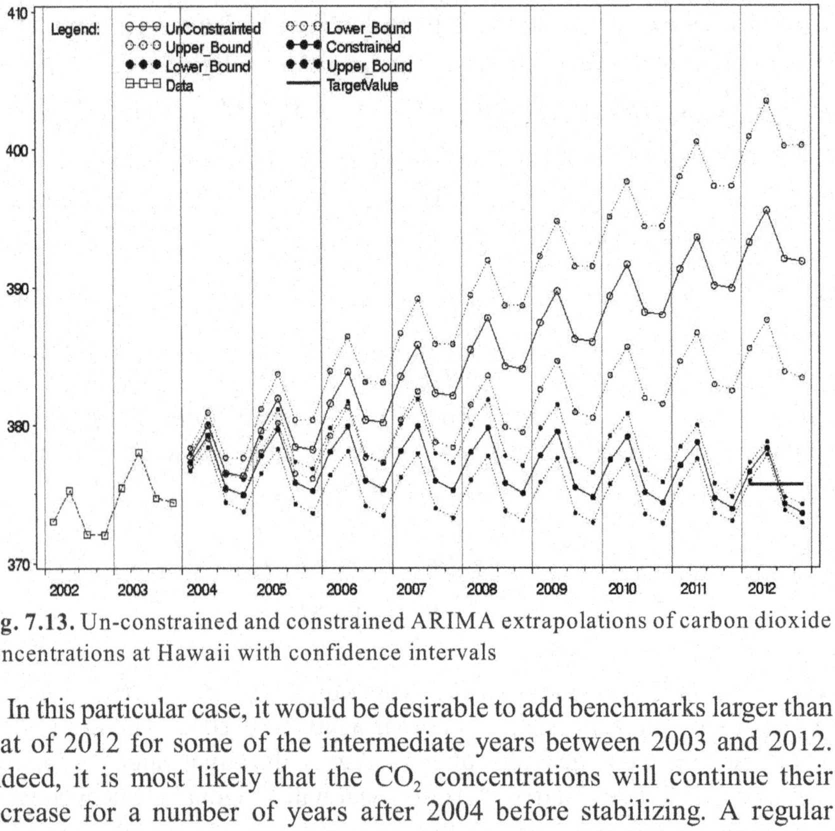

Fig. 7.13. Un-constrained and constrained ARIMA extrapolations of carbon dioxide concentrations at Hawaii with confidence intervals

In this particular case, it would be desirable to add benchmarks larger than that of 2012 for some of the intermediate years between 2003 and 2012. Indeed, it is most likely that the CO_2 concentrations will continue their increase for a number of years after 2004 before stabilizing. A regular moving average positive parameter would have accomplished that. However, we could not significantly estimate such a parameter from the data. Recall that ARIMA models describe the *temporal* behaviour of time series, but does not explain it in terms of causality. Nevertheless the tool remains very useful to anticipate and monitor the course of the CO_2 concentrations before reaching the 2012 target.

8 Signal Extraction and Benchmarking

8.1 Introduction

This chapter provides an overview of other benchmarking methods which are based on signal extraction. The latter is done by means of ARIMA models or of structural time series models cast in state space form. These methods have not been implemented for use in statistical agencies and are seldom applied to real life cases. Some of their shortcomings are associated with the lack of fit for dis-aggregated series, their high computational complexity and the large number of parameters to be estimated from relatively short series especially for structural models. The methods are still in a developmental stage.

The reader should be well familiarized with ARIMA and structural models identification and estimation for a full understanding, otherwise the chapter can be skipped without loss of continuity. We keep the notation used by their authors with few minor changes.

In time series analysis, it is often assumed that the observations y_t are contaminated with errors e_t such that

$$y_t = \eta_t + e_t, \quad t=1,...,n, \tag{8.1}$$

where η_t denotes the true un-observable values. Following from electrical engineering, η_t is called the "signal" and e_t the "noise".

The signal extraction problem is to find the "best" estimates of η_t where best is defined according to a given criterion, usually that of minimizing the mean square error.

Signal extraction can be made by means of parametric models or non-parametric procedures. The latter has a long standing and was used by actuaries at the beginning of the 1900's. The main assumption in non-parametric procedures is that η_t is a smooth function of time. Different types of smoothers are used depending on the series under question. The most common smoothers are the cubic splines originally applied by Whittaker (1923) and Whittaker and Robinson (1924) to smooth mortality tables. Other smoother are moving averages and high order kernels used in the context of seasonal adjustment and form the basis of methods such as Census X-11 (Shiskin et al. 1967), X-11-ARIMA (Dagum 1980 1988), X-12-ARIMA (Findley et al. 1998), STL (Cleveland et al. 1990).

Non-parametric signal extraction has also been very much applied to estimate the trend (non-stationary mean) of time series (see among others Henderson 1916; Macaulay 1931; Gray and Thomson 1966; Dagum and Luati 2000).

On the other hand, signal extraction by means of explicit models arrived much later. Under the assumption that the entire realization of y_t is observed from $-\infty$ to ∞ and η_t and e_t are both mutually independent and stationary,[1] Kolmogorov (1939, 1941) and Wiener (1949) independently proved that the minimum mean square estimator of the signal η_t is the conditional mean given the observations y_t, that is $\hat{\eta}_t = E(\eta_t \,|\, y_t, y_{t-1}, \ldots)$. This fundamental result was extended by several authors who provided approximative solutions to the non-stationary signal extraction, particularly Hannan (1967), Sobel (1967) and Cleveland and Tiao (1976). Finally, Bell (1984) provided exact solutions for the conditional mean and conditonal variance of vector $\boldsymbol{\eta}$ when non-stationarity can be removed by applying differences of a finite order. This author used two alternatives regarding the generation of vectors $\boldsymbol{y}$, $\boldsymbol{\eta}$ and $\boldsymbol{e}$.

Model-based signal extraction was also used in the context of seasonal adjustment where the signal η_t is assumed to follow an ARIMA model of the Box and Jenkins (1970) type, plus a regression model for the deterministic variations (see e.g. Burman 1980, Gómez and Maravall 1996, Findley et al. 1998). The latter is applied to estimate deterministic components, such as trading-day variations or moving-holiday effects and outliers. Kitagawa and Gersch (1984) and Koopman et al. (1995) also used signal extraction for seasonal adjustment where the signal η_t is assumed to follow a structural time series component model (Harvey 1989) cast in state-space representation.

Signal extraction, parametric and non-parametric, is also widely applied for forecasting purposes.

We now discuss model-based signal extraction in the context of benchmarking. It should be kept in mind, that signal extraction is a data adjustment independent of benchmarking. When modelling is feasible, it provides a genuine covariance function to be used in the benchmarking

[1] The mean and variance are constant through time and the auto-covariance function depends only on the time lag.

process, which is a major advantage. ARMA models jointly used with standard deviations should in principle reflect the design of sample surveys. Hillmer and Trabelsi (1987) illustrated their ARIMA signal extraction and benchmarking method using an AR(1) model with parameter value equal to 0.80. Trabelsi and Hillmer (1990) used an ARMA (2,1)(1,0) model to account for the composite estimation and sample design of the U.S. Retail Trade series. Bell and Wilcox (1993) used a similar model for the same purpose. Binder and Dick (1989) proposed an ARMA(3,6) model to account for the sample rotation design of the Canadian Labour Force survey.

The ARMA modelling of the survey error on the basis of the sample rotation design has not been pursued further because of the modelling difficulty and the fact that the sample design changes occasionally through time.

8.2 ARIMA Model-Based Signal Extraction and Benchmarking: The Hillmer and Trabelsi Method

Hillmer and Trabelsi (1987) wrote a pioneering paper on signal extraction and benchmarking on the basis of ARIMA modelling. These authors assumed a purely linear[2] ARIMA model for the non-stationary time series with zero mean and unbiased survey error. Their main contribution was to demonstrate that the extracted signal $\hat{\boldsymbol{\eta}}_0$ is the best estimate of $\boldsymbol{\eta}$ given the vector of observation $\boldsymbol{y}$ alone, and that the benchmarked estimates $\hat{\boldsymbol{\eta}}$ are the best unbiased estimates of $\boldsymbol{\eta}$ given $\boldsymbol{y}$ and the benchmarks $\boldsymbol{z}$. For implementation, their procedure requires knowledge of the covariance matrices $\boldsymbol{\Sigma}_\eta$ and $\boldsymbol{\Sigma}_e$ and the unconditional mean $\boldsymbol{\mu} = E(\boldsymbol{\eta})$. We shall next present a short description of this method.

Let us assume

$$y_t = \eta_t + e_t, \; t=1,...,n, \tag{8.2}$$

where y_t denote the observations, η_t the true values and e_t an unbiased sample survey error. Both η_t and e_t follow a linear stochastic process,

$$\varphi_\eta(B)(\eta_t - \mu) = \theta_\eta(B)\, b_t\,, \tag{8.3a}$$

and

$$\varphi_e(B) e_t = \theta_e(B)\, c_t\,, \tag{8.3b}$$

[2] purely linear in the sense that deterministic components like trading-day variations, moving holidays, outliers, are assumed absent.

where (8.3a) represents an ARIMA model and (8.3b) an ARMA one. The symbol $\varphi_\eta(B)$is the generalized autoregressive operator representing a polynomial in B (the backshift operator) such that the roots are on or outside the unit circle; $\theta_\eta(B)$is a moving average operator representing a polynomial in B such that its roots are all outside the unit circle and different from those of $\varphi_\eta(B)$. Finally $b_t \sim N(0,\sigma_b^2)$ and $c_t \sim N(0,\sigma_c^2)$and are mutually independent.

In (8.3b), the sample survey error e_t is assumed to follow an ARMA process, hence the roots of both polyninomials $\varphi_e(B)$ and $\theta_e(B)$have their roots outside the unit circle and c_t is Gaussian white noise. Since b_t and c_t are assumed mutually independent, the ARIMA model (8.2) is

$$\varphi_y(B)(y_t - \mu) = \theta_y(B)\, a_t,\ t=1,...,n, \tag{8.4}$$

where $\varphi_y(B) = \varphi_\eta(B)\,\varphi_e(B)$ and $\theta(B)$can be obtained from (8.3a) and (8.3b).

Model (8.2) can be written in matrix form as

$$\boldsymbol{y} = \boldsymbol{\eta} + \boldsymbol{e}, \tag{8.5}$$

where $\boldsymbol{y} = [y_1 ... y_n]'$, $\boldsymbol{\eta} = [\eta_1 ... \eta_n]'$ and $\boldsymbol{e} = [e_1 ... e_n]'$.

The estimates of $\boldsymbol{\eta}$ are linear function of $\boldsymbol{y}$ and given the assumption of normality, the minimum mean square error of $\hat{\boldsymbol{\eta}}$ depend only on the first and second moments. The series is assumed multivariate normal $\boldsymbol{\eta} \sim N(\boldsymbol{\mu}, \Sigma_\eta)$ and similarly for the error , $\boldsymbol{e} \sim N(\boldsymbol{0}, \Sigma_e)$.

Hence, the minimum mean square error estimator of $\boldsymbol{\eta}$, given $\boldsymbol{y}$ is

$$E(\boldsymbol{\eta} \mid \boldsymbol{y}) = \hat{\boldsymbol{\eta}}_0 = (\boldsymbol{\Sigma}_e^{-1} + \boldsymbol{\Sigma}_\eta^{-1})^{-1}(\boldsymbol{\Sigma}_e^{-1}\boldsymbol{y} + \boldsymbol{\Sigma}_\eta^{-1}\boldsymbol{\mu}) \tag{8.6a}$$

with

$$var[\boldsymbol{\eta} \mid \boldsymbol{y}] = \boldsymbol{\Omega}_0 = (\boldsymbol{\Sigma}_e^{-1} + \boldsymbol{\Sigma}_\eta^{-1})^{-1}. \tag{8.6b}$$

Equations (8.6) show that the best estimate of η given $\boldsymbol{y}$ depends on knowing the first and second moments of the unobserved components η and $\boldsymbol{e}$. However, in some occasions, the variances and covariances of $\boldsymbol{e}$ can be estimated from detailed information on the sample survey. Since $\boldsymbol{y}$ is observable, an ARIMA model can be built for it and used to estimate $\boldsymbol{\Sigma}_y = \boldsymbol{\Sigma}_\eta + \boldsymbol{\Sigma}_e$ and $\boldsymbol{\mu}$.

Hillmer and Trabelsi add to Eq. (8.5) that of the benchmarks

$$\boldsymbol{z} = \boldsymbol{L}\boldsymbol{\eta} + \boldsymbol{\varepsilon}, \tag{8.7}$$

where $\boldsymbol{z}$ is an $m \times 1$ vector of benchmarks, $\boldsymbol{L}$ is an $m \times n$ matrix of zeroes and ones equivalent to matrix $\boldsymbol{J}$ discussed in section 3.6.1 and $\boldsymbol{\varepsilon}$ is an $m \times 1$ error vector distributed as $N(\boldsymbol{0}, \boldsymbol{\Sigma}_\varepsilon)$, assumed independent of $\boldsymbol{\eta}$ and $\boldsymbol{e}$. The benchmarks are binding when $\boldsymbol{\Sigma}_\varepsilon$ is the null matrix.

Each row of (8.7) associates a benchmark z_m to a set of η_ts. For example, if the second benchmark z_2 is binding (ε_2=0), and

$$z_2 = \Sigma_{n=13}^{24} \eta_n \tag{8.8}$$

measures the sum of the η_ts over periods 13 to 24 (e.g. the second year).

Combining equations (8.5) and (8.7) we obtain

$$\boldsymbol{\tau} = \boldsymbol{X}\boldsymbol{\eta} + \boldsymbol{u}, \tag{8.9}$$

where $\boldsymbol{\tau} = [\boldsymbol{y}' \ \boldsymbol{z}']'$, $\boldsymbol{u} = [\boldsymbol{e}' \ \boldsymbol{\varepsilon}']'$, $\boldsymbol{X} = [\boldsymbol{I} \ \boldsymbol{L}']'$ and $\boldsymbol{I}$ denotes the identity matrix. Under the assumption that $\boldsymbol{\eta} \sim N(\boldsymbol{\mu}, \Sigma_\eta)$ and $\boldsymbol{u} \sim N(\boldsymbol{0}, \Sigma_u)$, the conditional expectation of $\boldsymbol{\eta}$ given $\boldsymbol{\tau}$ is

$$E(\boldsymbol{\eta} \,|\, \boldsymbol{\tau}) = \hat{\boldsymbol{\eta}} = (\boldsymbol{\Sigma}_\eta^{-1} + \boldsymbol{X}'\boldsymbol{\Sigma}_u^{-1}\boldsymbol{X})^{-1} (\boldsymbol{X}'\boldsymbol{\Sigma}_u^{-1}\boldsymbol{\tau} + \boldsymbol{\Sigma}_\eta^{-1}\boldsymbol{\mu}) \tag{8.10}$$

$$var[\boldsymbol{\eta} \,|\, \boldsymbol{\tau}] = \boldsymbol{\Omega} = (\boldsymbol{\Sigma}_\eta^{-1} + \boldsymbol{X}'\boldsymbol{\Sigma}_u^{-1}\boldsymbol{X})^{-1}. \tag{8.11}$$

It is useful to write the benchmarked estimates $\hat{\boldsymbol{\eta}}$ in terms of the signal extraction estimates $\hat{\boldsymbol{\eta}}_0$ and the corrections $\hat{\boldsymbol{\eta}}_c$ brought about by the benchmarks as follows.

$$\hat{\eta}_0 = E(\eta \mid y) = \Omega_0 (\Sigma_e^{-1} y + \Sigma_\eta^{-1} \mu) \tag{8.12}$$

$$\hat{\eta}_c = E(\eta \mid z) = \Omega_0 L' (L \Omega_0 L' + \Sigma_\varepsilon)^{-1} (z - L\hat{\eta}_0), \tag{8.13}$$

where

$$\Omega_0 = var[\eta \mid y] = (\Sigma_e^{-1} + \Sigma_\eta^{-1})^{-1}. \tag{8.14}$$

The final benchmarked series and its covariance matrix are then

$$\hat{\eta} = \hat{\eta}_0 + \hat{\eta}_c = \hat{\eta}_0 + \Omega_0 L' (L \Omega_0 L' + \Sigma_\varepsilon)^{-1} (z - L\hat{\eta}_0) \, , \tag{8.15a}$$

$$\Omega = var(\hat{\eta} - \eta) = (\Omega_0^{-1} + L' \Sigma_\varepsilon^{-1} L)^{-1} = (\Sigma_e^{-1} + \Sigma_\eta^{-1} + L' \Sigma_\varepsilon^{-1} L)^{-1}$$

$$= \Omega_0 L' (L \Omega_0 L' + \Sigma_\varepsilon)^{-1} L \Omega_0 . \tag{8.15b}$$

Note the resemblance between solution (8.15) and the solution of the Cholette-Dagum regression-based benchmarking model (3.21) repeated here for convenience:

$$\hat{\theta} = s + V_e J' V_d^{-1} [a - Js],$$

with
$$var[\hat{\theta}] = V_e - V_e J' V_d^{-1} J V_e \, ,$$

where $V_d = (J V_e J' + \Sigma_\varepsilon)$. Vectors s and a of the additive regression-based model, respectively correspond to $\hat{\eta}_0$ and z; and matrices V_e and J correspond to Ω_0 and L.

Hillmer and Trabelsi illustrated their method with the monthly U.S. Retail Sales of Hardware Stores. In the first stage on signal extraction, these authors fit to the data an ARIMA (1,1,0) plus deterministic seasonality and assume an AR(1) model with φ=**0.80** for the sample survey error. Since the ratio σ_c^2 / σ_a^2 is very small, the behaviour of Ω_0 is dictated entirely by the model of the survey error. In Chapter 10, we illustrate this method with the Canadian Retail Trade series assuming a more realistic RegARIMA model for the signal plus a (1,1,0)(1,1,0) model for the sample survey error.

8.3 State Space Signal Extraction and Benchmarking : The Durbin and Quenneville Method

Durbin and Quenneville (1997) article made an excellent contribution to the problem of signal extraction and benchmarking. These authors offered two solutions: (1) a two-stage method as in the previous ARIMA modelling where the first stage fits a structural component model to the monthly data, and the second stage combines the resulting smoothed values with the benchmark data to obtain the final adjusted estimates; and (2) a one-stage method which simultaneously fits the structural model and performs the benchmarking.

Like in Hillmer and Trabesi method, the authors use a stochastic model for the signal, to take into account the correlation structure of the underlying series. If the model is appropriate, they achieve more efficiency relative to methods without time series modelling of the signal.

However, Durbin and Quenneville improve on the ARIMA model-based approach, by assuming an heteroscedastic sample survey error and avoiding the unsustainable assumption of zero mean for non-stationary series. The authors present additive and multiplicative benchmarking models with and without constant bias correction. They also make a major contribution to state-space model estimation, by putting the observation error into the state-space vector to facilitate the calculation of the covariance matrix which is otherwise un-manageable. The main draw-backs of their method, particularly for the single-stage benchmarking, is the high computational complexity and the large number of parameters to estimate from relatively short series. These draw-backs drastically reduce the range of application.

8.3.1 State-Space Model for Signal Extraction

Assume that the monthly series y_t originates from a sample survey,

$$y_t = \eta_t + k_t u_t, \quad t=1,\ldots,n, \tag{8.17}$$

where $k_t u_t$ is the heteroscedastic sample survey error, u_t being a unit variance stationary ARMA (p,q) series and where the values of k_t, p and q are provided by survey experts.

The benchmarks relationships are given by

$$z = L\eta + \varepsilon, \tag{8.18}$$

where z is an $m\times 1$ benchmark vector, $\boldsymbol{L}$ is an $m\times n$ matrix of zeroes and ones, equivalent to matrix $\boldsymbol{J}$ (coverage fractions) discussed in Chapter 3, and $\boldsymbol{\varepsilon}$ is an $m\times 1$ error vector normally distributed with mean zero and covariance $\boldsymbol{\Sigma}_{\varepsilon}$.

The signal η_t is assumed to follow a structural component model given by

$$\eta_t = \mu_t + \gamma_t + \Sigma_{j=1}^{k} \delta_{jt} w_{jt} + e_t, \quad e_t \sim N(0,\sigma_e^2),\ t=1,...,n, \tag{8.19}$$

where μ_t denotes the trend, γ_t are the seasonal effects, δ_{jt} are the daily coefficients, w_{jt} is the number of occurrences of day j in month t, and e_t is an error term.

The model for the trend is

$$\Delta^2 \mu_t = \xi_t, \quad \xi_t \sim N(0,\sigma_\xi^2), \tag{8.20}$$

where $\Delta = (1-B)$ is the first order difference operator. If $\xi_t = 0$, the trend given in (8.20) follows a straight line, otherwise the term ξ_t allows the trend to change slope in time.

The model for the seasonal variations is

$$\gamma_t = -\Sigma_{j=1}^{11} \gamma_{t-j} + \omega_t, \quad \omega_t \sim N(0,\sigma_\omega^2). \tag{8.21}$$

If $\omega_t = 0$, the seasonal pattern is constant, and moving otherwise.

The model for the trading-day variations is

$$\Delta \delta_{jt} = \zeta_t, \quad \zeta_t \sim N(0,\sigma_\zeta^2), \quad j=1,...,k. \tag{8.22}$$

Usually there are six (k=6) coefficients, one for each day of the working week. The Sunday is equal to the negative sum of the other six, since this kind of variations cancel over a week.

Hence, the observation equation of the structural model (8.17) results in

$$y_t = \mu_t + \gamma_t + \Sigma_{j=1}^{k} \delta_{jt} w_{jt} + e_t + k_t u_t, \ t=1,...,n, \tag{8.23}$$

where $k_t u_t$ is the heteroscedastic survey error with standard deviation k_t, and $u_t = \varphi u_{t-1} + a_t$, $a_t \sim N[0, (1-\varphi^2)]$.

The state-space representation of model (8.23) is

$$y_t = \tilde{\boldsymbol{X}}_t \boldsymbol{\alpha}_t, \tag{8.24}$$

where
$$\tilde{\boldsymbol{X}}_t = [1\ 0\ 1\ 0\ 0\ \cdots 0\ \omega_{1t} \cdots \omega_{kt}\ 1\ k_t]; \tag{8.25}$$

and the state vector $\boldsymbol{\alpha}_t$ is

$$\boldsymbol{\alpha}_t = [\mu_t\ \mu_{t-1}\ \gamma_t\ \gamma_{t-1} \cdots \gamma_{t-11}\ \delta_{1t} \cdots \delta_{kt}\ e_t\ u_t]'. \tag{8.26}$$

Notice the unusual feature of the above formulation where the error term e_t of the observation equation (8.23) is part of the state vector (8.26). Durbin and Quenneville justify this change for the sake of facilitating the calculation of the covariance matrix of the estimation errors of the η_t's.

The state transition equation is

$$\boldsymbol{\alpha}_t = \boldsymbol{T}_t \boldsymbol{\alpha}_{t-1} + \boldsymbol{R}_t \boldsymbol{v}_t,\ t=1,\ldots,n, \tag{8.27}$$

where $\boldsymbol{T}_t$ is a time-invariant block diagonal matrix with blocks

$$\begin{bmatrix} 2 & -1 \\ 1 & 0 \end{bmatrix}, \begin{bmatrix} -\mathbf{1}_{1\times 10} & -1 \\ \boldsymbol{I}_{10} & \mathbf{0}_{10\times 1} \end{bmatrix}, \boldsymbol{I}_k, 0\ ;\ \varphi\ , \tag{8.28}$$

for the trend, seasonal effects, trading-day component, observation error and survey error respectively. Matrix $\boldsymbol{R}_t$ is a selection matrix composed of a subset of columns of the identity matrix and $\boldsymbol{v}_t = [\xi_t\ \omega_t\ \zeta_1 \cdots \zeta_k\ e_t\ a_t]'$. Matrix $\boldsymbol{I}_c$ denotes a $c\times c$ identity matrix, $\mathbf{1}_{1\times c}$ a $1\times c$ vector of ones, and $\mathbf{0}_{r\times 1}$ a $r\times 1$ vector of zeroes.

8.3.2 Two-Stage Benchmarking: the Additive Model

The purpose of benchmarking is to estimate

$$\eta_t = \boldsymbol{X}_t \boldsymbol{\alpha}_t, \quad t=1,...,n, \tag{8.29}$$

using the information from the monthly observations and the benchmarks and where the $\boldsymbol{X}_t$ matrix associated to the signal η_t is equal to $\tilde{\boldsymbol{X}}_t$ of Eq. (8.25) except that $k_t=0$. The best estimator of η_t is shown to be the mean of its posterior distribution given all the information,

$$\hat{\eta}_t = E(\eta_t \mid \boldsymbol{y}, \boldsymbol{z}), \quad t=1,...,n, \tag{8.30}$$

where $\boldsymbol{y} = [y_1\ y_2\ \cdots\ y_n]'$ and $\boldsymbol{z} = [z_1\ z_2\ \cdots\ z_m]'$. We discuss next the two-stage additive benchmarking model.

Similarly to Hillmer and Trabelsi (1987), one of the benchmarking solutions consists of two stages. First, $\tilde{\boldsymbol{\eta}} = E(\boldsymbol{\eta} \mid \boldsymbol{y})$ and $var[\tilde{\boldsymbol{\eta}}] = \tilde{\boldsymbol{\Omega}}$ are computed by applying standard Kalman filtering and smoothing only to the monthly observations $\boldsymbol{y}$. In the second stage, the information available in $\tilde{\boldsymbol{\eta}}$ is combined with the vector of benchmarks $\boldsymbol{z}$ to obtain the final benchmarked series $\hat{\boldsymbol{\eta}}$. Hence, the benchmarked series $\hat{\boldsymbol{\eta}}$ and its MSE matrix are given by

$$\hat{\boldsymbol{\eta}} = \tilde{\boldsymbol{\eta}} + \hat{\boldsymbol{\eta}}_c = \tilde{\boldsymbol{\eta}} + \tilde{\boldsymbol{\Omega}}\boldsymbol{L}'(\boldsymbol{L}\tilde{\boldsymbol{\Omega}}\boldsymbol{L}' + \boldsymbol{\Sigma}_\varepsilon)^{-1}(\boldsymbol{z} - \boldsymbol{L}\tilde{\boldsymbol{\eta}}), \tag{8.31a}$$

and

$$var[\hat{\boldsymbol{\eta}}] = \tilde{\boldsymbol{\Omega}} - \tilde{\boldsymbol{\Omega}}\boldsymbol{L}'(\boldsymbol{L}\tilde{\boldsymbol{\Omega}}\boldsymbol{L}' + \boldsymbol{\Sigma}_\varepsilon)^{-1}\boldsymbol{L}\tilde{\boldsymbol{\Omega}}. \tag{8.31b}$$

Assuming $\boldsymbol{\Sigma}_\varepsilon$ known, Eq. (8.31) requires only the knowledge of $\tilde{\boldsymbol{\eta}}$ and $\tilde{\boldsymbol{\Omega}}$. Formula (8.31) is due to Hillmer and Trabelsi (1987, Eq. 2.16).

Making use of standard regression theory, Durbin and Quenneville give a simpler proof. Since $\boldsymbol{z} \equiv \boldsymbol{L}\boldsymbol{\eta} + \boldsymbol{\varepsilon}$, $E(\boldsymbol{z} \mid \boldsymbol{y}) = \boldsymbol{L}\tilde{\boldsymbol{\eta}}$, they obtain $\hat{\boldsymbol{\eta}} = E(\boldsymbol{\eta} \mid \boldsymbol{y}, \boldsymbol{z}) = E(\boldsymbol{\eta} \mid \boldsymbol{y}, \boldsymbol{z} - \boldsymbol{L}\tilde{\boldsymbol{\eta}})$. Since $E(\boldsymbol{z} - \boldsymbol{L}\tilde{\boldsymbol{\eta}}) = \boldsymbol{0}$ and $\boldsymbol{y}$ and $\boldsymbol{L}\tilde{\boldsymbol{\eta}}$ are orthogonal, they derive:

$$E(\boldsymbol{\eta} \mid \boldsymbol{y}, \boldsymbol{z} - \boldsymbol{L}\tilde{\boldsymbol{\eta}}) = E(\boldsymbol{\eta} \mid \boldsymbol{y}) + cov[\boldsymbol{\eta}, \boldsymbol{z} - \boldsymbol{L}\tilde{\boldsymbol{\eta}}]\,(var[\boldsymbol{z} - \boldsymbol{L}\tilde{\boldsymbol{\eta}}])^{-1}\,(\boldsymbol{z} - \boldsymbol{L}\tilde{\boldsymbol{\eta}}) \tag{8.33a}$$

and

$$var[\hat{\boldsymbol{\eta}}] = var[\tilde{\boldsymbol{\eta}}] - cov[\boldsymbol{\eta}, \boldsymbol{z} - \boldsymbol{L}\tilde{\boldsymbol{\eta}}]\,[var(\boldsymbol{z} - \boldsymbol{L}\tilde{\boldsymbol{\eta}})]^{-1}\,cov[\boldsymbol{\eta}, \boldsymbol{z} - \boldsymbol{L}\tilde{\boldsymbol{\eta}}]'. \tag{8.33b}$$

Substituting $E(\boldsymbol{\eta}\,|\,\boldsymbol{y}) = \tilde{\boldsymbol{\eta}}$, $cov[\boldsymbol{\eta}, \boldsymbol{z} - \boldsymbol{L}\tilde{\boldsymbol{\eta}}] = \tilde{\boldsymbol{\Omega}}\boldsymbol{L}'$ and $var[\boldsymbol{z} - \boldsymbol{L}\tilde{\boldsymbol{\eta}}] = \boldsymbol{L}\tilde{\boldsymbol{\Omega}}\boldsymbol{L}' + \boldsymbol{\Sigma}_\varepsilon$ in (8.33) yields (8.31).

The estimates obtained from sample surveys are often biased due to non-response, lack of coverage and other factors. Since the benchmarks are unbiased, it is possible to estimate the bias in the sub-annual series.

Assuming

$$y_t = \eta_t + b + k_t u_t, \quad t=1,...,n, \tag{8.34}$$

where b denotes a constant bias. Then $\tilde{\eta}_t = \eta_t + b$, and applying Kalman filtering and smoothing to the series y_t, the authors obtain $\tilde{\boldsymbol{\eta}} = E(\boldsymbol{\eta}\,|\,\boldsymbol{y})$ where $\tilde{\boldsymbol{\eta}} = [\tilde{\eta}_1 \cdots \tilde{\eta}_n]'$.

8.3.3 Single-Stage Benchmarking: Additive Model

In the single-stage benchmarking solution, a single series is built by including the monthly observations $\boldsymbol{y}$ and the benchmarks $\boldsymbol{z}$ to estimate $\hat{\boldsymbol{\eta}}$. The series is constructed by inserting each benchmark into the monthly series immediately after the last monthly value covered by the benchmark. For example, if the yearly benchmarks starts in the first year, the series take the form

$$y_1, ..., y_{12}, z_1, y_{13}, \cdots y_{24}, z_2, y_{25}, \cdots \tag{8.35}$$

These observations are referred to with the following index: 1, ..., 12, (12×1)′, 13, ..., 24, (12×2)′, 25, ..., 36, (12×3)′, 37, ...

Assuming the same structural component model as (8.23), but with trading-day coefficients updated only once a year each January such that

$$\delta_{j,12i+1} = \delta_{j,12i} + \zeta_{j,12i+1}, \quad j=1,...,k; \; i=1,...,L,$$

and $$\delta_{j,t} = \delta_{j,t-1} \text{ otherwise,}$$

where L denotes the last year for which a January value is available. The time point in the series at which benchmarks z_i occur is denoted by $(12i)'$ and located between $t=12i$ and $t=12i+1$. Hence the state vector is redefined as,

$$\boldsymbol{a}_t = [\mu_t \cdots \mu_{t-11} \; \gamma_t \cdots \gamma_{t-11} \; \delta_{1t} \cdots \delta_{kt} \; e_t \cdots e_{t-11} \; u_t]' \text{ for } t=1,...,n,$$

and $$\boldsymbol{a}_t \equiv \boldsymbol{a}_{12i}, \quad t=(12i)' \text{ for } i=1, ..., L. \tag{8.37}$$

The transition matrix $\boldsymbol{T}_t$ is block diagonal with block elements

$$\boldsymbol{T}_t = block\left(\begin{bmatrix} 2 & -1 & \mathbf{0}_{1\times 10} \\ 1 & 0 & \mathbf{0}_{11\times 10} \end{bmatrix}, \begin{bmatrix} -\mathbf{1}_{1\times 10} & -1 \\ \boldsymbol{I}_{10} & \mathbf{0}_{10\times 1} \end{bmatrix}, \boldsymbol{I}_k, 0\,;\varphi\right),\ t=1,...,n, \tag{8.38a}$$

$$\boldsymbol{T}_t = \boldsymbol{I}_{37+k},\ t=(12i)',\ i=1,...,L. \tag{8.38b}$$

The state error vector is

$$\boldsymbol{v}_t = [\,\xi_t\ \omega_t\ e_t\ a_t\,]',\ t=12i+2,..., 12(i+1), \tag{8.39a}$$

$$\boldsymbol{v}_t = [\,\xi_t\ \omega_t\ \zeta_{1t},...,\zeta_{kt}\ e_t\ a_t\,]',\ t=12i+1;\ i=1,...,L, \tag{8.39b}$$

$$\boldsymbol{v}_t = 0,\quad t=(12i)';\ i=1,...,L. \tag{8.39c}$$

Hence, matrix of Eq. (8.25) becomes

$$\tilde{\boldsymbol{X}}_t = [1\ \mathbf{0}_{1\times 11}\ 1\ \mathbf{0}_{1\times 11}\ \omega_{1t} \cdots \omega_{kt}\ 1\ \mathbf{0}_{1\times 11}\ k_t],\quad t=1,...,n. \tag{8.40a}$$

The matrices used to estimate the signal $\boldsymbol{a}_t$ for the monthly and the yearly observations are respectively:

$$\boldsymbol{X}_t = [1\ \mathbf{0}_{1\times 11}\ 1\ \mathbf{0}_{1\times 11}\ \omega_{1t} \cdots \omega_{kt}\ 1\ \mathbf{0}_{1\times 11}\ 0],\quad t=1,...,n, \tag{8.40b}$$

$$\tilde{\boldsymbol{X}}_t = [\tilde{\boldsymbol{\ell}}_i\ \tilde{\boldsymbol{\ell}}_i\ \tilde{\boldsymbol{\ell}}_i\boldsymbol{\omega}_{1(i)} \cdots \tilde{\boldsymbol{\ell}}_i\boldsymbol{\omega}_{k(i)}\ \tilde{\boldsymbol{\ell}}_i\ 0],\quad t=(12i)';\ i=1,...,L, \tag{8.40c}$$

where $\tilde{\boldsymbol{\ell}}_i = [\,\ell_{i,12i},...,\ell_{i,12i-11}\,]$ with $\ell_{i,t}$ defined as the (i,t)-th element of the design matrix $\boldsymbol{L}$ consisting of zeroes and ones, and $\boldsymbol{\omega}_{j(i)} = [\omega_{12i} \cdots \omega_{12i-11}]$ for $j=1,..., k$ and $i=1,..., L$.

Hence, $$y_t = \tilde{\boldsymbol{X}}_t \boldsymbol{a}_t,\quad t=1,...,n, \tag{8.41}$$

$$\eta_t = \boldsymbol{X}_t \boldsymbol{\alpha}_t, \quad t=1,...,n, \tag{8.42}$$

and

$$z_i = \tilde{\boldsymbol{X}}_t \boldsymbol{\alpha}_t + \varepsilon_i, \quad t=(12i)'; \; i=1,...,L. \tag{8.43}$$

The state-space model has a new form where the observation error $\boldsymbol{\varepsilon}$ is part of the state vector for the benchmarks in Eq. (8.43) and is zero in Eqs. (8.41) and (8.42). It is assumed that $\boldsymbol{\Sigma}_\varepsilon$ is diagonal to reduce the size of the sate vector, which otherwise becomes too large.

The state transition equation is

$$\boldsymbol{\alpha}_t = \boldsymbol{T}_t \boldsymbol{\alpha}_{t-1} + \boldsymbol{R}_t \boldsymbol{\nu}_t, \; t=1,...,12,(12\times1)',13,...,24,(12\times2)',25,... \tag{8.44}$$

where $\boldsymbol{R}_t$ is a selection matrix made of the appropriate columns of $\boldsymbol{I}_{37+k}$.

The model is cast in standard state-space form, and the benchmarked estimates $\hat{\eta}_t$ can be obtained by standard Kalman filtering and smoothing for $t=1,...,12,\; (12\times1)',\; 13,...,24,\; (12\times2)'\; 25,...$ The most common benchmarking situation encountered in practice is that where the benchmarks are annual totals, $z_i = \Sigma_{t=12i-11}^{12i}\, \eta_t$. The solution is obtained by substituting $\tilde{\ell}_i = \tilde{\ell}_{12}$ in the general formula.

Durbin and Quenneville (1997) apply this benchmarking method to the Canadian Retail Trade series and illustrate using: (1) two additive models, with and without bias, and (2) a multiplicative model without bias.

In principle, state space models are indeed more versatile. However this extra versatility is not required in benchmarking and interpolation. The models are very complex compared to those use in the regression method. Moreover, the smoothing part of the state space approach requires the storage of a large number of covariance matrices, which largely overlap.

8.4 Non-Parametric Signal Extraction and Benchmarking: The Chen, Cholette and Dagum Method

The article by Chen, Cholette and Dagum (1997) introduces an innovative way to estimate non-parametrically the covariance matrix for the stationary part of the signal to enable benchmarking via signal extraction with ARIMA modelling. The authors point out that identifying a model for $\{\eta_t\}$ from the observations $\{y_t\}$ in model (8.1) poses the so-called problem of "modelling error-in-variables", even if the model of $\{e_t\}$ is given. This is also the case when both $\{e_t\}$ and $\{\eta_t\}$ of equation (8.1) are assumed to be stationary ARMA models like in Hillmer and Trabelsi (1987). To overcome the problem, they proposed a non-parametric method to estimate $\boldsymbol{\Sigma}_\eta^{-1}$ from the observed data, assuming that the signal $\{\eta_t\}$ follows a general model given by

$$\eta_t = \mu_t + \xi_t \tag{8.45a}$$

and

$$\Delta^d \Delta_s^D \xi_t = \zeta_t, \tag{8.45b}$$

where $\{\mu_t\}$ and $\{\xi_t\}$ are respectively the deterministic and stochastic components, ζ_t is a stationary process, and $\Delta = (1-B)$ and $\Delta_s = (1-B^s)$ are the regular difference and seasonal difference operators respectively. If $\mu_t = 0$, (8.45) reduces to the well known difference stationary (**DS**) model, and to the trend-stationary (**TS**) if d=D=0.

Denoting $d^* = d+D\times s$ and $\tilde{n} = n - d^*$, the $\tilde{n}\times 1$ vector $\boldsymbol{\zeta}$ of (8.45b) may be written as

$$\boldsymbol{\zeta} = \boldsymbol{D\eta} = \boldsymbol{D\xi}, \tag{8.46}$$

where $\boldsymbol{D}$ is an $\tilde{n}\times n$ matrix. The covariance matrix $\boldsymbol{\Sigma}_\eta$ of (8.46) is not well defined. Chen et al. use two common assumptions to obtain $\boldsymbol{\Sigma}_\eta^{-1}$ given by

$$\boldsymbol{\Sigma}_\eta^{-1} = \boldsymbol{D}'\boldsymbol{\Sigma}_\zeta^{-1}\boldsymbol{D}. \tag{8.47}$$

The two assumptions made are:

$$var[\boldsymbol{\eta}_i] = h\boldsymbol{I}_{d*}, \quad h \to \infty \tag{8.48}$$

and

$$cov[\boldsymbol{\eta}_i\, \zeta_t] = 0, \quad t = d^* + 1, ..., n, \tag{8.49}$$

where $\boldsymbol{\eta}_i$ contains the first $d^* = d + D \times s$ values of $\boldsymbol{\eta}$.

Eq. (8.48) means that the initial values are diffuse; and Eq. (8.49), that the series is not correlated with the initial values for any $t > d^*$.

On the other hand, the TS (trend stationary) model is given by

$$\eta_t = \mu_t + \zeta_t, \quad t = 1, ..., n, \tag{8.50}$$

where ζ_t a zero mean stationary process and μ_t is a deterministic component, usually assumed to follow a linear model

$$\mu_t = \Sigma_{j=1}^{k} x_{tj} \beta_j \quad = \boldsymbol{x}_t' \boldsymbol{\beta}. \tag{8.51}$$

Hence, model (8.1) becomes

$$y_t = \mu_t + w_t, \tag{8.52}$$

where $w_t = \zeta_t + e_t$, ζ_t and e_t are mutually uncorrelated zero mean stationary processes. Assuming for the moment μ_t known, $w_t = y_t + \mu_t$ is treated as data to estimated the signal ζ_t. Hence,

$$\hat{\zeta}_0 = E(\boldsymbol{\zeta} | \boldsymbol{w}) = (\boldsymbol{\Sigma}_e^{-1} + \boldsymbol{\Sigma}_\zeta^{-1})^{-1} \boldsymbol{\Sigma}_e^{-1} \boldsymbol{w} \tag{8.53}$$

and

$$var[\hat{\zeta}_0] = (\boldsymbol{\Sigma}_e^{-1} + \boldsymbol{\Sigma}_\zeta^{-1})^{-1}. \tag{8.54}$$

The mutual independence of η_t and e_t in Eq. (8.50) implies that η_t and ζ_t, and y_t and w_t have the same autocovariance function, that is $\boldsymbol{\Sigma}_w = \boldsymbol{\Sigma}_e^{-1} + \boldsymbol{\Sigma}_\zeta^{-1}$. Eq. (8.53) and (8.54) may then be written as

$$\hat{\zeta}_0 = \Sigma_\zeta \Sigma_w^{-1} w, \tag{8.55}$$

and

$$var[\hat{\zeta}_0] = \Omega_0 = \Sigma_\zeta \Sigma_w^{-1} \Sigma_e, \tag{8.56}$$

which requires fewer matrix inversions than (8.6).

Chen et al. propose to estimate Σ_ζ and Σ_w by means of their spectrum smoothed in an appropriate manner. They obtain estimates of η_c and η using the same equations (8.13) and (8.15) due to Hillmer and Trabelsi (1987). Chen et al. also carry out a simulation comparing the efficiency of the proposed non-parametric method (**NP**), the additive Cholette-Dagum (1994) regression model (**REG**) and the Hillmer and Trabelsi ARIMA benchmarking (1987) (**SE**), assuming for the latter that Σ_η is known. The summarized results follow.

(1) A smooth spectral estimate is needed to estimate Σ_ζ, which must be positive definite to obtain an estimate of η close to the one obtained under the true value of Σ_η.

(2) Using RMSE as a measure of efficiency, these authors show that SE is more efficient than REG and NP gives very close values to SE.

(3) If the signal-to-noise ratio (**S/N** $= \gamma_\zeta(0)/\gamma_e(0)$) is high, the gain of SE and NP relative to simple REG is small. If S/N is low, the gain of SE and NP relative to REG is large. However, ARIMA models seldom fit noisy data. In fact, Hillmer and Tabelsi (1987) assume Σ_η to be known; and this is also the case in the simulation. In practice, signal extraction is not very useful for series which need it the most.

(4) When S/N is medium or low and the patterns of the spectrum estimates, $f_\eta(\lambda)$ and $f_e(\lambda)$, are similar, the gains in efficiency relative to REG are small for the TS model. If the difference between the two patterns is large, the gains are large. For the DS models, the gains tend to be always large.

9 Calendarization

9.1 Introduction

Time series data do not always coincide with calendar periods. Indeed, to accommodate the respondents to surveys and censuses, statistical agencies often accept data with fiscal reporting periods as valid. For example, the Canadian Federal and Provincial governments have a common fiscal year ranging from April 1 to March 31 of the following year; banks have a common fiscal year ranging from Nov. 1 to Oct. 31; school boards and academic institutions, usually from Sept. 1 to Aug. 31.

The timing and duration of a reported value covering fiscal or calendar periods are a function of its starting and ending dates. These dates are critical pieces of information in benchmarking, temporal distribution, interpolation and even more so in calendarization. The dates must be in terms of day, month and year. The starting and ending dates may be processed mathematically by assigning an integer value to each day of the 20th and 21st centuries in an ascending manner, e.g. 1 for Jan. 1 1900, 2 for Jan. 2 1900, etc.[1] This mapping scheme thus stores the starting month, the starting day and the starting year in one single integer variable; and the ending month, the ending day and the ending year into another variable. With only two numerical variables, it is thus possible to economically store and retrieve six variables.

The *duration* can then be defined as the number of days covered, that is the number assigned to the ending date minus that assigned to the starting date plus one. The *timing* of the data can be crudely defined as the middle date of the reporting period, that is the (number assigned to the) starting date plus the ending date divided by 2. The timing reflects whether the data pertain to the recent past, to the not so recent past or to the remote past. Under this definition, the timing of a reporting period, with starting and ending dates respectively equal to Jan. 1 2005 (38717 in Excel) and Dec. 31 2005 (38353), is July 2 2005 (38535); that of a reporting period, with starting

[1] This is how spreadsheet program Excel stores date variables. In a similar manner, program SAS (Statistical Analysis System) assigns -1 to Dec. 31 1959, 0 to Jan. 1 1960, 1 to Jan. 2 1960, 2 to Jan. 3 1960, and so forth. In both cases these numerical values can be displayed under a conventional date format, e.g. 31/12/2000.

and ending dates April 1 2004 (38078) and March 31 2005 (38442), is Sept. 29 2004 (38260).

For stock variables, the duration is typically one day, usually the day of the ending date, and the timing coincides with the ending date. For some stock variables, e.g. employment and unemployment, the duration is typically one week and the measurement presumably pertains to the average of the week. For the time being, we assume that the data pertain to flow variables.

In cases of homogenous fiscal years within an industry or sector, e.g. banking, taking the cross-sectional sums of the fiscal year data with same timing produces valid *tabulated* values. These may be used as such for socio-economic analysis provided awareness of the timing of the data. They may also be used in benchmarking, as long as the coverage fractions reflect the correct duration and timing as described in Chapter 3 section 3.1 and 3.3. Calendar year values may then be obtained as a by-product of benchmarking, by merely taking the temporal sums of the benchmarked series over the calendar years. Such situations are ideal for optimal calendarization, if monthly or quarterly data are available for the same yearly variable of interest.

In the absence of such sub-yearly data, the calendarized estimates will lack some of the variation associated with the business cycles, as discussed in section 9.2. Despite that, the model-based calendarization methods presented perform better than ad hoc methods, namely the assignment calendarization and the fractional methods, respectively discussed in the next two sections.

A similar situation prevails in the calendarization of quarterly data. As shown by Fig. 7.11 of Chapter 7, the quarterly sum operator (and therefore quarterly data) reduces the monthly seasonal variations and completely eliminates the 3-month seasonal cycle. This implies monthly seasonality cannot be fully recovered from quarterly flow data, unless a genuine monthly seasonal indicator is used.

Some statistical agencies address the calendarization problem by simply requiring that the data pertaining to the target variable be provided on a calendar period basis. In principle, this is desirable given the limitations of calendarization methods in the absence of genuine sub-annual data for the same variable. Indeed, the absence of such data is more often the rule than the exception. If such sub-annual data were available, there would probably be no calendarization problem.

9.2 The Assignment Calendarization Procedure

In most industries or economic sectors, the fiscal years are *not* homogenous: they typically vary from unit to *unit* (respondent). As a result, the timing and sometimes the duration of the data differ from unit to unit and even for a given unit. Hence, fiscal data is not very useful. Indeed, it is not possible to cross-sectionally sum over the various fiscal years, for a given calendar year, because they do not have the same timing; and, similarly for other calendar periods.

In order to make the data usable, one practice has been to assign the fiscal year data to particular calendar years according to an arbitrary rule. For example, fiscal years with ending dates in 2005 are assigned to calendar year 2005. Assuming that the fiscal years cover whole months, there are twelve fiscal years ending in 2005: the fiscal year covering from Feb. 1 2004 to Jan. 31 2005, that covering from March 1 2004 to Feb. 28 2005, and so forth, that covering from Jan. 1 2005 to Dec. 31 2005. These twelve fiscal years jointly cover 23 months. The fiscal data assigned to the target year are then tabulated (i.e. cross-sectionally summed).[2] For convenience, we refer to this process as the *assignment calendarization procedure.*

Table 9.1. Monthly weights resulting from the assignment calendarization procedure under equal distribution of fiscal years over the twelve possible fiscal years, assuming common trend-cycle component

	2004											2005											
(1)	F	M	A	M	J	J	A	S	O	N	D	J	F	M	A	M	J	J	A	S	O	N	D
1	1	1	1	1	1	1	1	1	1	1	1	1											
2		1	1	1	1	1	1	1	1	1	1	1	1										
3			1	1	1	1	1	1	1	1	1	1	1	1									
4				1	1	1	1	1	1	1	1	1	1	1	1								
5					1	1	1	1	1	1	1	1	1	1	1	1							
6						1	1	1	1	1	1	1	1	1	1	1	1						
7							1	1	1	1	1	1	1	1	1	1	1	1					
8								1	1	1	1	1	1	1	1	1	1	1	1				
9									1	1	1	1	1	1	1	1	1	1	1	1			
10										1	1	1	1	1	1	1	1	1	1	1	1		
11											1	1	1	1	1	1	1	1	1	1	1	1	
12												1	1	1	1	1	1	1	1	1	1	1	1
(2):	1	2	3	4	5	6	7	8	9	10	11	12	11	10	9	8	7	6	5	4	3	2	1

(1): Fiscal year according to the ending month (2): Monthly weights

[2] A variant of this practice is to assign the data with ending date between April 1 2005 and March 31 2006 inclusively, to year 2005.

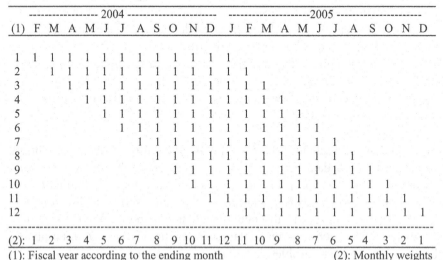

The twelve fiscal years assigned to year 2005 are represented in Table 9.1. The rows of the table can be seen as fiscal groups. To assess the impact of the assignment calendarization approach on the time series components, we now assume that all fiscal groups have the same monthly trend-cycle component, c_t, and that the other time series components (i.e. seasonality, trading-day variations and the irregulars) approximately sum to zero over any twelve consecutive months. Taking the cross-sectional (contemporaneous) sums for each month over the fiscal groups of Table 9.1 yields the values in the last row of the table. Given the above assumptions, we can factorize out c_t from these sums. As a result, the assignment calendarization procedure is equivalent to taking a weighted sum over 23 months, that is

$$\hat{c}_{2005} \approx c_{t-11} + 2c_{t-10} + \ldots + 11c_{t-1} + 12c_t + 11c_{t+1} + \ldots + 2c_{t+10} + c_{t+11}, \qquad (9.1a)$$

where $t-11$ stands for Jan. 2004, $t-10$ stands for Feb. 2004, and so forth, and $t+11$ stands for Dec. 2005.

If only the fiscal years with ending dates in Jan. 2005 (row 1 in Table 9.1), March 2005 (row 3), Oct. 2005 (row 10) and Dec. 2005 (row 12) are present in the industry or sector considered, the weighted sum becomes

$$\begin{aligned} \hat{c}_{2005} \approx\ & c_{t-11} + c_{t-10} + 2c_{t-9} + \ldots + 2c_{t-3} + 3c_{t-2} + 3c_{t-1} \\ & + 4c_t + 3c_{t+1} + 3c_{t+2} + 2c_{t+3} + \ldots + 2c_{t+9} + c_{t+10} + c_{t+11}. \end{aligned} \qquad (9.1b)$$

If in the same case, the first fiscal group ending in Jan. 2005 is twice as important as the other groups, the weighted sum becomes

$$\begin{aligned} \hat{c}_{2005} \approx\ & 2c_{t-11} + 2c_{t-10} + 3c_{t-9} + \ldots + 3c_{t-3} + 4c_{t-2} + 4c_{t-1} \\ & + 5c_t + 3c_{t+1} + 3c_{t+2} + 2c_{t+3} + \ldots + 2c_{t+9} + c_{t+10} + c_{t+11}. \end{aligned} \qquad (9.1c)$$

Dividing the weights in Eq. (9.1a) by their sum (144), we obtain a twelve by twelve average. Similarly dividing the weights in Eq. (9.1b) by their sum (48), we obtain a weighted average, which we shall call the Jan.-March-Oct.-Dec. average.

The properties of any linear operators, e.g. a weighted average, can be determined exactly by means of spectral analysis (e.g. Koopmans 1974). The gain function determines the fraction of a sine (or cosine) wave preserved by the operator. The phase-shift function determines the displacement in time of a sine wave. Ideally a filter should preserve 100% of the sine waves at the target frequencies with no phase-shift. In the case at hand, the target frequencies are those associated to the trend-cycle component. Note that this component is the one critical to most time series users and decisions makers.

Frequencies lower than 24/360, corresponding to 15-month cycle or longer, are usually associated with the trend-cycle component of the series to which the operator is applied. Note that the cycle periodicity is the reciprocal of the frequency. The frequency 0/360 corresponds to a sine wave of infinite periodicity (i.e. a constant).

The frequencies 30/360, 60/360, 90/360, 120/360 150/360 and 180/360 are the seasonal frequencies; the corresponding sine waves have periodicities of 12, 6, 4, 3, 2.4 and 2 months.

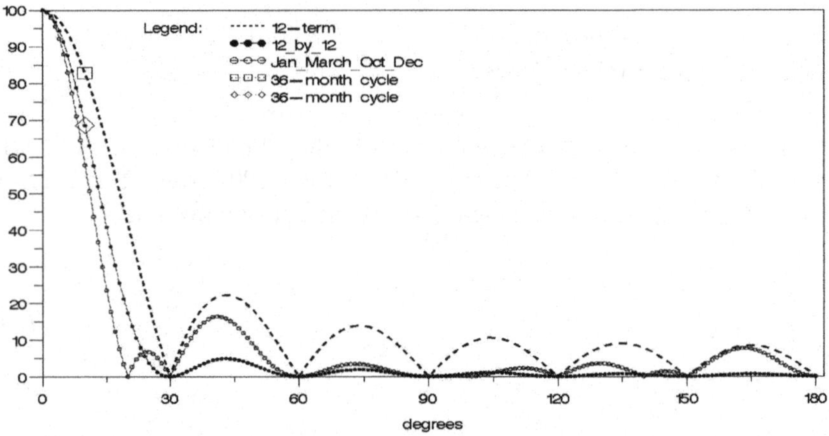

Fig. 9.1a. Gain functions of the 12-term average, the 12 by 12 average and the Jan., March, Oct. and Dec. average

As shown in Fig. 9.1a, the three linear operators eliminate seasonality, since the gains are all zero at the seasonal frequencies. Concerning the trend-cycle component, the performance of the 12-term average represents the target aimed at by the other two averages considered.[3] However, the 12 by 12 and the Jan.-March-Oct.-Dec. average further reduces the amplitude of business cycles compared to the 12-term average. In other words, the 12 by 12 average and the Jan.-March-Oct.-Dec. average under-estimate the peaks and over-estimates the throughs of business cycles. Indeed, Fig. 9.1a shows that the 12-term average preserves 82.8% of the 36-month (square "□" at frequency

[3] Yearly data contain reduced trend-cyclical fluctuations compared to monthly or quarterly data.

10/360) business cycle; the 12 by 12 average, 68.6% (diamond "◊"); and the Jan.-March-Oct.-Dec. average, only 14.2% of the same cycle.

According to the phase-shift functions exhibited in Fig. 9.1b, the 12-term average produces no phase shift. The 12 by 12 and Jan.-March-Oct.-Dec. averages produce important phase-shifts at the business cycle frequencies. They displace almost all business cycles by 5.5 months, particularly the 36-month cycle, which is a typical one for the American economy.

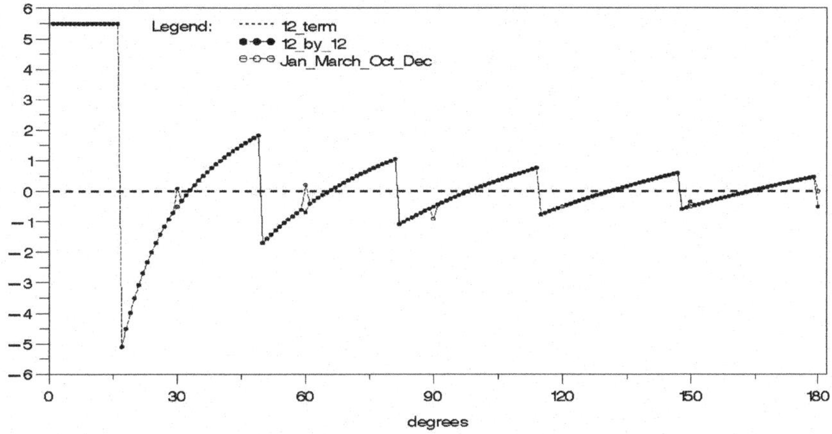

Fig. 9.1b. Phase shifts functions (in months) of the 12-term average, the 12 by 12 average and of the Jan., March, Oct. and Dec. average

Besides spectral analysis, there are simpler methods to assess the bias introduced by accepting fiscal year data as representing calendar year values. However, spectral analysis provides an exhaustive examination.

Table 9.2 displays the percentage errors in level introduced when a fiscal value pertaining to a unit is used as an estimate of its calendar year value. It is assumed that the variable is a flow, the fiscal periods cover twelve consecutive whole months and the underlying monthly series displays a constant growth rate, with no seasonality. The errors in level depend on the growth rates assumed and on the shift of the fiscal year with respect to the calendar year. A negative shift of 2 months, for instance, means that the fiscal year ends in Oct. 2005 instead of Dec. 2005. Fiscal year values covering 12 whole months under-estimate the true calendar year values in periods of growth (e.g. expansion) and over-estimate in periods of decline (e.g. recession). The larger the shift and the underlying growth rate (in absolute values), the greater is the error at the unit level. For instance if the underlying monthly series grows at an annual rate of 9% (monthly rate 0.72%), the percent error in level is −2.83% if the shift is −4 and −4.22% if the shift is − 6. Note that in leading-edge industries yearly growth rates often exceed 10%.

Table 9.2. Error in level when using the fiscal year value as an estimate of the true calendar year value, under the indicated true growth rates indicated

True yearly growth	corresp. monthly growth	Shift of the fiscal year in number of months with respect to the calendar year -10	-8	-6	-4	-2	-0
-9.00	-0.78	8.18	6.49	4.83	3.19	1.58	0.00
-6.00	-0.51	5.29	4.21	3.14	2.08	1.04	0.00
-3.00	-0.25	2.57	2.05	1.53	1.02	0.51	0.00
0.00	0.00	0.00	0.00	0.00	0.00	0.00	0.00
3.00	0.25	-2.43	-1.95	-1.47	-0.98	-0.49	0.00
6.00	0.49	-4.74	-3.81	-2.87	-1.92	-0.97	0.00
9.00	0.72	-6.93	-5.58	***-4.22***	***-2.83***	-1.43	0.00
12.00	0.95	-9.01	-7.28	-5.51	-3.71	-1.87	0.00
15.00	1.17	-10.99	-8.90	-6.75	-4.55	-2.30	0.00
18.00	1.39	-12.88	-10.45	-7.94	-5.37	-2.72	0.00
21.00	1.60	-14.69	-11.93	-9.09	-6.16	-3.13	0.00

To complicate matters, there are in fact more than twelve possible fiscal years, because some of them do not start on a first day of a month nor end on the last day of a month. For example a fiscal year may start on April 15 of one year and end on April 14 of the following year. In the United States, businesses may report on a 52-week basis and occasionally on a 53-week basis depending on the configuration of leap-years. Furthermore, some fiscal years cover more or less than twelve months (or 365 days). The first fiscal year of a new company often covers 15 to 18 months; and similarly, for the last fiscal year of company shutting down. When an existing company is taken over by another one, the absorbed company typically shortens or lengthens its fiscal year to synchronize it with that of the new parent company. Applying the assignment calendarization method to such fiscal year data may produces yearly estimates covering *more* than 23 months with unknown weights.

The assumptions made about the trend-cycle to produce Tables 9.1 and 9.2 are not likely to occur. Despite that, they clearly show that the assignment calendarization procedure seriously distorts the yearly estimates. Furthermore, the procedure *systematically* distorts the estimates differently in different industries, since different industries may have different fiscal year distributions. This is quite disturbing, since the assignment procedure generates many of the yearly benchmarks used in benchmarking, interpolation and temporal distribution. Furthermore, these yearly data are used in the building of Input-Output tables, which are the back-bone of National Accounts statistics.

A similar situation prevails with fiscal quarter data. The data typically cover any three consecutive whole months. In the case of homogeneous fiscal quarters, like in the Canadian banking sector, the fiscal quarter values may be directly tabulated and used for analysis (provided awareness of the timing) or as benchmarks with correctly specified coverage fractions (Cholette 1990).

In many cases, however, the fiscal quarters are not homogenous and may cover any consecutive 12 or 13 weeks, sometimes more than three months. For fiscal year data, the seasonal component largely cancels out for reporting periods covering approximately 365 days (except for stock series); this is not the case for fiscal quarter data. As a result, the assignment calendarization procedure results in wrong timing and duration of the calendar quarter estimates, in the presence of seasonality. The procedure severely distorts and reduces the amplitude of the underlying seasonal pattern, which accounts for most of the variations within the year. It is not rare to observe seasonal patterns with amplitude exceeding 50%.

The situation also prevails with respect to calendar months. Indeed many units, especially the important ones, send their data in the form of multi-weekly data, typically covering 4 or 5 weeks. The assignment calendarization procedure leads to serious biases regarding the timing, duration and amplitude of the monthly estimates.

In the case of flow variables, the assignment calendarization procedure applied to fiscal data distorts the timing and the duration of the estimates in different manner in different industries. This procedure mis-represents the "true" calendar values, blurs the relations and causality between socio-economic variables and leads to tainted analysis and decision making.

Fiscal data pertaining to stock variables (e.g. Inventories, Population) pose an even more serious calendarization problem, because they refer to one single day. As a result, even yearly stocks are inherently seasonal. Inventories or instance typically pertain to the last day of the fiscal year. In Canada, the yearly Population is measured on July 1 of every year.

Some stock data pertain to the average of a week. This is the case of the Canadian Labour Force Survey which measures employment, unemployment and other variables. The survey taken during the week containing the 15-th day of each month. The week of the 15-th may fall as early as the 9-th to the 15-th day of the month and as late as the 15-th to the 21-st day of the month. Furthermore, in some months the survey is taken before or after the week of the 15-th, in order to avoid the proximity of certain holidays, namely Christmas and the Canadian Thanksgiving Day on the second Monday of October.

The variability of the survey dates entails that four weeks, five weeks or six weeks elapse between the survey weeks, depending on the configuration of the calendar. During the months of seasonal increases, the reading is under-estimated if the "week of the 15-th" is early and over-estimated otherwise; and vice versa during the months of seasonal declines.

Fiscal data also occur in administrative data file. The Income Tax file pertaining to businesses contains fiscal and calendar year data. The Payroll Deductions file discussed in section 9.6 contains monthly and multi-weekly data. The Canadian Goods and Services Tax file contains a very large variety of fiscal periods.

9.3 The Fractional Calendarization Method and its Variants

The fractional calendarization method is widely used for flow data pertaining to fiscal years. This method consists of setting the calendarized value equal to a weighted sum of the two overlapping fiscal year values. Thus for fiscal years covering from April 1 of one year to March 31 of the following year, the calendar estimate of 2003 $\hat{a}_{2003}$ is set equal to 1/4 of the 2002-03 fiscal year value plus 3/4 the 2003-04 value, that is,

$$\hat{a}_{2003} = (1/4)\, a_{2002-03} + (3/4)\, a_{2003-04} \,. \tag{9.2}$$

The rationale is that 1/4 of the 2002-03 value belongs to 2003; and 3/4 of the 2003-04 value, to 2003. It is not necessary to take seasonality into account, because it cancels out over *any* 12 consecutive months. The method does not provide end estimates, for 2002 and 2004 in the example. The central estimates are acceptable for variables with linear behaviour.

The fractional method of Eq. (9.2) is equivalent to taking a 24-month weighted average of the underlying monthly series with weights 1/4 over the months of the first fiscal year and 3/4 over the months of the second. The average can be analysed spectrally, in terms of gain and phase-shift functions, like in section 9.2. The gain function of the 12-term average (yearly sum) preserves 82.8% of the 36-month business cycle against 54.8% for the weights of the fractional method of Eq. (9.2) standardized to sum 1. Recall that the performance of the 12-term average represents the target aimed by the calendarization of fiscal year data.

A similar approach is often applied to multi-weekly flow data covering four or five weeks to obtain calendar month value such that

$$\hat{a}_n = w_{1,n}\, a_{m-1} + w_{2,n}\, a_m\,, \; n=1,...,N, \tag{9.3a}$$

$$w_{1,n} = (\Sigma_{\tau \in n}\, \delta_\tau) / (\Sigma_{\tau \in (m-1)}\, \delta_\tau),\;\; w_{2,n} = (\Sigma_{\tau \in n}\, \delta_\tau) / (\Sigma_{\tau \in m}\, \delta_\tau), \tag{9.3b}$$

where $\hat{a}_n$ denotes the calendarized monthly value, n is the calendar month index, and m is the multi-weekly data index, which temporally overlap the month, and δ_τ is a daily pattern reflecting the relative importance of the days.

If the daily pattern is equal to 1, the weights $w_{1,n}$ and $w_{2,n}$ are respectively the fractions of the days covered by a_{m-1} and a_m in calendar month n. For example, if n stands for June 2005, and $m-1$ and m stand for data points covering from May 15 to June 18 and from June 19 to July 16 2005, the fractions $w_{1,n}$ and $w_{2,n}$ are respectively 18/35 and 12/28.

As such, the fractional method does not provide estimates at the ends. An extension to obtain monthly estimates for the first (n=1) and last (n=N) month is the following

$$\hat{a}_{n=1} = w_{n=1}\, a_{m=1},\;\; w_{n=1} = (\Sigma_{\tau \in (n=1)}\, \delta_\tau) / (\Sigma_{\tau \in (m=1)}\, \delta_\tau), \tag{9.3c}$$

$$\hat{a}_N = w_N\, a_M,\;\; w_N = (\Sigma_{\tau \in N}\, \delta_\tau) / (\Sigma_{\tau \in M}\, \delta_\tau), \tag{9.3d}$$

where the weight $w_{n=1}$ and w_N may be greater than 1.

The method characterized by Eq. (9.3) assumes that no multi-weekly data covering 4 weeks is embedded *within* a target calendar month. In such a case, that target month overlaps three multi-weekly data points and requires extra rules besides those of Eqs. (9.3a) to (9.3d).

The next section reveals the implicit interpolations (temporal distributions) associated with the fractional calendarization methods of Eqs. (9.2) and (9.3).

9.4 Model-Based Calendarization Methods

Calendarization can be seen as a benchmarking problem. The approach consists of benchmarking sub-annual series to fiscal data and taking the temporal sums of the benchmarked or interpolated series over the desired calendar periods (Cholette 1989). For example, a monthly series is benchmarked to fiscal year values, and the calendar year values are set equal to the sums of the benchmarked series over calendar years. The original sub-annual series may be an *indicator* (proxy) or preferably a genuine sub-annual series pertaining to the same target variable as the benchmarks.

9.4.1 Denton-Based Methods

The fractional method described above can be seen as a particular case of benchmarking with the modified variants of the Denton method described in Chapter 6. The solution given by Eq. (6.8) is repeated here for convenience

$$\begin{bmatrix} \hat{\boldsymbol{\theta}} \\ \hat{\boldsymbol{\gamma}} \end{bmatrix} = \begin{bmatrix} \boldsymbol{D}'\boldsymbol{D} & \boldsymbol{J}' \\ \boldsymbol{J} & \boldsymbol{0} \end{bmatrix}^{-1} \begin{bmatrix} \boldsymbol{D}'\boldsymbol{D} & \boldsymbol{0}_{T\times M} \\ \boldsymbol{J}_{M\times T} & \boldsymbol{I}_M \end{bmatrix} \begin{bmatrix} \boldsymbol{s} \\ (\boldsymbol{a}-\boldsymbol{J}\boldsymbol{s}) \end{bmatrix}, \tag{9.4a}$$

$$\Rightarrow \begin{bmatrix} \hat{\boldsymbol{\theta}} \\ \hat{\boldsymbol{\gamma}} \end{bmatrix} = \boldsymbol{W} \begin{bmatrix} \boldsymbol{s} \\ (\boldsymbol{a}-\boldsymbol{J}\boldsymbol{s}) \end{bmatrix} = \begin{bmatrix} \boldsymbol{I}_T & \boldsymbol{W}_\theta \\ \boldsymbol{0}_{M\times T} & \boldsymbol{W}_\gamma \end{bmatrix} \begin{bmatrix} \boldsymbol{s} \\ (\boldsymbol{a}-\boldsymbol{J}\boldsymbol{s}) \end{bmatrix}, \tag{9.4b}$$

$$\Rightarrow \hat{\boldsymbol{\theta}} = \boldsymbol{s} + \boldsymbol{W}_\theta\,(\boldsymbol{a}-\boldsymbol{J}\boldsymbol{s}), \tag{9.4c}$$

where $\boldsymbol{s}$ and $\hat{\boldsymbol{\theta}}$ respectively denote the original sub-annual series and the corresponding benchmarked-interpolated series, $\boldsymbol{a}$ contains the fiscal values to be calendarized, and $\hat{\boldsymbol{\gamma}}$ denotes the Lagrange multipliers associated with the linear constraints $\boldsymbol{a}-\boldsymbol{J}\boldsymbol{\theta}=\boldsymbol{0}$. Matrix $\boldsymbol{D}$ is either a difference operator or an identity matrix, and $\boldsymbol{J}$ is a temporal sum operator. Assuming that $\boldsymbol{s}$ covers M complete calendar years and that the fiscal years run from April to March, matrix $\boldsymbol{J}$ of dimension $M\times T$ is

$$\boldsymbol{J} = \left[\,\boldsymbol{0}_{M\times 3}\ \ \boldsymbol{I}_M \otimes \boldsymbol{1}_{1\times 12}\ \ \boldsymbol{0}_{M\times 9}\,\right],\ \boldsymbol{J} = \left[\,\boldsymbol{0}_{M\times 1}\ \ \boldsymbol{I}_M \otimes \boldsymbol{1}_{1\times 4}\ \ \boldsymbol{0}_{M\times 3}\,\right],$$

for monthly and quarterly indicators respectively.

The calendarized values are the appropriate temporal sums of the benchmarked or interpolated series $\hat{\boldsymbol{\theta}}$ obtained from (9.4c),

$$\boldsymbol{a}_c = \boldsymbol{J}_c \hat{\boldsymbol{\theta}} = \boldsymbol{J}_c \boldsymbol{s} + \boldsymbol{J}_c \boldsymbol{W}_\theta (\boldsymbol{a} - \boldsymbol{J}\boldsymbol{s}) \equiv \boldsymbol{J}_c \boldsymbol{s} + \boldsymbol{W}_c (\boldsymbol{a} - \boldsymbol{J}\boldsymbol{s}), \quad (9.5a)$$

where $\boldsymbol{J}_c = \boldsymbol{I}_{M+1} \otimes \boldsymbol{1}_{1\times 12}$ and $\boldsymbol{J}_c = \boldsymbol{I}_{M+1} \otimes \boldsymbol{1}_{1\times 4}$ for monthly and quarterly indicators respectively .

If vector $\boldsymbol{s}$ contains zeroes, the calendarized value are given by

$$\boldsymbol{a}_c = \boldsymbol{J}_c \boldsymbol{W}_\theta \, \boldsymbol{a} = \boldsymbol{W}_c \, \boldsymbol{a}, \quad (9.5b)$$

If $\boldsymbol{D} = \boldsymbol{I}_T$, M=4 and $\boldsymbol{s} = \boldsymbol{0}$, the weights $\boldsymbol{W}_c$ are:

0.75000	0.00000	0.00000	0.00000
0.25000	0.75000	0.00000	0.00000
0.00000	0.25000	0.75000	0.00000
0.00000	0.00000	0.25000	0.75000
0.00000	0.00000	0.00000	0.25000

whether the indicator is monthly or quarterly. If the first year is 2001, the weights in the second, third and fourth rows are used to estimate the calendar year values of 2002, 2003 and 2004; the weights 1/4 and 3/4 are those of the fractional method of Eq. (9.2). The weights in the first and last rows, used to estimate year 2001 and 2005, provide wrong estimates, namely 3/4 of the 2001-02 fiscal value and 1/4 of the 2004-05 fiscal value.

The interpolations or temporal distribution resulting from (9.4c) with $\boldsymbol{D} = \boldsymbol{I}_T$ and $\boldsymbol{s} = \boldsymbol{0}$ are constant within each fiscal year and equal to the corresponding fiscal year value divided by 12, or by 4 in the case of quarterly distributions. The extrapolations are equal to zero for the first calendar quarter of 2001 and the last three calendar quarters of 2005, which explains the poor first and last calendarized yearly estimates.

If M=2, $\boldsymbol{s} = \boldsymbol{0}$ and $\boldsymbol{D}$=$\boldsymbol{D}^2$, the second order difference operator from Eq. (6.19), the weights $\boldsymbol{W}_c$ of (9.5b) are

1.25000	-0.25000
0.25000	0.75000
-0.75000	1.75000

The weights in the second row also coincide with those of the fractional method of Eq. (9.2), i.e. 1/4 and 3/4. The weights in the first and last rows are used to estimate first and third calendar years. The underlying interpolations behave linearly. If $M>2$, the weights are no longer equal to 1/4 and 3/4, and the interpolations behave as linearly as possible.

If M=2, $\boldsymbol{s}=\boldsymbol{0}$ and $\boldsymbol{D}=\boldsymbol{D}^1$, the first order difference operator from Eq. (6.4), the $\boldsymbol{W}_c$ weights of (9.5b) are

1.14360	-0.14360
0.22664	0.77336
-0.24394	1.24394

Fig. 9.2a displays the first and second order difference monthly interpolations (or temporal disaggregations), obtained from two fiscal year values covering April to March of the following year, along with the corresponding calendarized yearly values for 2002, 2003 and 2004. Both the fiscal and calendarized values are represented divided by 12 to bring them to the level of the monthly interpolations. Note that with first differences, the interpolations are as horizontal as possible; and with second difference, as linear as possible. The second difference extrapolations and calendarized value for 2004 are rather extreme considering the fact that only three months of 2004 are covered by data, namely Jan., Feb. and March. The first difference interpolations and calendarized values provide more conservative values at both ends. For $\boldsymbol{D}=\boldsymbol{I}_T$, the fractional interpolations coincide with the fiscal values divided by 12.

Note that the second difference interpolations and calendarized values would be appropriate to calendarize fiscal year data, which behave monotonically, e.g. linearly. Perhaps the yearly sales of food would display such a behaviour.

Casting calendarization as a Denton-based benchmarking-interpolation problem, provides a more general framework to address the issue. As shown above, the benchmarking framework offers weights for all three calendar years covered by two fiscal year values of Fig. 9.2a and not just the middle year 2003. A further generalization is to apply the benchmarking approach to more fiscal years.

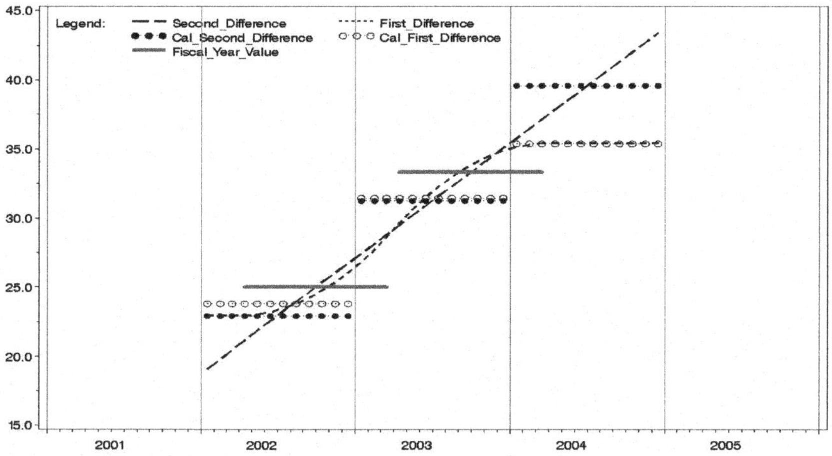

Fig. 9.2a. Monthly interpolations and corresponding calendarized yearly values obtained with the modified second and first difference additive Denton benchmarking methods applied to two fiscal values

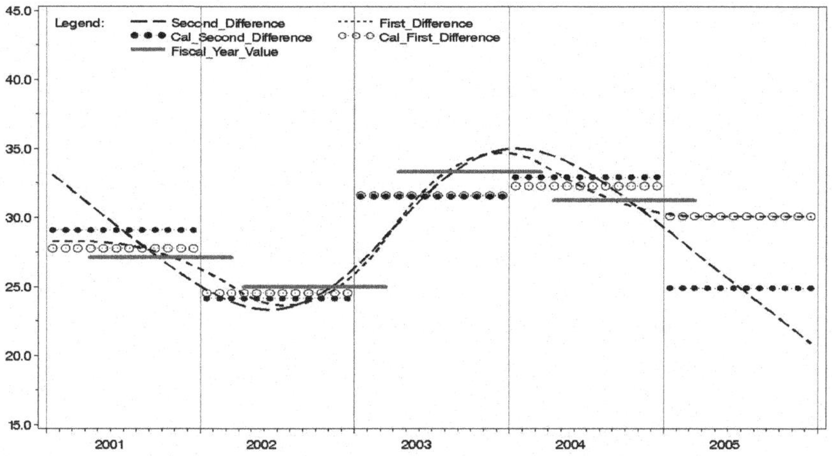

Fig. 9.2b. Monthly interpolations and corresponding calendarized yearly values obtained with the modified second and first difference additive Denton benchmarking methods applied to four fiscal values

Setting M=4, s=0 and D=D^2 from Eq. (6.19), the weights W_c are

1.35502	-0.49751	0.17998	-0.03748
0.15002	0.98155	-0.16315	0.03158
-0.04344	0.26629	0.84775	-0.07060
0.02859	-0.13947	0.44317	0.66771
-0.12127	0.58012	-1.54643	2.08758

Setting $\boldsymbol{D}=\boldsymbol{D}^1$ from (6.4), the weights for a monthly indicator, the above weights $\boldsymbol{W}_c$ of Eq. (9.5b) become

1.15361	-0.19410	0.05050	-0.01002
0.20197	0.89777	-0.12441	0.02468
-0.05018	0.26698	0.87316	-0.08996
0.01483	-0.07478	0.30143	0.75852
-0.01701	0.08579	-0.32973	1.26096

Fig 9.2b displays the modified Denton-based interpolations and the calendarized values over five years, obtained from four fiscal year values. Because, the calendarization involves four fiscal year (instead of two), the reliability of the estimates embedded in fiscal data (i.e. 2002 and 2003 in the figure) is increased. Again the second order difference interpolations and calendarized estimates at both ends are quite extreme, especially for 2005. As shown in Fig 9.2b, the 2002, 2003 and 2004 calendarized estimates of Fig 9.2a undergo revisions as a result of incorporating the 2001-02 and the 2004-05 fiscal values. The revision is most drastic for the 2004 second order difference estimate.

9.4.2 Regression-Based Method

Calendarization can also be achieved with other benchmarking methods, namely the Cholette-Dagum regression-based models of Chapter 3 and 5. The additive and the proportional models can be used.

Fig. 9.2c shows the monthly interpolations and calendarized values produce by the additive regression-based benchmark method, where a constant (matrix $\boldsymbol{R}=-1$) is used to capture the average level of the fiscal year values and an autoregressive error model $e_t = \mathbf{0.90}\, e_{t-1} + v_t$ to perform the residual adjustment necessary to satisfy the benchmarks. This error model causes the extrapolations to converge to the average level. In this particular case, the modified Denton first difference interpolations are very close to those obtained from the additive regression-based model with autoregressive errors. This is also the case for the end years 2001 and 2005, because of the particular configuration of the fiscal values. However, if the 2004-05 fiscal benchmark were at the level of the 2003-04 value, the modifed Denton first difference extrapolations would not converge to the average level. Indeed, the Denton extrapolations do not converge beyond the last sub-annual period covered by the last benchmark and before the first sub-annual period covered by the first benchmark.

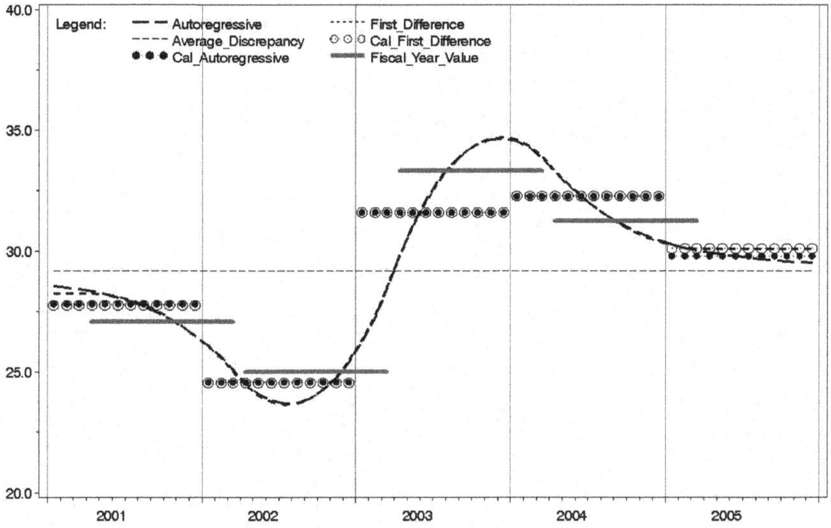

Fig. 9.2c. Monthly interpolations and corresponding calendarized yearly values obtained from the modified first difference additive Denton method and from the additive Cholette-Dagum benchmarking method under AR(1) errors model with φ=0.90

Note that the two methods produce 2002 calendarized values lower than both the 2001-02 and the 2002-03 fiscal values. This makes sense because the three first fiscal values imply a turning point most probably during the 2002-03 fiscal period. As a result the underlying monthly series must reach under the 2002-03 fiscal value. Recall that the fiscal and calendarized values are represented divided by 12.

The fractional method of Eq. (9.2), on the other hand, always produce values (not shown) between the two overlapping fiscal values, because it is a convex combination of the two fiscal values. This clearly underestimate the amplitude of the business cycle compared to the two other methods illustrated.

In the case of fiscal years covering 12 months, it is not necessary to generate *seasonal* interpolations, because seasonality approximately cancels out over *any* twelve consecutive months, in the fiscal year data as well as in the calendar year estimates.

Note that the resulting monthly (or quarterly) interpolations contain trend-cyclical variations; however the timing of the turning points may not be correct, and the amplitude may be underestimated. Indeed, the gain function of yearly sums in Fig. 9.1a (12-term curve) shows that short business cycles are partly *eliminated* in fiscal or calendar year data; as a result these cycles are not fully recoverable without an appropriate cyclical indicator series.

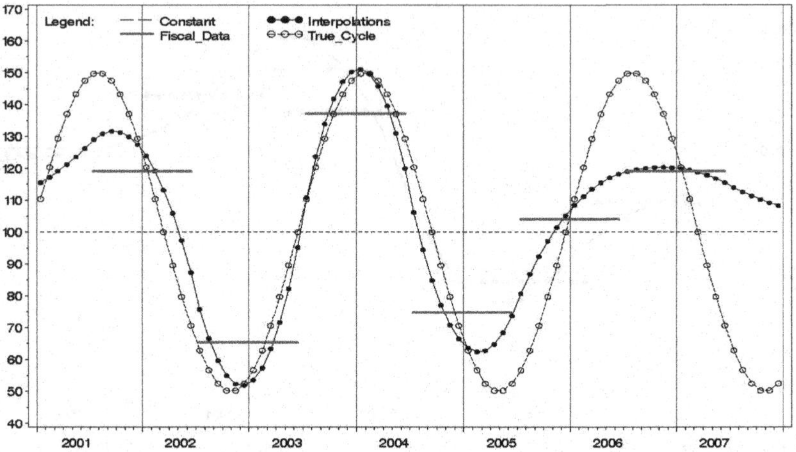

Fig. 9.3a. Cycle lasting 30 months and corresponding fiscal year values with interpolations obtained from the regression-based benchmarking method with an AR(1) error model with φ=0.90

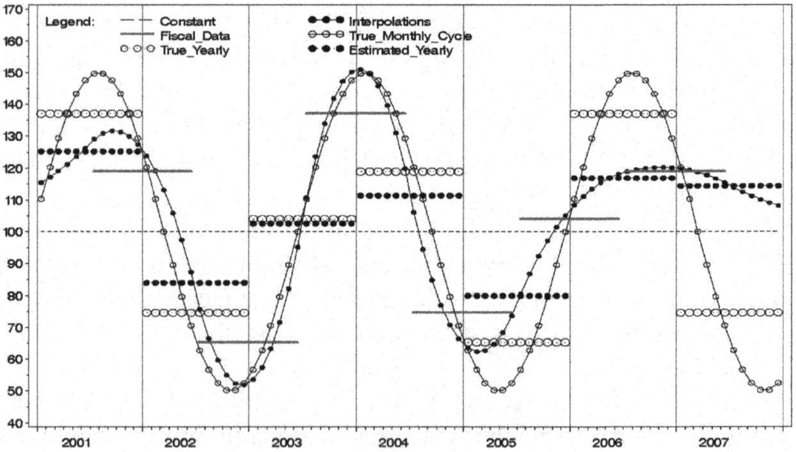

Fig. 9.3b. Corresponding true and estimated calendar year values

In order to illustrate that point, Fig. 9.3a displays a 30-month sine wave $c_t = 100 + 50 \times sin(2\pi t/30)$, the corresponding fiscal year values covering from July 1 to June 30 of the following year and the interpolations obtained from the additive regression-based method with autoregressive errors (φ=0.90). The interpolations reproduce the true trend-cycle relatively well for 2003 and 2004 but not for the other years. The performance would be better for longer cycles and worse for shorter ones.

Fig 9.3b also displays the true calendarized yearly values (i.e. the yearly sums of c_t) and the estimated calendar year estimates obtained from the interpolations. Recall that the calendar and fiscal year values are represented divided by 12 to bring them to the scale of the monthly interpolations.

Table 9.3 exhibits the estimated calendar year values obtained with the regression-based interpolation method and with the fractional method of Eq. (9.2) with fractions equal to 0.50 and 0.50. The estimation errors are smaller with the regression-based method, because a more appropriate model is applied and the calendarized value depends on all six fiscal values. The estimates of the fractional method, on the other hand, are based only on the two overlapping fiscal values. Moreover, the fractional method produces no estimate at the ends, for 2001 and 2007.

Table 9.3. Calendarized yearly values obtained from the regression-based interpolation method and the fractional method

Year	True Value	Reg.-based Estim.	Reg.-based Error	Fractional Estim.	Fractional Error
2001	1,645	1,503	-8.6%	.	.
2002	896	1,008	12.6%	1,106	23.4%
2003	1,248	1,231	-1.3%	1,215	2.7%
2004	1,427	1,337	-6.3%	1,271	-11.0%
2005	784	960	22.3%	1,072	36.7%
2006	1,645	1,403	-14.7%	1,338	-18.7%
2007	896	1,373	53.3%	.	.

If the goal is to obtain calendar quarter or calendar month estimates, these calendarized values will display no seasonality nor trading-day variations.

Casting the calendarization problem in the benchmarking-interpolation framework also allows for seasonality and trading-day variations if applicable. Indeed, if the original series $\boldsymbol{s}$ is a seasonal pattern, or a seasonal series related to the variable of interest, the resulting interpolations will display seasonality.

Fig. 9.4 illustrates such a case. The fiscal year values pertain to the Operation Profits series of the banking sector, with reporting periods ranging from Nov. 1 to Oct. 31 of the following year. The interpolations are obtained by benchmarking a fictitious seasonal pattern using the proportional benchmarking model with autoregressive errors described in Chapter 3 section 3.8. The autoregressive parameter equal to 0.90 causes the extrapolations to converge to the multiplicatively rescaled seasonal pattern. This is particularly apparent in 1992.

The resulting interpolations may be used directly as calendarized monthly values or transformed in calendar quarter values by taking appropriate temporal sums. It should be noted that the seasonal pattern of the estimates are appropriate only if the indicator series is derived from a variable highly correlated with the target variable.

In the case of stock series, the use of an AR(2) error model should be preferred to an AR(1), as shown in Chapter 4.

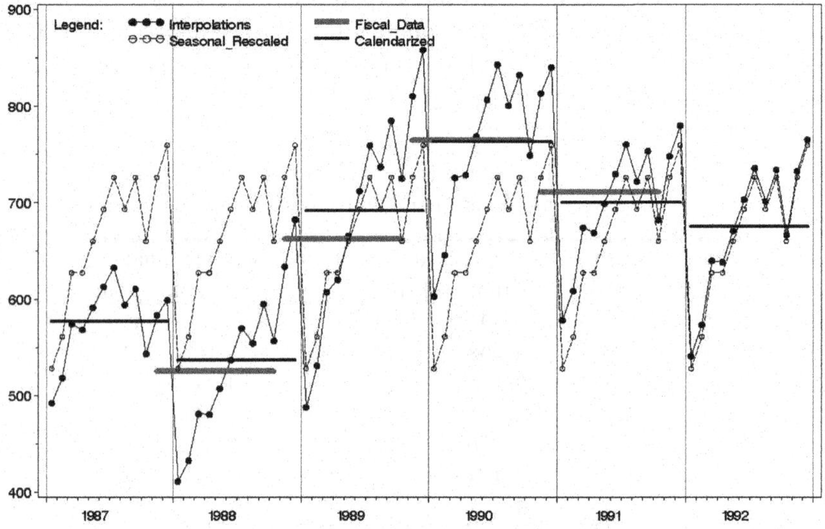

Fig. 9.4. Monthly interpolations and corresponding calendar year values obtained from the proportional Cholette-Dagum benchmarking method under an AR(1) error model with φ=0.90

In some cases, calendarization may be achieved as a by-product of genuine benchmarking, where both the annual and sub-annual data measure the same variable at different frequencies. Such situations are ideal for calendarization. This is potentially the case of the Canadian banking sector. All businesses in that sector have a common fiscal year, ranging from Nov. 1 to Oct 31, and common fiscal quarter consistent with the fiscal year: Nov. 1 to Jan. 31, Feb. 1 to April 30, and so forth. Calendarization can then be achieved by benchmarking monthly series to the fiscal year values and/or the fiscal quarter values. The calendar year and calendar quarter estimates are simply obtained by taking the temporal sums benchmarked series over the calendar years and the calendar quarters.

Fig. 7.8 of Chapter 7 illustrates such a case where the monthly interpolations (distributions) are obtained from fiscal quarter benchmarks and a constant indicator. The model used is the multiplicative benchmarking model of Chapter 5, with seasonal regressors, a spline trend levelling at the ends and autoregressive errors. These interpolations can be temporally summed into calendar quarter values.

The federal and provincial governments in Canada have common fiscal years ranging from April 1 to March 31 of the following year. Such situations are ideal for calendarization, if such sectors have genuine sub-annual data corresponding to the fiscal years.

9.5 Calendarizing Multi-Weekly Data Covering 4 or 5 Weeks

As already mentioned, some businesses report on a multi-weekly basis, i.e. every four or five weeks. These are typically the most important units in terms of activity. Such reports must be calendarized, i.e. transformed into monthly values. Under our approach to calendarization, there are two ways to calendarize reported values of short duration. One way is to generate daily interpolations (temporal distributions) followed by temporal sums; and the other, to directly generate monthly interpolations.

Under the daily alternative, a daily pattern $\{d_\tau\}$ is benchmarked to multi-weekly data. In this regard, Cholette and Chhab (1991) recommended the use of the modified proportional first difference Denton method. However, the proportional benchmarking model of Chapter 3 section 3.8, with autoregressive errors of order 1, gives better results. Since $\varphi = 0.90$ is appropriate for monthly series, the consistent choice for daily series is $\varphi = 0.90^{1/30.4375}$. The resulting model $e_\tau / d_\tau = 0.90^{1/30.4375} e_{\tau-1} / d_{\tau-1} + v_\tau$ achieves a convergence after a month (30 days) comparable to the corresponding monthly model: $e_t / s_t = 0.90\, e_{t-1} / s_{t-1} + v_t$.

Fig. 9.5 illustrates such daily interpolations. The daily weighting pattern used is equal to 700/5 for working days and 0.001 for Saturday and Sunday. The multi-weekly benchmarks are represented divided by the number of days covered, the actual values are displayed in Table 9.4. The error model causes the convergence of the extrapolations to the multiplicatively re-scaled daily pattern.

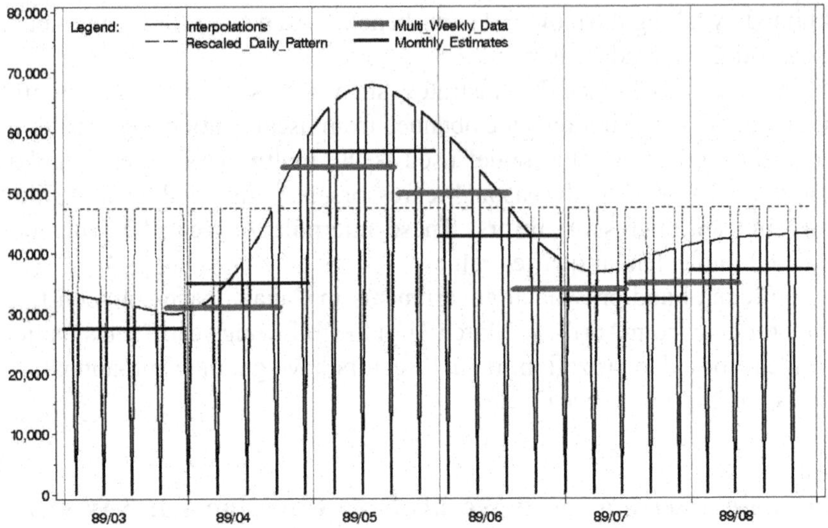

Fig. 9.5. Daily interpolations and the corresponding calendar month estimates obtained from the proportional Cholette-Dagum benchmarking method under an AR(1) error model with φ=0.90$^{(1/30.4375)}$

In the absence of an appropriate daily pattern any constant can be used, e.g. 100; this would produce monthly values without trading-day variations. Daily patterns cannot be estimated from data covering multiples of seven days, because the daily pattern cancels out over any seven consecutive days. This problem could be addressed if the multi-weekly reporters provided their daily pattern stating the relative importance the days of the week.

The calendarized monthly values and their covariance matrix are given by Eq. (3.24):

$$\hat{\boldsymbol{\theta}}_c = \boldsymbol{J}_c \hat{\boldsymbol{\theta}}, \tag{9.6a}$$

$$var[\hat{\boldsymbol{\theta}}_c] = \boldsymbol{J}_c \, var[\hat{\boldsymbol{\theta}}] \, \boldsymbol{J}_c', \tag{9.6b}$$

where $\hat{\boldsymbol{\theta}}$ denotes the daily interpolations, $\boldsymbol{J}_c$ is a monthly sum operator and $var[\hat{\boldsymbol{\theta}}]$ is the covariance matrix of the daily interpolations defined in Eq. (3.21). The calculation of the daily interpolations and particularly their covariance matrix is computer intensive. The formulae given in Appendix C reduce the computational burden.

The purpose of $var[\hat{\boldsymbol{\theta}}_c]$ is to provide a measure of relative reliability under the assumptions that both the error model and the daily pattern are appropriate. The resulting calendarized values and their CVs are displayed in Table 9.4. Note that the months less covered by data (March and August) display larger CVs, whereas those more covered by data (April to July) have smaller CVs.[4]

Table 9.4. Original multi-weekly values and corresponding monthly estimates

Reporting Periods		Reported Values			Calendarized Values	CVs
27/03/89	23/04/89	865,103	01/03/89	31/03/89	873,463	15.5
24/04/89	21/05/89	1,514,139	01/04/89	30/04/89	1,016,092	2.9
22/05/89	18/06/89	1,394,750	01/05/89	31/05/89	1,793,213	1.6
19/06/89	16/07/89	945,691	01/06/89	30/06/89	1,285,258	2.6
17/07/89	13/08/89	973,716	01/07/89	31/07/89	966,809	3.8
			01/08/89	31/08/89	1,168,915	9.1

The seasonality present in the monthly estimates originates from the multi-weekly data which are more frequent than the month. However, the amplitude and the location of the directional changes in the seasonal pattern may not be exact. Furthermore, the extrapolation of the daily values does not take seasonality into account.

A solution is to benchmark the product of a daily pattern multiplied by a seasonal pattern defined on a daily basis. However, it is not easy to find even a monthly seasonal pattern for the purpose of calendarization at the unit (micro) level. One alternative would be to borrow a monthly seasonal pattern, repeat each seasonal factor over the days of each month and standardize it for the length of month. The latter could be achieved by dividing the daily values of the seasonal pattern by the number of days in each month and multiplying by 30.4375 (the average number of days in a month).

A solution *not* requiring a seasonal pattern is to apply model (7.12) with $\boldsymbol{J}=\boldsymbol{I}_T$ (which is then equivalent to the smoothing model (10.7)):

$$\boldsymbol{s} = \boldsymbol{Z}\boldsymbol{\delta} + \boldsymbol{\eta} + \boldsymbol{e}, \qquad E(\boldsymbol{e}) = 0,\ E(\boldsymbol{e}\,\boldsymbol{e}') = \boldsymbol{V}_e \tag{9.7a}$$

$$\boldsymbol{D}\,\boldsymbol{\eta} = \boldsymbol{\upsilon}, \qquad E(\boldsymbol{\upsilon}) = 0,\ E(\boldsymbol{\upsilon}\,\boldsymbol{\upsilon}') = \boldsymbol{V}_\upsilon. \tag{9.7b}$$

where $\boldsymbol{s}$ now denotes the monthly calendarized values $\hat{\boldsymbol{\theta}}_c$ with their covariance matrix $\boldsymbol{V}_e = var[\hat{\boldsymbol{\theta}}_c]$ from (9.6). The matrix $\boldsymbol{Z}$ may contain trading-day regressors and $\boldsymbol{\delta}$ the corresponding daily coefficients. Matrix $\boldsymbol{D}$ of dimension $T-13\times T$ is a combined regular and monthly seasonal difference

[4] The calculation of the variances and the corresponding coefficients of variation entails the calculation of the residuals according to Eq. (3.22) and (3.23).

operator. (This operator is exemplified by (7.13c) for quarterly series.) As a result, the stochastic part η_t tends to behave in a seasonal manner

$$\eta_t - \eta_{t-1} = \eta_{t-12} - \eta_{t-13} + \upsilon_t \,, \tag{9.8a}$$

$$\Rightarrow \eta_t - \eta_{t-12} = \eta_{t-1} - \eta_{t-13} + \upsilon_t \,. \tag{9.8b}$$

Eq. (9.8a) states that the month-to-month change in η_t tends to largely repeat from year to year. Eq. (9.8b) states that the year-to-year change in η_t tends to largely repeat from month to month. The solution to model (9.7) is given by Eqs. (7.15). The new monthly estimates, $\tilde{\boldsymbol{\theta}} = \boldsymbol{Z}\tilde{\boldsymbol{\delta}} + \tilde{\boldsymbol{\eta}}$, are given by Eq. (7.15b) and their covariance matrix by Eq. (7.15d). The lower the signal-to-noise ratio, $\sigma_\upsilon^2 / \sigma_e^2$, the more η_t behaves as per Eq. (9.8).

Applied in this manner, model (9.7) can be viewed as a signal extraction model. The model provides structure to the monthly interpolations and extrapolations. However, it requires a continuous sequence of multi-weekly data covering at least two full years and may produce negative values.

Under the monthly alternative to calendarization, a monthly indicator with seasonality and trading-day variations is benchmarked to the multi-weekly reported values, specified as benchmarks covering fractions of adjacent months. For example, a 4-week reported value with starting and ending dates respectively equal to June 17 2005 and July 14 2005 covers 14/30 of June 2005 and 14/31 of July 2005. The corresponding benchmarking constraint is

$$a_m = j_{m,t-1}\, \theta_{t-1} + j_{m,t}\, \theta_t \,, \tag{9.9}$$

where a_m stands for the 4-week reported value, $t-1$ and t respectively denote the months of June and July 2005. The coverage fractions are then equal to $j_{m,t-1}$ = 14/30 for June and $j_{m,t}$ = 14/31 for July. A five-week reported value can span more than two months, in which case there would be a third coverage fraction.

Formally, let $\boldsymbol{A}_m$ be the set of dates (defined by their numerical values as explained in section 9.1) covered by the m-th benchmark and $\boldsymbol{B}_t$ the set of days covered by month t. The coverage fraction is then

$$j_{mt} = \frac{\textit{number of dates in } \boldsymbol{A}_m \cap \boldsymbol{B}_t}{\textit{number of dates in } \boldsymbol{B}_t} .$$

Better coverage fractions can be obtained by replacing the dates in each set $\boldsymbol{A}_m \cap \boldsymbol{B}_t$ and $\boldsymbol{B}_t$ with the corresponding daily weights and by taking the sum:

$$j_{mt} = \frac{\textit{sum of the daily weights in } \boldsymbol{A}_m \cap \boldsymbol{B}_t}{\textit{sum of the daily weights in } \boldsymbol{B}_t}.$$

The resulting coverage fractions take into accounts the relative importance of the days.

When the reporting periods are short, there may be more benchmarks than monthly interpolations to be produced. Specifying these benchmarks as binding causes a non-invertible matrix in the calculations due over-determination. One solution is to assign variances to some of the benchmarks. For example, in the case of the Canadian Goods and Services Tax, the problem is addressed by setting the variance of the reported. value σ_m^2 equal to $(cv \times y_m \times \pi_m)^2$ where cv stands for a pre-determined coefficient of variation parameter, e.g. cv =0.1%. The factor π_m is equal to $[\Sigma_t^T \min(j_{m,t}, 1-j_{m,t})] / (\Sigma_t^T j_{m,t})$, where the numerator is the number of months partially covered and the denominator is the total number of months covered by a_m. For example if the reporting period considered covers the last 10 days of May 1999, all the days of June 1999 and the 15 first days of July 1999, the number of months partially covered is equal to 25/31 (i.e. 10/31+0+15/31) and the total number of months covered $(\Sigma_t^T j_{m,t})$ is 10/31+30/30+15/31, hence π_m is equal to 0.4464.

The reported values are then specified as binding benchmarks when they cover only whole-months ($j_{m,t}$ equal to 0 or 1); and, as mildly non-binding otherwise.

The monthly approach provides satisfactory results when the reported values cover at least than 42 days (six weeks), except for the reported values covering one whole month. In the presence of non-monthly reported values covering less than 42 days, the coverage fractions may take value in the neighbourhood of 0.50, e.g. 14/30 and 14/31. Such fractions may produce negative values and large spurious fluctuations in the monthly interpolations. Note that monthly reported values (which cover less than 42 days) are not problematic, because their coverage fraction is equal to 1; on the contrary, they contribute the most valuable information to the calendarization, i.e to the estimation of monthly values.

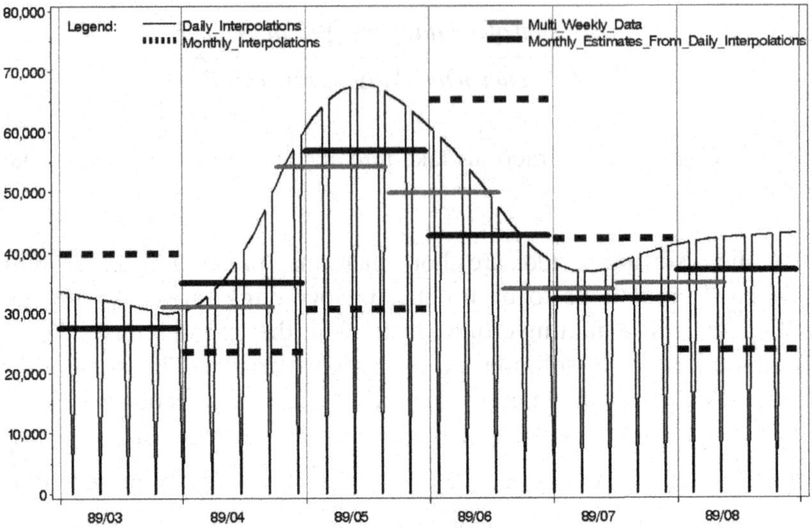

Fig. 9.6. Daily interpolations and corresponding calendar month estimates obtained from the proportional Cholette-Dagum benchmarking method under an AR(1) error model with $\varphi=0.90^{(1/30.4375)}$

Fig. 9.6 compares the monthly calendarized estimates obtained from daily interpolations (also displayed in Fig. 9.5) with those obtained from monthly interpolations. The latter estimates are produced by proportionally benchmarking a re-scaled monthly constant to the multi-weekly data; the same daily pattern is used in the calculation of the coverage fractions. The resulting monthly estimates exhibit a behaviour not supported by the data. The estimate for May 1989 lies very much below the overlapping April-May and May-June multi-weekly data points. On the contrary, the June estimate lies much above the data points overlapping June. The other estimates are also implausible.

The problem can be observed by additively benchmarking a constant monthly series to constant yearly benchmarks with reporting periods starting more or less in the middle of a month and ending in the middle of the following month, using model (3.15) with $\boldsymbol{V}_e = \boldsymbol{I}_T$, $\boldsymbol{V}_\varepsilon = \boldsymbol{0}$ and H=0. This experiment isolates the effect of coverage fractions in the neighbourhood of 0.50.

The estimates obtained from daily interpolations, on the other hand, are consistent with the movements in the multi-weekly data: when the data goes up so do the monthly estimates and vice versa.

9.6 Calendarizing Payroll Deductions

For certain variables, e.g. Income Tax, Superannuation and Employment Insurance premiums, the daily pattern is defined over two weeks instead of one. These variables are related to pay days and the Payroll Deduction file. Pay days typically take place every second week, often on Wednesday, Thursday or Friday. As a result, two or three times per year, some months contain three bi-weekly pay-days instead of two, which can causes an excess of "activity" in order of 50%. Similarly, some quarters may contain from six to eight pay-days instead of six, which can cause an excess of 17% or 25%. Note that for these lamented excesses in the monthly data imply that six weeks of data were assigned to a specific month.

Appendix E provides regressors for bi-weekly daily patterns. In order to use these regressors on monthly or quarterly data, the data must cover only the days in each month or quarter. Unfortunately, the data is forwarded to the statistical agency in the form of reported values with different and possibly unknown timings and durations. Each reported value is likely to be arbitrarily assigned a month, perhaps after some editing.

We propose a simpler method, which relies on the following assumptions:

(1) The reported values cover two or four weeks, well defined by their starting and ending dates. Reports covering two weeks would provide better timeliness.

(2) The Pay and the other flow variables pertain to the active days of the reporting periods. Subject matter expertise could determine whether a daily pattern is required.

The method proposed consist of benchmarking a daily pattern to the bi-weekly reported values over the days. The proportional benchmarking model described in section 3.8 of Chapter 3 is appropriate, with error model $e_\tau / d_\tau = 0.90^{1/30.4375}\, e_{\tau-1} / d_{\tau-1} + v_\tau$, where d_τ is the daily pattern. This approach distributes the reported values gradually over the days of the reporting periods.

In the absence of a daily pattern, the monthly sums of the resulting daily interpolations (distributions) yield the calendarized monthly values without trading-day variations. In the absence of a seasonal pattern, the daily extrapolations and the monthly values involving these extrapolations would subject to heavy revision, as more data are added. The calendarized monthly values retained should be embedded in bi-weekly data. For example, if the last reported values ends on Oct. 12 2006, the last usable calendarized monthly value would pertain to Sept. 2006, which would also be subject to a much smaller revision.

10 A Unified Regression-Based Framework for Signal Extraction, Benchmarking and Interpolation

10.1 Introduction

In 1998, Dagum, Cholette and Chen developed a generalized dynamic stochastic regression model that encompasses statistical adjustment, such as signal extraction, benchmarking, interpolation, temporal disaggregation and extrapolation. This dynamic model provides a unified regression-based framework for all the adjustments treated as separate problems in the previous chapters. ARIMA model-based techniques, e.g. benchmarking with ARIMA model signal extraction by Hillmer and Trabelsi (1987), is included as a special case of the generalized regression model. Other benchmarking methods, e.g. those by Denton (1971) and its variants, are also included. Several interpolation and temporal distribution are shown to be particular cases of the generalized model, e.g. the Boot, Feibes and Lisman (1967) method, the Chow-Lin (1971 1976) method and several of its variants. Restricted ARIMA forecasting methods are also comprised, e.g. Cholette (1982), Pankratz (1989), Guerrero (1989) Trabelsi and Hillmer (1989).

10.2 The Generalized Dynamic Stochastic Regression Model

The generalized dynamic stochastic regression model developed by Dagum, Cholette and Chen (1998) consists of three equations:

$$s = R\beta + Z\delta + \eta + e, \qquad E(e) = 0,\ E(e\,e') = V_e, \qquad (10.1a)$$

$$a = JZ\delta + J\eta + \varepsilon, \qquad E(\varepsilon) = 0,\ E(\varepsilon\,\varepsilon') = V_\varepsilon, \qquad (10.1b)$$

$$D\,\eta = \upsilon, \qquad E(\upsilon) = 0,\ E(\upsilon\,\upsilon') = V_\upsilon, \qquad (10.1c)$$

where $E(e\,\varepsilon') = 0$, $E(e\,\upsilon') = 0$, $E(\varepsilon\,\upsilon') = 0$. The notation from the article (*Ibid.*) is altered to conform with that used in previous chapters.

Equation (10.1a) states that the high frequency or *sub-annual* observations of the *signal* $\theta = Z\delta + \eta$ are contaminated by bias and by autocorrelated errors. Vector $\boldsymbol{s}$ denotes the T sub-annual observations of a socio-economic variable. Matrix $\boldsymbol{R}$ with dimension $T \times H$ contains regressors, such that the product $\boldsymbol{R\beta}$ captures the systematic difference between the low frequency or *annual* and the "sub-annual" data. This difference may be one of level, in which case the $\boldsymbol{R}$ is $T \times 1$ equal to -1. In the case of a trend, matrix $\boldsymbol{R}$ is $T \times 2$ and also includes a function of time. This function may be time itself or the conservative spline trend regressor discussed in section 3.4.

The signal to be estimated, $\boldsymbol{\theta}$, follows a mixed stochastic process, consisting of a *deterministic part* and a *stochastic part*. The deterministic part $\boldsymbol{Z\delta}$, where $\boldsymbol{Z}$ is a $T \times K$ matrix of known regressors and $\boldsymbol{\delta}$ a $K \times 1$ vector of unknown coefficients, captures seasonal and/or calendar variations in $\boldsymbol{s}$; it may also contain intervention variables, or regression variables related to $\boldsymbol{s}$ and $\boldsymbol{a}$ (e.g. Chow and Lin 1971). The regressors $\boldsymbol{R}$ and $\boldsymbol{Z}$ are typically not present at the same time. The stochastic part of $\boldsymbol{\eta}$ follows a stochastic process as described below.

The last term $\boldsymbol{e}$ denotes the errors affecting the observations; their covariance matrix is given by $\boldsymbol{V}_e$. This covariance matrix can originate from a statistical model previously identified and applied to the series considered.

In most cases, the sub-annual covariance matrix has to be built from known or assumed standard deviations of the sub-annual series and from a known or assumed autocorrelation function, as follows:

$$\boldsymbol{V}_e = \boldsymbol{\Xi}^{\lambda} \boldsymbol{\Omega} \, \boldsymbol{\Xi}^{\lambda}, \tag{10.2}$$

where $\boldsymbol{\Xi}$ is a diagonal matrix containing the standard deviations, matrix $\boldsymbol{\Omega}$ typically contains the autocorrelations of an AutoRegressive Moving Average (ARMA) process, and λ takes the value 0, 1/2 or 1.

When the series originates from a survey, the covariance matrix should ideally be supplied as a by-product of the survey. In practice however, the survey only produces the coefficients of variation. These coefficients can be converted in standard deviations and stored in $\boldsymbol{\Xi}$ of Eq. (10.2).

Eq. (10.1b) states that the observations of the annual sums of the signal are contaminated with errors. Vector $\boldsymbol{a}$ contains the M benchmark measurements of the variable considered. The $M \times T$ Matrix $\boldsymbol{J}$ (discussed in section 3.6.1) is typically an annual sum operator (e.g. $\boldsymbol{J} = \boldsymbol{I}_M \otimes \boldsymbol{1}_{1\times 12}$); but, formally is a design matrix with elements equal to 0 or 1. It can accommodate flow and stock series, calendar and fiscal year benchmarks, annual and sub-annual benchmarks, etc. When a benchmark corresponds to the average of the signal over some sub-annual periods, as is the case for index series, its value and standard deviation are multiplied by the number of periods covered.

The last term ε denotes the errors affecting the benchmarks with known covariance matrix V_ε. Usually, V_ε is diagonal with elements $\sigma_{\varepsilon,m}$, i.e. $\varepsilon_m / \sigma_{\varepsilon,m}$, m=1,...,M, is white noise. The annual errors are usually much smaller than the sub-annual errors. If a benchmark $\boldsymbol{a}$ is not subject to error, i.e. $\varepsilon_m = 0$ and $\sigma^2_{\varepsilon,m} = 0$, it is considered fully reliable and *binding*. *Non-binding* benchmarks are not benchmarks in a strict sense, but simply low frequency measurements of the target variable.

The errors $\boldsymbol{e}$ and $\boldsymbol{\varepsilon}$ are assumed to be mutually independent. This would be the case, for example, when the sub-annual and annual data come from two separate sources (e.g. from a survey and from administrative records) or when $\boldsymbol{\varepsilon} = \boldsymbol{0}$. There are cases where the assumption of mutual independence does not hold, For example, both the American and the Canadian Annual Retail Trade Surveys use a sample drawn from the corresponding monthly survey. In such cases, a new problem arises, that of defining the covariance between $\boldsymbol{e}$ and $\boldsymbol{\varepsilon}$.

Eq. (10.1c) specifies the dynamic behaviour of $\boldsymbol{\eta}$: the appropriate regular and seasonal differences of $\boldsymbol{\eta}$, $\boldsymbol{D\theta}$ follow a stationary stochastic process with mean 0 and covariance matrix V_υ. Often V_υ is such that $\boldsymbol{\upsilon}$ follows an ARMA model (Box and Jenkins 1970). McLeod (1975) provides an algorithm to build the covariance matrix of a stationary ARMA process. Process $\boldsymbol{\upsilon}$ is assumed independent of $\boldsymbol{e}$ and $\boldsymbol{\varepsilon}$.

In the case of quarterly series, a regular first order difference times a seasonal difference $(1-B)(1-B^4)$, the $N{\times}T$ ($N{\leq}T$) matrix $\boldsymbol{D}$ is:

$$\boldsymbol{D} = \begin{bmatrix} 1 & -1 & 0 & 0 & -1 & 1 & 0 & 0 & \cdots \\ 0 & 1 & -1 & 0 & 0 & -1 & 1 & 0 & \cdots \\ 0 & 0 & 1 & -1 & 0 & 0 & -1 & 1 & \cdots \\ \vdots & \vdots & \vdots & \vdots & \vdots & \vdots & \vdots & \vdots & \ddots \end{bmatrix}_{(T-5)\times T} . \tag{10.3}$$

We can write Eqs. (10.1) in the form of a single regression model as follows:

$$\begin{bmatrix} \boldsymbol{s} \\ \boldsymbol{a} \\ \boldsymbol{0} \end{bmatrix} = \begin{bmatrix} \boldsymbol{R} & \boldsymbol{Z} & \boldsymbol{I}_T \\ \boldsymbol{0} & \boldsymbol{JZ} & \boldsymbol{J} \\ \boldsymbol{0} & \boldsymbol{0} & \boldsymbol{D} \end{bmatrix} \begin{bmatrix} \boldsymbol{\beta} \\ \boldsymbol{\delta} \\ \boldsymbol{\eta} \end{bmatrix} + \begin{bmatrix} \boldsymbol{e} \\ \boldsymbol{\varepsilon} \\ -\boldsymbol{\upsilon} \end{bmatrix}, \tag{10.4a}$$

or $\boldsymbol{y} = \boldsymbol{X\alpha} + \boldsymbol{u},\ E(\boldsymbol{u}) = \boldsymbol{0},\ E(\boldsymbol{u}\boldsymbol{u}') = block(\boldsymbol{V}_e, \boldsymbol{V}_\varepsilon, \boldsymbol{V}_\upsilon) = \boldsymbol{V}_u$ (10.4b)
where $block(.)$ denotes a block-diagonal matrix.

For standard regression models of the form (10.4b) where $\boldsymbol{\alpha}$ is a vector of fixed parameters, the well-known Gauss-Markov theorem states that the generalized least square formula (10.5a) minimizes the variance of the estimation error given by Eq. (10.5b). The estimator $\hat{\boldsymbol{\alpha}}$ is then called the best linear unbiased estimator (BLUE).

If vector $\boldsymbol{\alpha}$ contains random parameters with covariance matrix $\boldsymbol{V}_\alpha$ and $\boldsymbol{V}_u$ is equal to $var[\boldsymbol{u}|\boldsymbol{\alpha}]$ (a sufficient condition being the independence of $\boldsymbol{u}$ and $\boldsymbol{\alpha}$), the minimum variance linear unbiased estimator of $\boldsymbol{\alpha}$ is also given by Eqs. (10.5a) and (10.5b). Furthermore, in the case of known mean of $\boldsymbol{\alpha}$, $\boldsymbol{X}'\boldsymbol{V}_u^{-1}\boldsymbol{X}$ in (10.5) can be replaced by $\boldsymbol{V}_\alpha^{-1} + \boldsymbol{X}'\boldsymbol{V}_u^{-1}\boldsymbol{X}$ (Rao 1965 p. 192), and the minimum variance unbiased estimator is also given by Eqs. (10.5a) and (10.5b), in which $\boldsymbol{X}'\boldsymbol{V}_u^{-1}\boldsymbol{X}$ is modified according to Robinson (1991).

However, none of the above situations apply directly to our model (10.4a) because: (1) some elements of vector $\boldsymbol{y}$ are not real observations but forced to be zero (with certainty), (2) the "error" $\boldsymbol{\upsilon}$ is a linear combination of $\boldsymbol{\eta}$, and (3) the stochastic parameter $\boldsymbol{\eta}$ is a non-stationary time series with unknown mean. In order to take into account these special characteristics of model (10.4a), the Dagum et al. (1998) provide a proof of the extended Gauss-Markov theorem shown in Appendix A.

The generalized least squares estimator of both the fixed and random parameters in model (10.4),

$$\hat{\boldsymbol{\alpha}} = (\boldsymbol{X}'\boldsymbol{V}_u^{-1}\boldsymbol{X})^{-1}\boldsymbol{X}'\boldsymbol{V}_u^{-1}\boldsymbol{y}, \tag{10.5a}$$

is minimum variance linear unbiased, with covariance matrix

$$E((\hat{\boldsymbol{\alpha}}-\boldsymbol{\alpha})(\hat{\boldsymbol{\alpha}}-\boldsymbol{\alpha})') = var[\hat{\boldsymbol{\alpha}}] = (\boldsymbol{X}'\boldsymbol{V}_u^{-1}\boldsymbol{X})^{-1}. \tag{10.5b}$$

When the parameters are random variables as in this case, some authors refer to the estimates as "predictions"; and to $\hat{\boldsymbol{a}}$, as best linear unbiased predictors (BLUP) instead of BLUE. For the general theory on BLUP, the reader is referred to the literature on (i) the prediction of random effects (e.g. Henderson 1975, Harville 1976 and Robinson 1991) and (ii) state space models (e.g. Tsimikas and Ledolter 1994).

The estimates of the benchmarked signal $\boldsymbol{\theta} = \boldsymbol{Z}\boldsymbol{\delta} + \boldsymbol{\eta}$ and its covariance matrix are respectively

$$\hat{\boldsymbol{\theta}} = \boldsymbol{X}^{\dagger}\hat{\boldsymbol{a}}\,, \tag{10.6a}$$

$$var[\hat{\boldsymbol{\theta}}] = \boldsymbol{X}^{\dagger}\, var[\hat{\boldsymbol{a}}]\, \boldsymbol{X}^{\dagger\prime}, \tag{10.6b}$$

where $\boldsymbol{X}^{\dagger} = \begin{bmatrix} \boldsymbol{0}_{T\times H} & \boldsymbol{Z} & \boldsymbol{I}_T \end{bmatrix}$.

As discussed in the following Section, model (10.4) and the results (10.5) and (10.6) can be used to obtain optimal estimates of the signal without benchmarking or of the benchmarked series with or without signal extraction; it can also be used for interpolation and extrapolation.

10.3 Signal Extraction

Signal extraction consists of finding the "best" estimator of the unobserved "true" series (the signal) given data corrupted by noise. "Best" is usually defined as minimum mean square error. The objective is to reduce the impact of noise in an observed series. The signal is often assumed to evolve through time according to a stochastic process. The estimated signal is then regarded as an estimate of the "true" series, with smaller error variance than the original observations.

A signal extraction model is obtained from the generalized model given in Eq. (10.1) by considering only the first and third equations:

$$\begin{bmatrix} \boldsymbol{s} \\ \boldsymbol{0} \end{bmatrix} = \begin{bmatrix} \boldsymbol{Z} & \boldsymbol{I}_T \\ \boldsymbol{0} & \boldsymbol{D} \end{bmatrix} \begin{bmatrix} \boldsymbol{\delta} \\ \boldsymbol{\eta} \end{bmatrix} + \begin{bmatrix} \boldsymbol{e} \\ -\boldsymbol{\upsilon} \end{bmatrix}, \quad \begin{matrix} E(\boldsymbol{e}) = \boldsymbol{0},\; E(\boldsymbol{e}\,\boldsymbol{e}') = \boldsymbol{V}_e, \\ E(\boldsymbol{\upsilon}) = \boldsymbol{0},\; E(\boldsymbol{\upsilon}\,\boldsymbol{\upsilon}') = \boldsymbol{V}_{\upsilon}, \end{matrix} \tag{10.7}$$

where the vectors and matrices are as in (10.1) except that $\boldsymbol{\eta}$ now includes the constant bias of Eq. (10.1a) and the same notation is kept for simplicity. Model (10.7) can be written as

$$y = X\alpha + u,\ E(u) = 0,\ E(u\,u') = block(V_e, V_\upsilon) = V_u \qquad (10.8)$$

and, as shown in Appendix A, the generalized least square solution is applicable:

$$\hat{\alpha} = (X' V_u^{-1} X)^{-1} X' V_u^{-1} y,\ \ var[\hat{\alpha}] = (X' V_u^{-1} X)^{-1}. \qquad (10.9)$$

By performing the matrix inversion in (10.9) by parts and using well-known matrix identities, the analytical solution in terms of the basic vectors and matrices in X and y is obtained as:

$$\tilde{\delta} = -(Z'D'V_d^{-1}DZ)^{-1} Z'D'V_d^{-1}[0 - Ds], \qquad (10.10a)$$

$$var[\tilde{\delta}] = (Z'D'V_d^{-1}DZ)^{-1}, \qquad (10.10b)$$

$$\tilde{\eta} = s^{\dagger} + V_e D' V_d^{-1}[0 - D s^{\dagger}], \qquad (10.10c)$$

$$var[\tilde{\eta}] = var[\tilde{\eta}_0] + W V_{\tilde{\delta}} W', \qquad (10.10d)$$

where $var[\tilde{\eta}_0] = V_e - V_e D' V_d^{-1} D V_e$, and $s^{\dagger} = s - Z\tilde{\delta}$, $W = Z - V_e D' V_d^{-1} DZ$ and $V_d = [D V_e D' + V_\upsilon]$. The $N{\times}N$ matrix V_d is the theoretical covariance matrix of Ds. This matrix is non-singular because V_υ is positive definite. Appendix B provides more details.

The estimator and the covariance matrix of the signal are respectively,

$$\tilde{\theta} = Z\tilde{\delta} + \tilde{\eta}, \qquad (10.11a)$$

$$var[\tilde{\theta}] = var[\tilde{\eta}_0] + (Z - W)\, var[\tilde{\delta}]\, (Z - W)'. \qquad (10.11b)$$

It should be noted that the covariance matrix of the signal estimates in Eq. (10.11b) has two components: that pertaining to the stochastic part, $var[\tilde{\eta}_0]$ of (10.10d), and that pertaining to the deterministic part $(Z - W)\, var[\tilde{\delta}]\, (Z - W)'$(and similarly for (10.10d)). In the absence of the deterministic part $Z\delta$, the terms and formulae involving $\tilde{\delta}$ and $var[\tilde{\delta}]$ are simply dropped, and $\tilde{\theta} = \tilde{\eta}$ and $var[\tilde{\theta}] = var[\tilde{\eta}_0]$, which is smaller than the original variance matrix V_e (see (10.10d)).

If δ is known, $Z\tilde{\delta}$ is replaced by μ and $var[\tilde{\delta}]$ is dropped in (10.10) and (10.11). The results then coincide with those given by Bell and Hillmer

(1990). Kohn and Ansley (1987) have solved the finite non-stationary time series signal extraction problem for the purely stochastic case without fixed effects, using their modified Kalman filter approach rather than the matrix regression approach used here. Formulae (10.8) to (10.11) extend the results given of Jones (1980) by incorporating the deterministic part $\boldsymbol{Z\delta}$ in the non-stationary model (p. 224). Binder et al. (1993) provide related readings on time series methods for survey estimation.

10.4 Benchmarking With and Without Signal Extraction

To the best of our knowledge, all the benchmarking carried out by statistical agencies is done without signal extraction. However, should signal extraction be used, Hillmer and Trabelsi (1987) and Durbin and Quenneville (1997) showed that signal extraction can be done in a first step and benchmarking (of the signal estimates) in a second step. The latter authors specified the signal with a structural model cast in state-space form whereas the former did it with an ARIMA model. The same final estimates are obtained as if the calculations had been done in a single step (namely by model (10.1)). As discussed in Hillmer and Trabelsi (1987), Chen, Cholette and Dagum (1993 1997) and Durbin and Quenneville (1997), benchmarking with signal extraction reduces the variance of the error of the estimates compared to benchmarking alone, because the latter does not use statistical information about the signal available in the sub-annual data.

The regression-based benchmarking method by Cholette and Dagum (1994), which does not use signal extraction, can be obtained by considering equations (10.1a) and (10.1b) as follows:

$$\begin{bmatrix} \boldsymbol{s} \\ \boldsymbol{a} \end{bmatrix} = \begin{bmatrix} \boldsymbol{R} & \boldsymbol{I}_T \\ \boldsymbol{0} & \boldsymbol{J} \end{bmatrix} \begin{bmatrix} \boldsymbol{\beta} \\ \boldsymbol{\theta} \end{bmatrix} + \begin{bmatrix} \boldsymbol{e} \\ \boldsymbol{\varepsilon} \end{bmatrix}, \quad \begin{matrix} E(\boldsymbol{e}) = \boldsymbol{0},\ E(\boldsymbol{e}\,\boldsymbol{e}') = \boldsymbol{V}_e, \\ E(\boldsymbol{\varepsilon}) = \boldsymbol{0},\ E(\boldsymbol{\varepsilon}\,\boldsymbol{\varepsilon}') = \boldsymbol{V}_\varepsilon. \end{matrix} \tag{10.12}$$

In (10.12) $\boldsymbol{s}$ is the series to be benchmarked with covariance matrix $\boldsymbol{V}_e$, $\boldsymbol{\theta}$ is the "true" benchmarked series, $\boldsymbol{a}$ contains the benchmarks with error covariance matrix $\boldsymbol{V}_\varepsilon$, $\boldsymbol{\beta}$ is the bias parameter(s), which can be viewed as the systematic discrepancy between $\boldsymbol{a}$ and the corresponding sums of $\boldsymbol{s}$.

Since model (10.12) has the same matrix form as model (10.7), the formulae derived for the latter apply, *mutatis mutandis*. The estimator of the bias parameter(s) and its covariance matrix are respectively

$$\hat{\boldsymbol{\beta}} = -(\boldsymbol{R}'\boldsymbol{J}'\boldsymbol{V}_d^{-1}\boldsymbol{J}\boldsymbol{R})^{-1}\,\boldsymbol{R}'\boldsymbol{J}'\boldsymbol{V}_d^{-1}[\boldsymbol{a}-\boldsymbol{J}\boldsymbol{s}], \qquad (10.13a)$$

$$var[\hat{\boldsymbol{\beta}}] = (\boldsymbol{R}'\boldsymbol{J}'\boldsymbol{V}_d^{-1}\boldsymbol{J}\boldsymbol{R})^{-1}, \qquad (10.13b)$$

and the estimator of the benchmarked series and its covariance matrix are

$$\hat{\boldsymbol{\theta}} = \boldsymbol{s}^{\dagger} + \boldsymbol{V}_e\,\boldsymbol{J}'\boldsymbol{V}_d^{-1}[\boldsymbol{a} - \boldsymbol{J}\boldsymbol{s}^{\dagger}], \qquad (10.13c)$$

$$var[\hat{\boldsymbol{\theta}}] = var[\hat{\boldsymbol{\theta}}_0] + \boldsymbol{W}\,var[\hat{\boldsymbol{\beta}}]\,\boldsymbol{W}', \qquad (10.13d)$$

where $var[\hat{\boldsymbol{\theta}}_0] = \boldsymbol{V}_e - \boldsymbol{V}_e\boldsymbol{J}'\boldsymbol{V}_d^{-1}\boldsymbol{J}\boldsymbol{V}_e$ and $\boldsymbol{s}^{\dagger} = \boldsymbol{s} - \boldsymbol{R}\hat{\boldsymbol{\beta}}$,

$\boldsymbol{W} = \boldsymbol{R} - \boldsymbol{V}_e\boldsymbol{J}'\boldsymbol{V}_d^{-1}\boldsymbol{J}\boldsymbol{R}$ and $\boldsymbol{V}_d = [\boldsymbol{J}\boldsymbol{V}_e\boldsymbol{J}' + \boldsymbol{V}_\varepsilon]$. Matrix $\boldsymbol{V}_d$ is the covariance matrix of the *discrepancies* $[\boldsymbol{a} - \boldsymbol{J}\boldsymbol{s}^{\dagger}]$.

If model (10.12) has no bias (i.e. $\boldsymbol{s}$ is not biased), the terms involving $\hat{\boldsymbol{\beta}}$ or $var[\hat{\boldsymbol{\beta}}]$ are ignored, and $var[\hat{\boldsymbol{\theta}}] = var[\hat{\boldsymbol{\theta}}_0]$. Then these results coincide with those in Hillmer and Trabelsi (1987, eq. (10.6) and (2.6)) for the benchmarking step.

If $\boldsymbol{s}$ and $\boldsymbol{V}_e$ in (10.12) are the signal extraction estimates and their covariance matrix, we are in the situation of benchmarking with signal extraction. (In other words, $\tilde{\boldsymbol{\theta}}$ and $var[\tilde{\boldsymbol{\theta}}]$ are given by (10.11).) Series $\hat{\boldsymbol{\theta}}$ in (10.13c) is then both smoothed (by signal extraction) and benchmarked and has covariance matrix $var[\hat{\boldsymbol{\theta}}]$ given by (10.13d). In other words, estimators $var[\hat{\boldsymbol{\theta}}]$ and $var[\hat{\boldsymbol{\theta}}]$ coincide with those of (10.6a) and (10.6b), despite the fact that $\boldsymbol{a}$ is not used in estimating $\boldsymbol{\delta}$ ($\tilde{\boldsymbol{\delta}}$ was implicitly updated as it is part of $\tilde{\boldsymbol{\theta}} = \boldsymbol{Z}\tilde{\boldsymbol{\delta}} + \tilde{\boldsymbol{\eta}}$ being benchmarked).

The signal extraction method of Section 3 combined with the benchmarking method of this Section extends that of Hillmer and Trabelsi (1987) by incorporating a deterministic component $\boldsymbol{Z\delta}$ in the signal and by allowing a bias term $\boldsymbol{R\beta}$ in the sub-annual series $\boldsymbol{s}$.

10.5 Interpolation, Temporal Distribution and Extrapolation

In some situations, sub-annual (high frequency) values are required for *all* the sub-annual time periods $t=1,...,T$, but are missing for some or even all of these periods; some annual (low frequency) data may also be available. This section addresses the problem of estimating the high frequency values based on some data available at both the low and high frequencies.

The sub-annual values may be interpolated, temporally distributed, and/or extrapolated from a subset of model (10.1), defined as follows:

$$\begin{bmatrix} \boldsymbol{a} \\ \boldsymbol{0} \end{bmatrix} = \begin{bmatrix} \boldsymbol{JZ} & \boldsymbol{J} \\ \boldsymbol{0} & \boldsymbol{D} \end{bmatrix} \begin{bmatrix} \boldsymbol{\delta} \\ \boldsymbol{\eta} \end{bmatrix} + \begin{bmatrix} \boldsymbol{\varepsilon} \\ -\boldsymbol{\upsilon} \end{bmatrix}, \tag{10.14}$$

where vector $\boldsymbol{s}$ now contains the M available high frequency values and $\boldsymbol{D}$ is a $T \times N$ $(N \leq T)$ difference operator. The available values are related to the $T>M$ unobservable "true" high frequency values, $\boldsymbol{\theta} = \boldsymbol{Z\delta} + \boldsymbol{\eta}$, by $\boldsymbol{s} = \boldsymbol{J\theta} + \boldsymbol{\varepsilon}$. The stochastic part $\boldsymbol{\eta}$ is defined as in model (10.1), and similarly the deterministic part $\boldsymbol{Z\delta}$ can stand for seasonal and calendar variations and a linear trend (in which case $\boldsymbol{D}$ is the identity matrix). The matrix $\boldsymbol{V}_\varepsilon$ typically tends to $\boldsymbol{0}$.

Model (10.14) can be written as $\boldsymbol{y} = \boldsymbol{Xa} + \boldsymbol{u}$, $E(\boldsymbol{u}) = \boldsymbol{0}$, $E(\boldsymbol{u}\boldsymbol{u}') =$ $block(\boldsymbol{V}_\varepsilon, \boldsymbol{V}_\upsilon) = \boldsymbol{V}_u$. The generalized least squares estimator of $\boldsymbol{a} = \left[\boldsymbol{\delta}' \ \boldsymbol{\eta}'\right]'$ and its covariance matrix are given by (10.5). The estimator of the high time frequency values and their covariance matrix are given by (10.6) with $\boldsymbol{X}^\dagger = [\boldsymbol{Z} \ \boldsymbol{I}_T]$.

Model (10.14) can be estimated under the following conditions

(i) $\boldsymbol{V}_\upsilon$ is assumed or obtained from a model fitted to the available sub-annual data,

(ii) the total number of observations $M+N$ exceeds the number of parameters $K+T$, where N and K are the number of rows in $\boldsymbol{D}$ and $\boldsymbol{\delta}$ respectively and

(iii) the columns of $\boldsymbol{X}$ are linearly independent.

For missing sub-annual observations, the corresponding elements of $\hat{\boldsymbol{\theta}}$ are the *estimated* interpolations; and for available observations, they are the fitted *smoothed* values.

For the particular case where $\boldsymbol{D} = \boldsymbol{I}_T$ (i.e. $\boldsymbol{\theta}$ consists of a deterministic part $\boldsymbol{Z\delta}$ plus an ARMA process $\boldsymbol{\eta}$), the following analytical solution is obtained:

$$\hat{\boldsymbol{\delta}} = (\boldsymbol{Z}'\boldsymbol{J}'\boldsymbol{V}_d^{-1}\boldsymbol{JZ})^{-1}\boldsymbol{Z}'\boldsymbol{J}'\boldsymbol{V}_d^{-1}\boldsymbol{a}, \tag{10.15a}$$

$$var[\hat{\boldsymbol{\delta}}] = (\boldsymbol{Z}'\boldsymbol{J}'\boldsymbol{V}_d^{-1}\boldsymbol{JZ})^{-1}, \tag{10.15b}$$

$$\hat{\boldsymbol{\eta}} = \boldsymbol{V}_\upsilon\boldsymbol{J}'\boldsymbol{V}_d^{-1}[\boldsymbol{a} - \boldsymbol{JZ}\hat{\boldsymbol{\delta}}], \tag{10.15c}$$

$$var[\hat{\boldsymbol{\eta}}] = var[\hat{\boldsymbol{\eta}}_0] + \boldsymbol{W} var[\hat{\boldsymbol{\delta}}]\boldsymbol{W}', \tag{10.15d}$$

where $var[\hat{\boldsymbol{\eta}}_0] = \boldsymbol{V}_\upsilon - \boldsymbol{V}_\upsilon\boldsymbol{J}'\boldsymbol{V}_d^{-1}\boldsymbol{JV}_\upsilon$, and $\boldsymbol{W} = \boldsymbol{V}_\upsilon\boldsymbol{J}'\boldsymbol{V}_d^{-1}\boldsymbol{JZ}$ and $\boldsymbol{V}_d = \boldsymbol{JV}_\upsilon\boldsymbol{J}' + \boldsymbol{V}_\varepsilon$.

The estimator $\hat{\boldsymbol{\theta}}$ and the covariance matrix $var[\hat{\boldsymbol{\theta}}]$ of the missing high frequency values (interpolations, temporal distributions and/or extrapolations) are respectively

$$\hat{\boldsymbol{\theta}} = \boldsymbol{Z}\hat{\boldsymbol{\delta}} + \hat{\boldsymbol{\eta}}, \tag{10.16a}$$

$$var[\hat{\boldsymbol{\theta}}] = var[\hat{\boldsymbol{\eta}}_0] + (\boldsymbol{Z} - \boldsymbol{W})\, var[\hat{\boldsymbol{\delta}}]\, (\boldsymbol{Z} - \boldsymbol{W})'. \tag{10.16b}$$

Several interpolations and temporal distribution methods in the literature are included as special cases of model (10.14). One such method is that by Chow and Lin (1971, 1976) which produces monthly values from quarterly data using related series as regressors. The regression coefficients are estimated from the quarterly data and the quarterly sums of the monthly regressors. The monthly regressors are then used to predict the target monthly values. Model (10.14) achieves these results by specifying matrix $\boldsymbol{Z}$ as the monthly regressors, by setting $\boldsymbol{J} = \boldsymbol{I}_M \otimes \boldsymbol{1}_3 = block(\boldsymbol{1}_{1\times 3}, \ldots, \boldsymbol{1}_{1\times 3})$, $\boldsymbol{\varepsilon} \to \boldsymbol{0}$, $\boldsymbol{D} = \boldsymbol{I}_T$, and by setting $\boldsymbol{V}_\upsilon$ equal to the covariance matrix of an autoregressive process of order 1 (i.e. $\boldsymbol{\theta} = \boldsymbol{\eta}$ follows an AR(1)). Furthermore, formulae (10.15) to (10.16) are the same as in Bournay and Laroque (1979). The extensions discussed by Fernandez (1981) and Litterman (1983) also emerge from model (10.14) if the appropriate matrices $\boldsymbol{D}$ and $\boldsymbol{V}_\upsilon$ are specified.

Other methods emerge from the following particular case of model (10.14):

$$\boldsymbol{a} = \boldsymbol{J\theta} + \boldsymbol{\varepsilon}, \quad E(\boldsymbol{\varepsilon} = \boldsymbol{0}, \; E(\boldsymbol{\varepsilon}\boldsymbol{\varepsilon}') = \boldsymbol{V}_{\varepsilon}\,, \tag{10.17a}$$

$$\boldsymbol{D\theta} = -\boldsymbol{\upsilon}, \quad E(\boldsymbol{\upsilon}) = \boldsymbol{0}, \; E(\boldsymbol{\upsilon}\boldsymbol{\upsilon}') = \boldsymbol{V}_{\upsilon}\,, \tag{10.17b}$$

where $\boldsymbol{V}_{\varepsilon}$ tends to $\boldsymbol{0}$. The Boot, Feibes and Lisman (1967) method, which interpolates quarterly values between yearly data, is obtained from (10.17), by specifying $\boldsymbol{J} = \boldsymbol{I}_M \otimes \boldsymbol{1}_{1\times 4} = \boldsymbol{block}(\boldsymbol{1}_{1\times 4},...,\boldsymbol{1}_{1\times 4})$ and $\boldsymbol{D}$ as the first (or second) regular difference operator with $\boldsymbol{V}_{\upsilon} = \sigma_{\upsilon}^2 \boldsymbol{I}_{T-1}$.

Similarly the Cohen, Müller and Padberg (1971) method, which interpolates monthly values between annual data, is obtained by specifying $\boldsymbol{J} = \boldsymbol{I}_M \otimes \boldsymbol{1}_{12\times 1}$ and $\boldsymbol{D}$ as the first (or second) regular difference operator with $\boldsymbol{V}_{\upsilon} = \sigma_{\upsilon}^2 \boldsymbol{I}_{T-1}$.

The Stram and Wei (1986) and Wei and Stram (1990) method, which generalizes both previous methods by specifying $\boldsymbol{D}$ as a regular and/or seasonal difference operator of any order and by letting $\boldsymbol{\upsilon}$ follow any ARMA model with autocovariance matrix $\boldsymbol{V}_{\upsilon}$, is also included.

Some extrapolation methods, with and without a deterministic part $\boldsymbol{Z\delta}$, are also particular cases of model (10.14). Indeed model (10.14) can generate ARIMA extrapolations (Box and Jenkins 1970) and restricted ARIMA extrapolations (Cholette 1982, Pankratz 1989, Guerrero 1989 and Trabelsi and Hillmer 1989). Note that in all these $\boldsymbol{Z\delta}$ can be specified as simpler variants of intervention variables (Box and Tiao 1975).

Whether model (10.14) generates interpolations, temporal distributions and/or extrapolations is governed by the configuration of matrix $\boldsymbol{J}$. For example, if quarterly observations are available only for periods 5 to 12 and 17 to 20 ; and one annual observation or forecast from another source, for periods 25 to 28, we have

$$\boldsymbol{J} = \begin{bmatrix} \boldsymbol{0}_{8\times 4} & \boldsymbol{I}_{8\times 8} & \boldsymbol{0}_{8\times 4} & \boldsymbol{0}_{8\times 4} & \boldsymbol{0}_{8\times 4} & \boldsymbol{0}_{8\times 4} \\ \boldsymbol{0}_{4\times 4} & \boldsymbol{0}_{4\times 8} & \boldsymbol{0}_{4\times 4} & \boldsymbol{I}_{4\times 4} & \boldsymbol{0}_{4\times 4} & \boldsymbol{0}_{4\times 4} \\ \boldsymbol{0}_{1\times 4} & \boldsymbol{0}_{1\times 8} & \boldsymbol{0}_{1\times 4} & \boldsymbol{0}_{1\times 4} & \boldsymbol{0}_{1\times 4} & \boldsymbol{1}_{1\times 4} \end{bmatrix}_{13\times 28}.$$

This matrix $\boldsymbol{J}$ generates values for the 15 non-available observations as follows: backcasts for periods 1 to 4 (corresponding to the first column of partitions of $\boldsymbol{J}$), interpolations for time periods 13 to 16 (corresponding to the third column of partitions of $\boldsymbol{J}$); forecasts, for periods 21 to 24 (fifth column); and restricted ARIMA forecasts, for time periods 25 to 28 (sixth column). Furthermore, fitted values are obtained for periods 5 to 12 and 17 to 20 (second and fourth columns of partitions of $\boldsymbol{J}$).

When part of the target series is available at the desired sub-annual frequency, the following procedure should be used: (1) specify and estimate the model for the available segment of the series, using standard ARIMA modelling, with or without intervention, and (2) generate the interpolations and extrapolations, using the formulae above.

10.6 Multiplicative Models for Signal Extraction and Benchmarking

An additive benchmarking model is appropriate when the variances of the sub-annual error are independent of the level of the series. However, for most time series, this condition is not present, and additive benchmarking is sub-optimal. When the error variances are proportional to the level of the series, a multiplicative model is more appropriate. In such cases, all the components of the benchmarking model and the resulting benchmarked estimates tend to have constant CVs, $\hat{\sigma}_t / \hat{\theta}_t \approx c$, as intended by the survey design of the original series.

The seasonal component of most time series display a seasonal amplitude which increases and shrinks with the level of the series. This is illustrated in Fig. 5.3a for the Canadian Total Retail Trade series. Additively benchmarking such series can produce negative benchmarked values. Since benchmarking basically alters the level of series, it is desirable to use proportional (section 3.8) or multiplicative benchmarking. Both methods tend to preserve growth rates, which are widely used to compare the behaviour of various socio-economic indicators. The basic assumption in the proportional and multiplicative benchmarking is that seasonally larger values account for more of the annual discrepancies than the smaller ones. This assumption is widely accepted by practitioners in statistical agencies. The assumption also reduces the possibility of negative benchmarked values.

What distinguish the multiplicative model from proportional benchmarking is that the former can include a multiplicative regression part, which the proportional model does not.

Another important reason for the wide application of the multiplicative benchmarking model is that the proportional discrepancies are standardized, thus facilitating comparisons among series.

We now prove that the actual empirical coefficients of variation of the sub-annual series play the same role in multiplicative benchmarking as standard deviations in additive benchmarking.

Let an observed time series $\{s_t\}$ be the sum of true values $\{\theta_t\}$ *plus* heteroscedastic errors $\{e_t^A\}$, i.e.

$$s_t = \theta_t + e_t^A, \qquad E(e_t^A) = 0, \;\; E((e_t^A)^2) = \sigma_t^2 . \tag{10.18}$$

The additive error e_t^A can be expressed as the product of a standardized error $e_t^\dagger$ and an heteroscedastic standard deviation σ_t:

$$e_t^A \equiv \sigma_t e_t^\dagger , \qquad E(e_t^\dagger) = 0, \;\; E(e_t^{\dagger 2}) = 1, \tag{10.19}$$

where $\sigma_t = c_t s_t$. Substituting $\sigma_t = c_t s_t$ in (10.19), and (10.19) in (10.18), yields $s_t = \theta_t + c_t s_t e_t^\dagger$, which after rearranging becomes

$$s_t = \theta_t \,/\, (1 - c_t e_t^\dagger). \tag{10.20}$$

Taking the logarithmic transformation yields:

$$ln\, s_t = ln\, \theta_t - ln\,(1 - c_t e_t^\dagger). \tag{10.21}$$

Assuming that $(1 - c_t e_t^\dagger) > 0$ and close to one (as is often the case for real data), we can apply the approximation $ln\, x \approx x - 1$ to the last term of (10.21), which becomes $c_t e_t^\dagger$. Substituting $c_t e_t^\dagger$ in (10.21) yields:

$$ln\, s_t \approx ln\, \theta_t + c_t e_t^\dagger = ln\, \theta_t + ln\, e_t , \tag{10.22a}$$

where $E(e_t) = 0$, $E(e_t^2) = c_t^2$, implying a multiplicative model $s_t \approx \theta_t \times e_t$.

Eq. (10.22a) may be written in matrix notation as,

$$\boldsymbol{s}^* \approx \boldsymbol{\theta}^* + \boldsymbol{e}^*,\ E(\boldsymbol{e}^*) = \boldsymbol{0},\quad E(\boldsymbol{e}^* \boldsymbol{e}^{*\prime}) = \boldsymbol{V}_{e^*} = \boldsymbol{\Xi}\,\boldsymbol{\Omega}\,\boldsymbol{\Xi}\,, \tag{10.22b}$$

where $\boldsymbol{s}^* = [\,ln\, s_t\,]$, $\boldsymbol{\theta}^* = [\,ln\, \theta_t\,]$ and $\boldsymbol{e}^* = [\,e_t^*\,] = [\,c_t e_t^\dagger\,] \approx [ln(1 - c_t e_t^\dagger)]$, and $\boldsymbol{\Xi}$ is a diagonal matrix containing the CVs and $\boldsymbol{\Omega}$ is the correlation matrix. For benchmarking without signal extraction and interpolation, matrix $\boldsymbol{\Xi}$ should contain only constant CVs for benchmarking and interpolation, as discussed in Chapter 4. The value used could be that of the target CV or the average of the empirical CVs. Indeed using moving CVs produces undesirable effects, illustrated in Fig. 5.4 and Fig 10.1b. For signal extraction, however the actual CVs should be used. The correlation matrix $\boldsymbol{\Omega}$ is typically assumed to be that of an ARMA process with unit variance; in other words, the standardized error $e_t^\dagger = e_t^* / c_t$ is assumed to follow an ARMA model.

The benchmarked estimates $\hat{\boldsymbol{\theta}}^* = [\,\hat{\theta}_t^*\,]$ and their covariance matrix $var[\,\hat{\boldsymbol{\theta}}^*\,] = [\,v_{ti}^*\,]$ are calculated as described below. The CVs of $\hat{\theta}_t$ are then given by the square root of the diagonal elements of $[\,v_{ti}^*\,]$, i.e. $(v_{tt}^*)^{1/2}$. The estimates and the covariances in the original scale are given by

$$\hat{\theta}_t = exp(\hat{\theta}_t^*), \tag{10.23a}$$

$$v_{ti} = \{exp(v_{ti}^*) - 1\}\ \hat{\theta}_t\, \hat{\theta}_i\,. \tag{10.23b}$$

When the signal θ_t is the product of two parts, b_t and η_t, the same line of reasoning leads to similar conclusions, namely

$$\boldsymbol{s}^* \approx \boldsymbol{b}^* + \boldsymbol{\eta}^* + \boldsymbol{e}^*,\ E(\boldsymbol{e}^*) = 0,\quad E(\boldsymbol{e}^* \boldsymbol{e}^{*\prime}) = \boldsymbol{V}_{e^*} = \boldsymbol{\Xi}\,\boldsymbol{\Omega}\,\boldsymbol{\Xi}\,. \tag{10.24}$$

Equations (10.22) and (10.24) justify the use of empirical CVs, in building the covariance matrix. Note that in the logarithmic equation (10.22a), the CVs play the role of the standard deviations in an additive model. In other words, the CVs of the sub-annual series $\boldsymbol{s}$ are the standard deviations of the logarithmically transformed series.

In the case of signal extraction, where $\theta_t = z_t \times \eta_t \times e_t$, multiplicative signal extraction is achieved by the log-additive variant of model (10.7), with $\boldsymbol{s}$, $\boldsymbol{\delta}$, $\boldsymbol{\eta}$, $\boldsymbol{e}$ and $\boldsymbol{\upsilon}$ replaced by $\boldsymbol{s}^*$, $\boldsymbol{\delta}^*$, $\boldsymbol{\eta}^*$, $\boldsymbol{e}^*$ and $\boldsymbol{\upsilon}^*$. In addition, matrix $\boldsymbol{V}_{e^*}$ is such that e_t^* / c_t follows a known ARMA process with unit variance, and $\boldsymbol{V}_{\upsilon^*}$ is such that $\boldsymbol{\upsilon}^*$ also follows an ARMA model (assumed known). The model is linear in the logs and solved using the formulae of Eqs (10.10).

In the case of benchmarking, the sub-annual observations are "multiplicative" in their components: $s_t = b_t \times \theta_t \times e_t$. However, the annual observations remain additive because they measure the annual sums (and not the products) of the signal: $y_{a,m} = \Sigma_{t \in m} \beta_t + e_{a,m}$.

Model (10.12) thus becomes non-linear in the parameters:

$$\boldsymbol{s}^* = \boldsymbol{R}\,\boldsymbol{\beta}^* + \boldsymbol{\theta}^* + \boldsymbol{e}^*,\; E(\boldsymbol{e}^*) = 0,\; E(\boldsymbol{e}^* \boldsymbol{e}^{*\prime}) = \boldsymbol{V}_{e^*}, \tag{10.25a}$$

$$\boldsymbol{a} = \boldsymbol{J}\, exp(\boldsymbol{\theta}^*) + \boldsymbol{\varepsilon},\; E(\boldsymbol{\varepsilon}) = 0,\; E(\boldsymbol{\varepsilon}\boldsymbol{\varepsilon}') = \boldsymbol{V}_{\varepsilon},\; E(\boldsymbol{e}^* \boldsymbol{\varepsilon}') = 0, \tag{10.25b}$$

where $\boldsymbol{s}^* = [\, ln\, s_t\,]$ (and similarly for $\boldsymbol{\beta}^*$, $\boldsymbol{\theta}^*$ and $\boldsymbol{e}^*$) and where matrix $\boldsymbol{V}_{e^*}$ is such that e_t^* / c_t follows a known ARMA process with unit variance as in model (10.1). If the series to be benchmarked is given by the signal extraction estimates, $\boldsymbol{s}^*$ and $\boldsymbol{V}_e$ above are replaced by $\tilde{\boldsymbol{\theta}}^*$ and $var[\tilde{\boldsymbol{\theta}}^*]$ from (10.11). Section 5.4 provides a non-linear algorithm to estimate the multiplicative benchmarking model, based on linearized regression.

However, if all the benchmarks refer to single time periods (e.g. stock series), Eq. (10.25b) may be replaced by $\boldsymbol{a}^* = \boldsymbol{J}\boldsymbol{\theta}^* + \boldsymbol{\varepsilon}^*$, $E(\boldsymbol{\varepsilon}^*) = \boldsymbol{0}$, $E(\boldsymbol{\varepsilon}^* \boldsymbol{\varepsilon}^{*\prime}) = \boldsymbol{V}_{\varepsilon^*}$. The model is then linear in the logs and may be solved using the formulae (10.13). The estimates and the covariances in the original scale are given by Eq. (10.23).

10.7 A Real Case Example: the Canadian Total Retail Trade Series

Next we discuss two applications of the generalized dynamic stochastic regression model to the Canadian Total Retail Trade series: *Method 1* refers to multiplicative benchmarking without signal extraction, and *Method 2*, to multiplicative benchmarking with signal extraction, interpolation and constrained extrapolation. Both Methods use the actual coefficients of variation c_t of the original series s_t, displayed in Table 10.1 (col. (2)).

In both cases the standardized logarithmic survey error e_t^* / c_t is assumed to follow a seasonal autoregressive model $(1,0)(1,0)_{12}$ with parameters equal to 0.938 and 0.896, respectively. These parameters were obtained from the autocorrelations available at lags 1, 3, 6, 9 and 12 (Mian and Laniel 1993), by finding the parameter values which minimize the mean squared difference between the observed and the corresponding theoretical autocorrelations. The autocorrelations and their fitted values were respectively 0.970 and 0.976; 0.940 and 0.939; 0.918 and 0.912; 0.914 and 0.919; and 0.962 and 0.961.

This model is justified by the following facts: The "take all" units of the survey, accounting for a large fraction of the activity, always stay in the sample, which generates strong first order autocorrelation in the survey error; and the "take some" units are organized in rotating groups. These groups stay in the sample for a maximum of two years and stay out of the sample for a minimum of one year, which generates seasonal autocorrelation. The model selected for e_t^* is used for illustrative purposes.

10.7.1 Method 1: Multiplicative Benchmarking Without Signal Extraction

Method 1 consists of applying the multiplicative benchmarking model (10.25), with a survey error following the model described above. Fig. 10.1a displays the original monthly data and its benchmarks. We should note that the four available yearly benchmarks (divided by 12 in the figure) cover *fiscal* periods from February to January of the following year. The three monthly benchmarks for October, November and December 1989 are values from the "new" Statistics Canada redesigned Retail Trade survey introduced in 1989. Their levels are comparable to that of the yearly benchmarks; this is why they are specified as benchmarks. Both the yearly and monthly benchmarks and their CVs are displayed in Table 10.2.

The matrix $\boldsymbol{J}$ necessary in model (6.2.1) to accommodate the mixture of fiscal year and monthly benchmarks is

$$J_{7\times120} = \begin{bmatrix} 0_{4\times61} & I_4 \otimes 1_{1\times12} & 0_{4\times8} & 0_{4\times3} \\ 0_{3\times61} & 0_{3\times48} & 0_{3\times8} & I_{3\times3} \end{bmatrix}. \tag{10.26}$$

The matrix associates each benchmark to a set months and thus accounts for the duration and the timing of the benchmarks with respect to the range of months covered (Jan. 1980 to Dec. 1989).

The first row of partitions of J pertains to the four yearly benchmarks; this is why all the partitions have 4 rows. The reference periods of the yearly benchmarks start in time period 62, which explains the 61 columns of zeroes $0_{4\times61}$ preceding the partition $I_4 \otimes 1_{1\times12}$. This partition is in fact $block(1,1,1,1)_{4\times48}$ where 1 is a 1×12 vector of ones. The first yearly benchmark covers periods 62 to 73; the second, 74 to 85; the third 86 to 97; and the fourth, 98 to 109; i.e the fiscal years Feb. 1984 to Jan. 1985, Feb. 1985 to Jan. 1986, and so on.

The second row of partitions of J pertains to the three monthly benchmarks; this is why all the partitions have 3 rows. The first monthly benchmark covers period 118; this explains the 117 columns of zeroes preceding partition $I_{3\times3}$. The three monthly benchmarks cover the consecutive periods 118 to 120, hence the presence of $I_{3\times3}$.

The resulting benchmarked series displayed in Fig. 10.1a raises the original monthly series to the level of the benchmarks and leads to the new survey values (the monthly benchmarks) at the end of the series. Thus the linking or historical continuity of the "old" series to the "new" series is achieved as a by-product of benchmarking as in Helfand et al. (1977). The data and these results are displayed in Table 10.1 and Table 10.2. The estimate of the multiplicative bias parameter is 1.1061 with standard deviation 0.0061.

The corresponding multiplicative corrections $\hat{\theta}_t / s_t$ displayed in Fig. 10.1b display a mild seasonal pattern, due to the seasonal autoregressive model. The corrections also display erratic fluctuations, in 1987 and 1988 especially, due to the moving CVs displayed in Table 10.1. Indeed the use of varying CVs can cause unwanted effects. For example, a large positive outlier with large CVs will often become even larger when the corresponding annual discrepancy is positive. In the context of the regression-based benchmarking method, it is then advisable to use constant CVs given by the target CV of the survey or by an average of the available CVs. The situation is different when signal extraction is performed in Method 2.

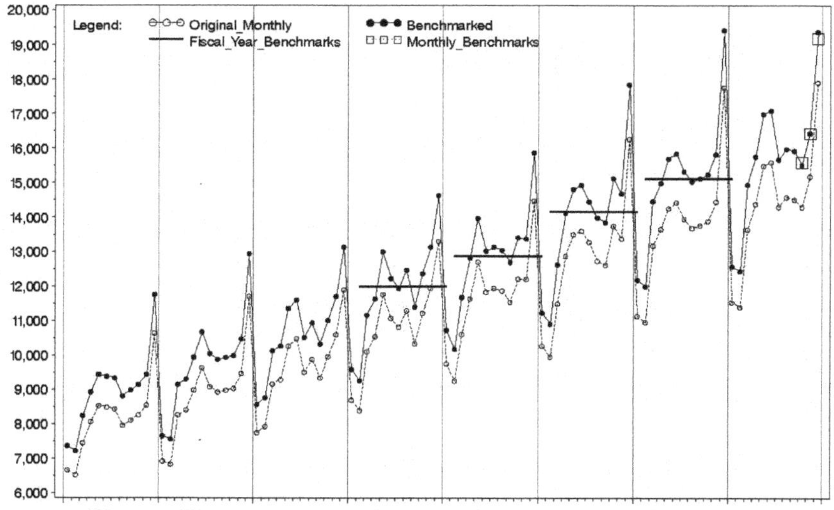

Fig. 10.1a. The Canadian Total Retail Trade series benchmarked from the multiplicative model without signal extraction (Method 1)

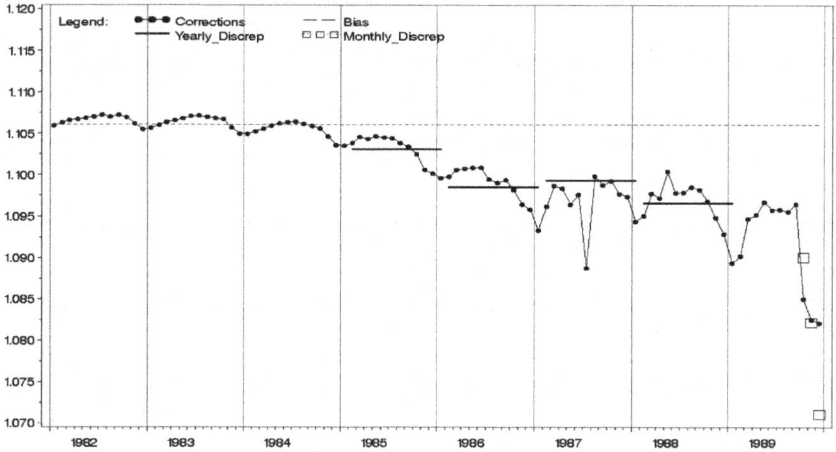

Fig. 10.1b. Multiplicative corrections for the Canadian Total Retail Trade Series, obtained from the multiplicative model (Method 1) with moving CVs

Table 10.1: Monthly data and estimates for the Canadian Total Retail Trade series
(1) and (2): Original monthly series and its empirical CVs
(3) and (4): Benchmarked series (Method 1) and estimated CVs
(5) and (6): Signal estimates (Method 2) and estimates CVs
(7) and (8): Benchmarked Signal Estimates Series and Estimates CVs

	(1)	(2)	(3)	(4)	(1)	(2)	(5)	(6)	(7)	(8)
01/80	5,651	0.80	6,251	0.65	5,651	0.80	5,665	0.74	6,236	0.64
02/80	5,761	0.70	6,374	0.59	5,761	0.70	5,774	0.64	6,354	0.59
03/80	6,128	0.80	6,782	0.64	6,128	0.80	6,145	0.73	6,766	0.64
04/80	6,585	0.70	7,289	0.58	6,585	0.70	6,602	0.64	7,266	0.58
05/80	7,362	0.70	8,151	0.57	7,362	0.70	7,381	0.64	8,125	0.58
06/80	6,952	0.70	7,697	0.57	6,952	0.70	6,968	0.64	7,672	0.57
07/80	7,070	0.90	7,830	0.68	7,070	0.90	7,091	0.82	7,813	0.68
08/80	7,148	0.90	7,917	0.68	7,148	0.90	7,168	0.82	7,898	0.67
09/80	7,048	0.80	7,804	0.61	7,048	0.80	7,065	0.73	7,780	0.61
10/80	7,614	0.70	8,429	0.56	7,614	0.70	7,630	0.64	8,400	0.55
11/80	7,625	0.70	8,437	0.55	7,625	0.70	7,641	0.64	8,411	0.54
12/80	9,081	0.70	10,043	0.54	9,081	0.70	9,101	0.64	10,017	0.54
01/81	6,684	0.90	7,393	0.65	6,684	0.90	6,703	0.82	7,383	0.64
02/81	6,258	0.80	6,924	0.58	6,258	0.80	6,275	0.72	6,910	0.58
03/81	7,098	0.90	7,856	0.64	7,098	0.90	7,122	0.81	7,846	0.63
04/81	7,796	0.70	8,629	0.52	7,796	0.70	7,816	0.63	8,606	0.52
05/81	8,256	0.90	9,143	0.62	8,256	0.90	8,284	0.81	9,129	0.62
06/81	8,257	0.80	9,144	0.56	8,257	0.80	8,281	0.71	9,123	0.56
07/81	8,110	1.00	8,984	0.67	8,110	1.00	8,139	0.89	8,975	0.66
08/81	7,663	0.90	8,488	0.60	7,663	0.90	7,688	0.80	8,473	0.60
09/81	7,748	0.90	8,581	0.60	7,748	0.90	7,772	0.80	8,566	0.59
10/81	8,249	0.80	9,134	0.54	8,249	0.80	8,273	0.71	9,113	0.53
11/81	8,199	0.70	9,072	0.48	8,199	0.70	8,219	0.62	9,049	0.48
12/81	9,973	0.70	11,027	0.48	9,973	0.70	9,999	0.62	11,007	0.47
01/82	6,666	0.80	7,372	0.52	6,666	0.80	6,686	0.71	7,363	0.51
02/82	6,526	0.80	7,220	0.51	6,526	0.80	6,548	0.71	7,211	0.51
03/82	7,449	0.80	8,244	0.50	7,449	0.80	7,475	0.71	8,234	0.50
04/82	8,065	0.60	8,925	0.42	8,065	0.60	8,086	0.53	8,899	0.42
05/82	8,528	0.60	9,440	0.41	8,528	0.60	8,550	0.53	9,412	0.42
06/82	8,480	0.60	9,388	0.41	8,480	0.60	8,501	0.52	9,358	0.41
07/82	8,428	0.70	9,332	0.43	8,428	0.70	8,452	0.61	9,310	0.43
08/82	7,956	0.60	8,807	0.40	7,956	0.60	7,976	0.52	8,780	0.40
09/82	8,107	0.80	8,977	0.46	8,107	0.80	8,135	0.70	8,962	0.45
10/82	8,256	0.70	9,139	0.41	8,256	0.70	8,281	0.61	9,118	0.41
11/82	8,538	0.60	9,445	0.38	8,538	0.60	8,560	0.52	9,420	0.38
12/82	10,640	0.60	11,762	0.37	10,640	0.60	10,667	0.52	11,739	0.37
01/83	6,922	0.80	7,653	0.43	.	.	6,995	1.80	7,697	1.74
02/83	6,832	0.70	7,556	0.38	.	.	7,089	1.93	7,803	1.87
03/83	8,269	0.90	9,148	0.46	.	.	8,302	2.07	9,144	2.01
04/83	8,406	0.70	9,302	0.37	.	.	8,890	2.15	9,790	2.10
05/83	8,978	0.70	9,937	0.36	.	.	9,181	2.19	10,116	2.15
06/83	9,643	0.80	10,676	0.38	.	.	9,432	2.25	10,387	2.19
07/83	9,077	0.80	10,050	0.37	.	.	9,066	2.28	9,985	2.18
08/83	8,928	0.80	9,884	0.36	.	.	8,999	2.21	9,922	2.14
09/83	8,974	0.80	9,933	0.35	.	.	9,132	2.18	10,063	2.09
10/83	9,030	0.90	9,994	0.38	.	.	9,074	2.07	9,998	1.99
11/83	9,477	0.70	10,479	0.31	.	.	9,722	1.95	10,707	1.86
12/83	11,708	0.60	12,937	0.29	.	.	11,788	1.83	12,978	1.72

Table 10.1: Continuation - 1
(1) and (2): Original monthly series and its empirical CVs
(3) and (4): Benchmarked series (Method 1) and estimated CVs
(5) and (6): Signal estimates (Method 2) and estimates CVs
(7) and (8): Benchmarked Signal Estimates Series and Estimates CVs

	(1)	(2)	(3)	(4)	(1)	(2)	(5)	(6)	(7)	(8)
01/84	7,752	0.80	8,566	0.32	7,752	0.80	7,783	0.68	8,568	0.32
02/84	7,929	0.90	8,764	0.35	7,929	0.90	7,966	0.76	8,773	0.35
03/84	9,165	1.00	10,133	0.40	9,165	1.00	9,212	0.84	10,150	0.39
04/84	9,289	0.80	10,273	0.29	9,289	0.80	9,326	0.67	10,269	0.29
05/84	10,268	0.80	11,359	0.28	10,268	0.80	10,310	0.66	11,353	0.28
06/84	10,486	0.90	11,601	0.31	10,486	0.90	10,532	0.74	11,602	0.30
07/84	9,509	1.00	10,521	0.34	9,509	1.00	9,557	0.81	10,531	0.33
08/84	9,885	0.90	10,935	0.28	9,885	0.90	9,931	0.74	10,936	0.28
09/84	9,337	0.80	10,326	0.24	9,337	0.80	9,376	0.66	10,319	0.24
10/84	9,961	0.80	11,013	0.22	9,961	0.80	10,003	0.66	11,006	0.23
11/84	10,598	0.70	11,708	0.21	10,598	0.70	10,638	0.58	11,698	0.21
12/84	11,901	0.70	13,134	0.20	11,901	0.70	11,947	0.57	13,134	0.20
01/85	8,690	0.80	9,589	0.21	8,690	0.80	8,728	0.66	9,597	0.22
02/85	8,390	0.80	9,262	0.21	8,390	0.80	8,428	0.65	9,265	0.21
03/85	10,107	0.60	11,165	0.21	10,107	0.60	10,141	0.49	11,143	0.22
04/85	10,541	0.80	11,641	0.20	10,541	0.80	10,585	0.64	11,636	0.20
05/85	11,764	0.70	12,995	0.19	11,764	0.70	11,807	0.56	12,976	0.19
06/85	11,067	0.80	12,225	0.19	11,067	0.80	11,113	0.63	12,213	0.19
07/85	10,811	0.80	11,940	0.18	10,811	0.80	10,856	0.62	11,929	0.19
08/85	11,290	0.90	12,462	0.20	11,290	0.90	11,344	0.71	12,460	0.20
09/85	10,337	0.90	11,405	0.20	10,337	0.90	10,387	0.71	11,404	0.20
10/85	11,214	1.00	12,363	0.22	11,214	1.00	11,276	0.79	12,372	0.24
11/85	11,935	1.00	13,136	0.23	11,935	1.00	12,003	0.79	13,161	0.25
12/85	13,300	0.80	14,633	0.18	13,300	0.80	13,362	0.63	14,647	0.18
01/86	9,753	0.90	10,725	0.21	9,753	0.90	9,806	0.71	10,741	0.21
02/86	9,249	0.90	10,172	0.22	9,249	0.90	9,299	0.71	10,182	0.22
03/86	10,610	0.80	11,678	0.20	10,610	0.80	10,658	0.63	11,670	0.20
04/86	11,638	0.80	12,811	0.20	11,638	0.80	11,688	0.62	12,796	0.21
05/86	12,695	0.80	13,976	0.20	12,695	0.80	12,751	0.61	13,956	0.21
06/86	11,826	0.80	13,019	0.19	11,826	0.80	11,877	0.61	12,997	0.20
07/86	11,941	1.00	13,129	0.23	11,941	1.00	12,005	0.75	13,124	0.23
08/86	11,867	1.00	13,042	0.23	11,867	1.00	11,932	0.75	13,036	0.23
09/86	11,540	0.90	12,687	0.20	11,540	0.90	11,599	0.69	12,674	0.20
10/86	12,209	1.00	13,408	0.21	12,209	1.00	12,280	0.77	13,405	0.23
11/86	12,201	1.00	13,378	0.22	12,201	1.00	12,274	0.77	13,393	0.23
12/86	14,479	0.90	15,867	0.20	14,479	0.90	14,559	0.69	15,892	0.21
01/87	10,272	1.20	11,230	0.32	10,272	1.20	10,349	0.92	11,275	0.33
02/87	9,951	1.00	10,908	0.25	9,951	1.00	10,013	0.77	10,924	0.25
03/87	11,492	0.80	12,627	0.21	11,492	0.80	11,546	0.61	12,613	0.22
04/87	12,867	0.90	14,133	0.23	12,867	0.90	12,931	0.68	14,121	0.23
05/87	13,508	1.20	14,811	0.34	13,508	1.20	13,598	0.90	14,828	0.32
06/87	13,608	1.10	14,937	0.28	13,608	1.10	13,688	0.81	14,936	0.27
07/87	13,278	2.30	14,457	0.95	13,278	2.30	13,443	1.66	14,578	0.88
08/87	12,728	0.80	13,999	0.22	12,728	0.80	12,784	0.59	13,969	0.23
09/87	12,616	0.90	13,862	0.23	12,616	0.90	12,680	0.68	13,845	0.23
10/87	13,761	0.80	15,127	0.23	13,761	0.80	13,824	0.61	15,099	0.23
11/87	13,380	0.80	14,688	0.23	13,380	0.80	13,443	0.61	14,679	0.23
12/87	16,270	0.70	17,854	0.23	16,270	0.70	16,338	0.54	17,850	0.24

Table 10.1: Continuation - 2
(1) and (2): Original monthly series and its empirical CVs
(3) and (4): Benchmarked series (Method 1) and estimated CVs
(5) and (6): Signal estimates (Method 2) and estimates CVs
(7) and (8): Benchmarked Signal Estimates Series and Estimates CVs

	(1)	(2)	(3)	(4)	(1)	(2)	(5)	(6)	(7)	(8)
01/88	11,134	1.00	12,185	0.26	11,134	1.00	11,202	0.77	12,215	0.28
02/88	10,959	1.00	12,001	0.27	10,959	1.00	11,025	0.77	12,023	0.28
03/88	13,178	0.80	14,466	0.22	13,178	0.80	13,237	0.62	14,456	0.24
04/88	13,666	0.90	14,995	0.24	13,666	0.90	13,733	0.69	14,990	0.25
05/88	14,268	0.60	15,700	0.25	14,268	0.60	14,314	0.46	15,656	0.27
06/88	14,433	0.90	15,845	0.24	14,433	0.90	14,501	0.68	15,833	0.25
07/88	13,961	0.90	15,327	0.24	13,961	0.90	14,027	0.67	15,317	0.25
08/88	13,691	0.80	15,040	0.23	13,691	0.80	13,750	0.61	15,019	0.25
09/88	13,773	0.80	15,126	0.23	13,773	0.80	13,834	0.62	15,108	0.25
10/88	13,901	0.90	15,247	0.22	13,901	0.90	13,969	0.71	15,243	0.27
11/88	14,453	0.90	15,824	0.22	14,453	0.90	14,527	0.71	15,848	0.28
12/88	17,773	0.90	19,423	0.22	17,773	0.90	17,865	0.72	19,486	0.28
01/89	11,537	1.20	12,568	0.33	.	.	11,705	2.00	12,638	1.59
02/89	11,402	1.20	12,431	0.35	.	.	11,851	2.21	12,741	1.86
03/89	13,653	0.90	14,946	0.28	.	.	14,186	2.43	15,186	2.00
04/89	14,392	0.90	15,762	0.29	.	.	14,373	2.64	15,308	2.08
05/89	15,487	0.80	16,985	0.29	.	.	15,630	2.82	16,582	2.11
06/89	15,595	0.90	17,089	0.30	.	.	16,030	2.99	16,913	2.10
07/89	14,305	0.90	15,676	0.30	.	.	14,688	3.21	15,408	2.05
08/89	14,584	0.90	15,978	0.30	.	.	15,192	3.33	15,882	1.94
09/89	14,522	0.80	15,923	0.30	.	.	15,006	3.47	15,608	1.79
10/89	14,297	1.70	15,513	0.37	.	.	15,030	3.62	15,581	0.48
11/89	15,183	1.60	16,435	0.34	.	.	16,035	3.77	16,441	0.57
12/89	17,910	1.40	19,380	0.32	.	.	18,858	3.88	19,202	0.48

Table 10.2: Benchmarks for the Canadian Total Retail Trade Series

Starting date	Ending date	Bench-marks	Cv	Temporal sum	Additiv. discrep.	Propor. discrep.
Yearly Benchmarks						
01/02/85	31/01/86	143,965	0.033	130,509	13,456	1.103
01/02/86	31/01/87	154,377	0.031	140,527	13,850	1.099
01/02/87	31/01/88	169,945	0.193	154,593	15,352	1.099
01/02/88	31/01/89	181,594	0.137	165,593	16,001	1.097
Monthly Benchmarks						
01/10/89	31/10/89	15,585	0.500	14,297	1,288	1.090
01/11/89	30/11/89	16,431	0.600	15,183	1,248	1.082
01/12/89	31/12/89	19,183	0.500	17,910	1,273	1.071

10.7.2 Method 2: Benchmarking with Signal Extraction, Interpolation and Extrapolation

This case of benchmarking with signal extraction, also includes interpolation and constrained extrapolation. In order to illustrate interpolation and extrapolation, we assume that the original monthly values are missing for years 1983 and 1989; they will be interpolated and extrapolated respectively. (Year 1989 was in fact a year of overlap between the "old" and the "new" redesigned survey. Method 2 thus illustrates how the costly maintenance of an overlap period could be avoided.

The desired estimates are obtained in two steps: (1) signal extraction with interpolation and *un*constrained extrapolation and (2) benchmarking which constrains the signal estimates, the interpolations and the extrapolations. The signal extraction model used is given by Eq. (10.14), which includes interpolation and extrapolation. The model is linear in the logs because because all the data in that step refer to one month. Indeed the sum operator $\boldsymbol{J}$ is trivial:

$$\boldsymbol{J}_{96\times 120} = \begin{bmatrix} \boldsymbol{I}_{36} & \boldsymbol{0}_{36\times 12} & \boldsymbol{0}_{36\times 60} & \boldsymbol{0}_{36\times 12} \\ \boldsymbol{0}_{60\times 36} & \boldsymbol{0}_{60\times 12} & \boldsymbol{I}_{60} & \boldsymbol{0}_{60\times 12} \end{bmatrix}, \tag{10.28}$$

where the first row of partitions pertains to the first stretch of 36 observations from 1980 to 1982; and the second row, to the second stretch of 60 observations covering from 1984 to 1988. Years 1983 and 1989 are interpolated and extrapolated respectively.

The model for signal extraction was identified on the log-transformed data $ln\, s_t$ from 1980 to 1989 using proc arima of SAS. The RegARIMA model found, a $(0,1,1)(0,1,1)_{12}$, with three intervention variables:

$$ln\, z_t = n_t^{-1} cos(\lambda_1 \tau)\, \delta_1 + n_t^{-1} sin(\lambda_1 \tau) \delta_2 + n_t^{-1} cos(\lambda_2 \tau) \delta_3 + e_t \,/\, (1-B)(1-B^{12}), \tag{10.29a}$$

$$e_t = (1-\psi_1 B)(1-\Psi_1 B^{12}) v_t = v_t - \psi_1 v_{t-1} - \Psi_1 v_{t-12} + \psi_1 \Psi_1 v_{t-12}, \quad E(v_t) = 0,\ E(v_t^2) = \sigma_v^2 \tag{10.29b}$$

where n_t is the number of days per month, $\lambda_j = 2\pi j \tau / 7$, the sine and cosine functions capture trading-day variation as explained in Appendix E.

The estimates of the RegARIMA parameters are displayed in Table 10.3 with other statistics. The regular and seasonal moving average parameters ψ_1 and Ψ_{12} are equal to 0.43619 and 0.78667 with *t*-values equal to 5.08 and 6.69 respectively; the standard deviation of the innovations σ_v is equal to 0.017864. The estimates of δ_1, δ_2 and δ_3 are equal to -0.29344, -0.23177 and -0.20863 with *t*-values equal to -12.81, -10.11 and -3.75. The resulting daily pattern in the original scale takes the values: 72.78, 102.77, 103.43 , 126.47 , 161.48, 104.57 and 60.53 percent, for Monday, Tuesday, and so forth.

Table 10.3. Reg-ARIMA model estimated on the observed monthly series

The ARIMA Procedure Maximum Likelihood Estimation

Parameter	Estimate	Standard Error	Approx. t Value	Pr > |t|	Lag	Variable	Shift
MA1,1	0.43619	0.08587	5.08	<.0001	1	LogOr	0
MA2,1	0.78667	0.11752	6.69	<.0001	12	LogOr	0
NUM1	-0.29344	0.02290	-12.81	<.0001	0	CosD1	0
NUM2	-0.23177	0.02293	-10.11	<.0001	0	SinD1	0
NUM3	-0.20863	0.05568	-3.75	0.0002	0	CosD2	0

Variance Estimate	0.000319	Std Error Estimate	0.017864
AIC	-541.133	SBC	-527.769
Number of Residuals	107		

Autocorrelation Check of Residuals

To Lag	Chi-Square	DF	Pr > ChiSq	Autocorrelations					
6	0.22	4	0.9945	0.009	-0.010	-0.015	0.015	0.009	0.034
12	2.29	10	0.9936	0.073	-0.008	-0.022	-0.045	0.096	0.009
18	11.04	16	0.8068	0.213	-0.041	-0.017	-0.092	-0.117	0.010
24	12.55	22	0.9449	0.037	-0.090	0.017	-0.010	0.033	0.017

Following Bell and Hillmer (1990), we assume that the model found for $\boldsymbol{ln}\ \boldsymbol{s}_t$ also holds for the signal. Matrix $\boldsymbol{D}$ of model is thus the monthly version of the regular and seasonal difference operator (10.3), with dimension 107×120. The covariance matrix $\boldsymbol{V}_{\upsilon^*}$ is such that $\boldsymbol{\upsilon}^*$ follows a seasonal moving average process (0,1)(0,1) with parameters 0.441 and 0.758. Matrix $\boldsymbol{Z}$ contains the sine and cosine functions for the trading-day variations; the signal-to-noise ratio $\sigma^2_{\eta^*}/\bar{\sigma}^2_{e^*}$ is equal to 8.282 ($\bar{\sigma}_{e^*}$ is the average CV of the monthly observations,).

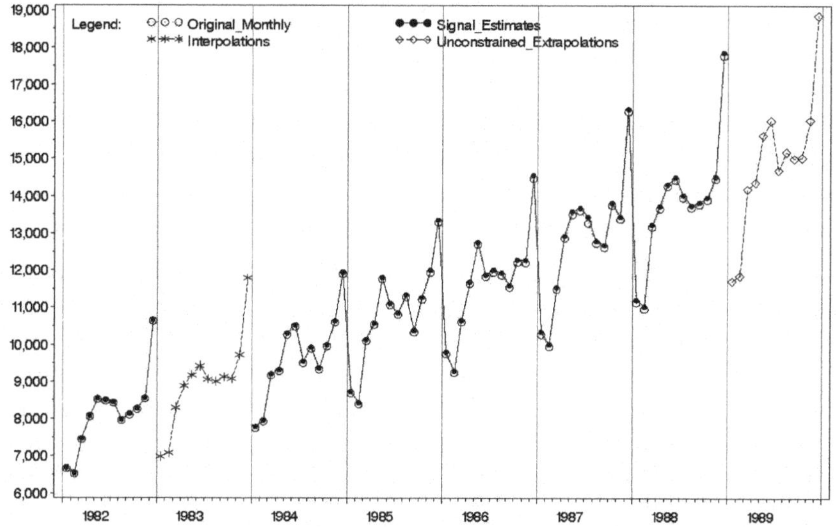

Fig. 10.2a. The Canadian Total Retail Trade Series: original series and signal extraction estimates obtained from multiplicative signal extraction (Method 2)

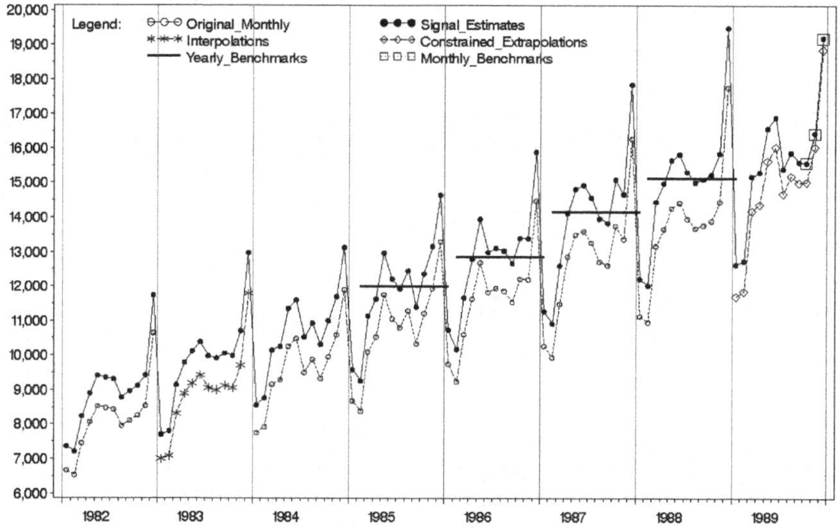

Fig. 10.2b. The Canadian Total Retail Trade Series: benchmarked signal estimates obtained from benchmarking with signal extraction estimates (Method 2)

The parameters δ_1, δ_2 and δ_3 re-estimated in the signal extraction model (10.7) are equal to -0.29867, -0.23924 and -0.27725 with t-values equal to -10.37, -8.32 and -4.25. The resulting daily pattern is 73.23, 108.66, 99.25, 122.14, 173.24, 106.45 and 56.22 percent. The un-constrained signal estimates, interpolations and extrapolations are illustrated in Fig. 10.2a and displayed in columns (5) and (6) of Table 10.1.

The second step consists of benchmarking both the signal estimates and the interpolations and extrapolations of the first step, by applying the multiplicative benchmarking model (10.25), where matrix $\boldsymbol{J}$ is as in (10.26) and $\boldsymbol{V}_{e^*}$ is the covariance matrix of the signal estimates of the first step. The final results are illustrated in Fig. 10.2b and displayed in columns (7) and (8) of Table 10.1.

As shown by Table 10.1, for both Methods 1 and 2, the CVs are lower than those of the original data which illustrates that benchmarking improves the reliability of the data, especially for periods covered by benchmarks. The CVs are lower with Method 2 (except for the months interpolated or extrapolated) showing its greater efficiency. The efficiency of Method 1 relative to Method 2 is proportional to the signal-to-noise ratio $\sigma^2_{\eta^*}/\bar{\sigma}^2_{e^*}$: the greater the survey error variance, the more the series stands to be improved by modelling the signal. Table 10.1 also shows that the CVs for Method 2 are much larger for 1983 and 1989, because the estimates of these two years are extrapolated since those two years were assumed to have no original monthly observations.

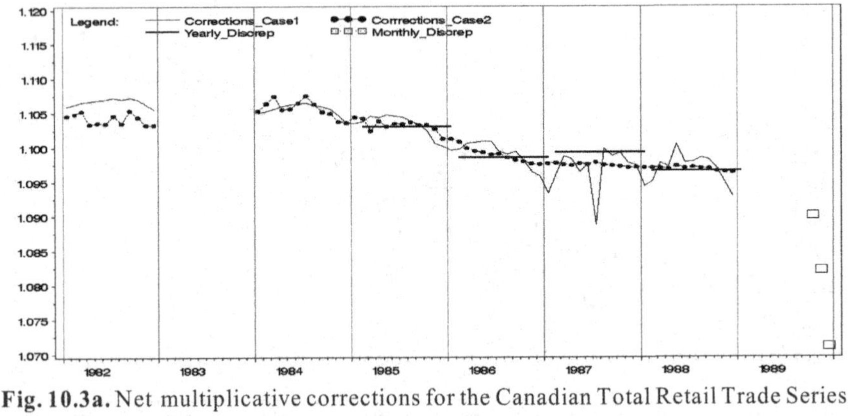

Fig. 10.3a. Net multiplicative corrections for the Canadian Total Retail Trade Series, obtained by multiplicative benchmarking (Method 1) and multiplicative benchmarking with signal extraction (Method 2)

Fig. 10.3a shows that the corrections obtained from Method 2 (with signal extraction and benchmarking) are smoother than those from Method 1 (with benchmarking alone), particularly for year 1987 which shows a very large CV. In the absence of signal extraction (Method 1), the observations with large CVs are by definition more modifiable, as a result they may become more outlier than they were in the first place. With signal extraction, on the other hand, Method 2 pulls the benchmarked series towards the fitted Reg-ARIMA value.

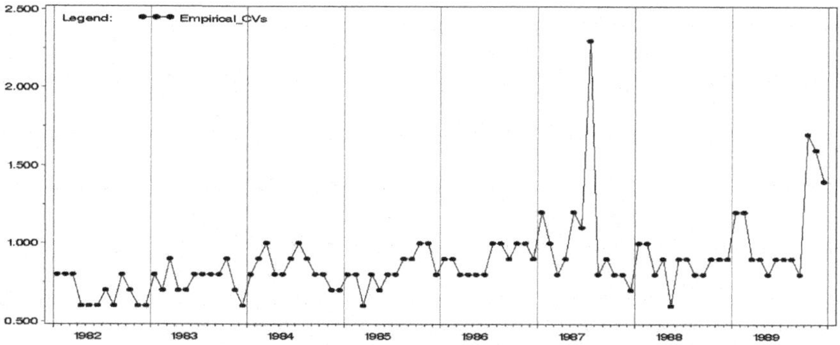

Fig. 10.3b. Actual CVs of the original Canadian Total Retail Trade series

Furthermore Method 2 with signal extraction homogenizes the CVs compared to Method 1; this is particularly true for the months of May, June and July of 1987. This desirable effect is due to the modelling of the signal, which can replace "out of line" values with high CVs by values closer to those suggested by the signal model.

Note that the fiscal benchmark covering Feb. 1987 to Jan. 1988 is not satisfied, which is obvious in Fig. 10.3a. That benchmark has a larger CV of 0.193% compared to the previous and following benchmarks with CVs equal to 0.031% and 0.137% respectively. Furthermore Fig. 10.3 shows a larger discrepancy for 1987-88 indicating that the corresponding benchmark is out of line.

Table 10.4 shows the estimates of the calendar year values

$$\hat{\boldsymbol{\theta}}_L = \boldsymbol{L}\,\hat{\boldsymbol{\theta}}, \quad var[\,\hat{\boldsymbol{\theta}}_L\,] = \boldsymbol{L}\,var[\,\hat{\boldsymbol{\theta}}\,]\,\boldsymbol{L}', \tag{10.27}$$

with their CVs, obtained from (10.23) with $\boldsymbol{L} = \boldsymbol{I}_{10} \otimes \boldsymbol{1}_{1\times 12}$. As a result, the fiscal year values (1985-86 to 1988-89) are now calendarized. Comparing the CVs of the calendarized values with those of the fiscal year data in Table 10.4 shows that calendarization does not necessarily entail a loss of accuracy. The larger CVs for Method 2 in 1983 and 1989 are due to the absence of monthly data for that method in those two years, which were respectively interpolated (italicized) and extrapolated (bold) respectively. Method 1 on the other hand had monthly data for those two years.

Table 10.4: Calendar Year Values and Corresponding CVs Estimated with the two Methods for the Canadian Total Retail Trade Series

Calendar periods		Without signal extraction		With signal extraction	
80-01	80-12	93,004	0.57	92,738	0.57
81-01	81-12	104,375	0.54	104,181	0.53
82-01	82-12	108,051	0.39	107,807	0.39
83-01	83-12	117,549	*0.31*	*118,588*	*1.27*
84-01	84-12	128,333	0.19	128,341	0.19
85-01	85-12	142,816	0.04	142,802	0.04
86-01	86-12	153,892	0.03	153,866	0.03
87-01	87-12	168,633	0.12	168,717	0.12
88-01	88-12	181,179	0.12	181,192	0.15
89-01	89-12	188,686	0.18	**187,489**	**0.94**

11 Reconciliation and Balancing Systems of Time Series

11.1 Introduction

The vast majority of time series data produced by statistical agencies are typically part of a system of series classified by attributes. For example, the Canadian Retail Trade series are classified by Trade Group (type of store) and Province; the Labour Force series are classified by Province, Age Group, Sex, and other attributes such as Part-Time and Full-Time employment. In such cases the series of the system must satisfy *cross-sectional aggregation constraints*, sometimes called contemporaneous constraints (e.g. Di Fonzo 1990). In many cases, each series must also temporally add up to its annual benchmarks; these requirements are referred to as *temporal aggregation constraints* (*Ibid.*), as discussed in the previous chapters.

Table 11.1. System of Time Series Classified by Industry and Province

Prov. / Ind.	Prov. 1	Prov. 2	...	Prov. $P-1$	Ind. Totals
Ind. 1	$\{s_{11,t}\}$	$\{s_{12,t}\}$	...	$\{s_{1,P-1,t}\}$	$\{s_{1P,t}\}$
Ind. 2	$\{s_{21,t}\}$	$\{s_{22,t}\}$	...	$\{s_{2,P-1,t}\}$	$\{s_{2P,t}\}$
⋮	⋮	⋮	⋱	⋮	⋮
Ind. $N-1$	$\{s_{N-1,1,t}\}$	$\{s_{N-1,2,t}\}$	...	$\{s_{N-1,P-1,t}\}$	$\{s_{N-1,P,t}\}$
Prov. Totals	$\{s_{N1,t}\}$	$\{s_{N2,t}\}$	...	$\{s_{N,P-1,t}\}$	$\{s_{NP,t}\}$

We focus on *one-way classified systems* of series, e.g. by Province; on *two-way classified systems* of series, e.g. by Industry and Province; and on *marginal two-way systems* of series, involving only the Industrial Totals, the Provincial Totals and the Grand Total of two-way classified systems. We henceforth refer to the corresponding reconciliation models as the *one-way reconciliation model*, the *two-way reconciliation model* and the *marginal two-way reconciliation model.*

The reconciliation problem is typically known as *balancing* in the context of National Accounts.

Table 11.1 represents a two-way classified system of series composed of $N-1$ Industries and $P-1$ Provinces. The system also contains the *marginal Totals* series, namely $N-1$ *Industrial Totals* in the last column of the table, $P-1$ *Provincial Totals* in the last row, and the *Grand Total* denoted by $s_{NP,t}$. The other entries of the table (i.e. for $n<N$ and $p<P$) are the *elementary cells* containing the *elementary series.* Each cell of the table contains the values of a "sub-annual" time series $s_{np,t}$ pertaining to a variable of interest (e.g. Sales) for Industry n and Province p. The table can also be viewed as containing vectors of time series or of scalars pertaining to period t. The subscript np should be interpreted as n, p.

The cross-sectional aggregation constraints require that the values of the series, $s_{np,t}$ add up to the appropriate marginal totals for each period of time t. More specifically, the system must satisfy N sets of *Industrial cross-sectional aggregation constraints*, over the Provinces in each Industry, and, P sets of *Provincial cross-sectional aggregation constraints*, over the Industries in each Province. The constraints can respectively be formalized as

$$s_{nP,t} = \Sigma_{p=1}^{P-1} s_{np,t}, \quad t=t_{1,nP},\ldots,t_{L,nP}, \quad n=1,\ldots,N, \tag{11.1a}$$

$$s_{Np,t} = \Sigma_{n=1}^{N-1} s_{np,t}, \quad t=t_{1,nP},\ldots,t_{L,nP}, \quad p=1,\ldots,P, \tag{11.1b}$$

where $t_{1,np}$ and $t_{L,np}$ are respectively the first and last available observations for series $s_{np,t}$. The number of observations for a given series, $s_{np,t}$, is thus $T_{np} = t_{L,np} - t_{1,np} + 1$. The number T_{np} may vary from cell to cell. In fact, T_{np} may be equal to 0 for some elementary series (i.e. $s_{np,t}$ for $n<N$ and $p<P$), because some Industries may not exist in some of the Provinces. Note that the number of observations for the Grand Total, T_{NP}, is greater than or equal to T_{np} for all n and p, because the sub-annual periods t covered by the Grand Total is the union of all sub-annual periods covered by all the other series in the system. Similarly, the number of observations for the Industrial Totals T_{nP} is greater than or equal to T_{np} for any given Industry n, and the number of observations for the Provincial Totals, T_{Np} is greater than or equal to T_{np}

for any given Province p. In other words, there are T_{nP} individual constraints in the n-th set of industrial conatraints; and T_{Np}, in the p-th set of provincial constraints.

For systems of series classified by one attribute, for example by Province, subscript n is omitted.

In many situations, the series in a system do not satisfy the cross-sectional aggregation constraints. This is for instance the case of the National Accounts, which bring together time series from a number of different sources, some series being derived (as opposed to observed) from such sources. The problem of reconciling large "accounting matrices" is central to the system of National Accounts. The main purpose of the latter is to remove cross-sectional and temporal discrepancies and residual errors, due to the fact that the observations can be obtained from income, expenditure or production measurements.

The problem also occurs for small area estimates which are not consistent with pre-existing values of larger areas (Pfeffermann and Bleuer 1993, Rao 2003).

An important case of reconciliation arises when systems of series are re-classified to maintain the relevance of statistical concepts. For example, the system of Retail and Wholesale Trade series was redefined in 1989 to adopt the 1980 Standard Industrial Classification (SIC), and in 2003 to conform to the North American Industrial Classification System (NAICS). This exercise entails performing domain estimation (Särndal et al. 1992), retro-actively adding the new industrial codes to the past survey records of the units, based on subject matter expertise, and retabulating according to the new industrial codes. The resulting re-classified system of series typically displays cross-sectional and temporal aggregation discrepancies. The rationale for reconciliation is that the Provincial Totals over the industries and the Grand Total usually refer to the same respective universe under both classifications.

In other situations, the system of series considered satisfies the cross-sectional constraints by construction; this is the case of series originating from surveys. However, even in such situations, the individual series often undergo *transformations* which cause *cross-sectional aggregation discrepancies* and temporal aggregation discrepancies (if applicable). Examples of such transformations are benchmarking, interpolation (temporal disaggregation), forecasting, seasonal adjustment.

The two-way classified system contains N sets of *Industrial cross-sectional aggregation discrepancies* and similarly P sets of *Provincial cross-sectional aggregation discrepancies*:

$$d_{n,t} = s_{nP,t} - \Sigma_{p=1}^{P-1} s_{np,t}, \quad t=t_{1,nP},...,t_{L,nP}, \quad n=1,...,N, \qquad (11.2a)$$

$$d_{p,t} = s_{Np,t} - \Sigma_{n=1}^{N-1} s_{np,t}, \quad t=t_{1,Np},...,t_{L,Np}, \quad p=1,...,P. \qquad (11.2b)$$

These aggregation discrepancies are more often expressed in terms of proportion to make them dimensionless. The Industrial and the Provincial *proportional cross-sectional aggregation discrepancies*, are

$$d_{n,t}^{(p)} = s_{nP,t} \,/\, \Sigma_{p=1}^{P-1} s_{np,t}, \quad t=t_{1,nP},...,t_{L,nP}, \quad n=1,...,N, \qquad (11.3a)$$

$$d_{p,t}^{(p)} = s_{Np,t} \,/\, \Sigma_{n=1}^{N-1} s_{np,t}, \quad t=t_{1,Np},...,t_{L,Np}, \quad p=1,...,P, \qquad (11.3b)$$

where the superscript *(p)* stands here for proportion and not province.

The cross-sectional additivity constraints are often implemented by means of the Iterative Proportional Fitting approach, also known as "raking". This entails the following steps:

(1) The elementary series of rows $n=1,...,N-1$ ($s_{np,t}$ for $n<N$ and $p<P$) are multiplied by the corresponding provincial proportional discrepancies $d_{p,t}^{(p)}$ of (11.3b). This step eliminates the provincial discrepancies.
(2) The industrial proportional discrepancies (11.3a) are recalculated, because step (1) alters these discrepancies.
(3) The elementary series of columns $p=1,...,P-1$ are multiplied by the corresponding industrial proportional discrepancies $d_{n,t}^{(p)}$ of step (2). This eliminates the industrial discrepancies.
(4) The provincial proportional discrepancies (11.3b) are recalculated.
(5) The steps (1) to (4) are repeated until the provincial proportional discrepancies are sufficiently close to one.

The Iterative Proportional Fitting technique implies that the Marginal Totals already satisfy the cross-sectional aggregation constraints and are imposed to the elementary cells (which is not the case of all methods).

The problem of reconciling systems of data has preoccupied economists and statisticians for more than sixty years. Deming and Stephan (1940) suggested the Iterative Proportional Fitting approach as an approximation to their "least square" numerical optimisation, probably justified by the lack of computing power at the time. According to Byron (1978), van der Ploeg (1982) and Weale (1988), the method of least squares was initially suggested by Stone et al. (1942). A major problem proved to be the amount of computations involved. Hence for the purpose of balancing input-output tables, the RAS[1] method developed by Bacharach (1965 1971) became very popular. However, it had the following shortcomings: no allowance for varying degrees of uncertainty in the initial estimates and constraints, dubious economic interpretation of pro rata adjustments to the accounts, and a very large number of iterations before convergence.

The attention reverted to the least square approach after Byron (1978), who developed alternative procedures for minimizing a constrained quadratic loss function based on the conjugate gradient algorithm.

The feasibility of balancing a large set of accounting matrices using least squares was demonstrated by van der Ploeg (1982) and Barker et al. (1984). Weale (1988) extended the approach to deal with the simultaneous adjustment of values, volumes and prices, by introducing non-linear constraints linearly approximated by means of logarithmic transformations.

Byron, van der Ploeg, Weale and others, observed that the balanced estimates based on the least squares approach are best linear unbiased. In his appendix, van der Ploeg (1982) claimed that "*the optimal* [least squares] *algorithm is not necessarily more expensive than the RAS approach*" and that "*the advantages of a more general treatment, allowing for uncertain row and column totals and relative degrees of reliability, are easily obtained*".

Theil (1967), Kapur et al. (1992) and Golan et al. (1994, 1996) proposed methods based on maximum entropy. MacGill (1977) showed that the entropy and the RAS methods converge to the same solution.

In the next section, we present a general regression-based method for the reconciliation of systems of time series. Following Weale (1988) and Solomou and Weale (1993), we refer to the model as a reconciliation model in order to distinguish it from Iterative Proportional Fitting or raking. This method has been used for many years at Statistics Canada and other statistical agencies, central banks and public institutions, to reconcile seasonally adjusted time series and reclassified systems of series.

The method has the following properties:

(1) It includes the Iterative Proportional Fitting (raking) technique as a particular case.

(2) Some series may be specified as *exogenous*, i.e. as pre-determined or unalterable. This can apply to some or all of the Totals, or to any elementary series ($s_{np,t}$ for $n<N$ and $p<P$).

(3) The Marginal Totals may be specified as *endogenous*, i.e. jointly estimated with the elementary series in the reconciliation process.

---

[1] The word refers to the product of matrices $\boldsymbol{R}$, $\boldsymbol{A}$ and $\boldsymbol{S}$ used in the document.

(4) The series may take negative values (e.g. profits).
(5) The method maintains the temporal aggregation constraints, so that the reconciled series still satisfy their benchmarks.
(6) Depending on the covariance matrices used for the series to be reconciled, different criteria are optimized, e.g. minimum proportional correction.
(7) When the covariance matrices are statistically genuine, the method modifies the original series according to their covariances and produces valid covariance matrices for the reconciled series.
(8) The method yields itself to hierarchical application.
(9) Finally, it can be used to impose identity constraints to stocks and contributing flows. For example the change in Inventories between time t and $t-1$ should be equal to the difference between Purchases and Sales at time t: $Inv_t - Inv_{t-1} = Purch_t - Sales_t + \varepsilon_t$

Properties (5) and (7) eliminate the need for a complex model which simultaneously includes both temporal and cross-sectional aggregation constraints. The incorporation of temporal constraints in a reconciliation model have a large impact on computing time and memory requirements.

In the context of the proposed reconciliation model, benchmarking can impose the temporal constraints to each series separately in a first stage. In a second stage, reconciliation can produce a system of series which satisfy both the temporal and the cross-sectional constraints, provided the same covariance matrices of the benchmarked series from the first stage are used in the reconciliation process. This is a property of recursive least squares.

The same holds for other transformations like interpolation (temporal disaggregation), forecasting (Pankratz 1989), seasonal adjustment, as long as these transformations provide positive semidefinite covariance matrices.

Property (8) facilitates timeliness in the production of large systems of series, by enabling a hierarchical approach. In a first stage of reconciliation, reconciled estimates can be produced at a relatively high level of aggregation e.g. the marginal totals. In a second stage, reconciled estimates can be produced at a lower (more detailed) level of aggregation, in a manner consistent with the previously obtained aggregates.

This strategy enables the publication of the more important aggregates within shorter deadlines; and, the less important series within longer deadlines. An example of this hierarchical strategy is to produce the Grand Total, the Provincial Totals and the Industrial Totals earlier; and, the industrial and provincial break down at later date.

11.2 General Regression-Based Reconciliation Method

The general regression-based reconciliation model consists of two linear equations:

$$s = I\,\theta + e, \quad E(e) = 0, \quad E(e\,e') = V_e, \tag{11.4a}$$

$$g = G\,\theta + \varepsilon, \quad E(\varepsilon) = 0, \; E(\varepsilon\,\varepsilon') = V_\varepsilon, \; E(\varepsilon\,e') = 0\,. \tag{11.4b}$$

Vector $\theta = [\,\theta_1' \cdots \theta_K'\,]'$ contains the K "true" series of interest, including the marginal totals; the true series satisfy the cross-sectional aggregation constraints. The estimates of θ obtained from the model are the reconciled series. Vector $s = [\,s_1' \cdots s_K'\,]'$ contains the K series to be reconciled, including the Marginal and the Grand Totals. Vector $e = [\,e_1' \cdots e_K'\,]'$ stands for the errors associated with s. Matrix V_e is usually block diagonal in practice $V_e = block(V_{e_1}, V_{e_2}, \ldots, V_{e_K})$. Section 11.3 discusses the choice of V_{e_k}.

Matrix G is the cross-sectional sum operator consisting of 0s, 1s and – 1s. For the moment, vector g reduces to a vector of zeroes, because all the series - including the Totals - are jointly estimated. Matrix V_ε is diagonal; the errors in ε may have a variance greater than zero for redundant constraints; and a variance equal to zero for the non-redundant ones. This issue will be discussed when specific models are presented, namely the one-way, the two-way and the marginal two-way reconciliation models.

Eq. (11.4a) states that the "true" series is observed with error. Eq. (11.4b) formalizes the cross-sectional aggregation constraints.

Model (11.4) can be written as a classical regression model,

$$\begin{bmatrix} s \\ g \end{bmatrix} = \begin{bmatrix} I \\ G \end{bmatrix} \theta + \begin{bmatrix} e \\ \varepsilon \end{bmatrix}, \tag{11.5}$$

or

$$y = X\,\theta + u, \quad E(u\,u') = V_u = block(V_e, V_\varepsilon), \tag{11.6}$$

where $y = \begin{bmatrix} s \\ g \end{bmatrix}$, $X = \begin{bmatrix} I \\ G \end{bmatrix}$ and $u = \begin{bmatrix} e \\ \varepsilon \end{bmatrix}$.

The generalized least square solution of model (11.6) is

$$\hat{\boldsymbol{\theta}} = (\boldsymbol{X}'\boldsymbol{V}_u^{-1}\boldsymbol{X})^{-1}\,\boldsymbol{X}'\boldsymbol{V}_u^{-1}\boldsymbol{y},\quad var[\hat{\boldsymbol{\theta}}] = (\boldsymbol{X}'\boldsymbol{V}_u^{-1}\boldsymbol{X})^{-1}. \qquad (11.7)$$

The solution (11.7) is achievable only if $\boldsymbol{V}_u$ is invertible, which requires that both $\boldsymbol{V}_e$ and $\boldsymbol{V}_\varepsilon$ be invertible. The implication is that all the diagonal elements of $\boldsymbol{V}_\varepsilon$ must be greater than zero, i.e. all the cross-sectional constraints must be non-binding. However, Appendix B shows that the solution (11.7) can be expressed as in Byron (1978); this solution admits binding constraints ($\boldsymbol{V}_\varepsilon = 0$):

$$\hat{\boldsymbol{\theta}} = \boldsymbol{s} + \boldsymbol{V}_e\,\boldsymbol{G}'(\boldsymbol{G}\boldsymbol{V}_e\boldsymbol{G}' + \boldsymbol{V}_\varepsilon)^{-1}\,[\boldsymbol{g} - \boldsymbol{G}\boldsymbol{s}]$$

$$= \boldsymbol{s} + \boldsymbol{V}_e\,\boldsymbol{G}'[\boldsymbol{V}_d^{-1}\,\boldsymbol{d}] \;\; = \boldsymbol{s} + \boldsymbol{V}_e\,\boldsymbol{G}'\,\hat{\boldsymbol{\gamma}}, \qquad (11.8a)$$

$$var[\hat{\boldsymbol{\theta}}] = \boldsymbol{V}_e - \boldsymbol{V}_e\boldsymbol{G}'\boldsymbol{V}_d^{-1}\boldsymbol{G}\boldsymbol{V}_e, \qquad (11.8b)$$

where $\boldsymbol{V}_d = \boldsymbol{G}\boldsymbol{V}_e\boldsymbol{G}' + \boldsymbol{V}_\varepsilon$ is the covariance matrix of the original cross-sectional aggregation discrepancies, and $\hat{\boldsymbol{\gamma}}$ contains the Lagrange multipliers. If vector $\boldsymbol{s}$ is considered as an initial estimate of $\boldsymbol{\theta}$ with covariance matrix $\boldsymbol{V}_e$, which ignores the observations in (11.4b), the final estimator (11.8) is the recursive least square estimator, which takes these observations into account. Alternatively, Eqs. (11.8a) and (11.8b) can be seen as the solution of the restricted generalized least square regression, where vector $\boldsymbol{s}$ stands for the unrestricted estimate with covariance matrix $\boldsymbol{V}_e$ and $\hat{\boldsymbol{\theta}}$ denotes the restricted (constrained) estimate with covariance matrix $var[\hat{\boldsymbol{\theta}}]$. The values of the latter covariance matrix are smaller than those of the original covariance $\boldsymbol{V}_e$, since the cross-sectional aggregation constraints implicitly increase the number of degrees of freedom by reducing the number of free parameters in vector $\hat{\boldsymbol{\theta}}$.

The dimension of matrix $\boldsymbol{V}_d$ is given by the number of cross-sectional constraints. For a two-way classified system of series comprising 13 Provinces by 19 Industries (including the Grand Total) over 23 quarters, there are 437 industrial and 299 provincial constraints, totalling 736 cross-sectional constraints. The size of $\boldsymbol{V}_d$ is then 736 by 736. The matrix may be conveniently inverted by parts. Furthermore, under the assumption of cross-sectional independence, the block diagonal parts of $\boldsymbol{V}_d$ are themselves block diagonal, which accelerates the calculations and preserves accuracy. (The inverse of a block diagonal matrix is the block diagonal matrix containing the inverted blocks.) If five temporal constraints per series were to be explicitly included in the model, the number of constraints would rise to 1,971.

It is possible to formulate the reconciliation problem in terms of a more conventional regression model, as proposed by Quenneville and Rancourt (2005). For illustration purposes, let a system contain three series, the Total s_3 and two elementary series s_1 and s_2, each containing one observation. The following model

$$\begin{bmatrix} s_1 \\ s_2 \\ s_3 \end{bmatrix} = \begin{bmatrix} 1 & 0 & 0 \\ 0 & 1 & 0 \\ 0 & 0 & 1 \end{bmatrix} \begin{bmatrix} \theta_1 \\ \theta_2 \\ \theta_3 \end{bmatrix} + \begin{bmatrix} e_1 \\ e_2 \\ e_3 \end{bmatrix} \quad \text{or} \quad \boldsymbol{s} = \boldsymbol{Z} + \boldsymbol{e},\ E(\boldsymbol{e}\,\boldsymbol{e}') = \boldsymbol{V}_e$$

can be estimated by means of standard regression software, which also offer the possibility of restrictions on the parameters. In this case, the restriction would be $\theta_3 = \theta_2 + \theta_1$. If the Total series is pre-determined, for example, another restriction is required, that is $\theta_3 = s_3$. If the Total is endogenous, θ_3 can be remove from the model; this also entail removing the third column of $\boldsymbol{Z}$ and replacing the third row of $\boldsymbol{Z}$ by 1s (for the two remaining columns). The model can then be estimated without restriction, but $\hat{\theta}_3$ must be obtained afterwards as the sum of $\hat{\theta}_1$ and $\hat{\theta}_2$.

The advantage of this approach is to allow the use of more readily available standard regression software. One disadvantage is that the specification of the restrictions may become cumbersome, especially if temporal constrains are also involved. Another limitation is that the matrix product $(\boldsymbol{Z}\,\boldsymbol{V}_e^{-1}\boldsymbol{Z})^{-1}$ will be calculated (probably in a recursive manner) and the size of the matrix may be prohibitive. Indeed this matrix includes the covariance matrices of each series (block diagonal partitions) and the cross-covariance matrices which are typically not needed. Hence, assuming a system of series of N Industries and P Provinces over T time periods, the dimension of the matrix would be $NRT \times NRT$. On the other hand, under model (11.4) with solution (11.8), the cross-correlation matrices, i.e. the off-diagonal partitions of $var[\hat{\boldsymbol{\theta}}]$, need not be calculated. Furthermore the matrix V_d to be inverted is much smaller with maximum dimension $(NT+RT)\times(NT+RT)$.

Vector s may contain benchmarked series, interpolated series, seasonally adjusted series, etc.; and V_e, the corresponding covariance matrices of these series. In the case of benchmarking and interpolation, reconciliation would preserve the temporal constraints which are implicit in the covariance matrices. Indeed, the covariances matrix of the benchmarked series V_{e_k} are such that the changes made by reconciliation must sum to zero over each "year" with binding benchmark.

According to van der Ploeg (1982), the estimator (11.8) is best linear unbiased, if the covariance matrices are statistically genuine.

In many applications however, the covariance matrices of the series to be reconciled are not derived from statistical models but built as described in the next section.

11.3 Choosing the Covariance Matrices

The cross-covariance matrix is often assumed to be block diagonal:

$$V_e = block(V_{e_1}, V_{e_2}, ..., V_{e_K}) = \begin{bmatrix} V_{e_1} & \cdots & 0 \\ \vdots & \ddots & \vdots \\ 0 & \cdots & V_{e_K} \end{bmatrix}, \qquad (11.9)$$

where V_{e_k} is the covariance matrix of the k-th series in the system of series. If the Totals are exogenous, i.e. pre-determined, the matrix is block diagonal.

When the totals are endogenous, i.e. jointly estimated with the elementary series, it would be desirable to have a cross-covariance matrices between the Totals and the elementary series. However, this raises the issue of choosing these cross-covariance matrices. Assuming these were available, this would dramatically increase the magnitude of the calculations. Furthermore, it should be noted that as the number of elementary series increases, the effect of cross-covariance between these series and the Totals becomes less important in the reconciliation process.

When cross-covariance matrices between the Totals and the elementary series are available and the system of series is not too large, the general solution (11.8) can be used.

Note that in a survey situation, the Totals are the cross-sectional sums of the appropriate elementary series. Consequently, the cross-covariance matrix between a given Total and its elementary series is given by the covariance matrix of the elementary series. In such situations there is simply no reconciliation problem. If this system of series were then to be processed by a common linear filter, the resulting system would still satisfy cross-sectional aggregation constraints. If a specific linear filter were used for each series, the system would no longer satisfy the constraints. The cross-covariance matrices could possibly be determined exactly, which is another research topic.

The covariance matrix $\boldsymbol{V}_{e_k}$ of each series is chosen as discussed in Chapter 4.

The situation most likely to supply covariance matrices, applicable as such in reconciliation, occurs when the sub-annual series originate from a previous statistical process which generates covariance matrices. In reconciliation, benchmarking becomes another such statistical process. The use of covariance matrices originating from benchmarking preserves the temporal aggregation constraints imposed by benchmarking (e.g. the annual benchmarks). If the benchmarks are specified as binding, the temporal constraints remain satisfied after reconciliation. In other words, the temporal constraints are implicit in the covariance matrices used in the reconciliation model. Indeed, these covariance matrices are such that the modifications brought about by reconciliation for a "year" (with benchmark) must sum to 0.

Another situation providing covariance matrices is small area estimation (Rao 1999 and 2003, Pfeffermann and Bleuer 1993). In some of the literature, the cross-sectional additivity is indeed imposed by means of Eq. (11.8).

In cases of genuine covariance matrices, the reconciliation model produces optimal reconciled series with valid covariance matrices.

In the absence of existing covariance matrices, it is necessary to build them according to the following structure:

$$\boldsymbol{V}_{e_k} = \alpha_k \, \boldsymbol{\Xi}_k^{\lambda} \boldsymbol{\Omega}_k \boldsymbol{\Xi}_k^{\lambda}, \quad k=1, ..., K, \tag{11.10}$$

where matrix $\boldsymbol{\Xi}_k$ is a diagonal matrix containing the standard deviations, $\sigma_{k,t}, ..., \sigma_{k,T}$, $\boldsymbol{\Omega}_k$ contains the autocorrelations corresponding to an ARMA process, and α_k is the alterability coefficient for the k-th series. Typically the standard deviations are given in the form of a constant coefficient of variation, $\sigma_{k,t} = c_k \times s_{k,t}$, for each series or for all series ($c_{k,t} = c$).

The alterability coefficients artificially increase or reduce the covariance matrices of some of the series in the system, so that these series are more or less affected by reconciliation. In particular, setting the alterability coefficient of the Grand Total to zero causes it to be un-*altered*, i.e. pre-determined, in the reconciliation process. Setting the alterability coefficient of any series to zero - whether Total or not - causes it to be unaltered in the reconciliation. On the other hand, setting the alterability coefficient of a series equal to relatively large values causes the series to be altered more than those with lower alterability coefficients. The alterability coefficients must take non-negative values. Of course if too many series have alterability coefficients equal to 0, there is no solution to the reconciliation problem.

Both the alterability coefficients and the coefficients of variations affect corrections made to each observation of each series in the system. The general practice consists of setting the coefficients of variation equal to 1 for all series and 0 for the un-alterable series. The alterability coefficients could also reflect the relative reliability of the various series. Note that volatile series may be measured precisely.

For variables of interest which may take negative values (e.g. Profits), the absolute value of the standard deviations must be taken before substituting them in matrix $\boldsymbol{\Xi}$ of Eq. (11.10), i.e. $\sigma_{k,t} = c_k \times |s_{k,t}|$. As a result, the objective function is specified in terms of absolute values (Junius and Oosterhaven 2003). It is advisable to exclude zero values from the reconciliation process and reintroduce them in the system of series after reconciliation. Indeed, they do not contribute to the discrepancies; and, under proportional reconciliation, the results for the zero observations are already known in advance to be zero. This is especially appropriate if the errors are independently distributed, i.e. $\boldsymbol{\Omega}_k = \boldsymbol{I}_{T_k}$.

In order to achieve proportional reconciliation, λ must be set equal to 1/2, the coefficients of variation must be the same for all series ($c_{k,t} = c$), and the alterability coefficients must be equal to a constant, e.g. 1. In *proportional reconciliation*, the proportional corrections to the original series, $|(\hat{\theta}_{k,t} - s_{k,t})| / s_{k,t}$, tend to be uniform for all series in the system, at least when all series take positive values. This can be exactly verified for one-dimensional systems of series with positive values, in the absence of temporal constraints.

Using different alterability coefficients for different series leads to a more general concept of proportional reconciliation: the proportional corrections to the k-th original series tend to be proportional to the alterability coefficients, α_k. For example, the proportional corrections of a series with alterability coefficients equal to 2 tend to be twice as large as those other series with alterability coefficients equal 1.

In order to see why uniform proportional reconciliation requires that parameter λ be set equal to 1/2, let us assume that

(a) the system of series is classified by P Provinces where the P-th Province is the Total,

(b) all the alterability coefficients and all the coefficients of variations are equal to 1 and

(c) the errors are independently and identically distributed, i.e. $\boldsymbol{\Omega}_k = \boldsymbol{I}_{T_k}$, for simplicity.

The objective function is then:

$$min \; \{ \Sigma_{t=1}^{T} \, e_{P,t}^{2} \, / s_{P,t}^{2\lambda} \; + \; \Sigma_{p=1}^{P-1} \, \Sigma_{t=1}^{T} \, e_{p,t}^{2} \, / s_{p,t}^{2\lambda} \} \; ,$$

where $e_{p,t} = s_{p,t} - \theta_{p,t}$. For $\lambda=1$ and $\lambda=1/2$, the objective functions are respectively

$$min \; \{ \Sigma_{t=1}^{T} \, e_{P,t}^{2} \, / s_{P,t}^{2} \; + \; \Sigma_{p=1}^{P-1} \, \Sigma_{t=1}^{T} \, e_{p,t}^{2} \, / s_{p,t}^{2} \} ,$$

$$min \; \{ \Sigma_{t=1}^{T} \, e_{P,t}^{2} \, / |s_{P,t}| \; + \; \Sigma_{p=1}^{P-1} \, \Sigma_{t=1}^{T} \, e_{p,t}^{2} \, / |s_{p,t}| \} .$$

With $\lambda=1$ each squared error is divided by the *square* of the corresponding observed value $s_{p,t}^2$. As a result, the errors $e_{p,t} = s_{p,t} - \theta_{p,t}$ are proportional to the square of the observed values, and the large series, namely the Total and the large Provinces, absorb disproportionate fractions of the cross-sectional aggregation discrepancies, and the small Provinces remain almost unaltered.

With $\lambda=1/2$, each error is divided by the observed absolute value $s_{p,t}$. As a result, all series in the system tend to display uniform proportional corrections $\hat{\theta}_{p,t} / s_{p,t}$. Note that denominator of the Total $s_{P,t}$ for time t is comparable to the sum of the denominators of the elementary series $\Sigma_{p=1}^{P-1} s_{p,t}$. Consequently, the Total tends to absorb half the cross-sectional aggregation discrepancy and the elementary series, the other half.

Note that the role of parameter λ holds for any correlation matrix, not only for $\boldsymbol{\Omega}_k = \boldsymbol{I}_{T_k}$.

Aware that reconciliation (or raking) distorts the movement of the various series, many practitioners advocate using the strong Denton movement preservation criterion in reconciliation.[2] However, movement preservation cannot achieved by reconciliation, because the cross-sectional discrepancies typically change sign from period to period. Each discrepancy *must* be cross-sectionally distributed over the various series in order to satisfy the cross-sectional constraints. As a result, attempts to preserve movement in reconciliation prove futile. Furthermore, in the absence of temporal constraints, large modifications and negative values may emerge in the reconciled series, because the movement preservation specifies nothing about the level of the individual reconciled series. This point will be illustrated in section 12.5.

Regarding the failure to preserve movement in reconciliation, Solomou and Weale (1993, p. 90) wrote that "*if all the variables have the same serial correlation, then the problem reduces to the simple case in which the data can be balanced one observation at a time* [i.e. $\boldsymbol{\Omega}_k = \boldsymbol{I}$]". The authors further observe (p. 103) that "*Experiments suggested that sharp reduction in the serial correlation parameters do not make much difference*". This interpretation is compatible with that of the previous paragraph.

If the movement preservation criterion is specified for only one of the series, that reconciled series does preserve movement, at the expense of distorting the other series even more; furthermore, in the absence of temporal constraints that same series may change level.

Furthermore, Weale (1988) pointed out that *"If a time series of data is to be balanced* [reconciled], *the presence of error persistence is inevitable and that the CSO* [Central Statistical Office of the U.K.], *suggests that the difference from one year to the next are measured more accurately than the levels themselves. This implies that the errors are positively autocorrelated; the results of Stone (1982) confirm this view. Stone (1980) argued that, provided all the data errors show the same autocorrelation, it is perfectly permissible to balance each observation separately, as though there were no autocorrelation."*

Under the strong assumption of common autocorrelation for all series, a system of series subject to yearly constraints can be reconciled one year at a time with no penalty in terms of movement discontinuities between the

[2] In the context of model (11.4), movement preservation would be attempted by means of an AR(1) process embodied in $\boldsymbol{\Omega}_k$ of Eq. (11.10), with an autoregressive parameter close to 1.0.

years; or even, one period at a time, if there are no temporal constraints. In particular, we have verified that if the Denton additive or proportional movement preservation criterion is used for all series, no gain in movement preservation is achieved.

The best way to preserve movement in reconciliation is to minimize the size of the corrections brought about by reconciliation; and more importantly, to spread the cross-sectional aggregation discrepancies over as many series as possible. In other words, the system of series should be quite large in terms of (number of) classes and contain mostly alterable series.

11.4 Data Problems

This section reviews some of the data problems encountered in reconciliation. We start with a few simple processes designed to identify problems with the data to be reconciled. We assume a classification by Industry and Province.

(1) A very useful process is to calculate and examine the number of dates, the minimum starting dates and maximum ending dates, and the minimum and maximum values for the variables of interest considered, for each combination of Industrial and Provincial identifiers. Ideally this process should also determine whether some starting and ending dates are absent from the target range of dates (e.g. Jan. 2001 to Nov 2004). Dates and values outside expected ranges, and dates missing from the target range of dates, should be discussed with the *client* or the subject matter expert providing the data.

(2) Another very useful process is to take the union of the Industrial codes and the union of the Provincial codes on the file produced by process (1).The number of Industries and Provinces in the file are then the number of elements in the respective unions. (The union operation usually eliminates duplicates and alpha-numerically sorts the elements.) The unions should be examined to determine whether all expected Industries and Provinces are present in the file and whether unexpected Industries or Provinces are also present. Missing Industries and Provinces should be discussed with the client providing the data.

(3) After fixing the above problems, e.g. ignoring the unwanted series and dates, the next process should be to calculate cross-sectional additive and proportional aggregation discrepancies of the system of series. In some cases, the proportional discrepancies may be questionably large;

or the Marginal Totals do not cross-sectionally sum to the Grand Total when they are supposed to; etc.

(4) If benchmarks are supposed to be available for each series, processes (1) to (3) should be applied to the benchmarks. Furthermore, the proportional temporal aggregation discrepancies should be calculated, temporal discrepancies outside a pre-established range should be discussed with the client.

Large Cross-Sectional Aggregation Discrepancies

In general, a system of series to be reconciled should display small or moderate Industrial and Provincial proportional cross-sectional aggregation discrepancies (11.3), and small proportional temporal aggregation discrepancies if applicable. In other words, the proportional discrepancies should be in the neighbourhood of 1.00. The presence of proportional discrepancies outside a certain range probably indicates something wrong with the system of series. This critical range would vary with the type of variable of interest. For inherently volatile variables such as Exports, the critical range could be relatively large, e.g. 0.90 to 1.10; for inherently predictable variables such as population, relatively low, e.g. 0.98 to 1.02.

Systems of seasonally adjusted series should also display small proportional discrepancies. In the case of the seasonally adjusted Canadian Retail Trade series, classified by Trade Group and by Province, the proportional discrepancies vary between 0.975 and 1.025; for the more volatile Wholesale Trade series (identically classified), between 0.95 and 1.05.

Large proportional cross-sectional discrepancies may indicate the presence of subtotals in the data file, leading to partial double counting. For example, the file supposed to contain data classified by Industry and Province may include data pertaining to larger metropolitan areas and to related industries. This translates into proportional discrepancies much lower than 1. The unions recommended in (2) above would have identified this problem.

Another problem may be the absence in the data file of some of the Marginal Totals, e.g. the Industrial Totals, the Provincial Totals or the Grand Total. Sometimes the Industrial or Provincial codes change after a certain date. Again the unions of the Industrial and Provincial codes proposed in (2) above could have detected the problem earlier.

After reconciliation, the proportional cross-sectional aggregation discrepancies of the reconciled series should be equal to 1.0; and the additive ones, equal to 0, except for rounding errors; similarly for the temporal discrepancies, if applicable. The presence of significant residuals discrepancies indicates that some time series data are contradictory.

The size of the proportional corrections $\hat{\theta}_{k,t}/s_{k,t}$ made to the series to achieve reconciliation is also informative. Simple statistics on these corrections, namely the minimum, maximum and standard deviation, for each series are very useful. Large corrections confirms the presence of large cross-sectional discrepancies and/or temporal discrepancies in the original system of series. Furthermore the statistics on the corrections can be used to verify that the unalterable series remained unaltered. The presence of negative corrections indicates the presence of negative reconciled values.

Contradictory Data

Contradictory data can occur in different manners. For example, consider a two-way system of series classified by Industry and Province where the Totals are pre-determined and imposed onto the elementary (cell) series. If the Industrial or the Provincial Totals do not cross-sectionally add up to the Grand Total, it is impossible to restore additivity to the system of series. Process (3) above would reveal the situation.

Contradictory data can also occur in the temporal benchmarks. If the series are subject to yearly temporal constraints (say) and the benchmarks themselves do not satisfy cross-sectional aggregation constraints, it is again impossible to obtain a system of series satisfying both temporal and cross-sectional constraints. The cross-sectional discrepancies of the benchmarks are imposed to the reconciled system of series. The solution is to rectify the benchmarks or make them non-binding. Process (3) applied to the benchmarks above reveals the problem beforehand.

For example, in the case of seasonally adjusted series, the yearly totals of the *raw* (seasonally *un*-adjusted) series are used as yearly benchmarks. If the raw data does not satisfy cross-sectional aggregation constraints, the benchmarks do not satisfy the cross-sectional constraints either, and it is impossible (contradictory) to obtain a system of series satisfying both the benchmarks and cross-sectional constraints.

Contradictory data can also occur in conjunction with the alterability coefficients. This would be the case if all the alterability coefficients in one Industry or one Province are equal to 0. This is a contradiction in the sense that no series in the Industry or Province in question can be modified to satisfy the cross-sectional constraints.

Problems with Files

Several of the problems mentioned above originate from the variability of the files sent by clients or subject matter experts; the content and record layout can vary from occasion to occasion - even within a day. The "corrected" file may contain new problems. As a result, an inordinate amount of time and effort is spent figuring out the data: determining which series is which, whether some Totals are missing, whether there is double counting; some files are plainly not machine readable. This kind of problem also exists for micro longitudinal data and survey data in general.

Such problems illustrate the need of protocols and standards, for storing and sending data - time series data especially - between various sections of statistical agencies and between statistical agencies.

Some of the problems mentioned above are exemplified below.

11.5 Strategies for Reconciliation

This section provides strategies to approach the reconciliation problem. The choice of strategy becomes much clearer when systems of series are seen as sets of two-way tables, some at the finest detail of classification by attributes and some at the more aggregate level.

The Canadian monthly Labour Force Survey series are classifed by Sex; Age Groups and by "Duration", i.e. Full-Time and Part-Time employment. The example focuses on the seasonally adjusted Employment series.

Recall that seasonal adjustment typically produces mild cross-sectional discrepancies in systems of series which satisfied cross-sectional constraints at the outset. The reason is that different seasonal adjustment options are used for each series and that seasonal adjustment methods are generally non-linear.

As shown in Table 11.2, the seasonally adjusted Employment estimates are published by Major Age-Sex Groups, by Province and by employment Duration.

The seasonally adjusted Employment estimates are also published by Major Age-Sex Groups by Province as displayed in Table 11.3.

At the Canada level, the series are also published at a more *detailed* Age-Sex Groups, as shown in Table 11.4.

Table 11.2. Canadian Labour Force Employment series by Major Age-Sex-Duration Groups and Province

Province / Age-Sex-Duration	NF	NS	···	BC	Canada Total
Male 15-24 years full-time			···		
Male 15-24 years part-time			···		
Male 25 years + full-time			···		
Male 25 years + part-time			···		
Female 15-24 years full-time			···		
Female 15-24 years part-time			···		
Female 25 years + full-time			···		
Female 25 years + part-time			···		
Provincial Total			···		

Table 11.3. Canadian Labour Force Employment series by Major Age-Sex Groups and Province

Province / Major Age-Sex	NF	···	BC	Canada Major Age-Sex Total
Male 15-24 years		···		
Male 25 years +		···		
Female 15-24 years		···		
Female 25 years +		···		
Provincial Totals		···		

The problem is to find an appropriate manner to perform hierarchical reconciliation in order to minimize the distortions of the seasonally adjusted series. In this particular case, we advance two strategies. The strategy chosen should depend on the reliability of the seasonally adjusted series to be reconciled.

Table 11.4. Canadian Labour Force Employment series by detailed Age-Sex Groups at the Canada level

Sub-Age Group / Major Age-Sex Group	Sub-Age Group 1	Sub-Age Group 2	Canada Major Age-Sex Totals
Male 15-24 years	M 15-19	M 20-24	
Male 25 years +	M 25-54	M 55 +	
Female 15-24 years	F 15-19	F 20-24	
Female 25 years +	F 25-54	F 55 +	

Strategy One

Table 11.3 can obviously be obtained from Table 11.2 by collapsing the Full-Time and Part-Time groups for each Major Age-Sex Group. This calls for the following hierarchical solution, which assumes that *all* series to be reconciled were directly seasonally adjusted.

(1) Reconcile Table 11.2, where all the series are treated as endogenous. This is achieved by assigning alterability coefficients greater than 0 to all series. As a result, all series are jointly estimated, including the Provincial Totals and the Canada Totals by Group.

(2) Generate Table 11.3 from Table 11.2 by taking the cross-sectional sums over the Full-Time and Part-Time grups for each major Age-Sex Group.

(3) Reconcile Table 11.4, where the Canada Major Age-Sex Totals originate from the last column of Table 11.3 produced in step (2) and are treated as exogenous by means of alterability coefficients equal to 0. Note that Table 11.4 is a pseudo two-way system of series, in the sense that it has no column totals. Each row is in fact a one-way system.

The two-way reconciliation model of Chapter 14 can be used for both Tables 11.2 and 11.4. Indeed, the two-way model can also process pseudo two-way systems of series, which simplifies the management of files. Alternatively, the one-way model can be used for each row of Table 11.4.

The more uniform the *non*-zero alterability coefficients (e.g. $\alpha_k=1$), the more the cross-sectional aggregation discrepancies are absorbed by a larger number of series, thus minimizing the distortion of the proportional movement. This would be most important for Table 11.2 in step (1). In order to perform proportional reconciliation, the parameter λ should be set to ½. Section 11.3 discusses the covariance matrices, the alterability coefficients and parameter λ in more details.

Strategy One is attractive if the seasonally adjusted series of Table 11.2 are sufficiently reliable for publication. However, in small provinces this is not likely to be the case for all Age-Sex-Duration Groups.

Strategy Two

Another hierarchical strategy is to start from Table 11.3, where the seasonally adjusted series are more likely to be reliable, because they are more aggregated. This calls for the following strategy, which assumes that all series are directly seasonally adjusted.

(1) Reconcile Table 11.3 where all series are specified as endogenous by means of alterability greater than 0. The resulting table contains the reconciled series by Major Age-Group and by Province, including Canada.

(2) Reconcile Table 11.4, where the Canada Major Age-Sex Totals originate from the last column of Table 11.3 produced in step (1) and specified as exogenous by means of alterability coefficients equal to 0. The resulting table contains the more *detailed* Age-Group estimates at the Canada level. Again note that Table 11.4 is a pseudo two-way system of series without column totals; each row is in fact a one-way system.

(3) Reconcile Table 11.5 where the Major Age-Sex Totals originate from Table 11.3 produced in step (1) and specifed as exogenous. There are eleven such tables: one for each Province and one for Canada. The resulting tables jointly contain the same information as Table 11.2 albeit obtained and presented in a different manner. Note that Table 11.5 is also a pseudo two-way system of series, each row being a one-way system.

The two-way reconciliation model can be used to reconcile Tables 11.3, 11.4 and 11.5. Note that the eleven Tables 11.5 of step (3) can be appended into one single table and reconciled as such by means of the two-way model. Alternatively each row of the appended table can be reconciled by the one-way model. This alternative would complicate file management.

Table 11.5. Canadian Labour Force Employment series by Age-Sex-Duration Groups at the Province level

Duration / Major Age-Sex Group	Full-Time	Part-Time	Major Age-Sex Total for a given Province
Male 15-24 years			
Male 25 years +			
Female 15-24 years			
Female 25 years +			

Strategy One has the advantage of spreading the cross-sectional aggregation discrepancies over more series, which tends to minimize changes in movement. Solution Two starts the hierarchical reconciliation process with more reliable series, however probably at the cost of larger movement distortions.

Note that Strategy One can be used even if some of the series of Table 11.2 are not reliable. Those series could remain unpublished.

Under both strategies, a case can be made to maintain the yearly sums of the raw series, despite the fact they are stock series. Indeed, the yearly level of employment and unemployment is usually represented by the yearly average.

Recall that if all the alterability coefficients in one Industry or one Province are equal to 0 in a two-way table, there is no solution to the reconciliation problem. A related problem is that of redundant constraints discussed in sections 12.4, 13.4 and 14.6, for the one-way, the marginal two-way and the two-way models.

12 Reconciling One-Way Classified Systems of Time Series

12.1 Introduction

This chapter focuses on *one-way classified systems* of series. We shall henceforth refer to the corresponding reconciliation model as the *one-way reconciliation model.*

Without loss of generality, we assume a system classified by Province. Such a system must satisfy one set of *cross-sectional aggregation constraints*:

$$s_{P,t} = \Sigma_{p=1}^{P-1} s_{p,t}, \quad t=t_{1,P},...,t_{L,P} \text{ ,} \tag{12.1}$$

where P denotes the Total (the "all" Province), and $t_{1,p}$ and $t_{L,p}$ are respectively the first and last available observations for series $s_{p,t}$. The number of observations for a given series, $s_{p,t}$, is thus $T_p = t_{L,p} - t_{1,p} + 1$. The number T_p may vary from province to province. The number of constraints in the (12.1) is equal to the number of observations T_P for the Total series.

The system contains one set of *cross-sectional aggregation discrepancies* and one set of *proportional cross-sectional aggregation discrepancies*:

$$d_t = s_{P,t} - \Sigma_{p=1}^{P-1} s_{p,t}, \quad t=t_{1,nP},...,t_{L,nP}, \quad n=1,...,N, \tag{12.2a}$$

$$d_t^{(p)} = s_{P,t} \,/\, \Sigma_{p=1}^{P-1} s_{p,t}, \quad t=t_{1,nP},...,t_{L,nP}, \quad n=1,...,N, \tag{12.2b}$$

where the superscript *(p)* stands here for proportion and not province.

The cross-sectional additivity constraints of Eq. (12.1) are often implemented by means of the Iterative Proportional Fitting approach, also known as "raking". In case of one-way systems raking merely consist of multiplying the elementary series ($s_{p,t}$, $p<P$) by the corresponding proportional discrepancy $d_t^{(p)}$ of Eq. (12.2b). This method implies that the Total Series is exogenous, i.e. pre-determined.

12.2 The Reconciliation Model for One-Way Classified Systems of Series

This section provides an analytical solution for the reconciliation model for one-way classified systems of series. For a system of K=P "Provincial" series, model (11.4) consists of the two following equations:

$$\begin{bmatrix} s_1 \\ s_2 \\ \vdots \\ s_P \end{bmatrix} = \begin{bmatrix} I_{T_1} & 0 & \cdots & 0 \\ 0 & I_{T_2} & \cdots & 0 \\ \vdots & \vdots & \ddots & \vdots \\ 0 & 0 & \cdots & I_{T_P} \end{bmatrix} \begin{bmatrix} \theta_1 \\ \theta_2 \\ \vdots \\ \theta_P \end{bmatrix} + \begin{bmatrix} e_1 \\ e_2 \\ \vdots \\ e_P \end{bmatrix}, \tag{12.3a}$$

where I_{T_p} is an identity matrix of dimension $T_p \times T_p$, s_P is the original Total series and $E(e_p) = 0$, $E(e_p e_p') = V_{e_p}$, p=1,...,P and $E(e_i e_j') = 0$, for all $i \neq j$; and

$$g = \begin{bmatrix} G_1 & G_2 & \cdots & -G_P \end{bmatrix} \begin{bmatrix} \theta_1 \\ \theta_2 \\ \vdots \\ \theta_P \end{bmatrix} + \varepsilon, \tag{12.3b}$$

where $E(\varepsilon) = 0$, $E(e\,\varepsilon') = 0$ and $E(\varepsilon\,\varepsilon') = V_\varepsilon$ is typically equal to 0. The matrices G_p are usually selector matrices, i.e. identities matrices with T_P rows and $T_p \leq T_P$ columns; the columns removed correspond to periods t where the series does not exist. These matrices thus allow some of the elementary series (s_p for $p < P$) to begin later and/or end earlier than the total s_P. For example, let the Total s_P start in 2001 and end in 2005 (5 observations); and, the second series s_2 start in 2002 and end in 2004 (3 observations). Matrix G_P equal to I_5; and, G_2 is equal to I_5 with the first and fifth columns removed.

For the moment, all the series in model (12.3) – including the Total – are considered endogenous, that is jointly estimated with the elementary series. As a result, vector g of dimension T_P×1 contain zeroes. Specifying all series as endogenous leads to a more general solution. It will then be easy to treat some of the series as exogenous, that is pre-determined.

If all series start at the same date and have the same length, i.e. $T_p = T$, model (12.3) can be written more compactly in terms of Kronecker products[1] as follows

$$s = I_P \otimes I_T\, \theta + e\,, \quad E(e) = 0, \; E(e\,e') = V_e\,,$$
$$g = \iota \otimes I_T\, \theta + \varepsilon, \quad E(\varepsilon) = 0, \; E(\varepsilon\,\varepsilon') = V_\varepsilon\,,$$

where I_P and I_T are identity matrices of dimension $P{\times}P$ and $T{\times}T$, ι is a row vector of dimension 1 by P equal to $[1 \;\; 1 \,...\, 1 \; {-1}]$, V_e is block diagonal, and V_ε is diagonal and e and ε are mutually independent. However, since series often vary in length especially within a Province, we continue with the notation in Eq. (12.3). This notation is more cumbersome but leads to a more general analytic solution.

The general solution (11.8) is repeated here for convenience:

$$\begin{aligned}\hat{\theta} &= s + V_e\, G'(G\, V_e G' + V_\varepsilon)^{-1}\,[g - G s] \\ &= s + V_e\, G'[V_d^{-1}\, d] \;=\; s + V_e\, G'\, \hat{\gamma}\,,\end{aligned} \qquad (12.4a)$$

$$var[\hat{\theta}] = V_e - V_e G' V_d^{-1}\, G\, V_e\,, \qquad (12.4b)$$

where $G = \begin{bmatrix} G_1 & G_2 & \dots & -G_P \end{bmatrix}$ and g is a $T_P{\times}1$ vector of zeroes.

Our main purpose is now to obtain a specific analytical solution for Eq. (12.4). To achieve this, we need to determine the contents of some of the matrix products of (12.4), namely $V_e G'$, $V_d = (G\, V_e G' + V_\varepsilon)$, d and $V_e G' V_d^{-1}\, G\, V_e$, in terms of the partitions of model. (12.3).

Given $G = \begin{bmatrix} G_1 & G_2 & \cdots & -G_P \end{bmatrix}$ and the block diagonal partitions of V_e, and denoting $G_p V_{e_p}$ by L_p, we obtain:

---

[1] The elements of Kronecker product $A \otimes B$ of matrices A and B respectively of dimension $K{\times}L$ and $M{\times}N$ are given by $a_{ij}\, B$, where a_{ij} are the elements of matrix A. The resulting product has dimension $KM{\times}LN$.

$$\boldsymbol{V}_e \, \boldsymbol{G}' = \begin{bmatrix} \boldsymbol{V}_{e_1} & 0 & \cdots & 0 \\ 0 & \boldsymbol{V}_{e_2} & \cdots & 0 \\ \vdots & \vdots & \ddots & \vdots \\ 0 & 0 & \cdots & \boldsymbol{V}_{e_P} \end{bmatrix} \begin{bmatrix} \boldsymbol{G}_1' \\ \boldsymbol{G}_2' \\ \vdots \\ -\boldsymbol{G}_P' \end{bmatrix} = \begin{bmatrix} \boldsymbol{V}_{e_1} \boldsymbol{G}_1' \\ \boldsymbol{V}_{e_2} \boldsymbol{G}_2' \\ \vdots \\ -\boldsymbol{V}_{e_P} \boldsymbol{G}_P' \end{bmatrix} = \begin{bmatrix} \boldsymbol{L}_1' \\ \boldsymbol{L}_2' \\ \vdots \\ -\boldsymbol{L}_P' \end{bmatrix}, \tag{12.5a}$$

$$\boldsymbol{G}(\boldsymbol{V}_e \boldsymbol{G}') = \left[\Sigma_{p=1}^{P} \, \boldsymbol{G}_p \boldsymbol{V}_{e_p} \boldsymbol{G}_p' \right], \tag{12.5b}$$

$$\boldsymbol{V}_d = \left[\Sigma_{p=1}^{P} \, \boldsymbol{G}_p \boldsymbol{V}_{e_p} \boldsymbol{G}_p' + \boldsymbol{V}_{\varepsilon} \right], \tag{12.5c}$$

$$\boldsymbol{d} = [\boldsymbol{g} - \boldsymbol{G}\boldsymbol{s}] \equiv \boldsymbol{g} - \boldsymbol{G}_1 \boldsymbol{s}_1 - \boldsymbol{G}_2 \boldsymbol{s}_2 - \ldots + \boldsymbol{G}_P \boldsymbol{s}_P, \tag{12.5d}$$

$$\boldsymbol{V}_e \, \boldsymbol{G}' \boldsymbol{V}_d^{-1} = \begin{bmatrix} \boldsymbol{V}_{e_1} \boldsymbol{G}_1' \boldsymbol{V}_d^{-1} \\ \boldsymbol{V}_{e_2} \boldsymbol{G}_2' \boldsymbol{V}_d^{-1} \\ \vdots \\ -\boldsymbol{V}_{e_P} \boldsymbol{G}_P' \boldsymbol{V}_d^{-1} \end{bmatrix} \begin{bmatrix} \boldsymbol{L}_1' \boldsymbol{V}_d^{-1} \\ \boldsymbol{L}_2' \boldsymbol{V}_d^{-1} \\ \vdots \\ -\boldsymbol{L}_P' \boldsymbol{V}_d^{-1} \end{bmatrix}, \tag{12.5e}$$

$$(\boldsymbol{V}_e \, \boldsymbol{G}' \boldsymbol{V}_d^{-1}) \, \boldsymbol{G} \, \boldsymbol{V}_e = \begin{bmatrix} \boldsymbol{E}_{11} & \boldsymbol{E}_{12} & \cdots & -\boldsymbol{E}_{1P} \\ \boldsymbol{E}_{21} & \boldsymbol{E}_{22} & \cdots & -\boldsymbol{E}_{2P} \\ \vdots & \vdots & \ddots & \vdots \\ -\boldsymbol{E}_{P1} & -\boldsymbol{E}_{P2} & \cdots & +\boldsymbol{E}_{PP} \end{bmatrix}, \tag{12.5f}$$

where $\boldsymbol{L}_p = \boldsymbol{G}_p \boldsymbol{V}_{e_p}$ and $\boldsymbol{E}_{ij} = \boldsymbol{L}_i' \boldsymbol{V}_d^{-1} \boldsymbol{L}_j$.

Substituting (12.5a), (12.5c), (12.5d) and (12.5f) in (12.4) yields:

$$\begin{bmatrix} \hat{\boldsymbol{\theta}}_1 \\ \hat{\boldsymbol{\theta}}_2 \\ \vdots \\ \hat{\boldsymbol{\theta}}_P \end{bmatrix} = \begin{bmatrix} \boldsymbol{s}_1 \\ \boldsymbol{s}_2 \\ \vdots \\ \boldsymbol{s}_P \end{bmatrix} + \begin{bmatrix} \boldsymbol{L}_1{}' \\ \boldsymbol{L}_2{}' \\ \vdots \\ -\boldsymbol{L}_P{}' \end{bmatrix} [\, \boldsymbol{V}_d^{-1} \boldsymbol{d} \,], \qquad (12.6a)$$

and

$$var \begin{bmatrix} \hat{\boldsymbol{\theta}}_1 \\ \hat{\boldsymbol{\theta}}_2 \\ \vdots \\ \hat{\boldsymbol{\theta}}_P \end{bmatrix} = \begin{bmatrix} \boldsymbol{V}_{e_1} & 0 & \cdots & 0 \\ 0 & \boldsymbol{V}_{e_2} & \cdots & 0 \\ \vdots & \vdots & \ddots & \vdots \\ 0 & 0 & \cdots & \boldsymbol{V}_{e_P} \end{bmatrix} - \begin{bmatrix} \boldsymbol{E}_{11} & \boldsymbol{E}_{12} & \cdots & -\boldsymbol{E}_{1P} \\ \boldsymbol{E}_{21} & \boldsymbol{E}_{22} & \cdots & -\boldsymbol{E}_{2P} \\ \vdots & \vdots & \ddots & \vdots \\ -\boldsymbol{E}_{P1} & -\boldsymbol{E}_{P2} & \cdots & +\boldsymbol{E}_{PP} \end{bmatrix}, \qquad (12.6b)$$

where $\boldsymbol{E}_{ij} = \boldsymbol{L}_i{}' \boldsymbol{V}_d^{-1} \boldsymbol{L}_j = \boldsymbol{V}_{e_i} \boldsymbol{G}_i{}' \boldsymbol{V}_d^{-1} \boldsymbol{G}_j \boldsymbol{V}_{e_j}$.

From the above results, the analytical solution is

$$\begin{aligned} \hat{\boldsymbol{\theta}}_p &= \boldsymbol{s}_p + \delta_{pP} \boldsymbol{V}_{e_p} \boldsymbol{G}_p{}' [\, \boldsymbol{V}_d^{-1} \boldsymbol{d} \,] \\ &\equiv \boldsymbol{s}_p + \delta_{pP} \boldsymbol{V}_{e_p} \boldsymbol{G}_p{}' \hat{\boldsymbol{\gamma}}, \quad p=1,\ldots,P, \end{aligned} \qquad (12.7a)$$

$$var[\hat{\boldsymbol{\theta}}_p] = \boldsymbol{V}_{e_p} - \boldsymbol{V}_{e_p} \boldsymbol{G}_p{}' \boldsymbol{V}_d^{-1} \boldsymbol{G}_p \boldsymbol{V}_{e_p}, \quad p=1,\ldots,P, \qquad (12.7b)$$

$$cov[\hat{\boldsymbol{\theta}}_p \, \hat{\boldsymbol{\theta}}_{p'}] = -\, \delta_{pP} \delta_{p'P} \boldsymbol{V}_{e_p} \boldsymbol{G}_p{}' \boldsymbol{V}_d^{-1} \boldsymbol{G}_{p'} \boldsymbol{V}_{e_{p'}}, \quad p \neq p', \qquad (12.7c)$$

where the scalar δ_{pP} is equal to -1 for $p=P$; and to 1, for $p \neq P$. The vector of cross-sectional aggregation discrepancies is given by (12.5d) $\boldsymbol{d} = \boldsymbol{g} - \Sigma_{p=1}^{P} \delta_{pP} \boldsymbol{G}_p \boldsymbol{s}_p$, with $\boldsymbol{g} = \boldsymbol{0}$.

Any series $\boldsymbol{s}_p$ - the Total in particular - can be specified as exogenous, i.e. pre-determined, by setting its alterability coefficient α_p equal to 0, which implies that $\boldsymbol{V}_{e_p} = \boldsymbol{0}$. (See Eq. (11.10) of Chapter 11.)

The Total can also be made exogenous (i.e. $\boldsymbol{\theta}_P = \boldsymbol{s}_P$) by storing it in vector $\boldsymbol{g}$ and by letting p range from 1 to $P-1$ in Eqs. (12.7). Note that the vector of cross-sectional discrepancies $\boldsymbol{d}$ remains unchanged. This implicitly entails removing the last row of partitions from Eq. (12.3a), and removing vector $\boldsymbol{\theta}_P$, and matrix $-\boldsymbol{G}_P$ from Eq. (12.3b).

Note that if only one Province is alterable, Province 2 for example, Eq. (12.7a) becomes $\hat{\boldsymbol{\theta}}_2 = \boldsymbol{s}_2 + \boldsymbol{V}_{e_2} \boldsymbol{V}_{e_2}^{-1} \boldsymbol{d} = \boldsymbol{s}_2 + \boldsymbol{d}$ (assuming that $\boldsymbol{s}_2$ and $\boldsymbol{s}_P$ have same length). In other words, Province 2 absorbs all the cross-sectional discrepancies.

Solution (12.7) is analytical in the sense that each reconciled series $\hat{\boldsymbol{\theta}}_p$ is expressed in terms of: (a) its "own" data, $\boldsymbol{s}_p$ and $\boldsymbol{V}_{e_p}$, and (b) the minimum set of "common" data contained in $\boldsymbol{d}$ and $\boldsymbol{V}_d^{-1}$.

12.3 Implementation of the Analytical Solution

This section is mainly intended for readers who consider programming the one-way reconciliation model. If the cross-covariance matrices are not required, the implementation of solution (12.7) requires two runs through the data, i.e. each series is processed twice.

Note that if the Total is exogenous, and therefore contained in vector $\boldsymbol{g}$, the index p ranges from 1 P-1.

First Run

The first run performs the following calculations for each series separately.

(1a) The series $\boldsymbol{s}_p$ is read and its covariance matrix $\boldsymbol{V}_{e_p}$ is generated according to Eq. (11.10) of Chapter 11. Alternatively, $\boldsymbol{V}_{e_p}$ is read from a file if a previous adjustment generated the covariance matrices. In particular this adjustment may be benchmarking, in which case the series would also be subject to temporal aggregation constraints. (The temporal constraints are then implicit in the covariance matrix.)

(1b) To avoid potential numerical problems, the series are re-scaled. Each series is divided by

$$\bar{s} = (max(abs(s_{p,t})) - min(abs(s_{p,t})) / 2, \tag{12.8}$$

where $max(abs(s_{p,t}))$ is the maximum absolute value of the various series and $min(abs(s_{p,t}))$ the minimum absolute value. The covariance matrix of the series is divided by $\bar{s}^2$. The resulting covariance matrix is then multiplied by the alterability coefficient selected for the series. This product becomes the covariance matrix used in the reconciliation model.

(1c) Matrix $V_d = (\Sigma_{p=1}^{P} G_p V_{e_p} G_p' + V_\varepsilon)$ is accumulated, each series contributing one term to the summation. Similarly, the cross-sectional discrepancies are accumulated in vector $d = g - \Sigma_{p=1}^{P} \delta_{pP} G_p s_p$. The proportional discrepancies could also be calculated, to better assess the need for reconciliation.

Steps (1a) to (1c) are repeated for each series in the system.

After the first run is completed, the following calculations are performed.

(a) Matrix V_d of dimension $T_P \times T_P$ is inverted. In order to deal with the problem of potential redundant constraints (discussed in the next section), the Moore-Penrose generalized matrix inversion can be used.

(b) The Lagrange multipliers of (12.7a) are calculated: $\hat{\gamma} = [V_d^{-1} d]$.

Second Run

The second run performs the following calculations for each series separately.

(2a) This step is identical to (1a).

(2b) This step is identical to (1b).

(2c) The reconciled series $\hat{\theta}_p$ is calculated according to the last expression of Eq. (12.7a). If applicable, its covariance matrix, $var[\hat{\theta}_p]$, is calculated according to Eq. (12.7b).

(2d) The series considered and its covariance matrix are converted into the original scale by multiplying them by $\bar{s}$ of Eq. (12.8) and by $\bar{s}^2$, respectively. Both $\hat{\theta}_p$ and $var[\hat{\theta}_p]$ are saved on a file. Since $var[\hat{\theta}_p]$ is symmetric, it is sufficient to store only the elements on the main diagonal and on the lower diagonals of the matrix, i.e. $T \times (T+1)/2$ elements instead of T^2. In most cases, only the diagonal elements are of interest.

(2e) The residual cross-sectional discrepancies are accumulated in vector $\boldsymbol{d}^{(r)} = \boldsymbol{g} - \Sigma_{p=1}^{P} \delta_{pP} \boldsymbol{G}_p \hat{\boldsymbol{\theta}}_p$. These residual discrepancies may be used to assess whether the reconciliation satisfied the cross-sectional constraints.

Under the solution and implementation discussed, it is not necessary to generate the large matrices in model (12.4), namely $\boldsymbol{V}_e$ and $\boldsymbol{G}$; and similarly, for $\boldsymbol{G}_p$. For example, let the Total $\boldsymbol{s}_P$ start in 2001 and end in 2005 (5 observations); and, the second series $\boldsymbol{s}_2$ start in 2002 and end in 2004 (3 observations). The first and last observations of $\boldsymbol{s}_2$ are then $t_{1,2}=2$ and $t_{L,2}=4$. Matrix $\boldsymbol{G}_2$ and matrix product $\boldsymbol{G}_2\,\boldsymbol{s}_2$ are then respectively

$$\boldsymbol{G}_2 = \begin{bmatrix} 0 & 0 & 0 \\ 1 & 0 & 0 \\ 0 & 1 & 0 \\ 0 & 0 & 1 \\ 0 & 0 & 0 \end{bmatrix}, \quad \boldsymbol{G}_2\,\boldsymbol{s}_2 = \begin{bmatrix} 0 & 0 & 0 \\ 1 & 0 & 0 \\ 0 & 1 & 0 \\ 0 & 0 & 1 \\ 0 & 0 & 0 \end{bmatrix} \begin{bmatrix} s_{2,2} \\ s_{2,3} \\ s_{2,4} \end{bmatrix} = \begin{bmatrix} 0 \\ s_{2,2} \\ s_{2,3} \\ s_{2,4} \\ 0 \end{bmatrix}.$$

As a result, subtracting $\boldsymbol{G}_2\,\boldsymbol{s}_2$ (dimension 5×1) from vector $\boldsymbol{d}$ (5×1) amounts to subtracting $\boldsymbol{s}_2$ from rows 2 to 4 of vector $\boldsymbol{d}$. Similarly adding $\boldsymbol{G}_2\,\boldsymbol{V}_{e_2}\,\boldsymbol{G}_2'$ to $\boldsymbol{V}_d$ (5×5) amounts to adding $\boldsymbol{V}_{e_2}$ (3×3) to rows 2 to 4 and columns 2 to 4 of $\boldsymbol{V}_d$. Operationally, matrix $\boldsymbol{G}_2$ is thus replaced by the use of row and/or column indexes. (Indices can contain non-contiguous values, e.g. 1, 2, 5.)

Note that some of the calculations may be avoided, namely for series with alterability coefficients equal to 0. Indeed, the results for these series are already known to be $\hat{\boldsymbol{\theta}}_p = \boldsymbol{s}_p$ and $var[\hat{\boldsymbol{\theta}}_p] = \boldsymbol{0}$. These series contribute to $\boldsymbol{d}$ and to $\boldsymbol{d}^{(r)}$, but not to matrix $\boldsymbol{V}_d$.

12.4 Redundant Constraints in the One-Way Reconciliation Model

The one-way model has one set of cross-sectional constraints. This set contains T_P such constraints, one for each observation of the Total:

$$\theta_{P,t} = \Sigma_{p=1}^{P-1} \theta_{p,t}, \quad t=1,...,T_P.$$

We distinguish two situations regarding the temporal constraints:

(a) the absence of any temporal constraints and

(b) the presence of temporal constraints. In this case, we assume that (i) all series start at the same date and have equal length, (ii) all series have annual benchmarks for each complete "year" and (iii) all the temporal constraints are binding and implicit in the covariance matrices V_{e_p} of the series to be reconciled.

In the absence of temporal constraints (case (a) above), the one-way model (12.3) has no redundant cross-sectional constraints.

In the presence of temporal constraints (case (b) above), the model has one redundant cross-sectional constraint per "year" with benchmark *or* has one set of temporal constraints for any one of the P series. In fact choosing which constraints are redundant is arbitrary, because redundancy is an attribute of the whole set of constraints. The problem of redundancy can therefore be addressed in a number of different ways:

(1) Omit one cross-sectional constraint per "year" with benchmark.

(2) Make one cross-sectional constraint per "year" non-binding, by setting the corresponding diagonal element of V_ε equal to 1 (say).

(3) Omit all the temporal constraints for one of the series.

(4) Make the temporal constraints non-binding for any one of the series.

(5) Invert V_d using the Moore-Penrose generalized matrix inversion.

Under conditions (1) to (4), the omitted or non-binding constraints are satisfied through the remaining ones. If some constraints are not satisfied, it means some of the data is contradictory. For example, the benchmarks imposed on the system of series do not satisfy the cross-sectional aggregation constraints.

In practical situations, some series may have temporal constraints for each "year", and others fewer temporal constraints or none at all. This makes it hard to determine which constraints should be omitted or made non-binding.

The generalized matrix inversion was used by Theil (1971) in a more general context and specifically applied by Di Fonzo (1990) in reconciliation. This approach offers a valid and practical solution. Failure of the resulting reconciled series to satisfy all constraints indicates contradictory data, which should be fixed by the subject matter experts.

12.5 An Example of One-Way Reconciliation

The one-way reconciliation model lends itself to a better understanding of the alterability coefficients and of the criteria embodied in the covariance matrices $\boldsymbol{V}_{e_p} = \alpha_p\, \boldsymbol{\Xi}_p^{\lambda} \boldsymbol{\Omega}_p\, \boldsymbol{\Xi}_p^{\lambda}$.

The system of series exemplified comprises three simulated quarterly series pertaining to Provinces over two years 2001 and 2002, with benchmarks for year 2001 only. Province 3 is the Total of the other two Provinces. The benchmark is 444 for the first Province; 843 for the second; and 1,287 for the third (the Total). The system of series will be reconciled considering six different sets of options, referred to as case (1) to case (6). The resulting reconciled series, the corresponding *proportional corrections* $\hat{\theta}_{p,t} / s_{p,t}$ and *change in proportional movement* $\hat{\theta}_{p,t} / s_{p,t}$ - $\hat{\theta}_{p,t-1} / s_{p,t-1}$ are displayed in Tables 12.1 to 12.5. The additive and proportional cross-sectional discrepancies also appear in the tables.

Case (1) contains no temporal constraints (no benchmarks). The alterability coefficients α_p are equal to 1 so that all series are equally alterable. The coefficients of variations are equal to 0.01, so that all observations of a given series are equally alterable. The covariances matrices are equal to $\boldsymbol{V}_{e_p}$ = $\alpha_p\, \boldsymbol{\Xi}_p^{\lambda} \boldsymbol{\Omega}_p\, \boldsymbol{\Xi}_p^{\lambda}$ with $\alpha_p = 1$, $\boldsymbol{\Xi}_p = diag(0.01 \times \boldsymbol{s}_p)$, $\lambda = 1/2$ and $\boldsymbol{\Omega}_p = \boldsymbol{I}_{T_p}$ for all series. The resulting objective function minimizes the proportional errors,

$$(1/0.01) \times min\, \{ \Sigma_{p=1}^{P}\, \Sigma_{t=1}^{T_p}\, e_{p,t}^{2} / s_{p,t} \},$$

where $e_{p,t} = s_{p,t} - \theta_{p,t}$, P=3 and $T_p = 8$.

Case (2) is identical to case (1) except that the Total series is un-alterable with $\alpha_3 = 0$.

Table 12.1a displays the original and the reconciled series obtained under the options used for cases (1) and (2), which illustrates the role of the alterability coefficients.

Case (1) (i.e. the associated set of options) spreads each proportional cross-sectional discrepancy pertaining to period t uniformally on *all* the series for the same period. As shown in the table, the corrections of period t=8 (say) are equal to −3.4% for the two elementary series (corrections equal to 0.966) and +3.4 for the Total series (correction equal to 1.034). The two elementary series absorb half of each proportional discrepancies, and the Total series the other half.

Case (2), on the other hand, spreads each proportional cross-sectional discrepancy pertaining to period t, uniformally on the elementary series *only*. As shown in Table 12.1a, the corrections for period t=8, are equal to −6.6% for the two elementary series and 0% for the Total series. Each correction of the elementary series for period t is equal to the proportional aggregation discrepancy for that period displayed in the fifth column. The elementary series absorbs all of the discrepancies and the Total series none. As a result, case (2) produces larger corrections because the discrepancies are distributed over fewer series.

Case (1) produces smaller corrections - and therefore smaller changes in the proportional movement of the original series - than case (2). The standard deviations of the proportional corrections exhibited in Table 12.1b are indeed smaller for case (1) than for case (2), 0.029 for all series against 0.046.

Note that case (2) produces results identical to those of Iterative Proportional Fitting (raking), in the absence of benchmarks. For one-dimensional systems raking converges in one iteration: each elementary series $s_{p,t}$ is multiplied by the proportional aggregation discrepancy $d_t^{(p)}$ for the same period. In fact, raking is a particular case of case (2). Case (1) can be seen as a generalization of case (2) where the Total becomes alterable.

Comparing the results of cases (1) and (2) leads to the following conclusion. The higher the number of alterable series in a system of time series, the smaller are the corrections and the changes in movement brought about by reconciliation.

Case (3) is identical to case (1) except that λ=1. The resulting objective function minimizes the corrections proportional to the square of the original values:

$$(1/0.01^2) \times min\,\{\, \Sigma_{p=1}^{P}\ \Sigma_{t=1}^{T_p}\ e_{p,t}^2\,/s_{p,t}^2 \,\}\,.$$

Table 12.2a displays the original and the reconciled series for cases (1) and (3), which illustrates the role of parameter λ. As shown in the table, case (1) corrects all series by the same proportion. Case (3), on the other hand, corrects the series proportionally to the *square* of the original values of each series. Indeed the above objective function proportionally penalizes larger series less than smaller ones. As a result, case (3) produces larger corrections for the important series, namely for the Total (p=3), and smaller corrections for the less important series. The standard deviations of the proportional corrections exhibited in Table 12.2b are the same for the three series under (the options of) case (1); and, higher for the more important series under case (3).

Comparing the results leads to the following conclusion: if proportional corrections are desired, parameter λ should be set equal to 1/2.

Case (4) is identical to case (1) except that the autocorrelations in $\boldsymbol{\Omega}$ are those of an AR(1) with φ equal to 0.999. The resulting objective function specifies that the proportional movement of the original series should be preserved, by basically repeating the proportional error from one period to the next,

$$((1-0.999^2)/0.01) \times min\,\{\ (e_1/s_1^{1/2})^2/(1-0.999^2)\ +$$
$$\Sigma_{t=2}^{T}\,(\,e_t/s_t^{1/2}\ -\ 0.999\,e_{t-1}/s_{t-1}^{1/2}\,)^2\ \}$$

Table 12.3a shows the original and the reconciled series for cases (1) and (4), the latter attempting to preserve proportional movement.

As shown in the table, (the options of) case (4) dramatically change the level of the series, to the point that two of the series take only negative values. Indeed, in the absence of temporal benchmarks, nothing restricts the level of the series in the achievement of movement preservation. The standard deviations of the change in movement exhibited in Table 12.3b show that case (4) does not achieve better movement preservation (lower standard deviations) for any of the series.

The results lead to the following conclusion: strong movement preservation should not be attempted in reconciliation, especially in the absence of temporal benchmarks.

Setting $\lambda=1$ instead of $\lambda=1/2$ also leads to drastic changes in level, with no negative values however.

Applying movement preservation to only one of the series (case not displayed) makes that series rather unalterable, thus forcing the discrepancies into the other series.

Case (5) is identical to case (1), except for binding temporal benchmarks, available for year 2001, i.e. for t=1,...,4. The benchmarks (not shown in the tables) are 444 for the first Province, 843 for the second and 1,287 for the third (the Total). Otherwise, the two cases minimizes the same objective function.

Table 12.4a displays the original and the reconciled series obtained for both cases. Contrary to case (1), case (5) produces corrections which are different for a given t across the three series. The reason is that the proportional yearly discrepancies given by

$$444/(103.0+108.0+110.0+117.0) = 1.014$$
$$843/(207.0+207.0+207.0+229.0) = 0.992$$
$$1{,}287/(313.0+283.0+299.0+362.0) = 1.024$$

vary from Province to Province. For Provinces 1 and 3 the benchmarks pull the series upward; and for Province 2, downward.

Comparing the results leads to the following conclusion: in the presence of temporal constraints (benchmarks), the result of one-way reconciliation are no longer intuitively predictable. Indeed the temporal constraints in case (5) introduce the dimension of time which is technically absent from case (1). (Since $\boldsymbol{\Omega} = \boldsymbol{I}$ for case (1), there is temporal dependence. Reconciling each of the eight quarters separately would yield the same result.)

Case (6) is the same as case (5) except that the autocorrelation matrices are those corresponding to an AR(1) process with φ equal to 0.999. Both cases entail temporal constraints. However, case (1) makes no attempt at movement preservation (φ=0); and case (2), makes a strong attempt (φ=0.999).

Table 12.5a displays the original and the reconciled series obtained for both cases. Both cases produce very similar reconciled series, especially for the year 2001 (i.e. for t=1,...,4). The results are less similar for year 2002 without benchmarks (i.e. for t=5,...,8). According to the standard deviations of the change in movement in Table 12.5b, case (6) does not achieve convincingly superior movement preservation.

Comparing the result of cases (6) and (5) leads to the following conclusion. In the presence of temporal constraints (benchmarks), a movement preservation criterion (an AR(1) with φ close to 1) has very negligible effects on actual movement preservation. Our experience suggests strong movement preservation may cause spurious change of level in the current year without benchmark, which is not supported by data.

Table 12.1a. Simulated system of quarterly series reconciled under the options used for cases (1) and (2)

			Discrepancy		Reconciled		Prop. correction		Change mov.	
Prov	t	Orig.	add.	prop.	(1)	(2)	(1)	(2)	(1)	(2)
1	1	103.0	3.0	1.010	103.5	104.0	1.005	1.010	.	.
1	2	108.0	-32.0	0.898	102.2	97.0	0.946	0.898	-0.058	-0.111
1	3	110.0	-18.0	0.943	106.8	103.8	0.971	0.943	0.024	0.045
1	4	117.0	16.0	1.046	119.6	122.4	1.023	1.046	0.052	0.103
1	5	133.0	-3.0	0.992	132.4	131.9	0.996	0.992	-0.027	-0.055
1	6	118.0	24.0	1.069	121.9	126.1	1.033	1.069	0.037	0.077
1	7	136.0	0.0	1.000	136.0	136.0	1.000	1.000	-0.033	-0.069
1	8	129.0	-26.0	0.934	124.6	120.5	0.966	0.934	-0.034	-0.066
2	1	207.0	3.0	1.010	208.0	209.0	1.005	1.010	.	.
2	2	207.0	-32.0	0.898	195.9	186.0	0.946	0.898	-0.058	-0.111
2	3	207.0	-18.0	0.943	201.0	195.2	0.971	0.943	0.024	0.045
2	4	229.0	16.0	1.046	234.2	239.6	1.023	1.046	0.052	0.103
2	5	224.0	-3.0	0.992	223.1	222.1	0.996	0.992	-0.027	-0.055
2	6	232.0	24.0	1.069	239.7	247.9	1.033	1.069	0.037	0.077
2	7	244.0	0.0	1.000	244.0	244.0	1.000	1.000	-0.033	-0.069
2	8	264.0	-26.0	0.934	255.0	246.5	0.966	0.934	-0.034	-0.066
3	1	313.0	3.0	1.010	311.5	313.0	0.995	1.000	.	.
3	2	283.0	-32.0	0.898	298.1	283.0	1.054	1.000	0.058	0.000
3	3	299.0	-18.0	0.943	307.7	299.0	1.029	1.000	-0.024	0.000
3	4	362.0	16.0	1.046	353.8	362.0	0.977	1.000	-0.052	0.000
3	5	354.0	-3.0	0.992	355.5	354.0	1.004	1.000	0.027	0.000
3	6	374.0	24.0	1.069	361.6	374.0	0.967	1.000	-0.037	0.000
3	7	380.0	0.0	1.000	380.0	380.0	1.000	1.000	0.033	0.000
3	8	367.0	-26.0	0.934	379.6	367.0	1.034	1.000	0.034	0.000

(1): No temporal constraints, $\alpha_1 = \alpha_2 = \alpha_3 = 1$, $\boldsymbol{\Omega}_1 = \boldsymbol{\Omega}_2 = \boldsymbol{\Omega}_3 = \boldsymbol{I}_T$, $\lambda = 1/2$, $c_{p,t} = 0.01$

(2): Same as (1) except $\alpha_3 = 0$

Table 12.1b. Some statistics on the corrections and movement for the series exhibited in Table 12.1a

Prov	Alter. coef. (1)	Std. dev. cor. (1)	Std. change mov. (1)	Alter. coef. (2)	Std. dev. cor. (2)	Std. change mov. (2)
1	1.0	0.029	0.043	1.0	0.058	0.084
2	1.0	0.029	0.043	1.0	0.058	0.084
3	1.0	0.029	0.043	0.0	0.000	0.000
All	.	***0.029***	*0.041*	.	***0.046***	*0.065*

Table 12.2a. Simulated system of quarterly series reconciled under the options used for cases (1) and (3)

			Discrepancy		Reconciled		Prop. correction		Change mov.	
Prov	t	Orig.	add.	prop.	(1)	(3)	(1)	(3)	(1)	(3)
1	1	103.0	3.0	1.010	103.5	103.2	1.005	1.002	.	.
1	2	108.0	-32.0	0.898	102.2	105.2	0.946	0.974	-0.058	-0.028
1	3	110.0	-18.0	0.943	106.8	108.5	0.971	0.986	0.024	0.012
1	4	117.0	16.0	1.046	119.6	118.1	1.023	1.009	0.052	0.023
1	5	133.0	-3.0	0.992	132.4	132.7	0.996	0.998	-0.027	-0.012
1	6	118.0	24.0	1.069	121.9	119.6	1.033	1.014	0.037	0.016
1	7	136.0	0.0	1.000	136.0	136.0	1.000	1.000	-0.033	-0.014
1	8	129.0	-26.0	0.934	124.6	127.0	0.966	0.985	-0.034	-0.015
2	1	207.0	3.0	1.010	208.0	207.8	1.005	1.004	.	.
2	2	207.0	-32.0	0.898	195.9	196.8	0.946	0.951	-0.058	-0.053
2	3	207.0	-18.0	0.943	201.0	201.7	0.971	0.974	0.024	0.023
2	4	229.0	16.0	1.046	234.2	233.3	1.023	1.019	0.052	0.044
2	5	224.0	-3.0	0.992	223.1	223.2	0.996	0.997	-0.027	-0.022
2	6	232.0	24.0	1.069	239.7	238.2	1.033	1.027	0.037	0.030
2	7	244.0	0.0	1.000	244.0	244.0	1.000	1.000	-0.033	-0.027
2	8	264.0	-26.0	0.934	255.0	255.8	0.966	0.969	-0.034	-0.031
3	1	313.0	3.0	1.010	311.5	311.1	0.995	0.994	.	.
3	2	283.0	-32.0	0.898	298.1	302.0	1.054	1.067	0.058	0.073
3	3	299.0	-18.0	0.943	307.7	310.1	1.029	1.037	-0.024	-0.030
3	4	362.0	16.0	1.046	353.8	351.4	0.977	0.971	-0.052	-0.067
3	5	354.0	-3.0	0.992	355.5	355.9	1.004	1.005	0.027	0.035
3	6	374.0	24.0	1.069	361.6	357.8	0.967	0.957	-0.037	-0.049
3	7	380.0	0.0	1.000	380.0	380.0	1.000	1.000	0.033	0.043
3	8	367.0	-26.0	0.934	379.6	382.8	1.034	1.043	0.034	0.043

(1): No temporal constraints, $\alpha_1=\alpha_2=\alpha_3=1$, $\boldsymbol{\Omega}_1=\boldsymbol{\Omega}_2=\boldsymbol{\Omega}_3=\boldsymbol{I}_T$, $\lambda=1/2$, $c_{p,t}=1$

(3): Same as (1) except $\lambda=1$

Table 12.2b. Some statistics on the corrections and movement for the series exhibited in Table 12.2a

Prov	Alter. coef. (1)	Std. dev. cor. (1)	Std. change mov. (1)	Alter. coef. (3)	Std. dev. cor. (3)	Std. change mov. (3)
1	1.0	0.029	0.043	1.0	0.013	0.019
2	1.0	0.029	0.043	1.0	0.026	0.037
3	1.0	0.029	0.043	1.0	0.038	0.054
All	.	0.029	0.041	.	0.027	0.038

Table 12.3a. Simulated system of quarterly series reconciled under the options used for cases (1) and (4)

			Discrepancy		Reconciled		Prop. correction		Change mov.	
Prov	t	Orig.	add.	prop.	(1)	(4)	(1)	(4)	(1)	(4)
1	1	103.0	3.0	1.010	103.5	-66.4	1.005	-0.645	.	.
1	2	108.0	-32.0	0.898	102.2	-68.0	0.946	-0.630	-0.058	0.015
1	3	110.0	-18.0	0.943	106.8	-66.5	0.971	-0.605	0.024	0.025
1	4	117.0	16.0	1.046	119.6	-62.6	1.023	-0.535	0.052	0.070
1	5	133.0	-3.0	0.992	132.4	-59.4	0.996	-0.446	-0.027	0.089
1	6	118.0	24.0	1.069	121.9	-61.6	1.033	-0.522	0.037	-0.076
1	7	136.0	0.0	1.000	136.0	-58.2	1.000	-0.428	-0.033	0.094
1	8	129.0	-26.0	0.934	124.6	-62.7	0.966	-0.486	-0.034	-0.058
2	1	207.0	3.0	1.010	208.0	23.0	1.005	0.111	.	.
2	2	207.0	-32.0	0.898	195.9	18.2	0.946	0.088	-0.058	-0.023
2	3	207.0	-18.0	0.943	201.0	20.3	0.971	0.098	0.024	0.010
2	4	229.0	16.0	1.046	234.2	37.4	1.023	0.163	0.052	0.065
2	5	224.0	-3.0	0.992	223.1	33.0	0.996	0.147	-0.027	-0.016
2	6	232.0	24.0	1.069	239.7	40.3	1.033	0.174	0.037	0.027
2	7	244.0	0.0	1.000	244.0	44.5	1.000	0.182	-0.033	0.009
2	8	264.0	-26.0	0.934	255.0	51.1	0.966	0.193	-0.034	0.011
3	1	313.0	3.0	1.010	311.5	-43.5	0.995	-0.139	.	.
3	2	283.0	-32.0	0.898	298.1	-49.8	1.054	-0.176	0.058	-0.037
3	3	299.0	-18.0	0.943	307.7	-46.2	1.029	-0.155	-0.024	0.021
3	4	362.0	16.0	1.046	353.8	-25.2	0.977	-0.070	-0.052	0.085
3	5	354.0	-3.0	0.992	355.5	-26.4	1.004	-0.075	0.027	-0.005
3	6	374.0	24.0	1.069	361.6	-21.3	0.967	-0.057	-0.037	0.018
3	7	380.0	0.0	1.000	380.0	-13.6	1.000	-0.036	0.033	0.021
3	8	367.0	-26.0	0.934	379.6	-11.6	1.034	-0.032	0.034	0.004

(1): No temporal constraints, $\alpha_1 = \alpha_2 = \alpha_3 = 1$, $\boldsymbol{\Omega}_1 = \boldsymbol{\Omega}_2 = \boldsymbol{\Omega}_3 = \boldsymbol{I}_T$, $\lambda = 1/2$, $c_{p,t} = 1$

(4): Same as (1) except $\boldsymbol{\Omega}_1 = \boldsymbol{\Omega}_2 = \boldsymbol{\Omega}_3$ correspond to an AR(1) model with $\varphi = 0.999$

Table 12.3b. Some statistics on the corrections and movement for the series exhibited in Table 12.3a

Prov	Alter. coef. (1)	Std. dev. cor. (1)	Std. change mov. (1)	Alter. coef. (4)	Std. dev. cor. (4)	Std. change mov. (4)
1	1.0	0.029	0.043	1.0	0.083	0.068
2	1.0	0.029	0.043	1.0	0.041	0.029
3	1.0	0.029	0.043	1.0	0.056	0.037
All	.	0.029	0.041	.	0.295	0.046

Table 12.4a. Simulated system of quarterly series reconciled under the options used for cases (1) and (5)

Prov	t	Orig.	Discrepancy add.	Discrepancy prop.	Reconciled (1)	Reconciled (5)	Prop. correction (1)	Prop. correction (5)	Change mov. (1)	Change mov. (5)
1	1	103.0	3.0	1.010	103.5	106.3	1.005	1.018	.	.
1	2	108.0	-32.0	0.898	102.2	105.1	0.946	0.960	-0.058	-0.058
1	3	110.0	-18.0	0.943	106.8	109.7	0.971	0.984	0.024	0.024
1	4	117.0	16.0	1.046	119.6	122.9	1.023	1.036	0.052	0.052
1	5	133.0	-3.0	0.992	132.4	135.8	0.996	1.008	-0.027	-0.028
1	6	118.0	24.0	1.069	121.9	125.0	1.033	1.045	0.037	0.038
1	7	136.0	0.0	1.000	136.0	139.5	1.000	1.012	-0.033	-0.033
1	8	129.0	-26.0	0.934	124.6	127.9	0.966	0.978	-0.034	-0.034
2	1	207.0	3.0	1.010	208.0	209.0	1.005	1.018	.	.
2	2	207.0	-32.0	0.898	195.9	196.8	0.946	0.958	-0.058	-0.060
2	3	207.0	-18.0	0.943	201.0	201.9	0.971	0.983	0.024	0.025
2	4	229.0	16.0	1.046	234.2	235.3	1.023	1.036	0.052	0.053
2	5	224.0	-3.0	0.992	223.1	223.9	0.996	1.008	-0.027	-0.028
2	6	232.0	24.0	1.069	239.7	240.7	1.033	1.046	0.037	0.039
2	7	244.0	0.0	1.000	244.0	244.9	1.000	1.012	-0.033	-0.034
2	8	264.0	-26.0	0.934	255.0	256.0	0.966	0.978	-0.034	-0.035
3	1	313.0	3.0	1.010	311.5	315.3	0.995	0.984	.	.
3	2	283.0	-32.0	0.898	298.1	301.9	1.054	1.042	0.058	0.058
3	3	299.0	-18.0	0.943	307.7	311.6	1.029	1.018	-0.024	-0.024
3	4	362.0	16.0	1.046	353.8	358.2	0.977	0.966	-0.052	-0.051
3	5	354.0	-3.0	0.992	355.5	359.7	1.004	0.992	0.027	0.026
3	6	374.0	24.0	1.069	361.6	365.8	0.967	0.955	-0.037	-0.037
3	7	380.0	0.0	1.000	380.0	384.5	1.000	0.988	0.033	0.033
3	8	367.0	-26.0	0.934	379.6	383.9	1.034	1.022	0.034	0.033

(1): No temporal constraints, $\alpha_1=\alpha_2=\alpha_3=1$, $\boldsymbol{\Omega}_1=\boldsymbol{\Omega}_2=\boldsymbol{\Omega}_3=\boldsymbol{I}_T$, $\lambda=1/2$, $c_{p,t}=1$

(5): Same as (1) except for yearly benchmarking constraints in 2001

Table 12.4b. Some statistics on the corrections and movement for the series exhibited in Table 12.4a

Prov	Alter. coef. (1)	Std. dev. cor. (1)	Std. change mov. (1)	Alter. coef. (5)	Std. dev. cor. (5)	Std. change mov. (5)
1	1.0	0.029	0.043	1.0	0.029	0.043
2	1.0	0.029	0.043	1.0	0.030	0.044
3	1.0	0.029	0.043	1.0	0.029	0.042
All	.	0.029	0.041	.	0.029	0.041

Table 12.5a. Simulated system of quarterly series reconciled under the option sused for cases (6) and (5)

Prov	t	Orig.	Discrepancy add.	prop.	Reconciled (6)	(5)	Prop. correction (6)	(5)	Change mov. (6)	(5)
1	1	103.0	3.0	1.010	106.6	106.3	1.021	1.018	.	.
1	2	108.0	-32.0	0.898	105.3	105.1	0.962	0.960	-0.059	-0.058
1	3	110.0	-18.0	0.943	109.7	109.7	0.984	0.984	0.022	0.024
1	4	117.0	16.0	1.046	122.4	122.9	1.032	1.036	0.049	0.052
1	5	133.0	-3.0	0.992	134.8	135.8	1.000	1.008	-0.032	-0.028
1	6	118.0	24.0	1.069	124.1	125.0	1.037	1.045	0.037	0.038
1	7	136.0	0.0	1.000	138.3	139.5	1.003	1.012	-0.034	-0.033
1	8	129.0	-26.0	0.934	126.8	127.9	0.970	0.978	-0.034	-0.034
2	1	207.0	3.0	1.010	209.1	209.0	1.019	1.018	.	.
2	2	207.0	-32.0	0.898	196.9	196.8	0.959	0.958	-0.060	-0.060
2	3	207.0	-18.0	0.943	201.8	201.9	0.983	0.983	0.024	0.025
2	4	229.0	16.0	1.046	235.2	235.3	1.036	1.036	0.053	0.053
2	5	224.0	-3.0	0.992	223.8	223.9	1.007	1.008	-0.028	-0.028
2	6	232.0	24.0	1.069	240.3	240.7	1.044	1.046	0.037	0.039
2	7	244.0	0.0	1.000	244.1	244.9	1.009	1.012	-0.035	-0.034
2	8	264.0	-26.0	0.934	255.0	256.0	0.974	0.978	-0.035	-0.035
3	1	313.0	3.0	1.010	315.7	315.3	0.985	0.984	.	.
3	2	283.0	-32.0	0.898	302.2	301.9	1.043	1.042	0.058	0.058
3	3	299.0	-18.0	0.943	311.5	311.6	1.017	1.018	-0.025	-0.024
3	4	362.0	16.0	1.046	357.6	358.2	0.965	0.966	-0.053	-0.051
3	5	354.0	-3.0	0.992	358.6	359.7	0.989	0.992	0.025	0.026
3	6	374.0	24.0	1.069	364.3	365.8	0.951	0.955	-0.038	-0.037
3	7	380.0	0.0	1.000	382.5	384.5	0.983	0.988	0.032	0.033
3	8	367.0	-26.0	0.934	381.8	383.9	1.016	1.022	0.033	0.033

(5): Temporal constraints, $\alpha_1=\alpha_2=\alpha_3=1$, $\boldsymbol{\Omega}_1=\boldsymbol{\Omega}_2=\boldsymbol{\Omega}_3=\boldsymbol{I}_T$, $\lambda=1/2$, $c_{p,t}=1$

(6): Same as (5) except $\boldsymbol{\Omega}_1=\boldsymbol{\Omega}_2=\boldsymbol{\Omega}_3$ correspond to an AR(1) model with $\varphi=0.999$

Table 12.5b. Some statistics on the corrections and movement for the series exhibited in Table 12.5a

Prov	Alter. coef. (6)	Std. dev. cor. (6)	Std. change mov. (6)	Alter. coef. (5)	Std. dev. cor. (5)	Std. change mov. (5)
1	1.0	0.028	0.042	1.0	0.029	0.043
2	1.0	0.030	0.043	1.0	0.030	0.044
3	1.0	0.030	0.042	1.0	0.029	0.042
All	.	0.028	0.041	.	0.029	0.041

12.6 A Real Data Example: One-Way Reconciliation Model of The Seasonally Adjusted Canadian Retail Trade Series

The Canadian system of Retail Trade series (Statistics Canada 2001) is classified by Trade Group (kinds of stores) and by Province (Statistics Canada 2001). The system contains 18 Trade Group Totals, 12 Provincial Totals and the Grand Total and ranges from Jan. 1991 to Nov. 2002. In this case, seasonal adjustment is the source of the cross-sectional aggregation discrepancies;[2] moreover, the method produces no covariance matrix for the seasonally adjusted series. In our example, the seasonally adjusted Grand Total is treated as exogenous or pre-determined, namely defined as the cross-sectional sum of the previously seasonally adjusted Trade Group Totals. The pre-determined Grand Total is imposed onto the Provincial Totals, by means of the one-way reconciliation model, where the alterability coefficient of the Grand Total is equal to 0 and the alterability coefficients of the Provincial Totals are equal to 1.

Table 12.6. Provincial Codes for the Canadian Retail Trade Series

p	Code	Province	p	Code	Province
1	10	Newfoundland	8	47	Saskatchewan
2	11	Prince Edward Island	9	48	Alberta
3	12	Nova Scotia	10	52	British Columbia
4	13	New Brunswick	11	60	Yukon
5	22	Québec	12	64	North-West Territories and Nunavut
6	32	Ontario			
7	42	Manitoba	13	00	"All" Province (Canada)

All the series must satisfy yearly benchmarks which are equal to the yearly sums of the *raw* (seasonally unadjusted) series. Each series is benchmarked as it is read in the reconciliation software and its covariance matrix is calculated. The method applied is proportional benchmarking described in section 3.8 with φ=0.90, and with λ equal to 1/2 instead of 1 to achieve a proportional adjustment across series.

The one-way reconciliation model is applied to the benchmarked series $\tilde{\theta}_p$ using their covariance matrix $var[\tilde{\theta}_p]$. These covariance matrices are multiplied by the alterability coefficient assigned to the series: $\boldsymbol{V}_{e_p}$ =

[2] The seasonal adjustment method used is non linear, and different series required different options, e.g. different moving averages.

$\alpha_p\, var[\tilde{\theta}_p]$. The problem of redundancies in the constraints, due to the presence of benchmarks, is addressed by making the January cross-sectional constraints non-binding for years with benchmark (1991 to 2001).[3] This strategy implies that the temporal binding constraints will be maintained; but not necessarily the cross-sectional constraints if the benchmarks do not satisfy the cross-sectional aggregation constraints themselves. This is indeed the case in the example.

In order to demonstrate the feasibility of the reconciliation methodology, we applied reconciliation to almost twelve years of monthly data. In many applications, one could process each year separately. However in many cases, it may be more convenient to perform reconciliation on many years at one time, in order to minimize the number of files to handle.

Table 12.7 displays the additive and the proportional cross-sectional discrepancies of the yearly benchmarks. These benchmarks do not quite add up, in additive terms at least. Indeed the benchmarks are the yearly sums of the monthly raw data which do not add up, albeit by small amounts. This common data problem is a contradiction in the data which will be imposed onto the system of seasonally adjusted series. Indeed if the series satisfy these benchmarks, they can never cross-sectionally add up. The cross-sectional discrepancies in the benchmarks are a measure of the contradiction.

Table 12.7. Cross-sectional aggregation discrepancies in the yearly benchmarks

Starting date	Ending date	Prop. disc.	Add. disc
01/01/91	*31/12/91*	*1.0000*	*-6*
01/01/92	*31/12/92*	*1.0000*	*4*
01/01/93	31/12/93	1.0000	0
01/01/94	31/12/94	1.0000	-3
01/01/95	31/12/95	1.0000	0
01/01/96	31/12/96	1.0000	2
01/01/97	31/12/97	1.0000	-2
01/01/98	31/12/98	1.0000	-5
01/01/99	31/12/99	1.0000	2
01/01/00	31/12/00	1.0000	-1
01/01/01	31/12/01	1.0000	2

Table 12.8 shows statistics on the temporal discrepancies between the yearly benchmarks and the yearly sums of the seasonally adjusted series to be reconciled. In principle these series should already satisfy the yearly benchmarks, because they were benchmarked by the seasonal adjustment program. Some of the series do not satisfy their benchmarks, because the version of the seasonal adjustment program used at the time was in single precision. This implies that digits beyond the seventh are not significant. The Grand Total (Prov=00) has more than seven digits, exceeding 10,000,000

[3] Any month can be chosen, because redundancy is an attribute of the whole set of constraints.

Table 12.8. Some statistics on the additive temporal discrepancies before reconciliation

Prov.	Min. Discrep.	Max. Discrep.
00	*-27*	*9*
10	-2	0
11	-1	2
12	-2	1
13	-1	1
22	-4	-1
32	-5	1
42	-2	1
47	-2	2
48	-4	1
52	-3	0
60	-2	1
64	-2	2

($ 000) since 1991. As a result, its temporal discrepancies are more noticeable. This illustrates another data problem. Note that even if there were no temporal discrepancies, it would still be necessary to generate their covariance matrices, in order to preserve the temporal constraints in the reconciliation.

Fig. 12.1 shows that the actual proportional cross-sectional discrepancies between the Grand total and the Provinces fluctuate in a rather erratic manner from month to month. Each discrepancy must be distributed over the Provincial Totals. The minimum and maximum additive discrepancies are respectively −188,585 and 333,352; and the minimum and maximum proportional, 0.9877 and 1.0163, i.e. −1.23% and 1.63%.

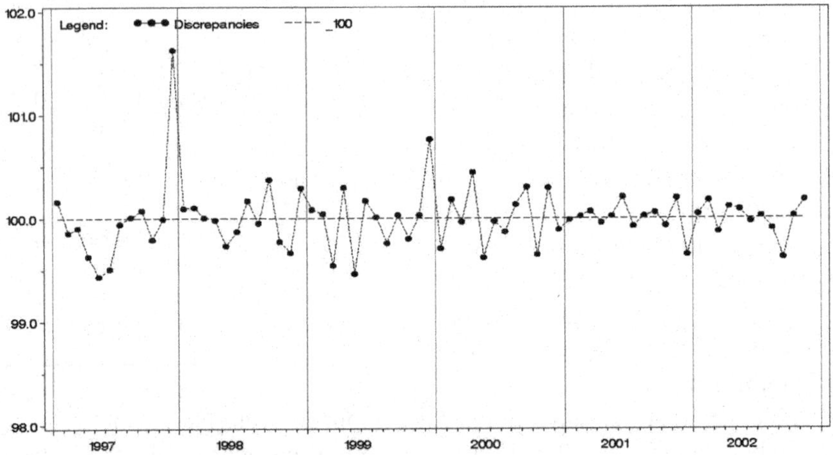

Fig. 12.1. Proportional cross-sectional aggregation discrepancies between the Grand Total and the sum of the Provinces

After reconciliation, the minimum and maximum *temporal* discrepancies reduce to −3 and +3, instead of −27 and 9 shown in Table 12.8. These residual additive temporal discrepancies are attributed to rounding errors. The reconciled series were rounded to the nearest unit by the reconciliation program before the calculation of the temporal and cross-sectional discrepancies. The purpose of such rounding is to anticipate the discrepancies attributable to rounding in the publication.

The minimum and maximum *cross-sectional* aggregation discrepancies after reconciliation are respectively -6 and 4, which coincide with the minimum and maximum of the additive cross-sectional discrepancies *of the yearly benchmarks* displayed italicized in Table 12.7. These cross-sectional discrepancies in the yearly benchmarks are forced onto the system of series, making it impossible to simultaneously satisfy all the temporal and cross-sectional constraints. The unsatisfied cross-sectional constraints are those of Januaries, which were specified as non-binding to avoid redundant constraints. Perhaps a seasonal peak month could have been chosen.

Table 12.9. Some statistics on the proportional corrections made by reconciliation and on the proportional movement

Prov.	Alter. coef.	Min. cor.	Max. cor.	Std.Dev. of cor.	Min. change mov.	Max. change mov.	Std. dev. of ch. mov.
00	0	1.0000	1.0000	0.0000	0.0000	0.0000	0.0000
10	1	0.9876	1.0163	0.0033	-0.0154	0.0189	0.0046
11	1	0.9878	1.0162	0.0033	-0.0153	0.0190	0.0046
12	1	0.9884	1.0162	0.0033	-0.0154	0.0191	0.0046
13	1	0.9879	1.0160	0.0033	-0.0154	0.0190	0.0046
22	1	0.9872	1.0169	0.0033	-0.0150	0.0188	0.0046
32	1	0.9878	1.0161	0.0033	-0.0154	0.0190	0.0046
42	1	0.9876	1.0160	0.0033	-0.0154	0.0190	0.0046
47	1	0.9881	1.0159	0.0032	-0.0155	0.0190	0.0046
48	1	0.9880	1.0160	0.0033	-0.0154	0.0190	0.0046
52	1	0.9876	1.0161	0.0033	-0.0154	0.0190	0.0046
60	1	0.9886	1.0164	0.0033	-0.0154	0.0191	0.0046
64	1	0.9879	1.0164	0.0033	-0.0154	0.0189	0.0046

We also reconciled the system of series using the generalized Moore-Penrose matrix inversion to handle the redundant constraints, instead of making the January cross-sectional constraints non-binding. The contradictory benchmark data translated into all the cross-sectional constraints of the years with benchmarks not being satisfied, but by smaller amounts which can be more easily attributed to rounding. More specifically, the January discrepancies of the previous method were equally distributed over all the months of each year with benchmark. The resulting residual minimum and maximum cross-sectional discrepancies were -2 and 3, instead of -6 and 4.

The Moore-Penrose approach distributes the contradictions in the data, i.e. the small cross-sectional aggregation discrepancies in the benchmarks, over more observations; whereas making one month per year non-binding concentrates the contradictions into the month chosen.

Continuing with the previous method of handling redundancies, Table 12.9 displays statistics on the proportional movement $\hat{\theta}_{p,t}/s_{p,t}$ - $\hat{\theta}_{p,t-1}/s_{p,t-1}$ and the proportional corrections $\hat{\theta}_{p,t}/s_{p,t}$ made to the benchmarked series to achieve reconciliation. The Grand Total with alterability equal to 0 is not corrected (except for benchmarking), and the Provincial totals are corrected in equal proportions. The standard deviations of the corrections made to the Provincial Totals hover between 0.0032 and 0.0033. In other words, the proportional corrections are very uniformally distributed over the Provincial Totals.

This example will be revisited in the next chapter which deals with the reconciliation of the marginal totals of a two-way system of series.

13 Reconciling the Marginal Totals of Two-Way Classified Systems of Series

13.1 Introduction

Two-way classified systems of series result from survey (say) classified by two attributes. As is often the case, only the marginal totals have sufficient reliability to deserve publication. This is the situation of the Canadian Retail and Wholesale Trade Series, which are classified by Province and Trade Group (kind of store). We will refer to such systems as *marginal two-way systems*.

There may be operational reasons to publish the marginal totals first, and the elementary series a few weeks later. This hierarchical strategy enables the production of the more important series, within shorter deadlines; and of the less important series, with longer deadlines.

We henforth refer to the model for marginal two-way systems of series, as the *marginal two-way reconciliation model*.

Table 13.1. System of Time Series Classified by Industry and Province

Prov. / Ind.	Prov. 1	Prov. 2	Ind. Totals
Ind. 1	$\{s_{11,t}\}$	$\{s_{12,t}\}$	$\{s_{13,t}\}$
Ind. 2	$\{s_{21,t}\}$	$\{s_{22,t}\}$	$\{s_{23,t}\}$
Ind. 3	$\{s_{31,t}\}$	$\{s_{32,t}\}$	$\{s_{33,t}\}$
Prov. Totals	$\{s_{41,t}\}$	$\{s_{42,t}\}$	$\{s_{43,t}\}$

Table 13.1 represents such a system of series, classified by two Provinces and three Industries. The system also contains the *Industrial Totals* in the last column, the *Provincial Totals* in the last row and the *Grand Total* denoted by $s_{43,t}$. The marginal two-way reconciliation model deals only with the six *marginal totals*: the two Provincial Totals $s_{41,t}$ and $s_{42,t}$ the three Industrial Totals $s_{13,t}$ $s_{23,t}$ and $s_{33,t}$ and the Grand Total $s_{43,t}$. The *elementary series*

(i.e. for $n<4$ and $p<3$) are excluded from the model. Note that the subscript np should be treated as a single index and not as the product of n by p.

This chapter provides an analytical solution for the marginal two-way reconciliation model. In order to minimize the typographical size of the matrices involved, we keep the number of series to six. It will then be easy to generalize the results.

The marginal two-way system contains one set of *Industrial cross-sectional aggregation constraints*, over the Provincial Totals; and one set of *Provincial cross-sectional aggregation constraints*, over the Industrial Totals. The two sets of constraints are respectively

$$s_{NP,t} = \Sigma_{p=1}^{P-1} s_{Np,t}, \quad t=1,...,T_{NP} , \tag{13.1a}$$

$$s_{NP,t} = \Sigma_{n=1}^{N-1} s_{nP,t}, \quad t=1,...,T_{NP} , \tag{13.1b}$$

where N=4 and P=3 in the example shown in Table 13.1. The number T_{NP} is the number of observations for the Grand Total. The summation over $s_{Np,t}$ is skipped if that series does not exist at time t, i.e. $t<t_{1,Np}$ or $t>t_{L,Np}$, $t_{1,Np}$ and $t_{L,Np}$ being the first and the last observations of $s_{Np,t}$; and, similarly for the summation over $s_{nP,t}$.

The Industrial and the Provincial *proportional cross-sectional aggregation discrepancies*, are

$$d_{N,t}^{(p)} = s_{NP,t} \,/\, \Sigma_{p=1}^{P-1} s_{Np,t}, \quad t=t_{1,NP},...,t_{L,NP}, \tag{13.2a}$$

$$d_{P,t}^{(p)} = s_{NP,t} \,/\, \Sigma_{n=1}^{N-1} s_{nP,t}, \quad t=t_{1,NP},...,t_{L,NP}, \tag{13.2b}$$

where the exponent *(p)* stands here for proportion and not province.

It is possible to impose the Grand Total onto the Industrial and Provincial Totals by means of the Iterative Proportional Fitting approach ("raking"). This entails the following steps:

(1) The Industrial Totals, $s_{nP,t}$, $n=1,...,N-1$, are multiplied by the Provincial proportional cross-sectional discrepancies $d_{p,t}^{(p)}$ of (13.2b). This step eliminates the provincial discrepancies.

(2) The industrial proportional cross-sectional discrepancies (13.2a) are recalculated, because step (1) alters these discrepancies.

(3) The Provincial Totals, $s_{Np,t}, p=1,...,P-1$ are multiplied by the industrial proportional discrepancies $d_{n,t}^{(p)}$ of step (2). This eliminates the industrial discrepancies.
(4) The provincial proportional discrepancies (13.2b) are recalculated.
(5) The steps (1) to (4) are repeated until the provincial proportional cross-sectional discrepancies are sufficiently close to one.

The Iterative Proportional Fitting technique implies that the Grand Total is exogenous, i.e. pre-determined. The marginal two-way model presented in the next section offers more flexibility in that regard.

13.2 The Marginal Two-Way Reconciliation Model

Following the general regression-based reconciliation model of Eqs (11.4), we can represent the marginal two-way model with three Industrial Totals, Two Provincial Totals and the Grand Total by

$$\begin{bmatrix} \boldsymbol{s}_{41} \\ \boldsymbol{s}_{42} \\ \boldsymbol{s}_{13} \\ \boldsymbol{s}_{23} \\ \boldsymbol{s}_{33} \\ \boldsymbol{s}_{43} \end{bmatrix} = \begin{bmatrix} \boldsymbol{I}_{T_{41}} & \boldsymbol{0} & \boldsymbol{0} & \boldsymbol{0} & \boldsymbol{0} & \boldsymbol{0} \\ \boldsymbol{0} & \boldsymbol{I}_{T_{42}} & \boldsymbol{0} & \boldsymbol{0} & \boldsymbol{0} & \boldsymbol{0} \\ \boldsymbol{0} & \boldsymbol{0} & \boldsymbol{I}_{T_{13}} & \boldsymbol{0} & \boldsymbol{0} & \boldsymbol{0} \\ \boldsymbol{0} & \boldsymbol{0} & \boldsymbol{0} & \boldsymbol{I}_{T_{23}} & \boldsymbol{0} & \boldsymbol{0} \\ \boldsymbol{0} & \boldsymbol{0} & \boldsymbol{0} & \boldsymbol{0} & \boldsymbol{I}_{T_{33}} & \boldsymbol{0} \\ \boldsymbol{0} & \boldsymbol{0} & \boldsymbol{0} & \boldsymbol{0} & \boldsymbol{0} & \boldsymbol{I}_{T_{43}} \end{bmatrix} \begin{bmatrix} \boldsymbol{\theta}_{41} \\ \boldsymbol{\theta}_{42} \\ \boldsymbol{\theta}_{13} \\ \boldsymbol{\theta}_{23} \\ \boldsymbol{\theta}_{33} \\ \boldsymbol{\theta}_{43} \end{bmatrix} + \begin{bmatrix} \boldsymbol{e}_{41} \\ \boldsymbol{e}_{42} \\ \boldsymbol{e}_{13} \\ \boldsymbol{e}_{23} \\ \boldsymbol{e}_{33} \\ \boldsymbol{e}_{43} \end{bmatrix}, \quad (13.3a)$$

where $E(\boldsymbol{e}_{np}) = \boldsymbol{0}$, $E(\boldsymbol{e}_{np}\, \boldsymbol{e}_{np}^{\prime}) = \boldsymbol{V}_{e_{np}}$ and $E(\boldsymbol{e}_{ij}\, \boldsymbol{e}_{np}^{\prime}) = \boldsymbol{0}$ for $i \neq n$ or $j \neq p$, and where $T_{np} \le T_{43} = T_{NP}$;

$$\begin{bmatrix} \boldsymbol{g}_A \\ \boldsymbol{g}_B \end{bmatrix} = \begin{bmatrix} \mathbf{G}_{41} & \mathbf{G}_{42} & \boldsymbol{0} & \boldsymbol{0} & \boldsymbol{0} & -\mathbf{G}_{43} \\ \boldsymbol{0} & \boldsymbol{0} & \mathbf{G}_{13} & \mathbf{G}_{23} & \mathbf{G}_{33} & -\mathbf{G}_{43} \end{bmatrix} \begin{bmatrix} \boldsymbol{\theta}_{41} \\ \boldsymbol{\theta}_{42} \\ \boldsymbol{\theta}_{13} \\ \boldsymbol{\theta}_{23} \\ \boldsymbol{\theta}_{33} \\ \boldsymbol{\theta}_{43} \end{bmatrix} + \begin{bmatrix} \boldsymbol{\varepsilon}_{4\bullet} \\ \boldsymbol{\varepsilon}_{\bullet 3} \end{bmatrix}, \quad (13.3b)$$

where $E(\varepsilon_{4\bullet}) = \boldsymbol{0}$, $E(\varepsilon_{\bullet 3}) = \boldsymbol{0}$, $E(\varepsilon_{4\bullet}\varepsilon_{4\bullet}') = \boldsymbol{V}_{\varepsilon_{4\bullet}}$, $E(\varepsilon_{\bullet 3}\varepsilon_{\bullet 3}') = \boldsymbol{V}_{\varepsilon_{\bullet 3}}$ and $E(\varepsilon_{4\bullet}\varepsilon_{\bullet 3}') = \boldsymbol{0}$. Matrices $\boldsymbol{V}_{\varepsilon_{4\bullet}}$ and $\boldsymbol{V}_{\varepsilon_{\bullet 3}}$ are diagonal. Typically these matrices are equal to zero. However, to make specific constraints non-binding, it is possible to set the corresponding diagonal elements to a value greater than zero. This is one way to deal with redundant constraints. Note again that the subscript *np* should be treated as a single index and not as the product of *n* by *p*.

The matrices $\boldsymbol{G}_{np}$ are usually selector matrices, i.e. identities matrices with T_{43} rows with some of the columns removed; the columns removed correspond to periods t where the series $\boldsymbol{s}_{Np}$ or $\boldsymbol{s}_{nP}$ does not exist. These matrices thus allow some of the series to begin later and/or end earlier than the Grand Total $\boldsymbol{s}_{NP}$. For example, let the system of series with Total $\boldsymbol{s}_{43}$ starting in 2001 and ending in 2005 (5 yearly observations); and, the second Provincial Total series $\boldsymbol{s}_{42}$ starting in 2002 and ending in 2004 (3 observations). The first and last of observations of $\boldsymbol{s}_{N2}$ are then $t_{1,2}=2$ and $t_{L,2}=4$, and matrix $\boldsymbol{G}_{42}$ is then

$$\boldsymbol{G}_{42} = \begin{bmatrix} 0 & 0 & 0 \\ 1 & 0 & 0 \\ 0 & 1 & 0 \\ 0 & 0 & 1 \\ 0 & 0 & 0 \end{bmatrix}.$$

The constraints (13.3b) may be rewritten as:

$$\begin{bmatrix} \boldsymbol{g}_A \\ \boldsymbol{g}_B \end{bmatrix} = \begin{bmatrix} \boldsymbol{G}_A \\ \boldsymbol{G}_B \end{bmatrix} \boldsymbol{\theta} + \begin{bmatrix} \varepsilon_A \\ \varepsilon_B \end{bmatrix}, \tag{13.3c}$$

where $E(\varepsilon_A) = \boldsymbol{0}$, $E(\varepsilon_A \varepsilon_A') = \boldsymbol{V}_{\varepsilon_A}$, $E(\varepsilon_B) = \boldsymbol{0}$, $E(\varepsilon_B \varepsilon_B') = \boldsymbol{V}_{\varepsilon_B}$, $E(\varepsilon_A \varepsilon_B') = \boldsymbol{0}$.

Matrix $\boldsymbol{G}_A$ is the Industrial cross-sectional sum operator over the three Provinces, and $\boldsymbol{G}_B$ is the Provincial cross-sectional sum operator over the four Industries. For the moment, all the series in model (13.3) - including the Grand Total - are endogenous, that is jointly estimated. As a result, vectors $\boldsymbol{g}_A$

and g_B contain zeroes. Specifying all series as endogenous leads to a more general solution. It will then be easy to treat some of the series as exogenous, that is pre-determined.

In the coming sub-sections, we provide analytical solutions to the marginal two-way reconciliation problem, considering first the main matrix partitions (A and B) and then the partitions pertaining to each series. The general solution (11.8) is repeated here for convenience:

$$\hat{\theta} = s + V_e\, G'(G V_e G' + V_\varepsilon)^{-1} [g - G s]$$
$$= s + V_e\, G' [V_d^{-1}\, d] = s + V_e\, G'\, \hat{\gamma}, \qquad (13.4a)$$
$$var[\hat{\theta}] = V_e - V_e G' V_d^{-1} G V_e, \qquad (13.4b)$$

where many of the matrices are large and sparse (i.e. with many partitions equal to 0).

The strategy used to obtain an analytic solution consists of developing the various matrix products in terms of their main partitions A and B, e.g. G_A, G_B; and then, in terms of the sub-partitions pertaining to each individual series, e.g. G_{23}, $V_{e_{23}}$.

13.2.1 Deriving an Analytical Solution in Terms of the Main Partitions

Given that $G' = \begin{bmatrix} G_A' & G_B' \end{bmatrix}$, we obtain:

$$V_e G' = V_e \begin{bmatrix} G_A' & G_B' \end{bmatrix} = \begin{bmatrix} V_e G_A' & V_e G_B' \end{bmatrix}, \qquad (13.5a)$$

$$V_d = (G V_e G' + V_\varepsilon) = \begin{bmatrix} G_A \\ G_B \end{bmatrix} \begin{bmatrix} V_e G_A' & V_e G_B' \end{bmatrix} + \begin{bmatrix} V_{\varepsilon_A} & 0 \\ 0 & V_{\varepsilon_B} \end{bmatrix}$$

$$= \begin{bmatrix} G_A V_e G_A' + V_{\varepsilon_A} & G_A V_e G_B' \\ G_B V_e G_A' & G_B V_e G_B' + V_{\varepsilon_B} \end{bmatrix} = \begin{bmatrix} V_{AA} & V_{AB} \\ V_{AB}' & V_{BB} \end{bmatrix}. \qquad (13.5b)$$

We partition the inverse of matrix V_d like the original matrix as follows:

$$V_d^{-1} = \begin{bmatrix} V^{AA} & V^{AB} \\ V^{AB\prime} & V^{BB} \end{bmatrix}, \qquad (13.5c)$$

where the four partitions have dimension $T_{43} \times T_{43}$ (i.e. $T_{NP} \times T_{NP}$.).

Continuing with the derivation of the analytical solution, we obtain,

$$\boldsymbol{d} = (\boldsymbol{g} - \boldsymbol{G}\boldsymbol{s}) \Rightarrow \begin{bmatrix} \boldsymbol{d}_A \\ \boldsymbol{d}_B \end{bmatrix} = \begin{bmatrix} \boldsymbol{g}_A \\ \boldsymbol{g}_B \end{bmatrix} - \begin{bmatrix} \boldsymbol{G}_A \\ \boldsymbol{G}_B \end{bmatrix} \boldsymbol{s} \Rightarrow \begin{matrix} \boldsymbol{d}_A = \boldsymbol{g}_A - \boldsymbol{G}_A \boldsymbol{s} \\ \boldsymbol{d}_B = \boldsymbol{g}_B - \boldsymbol{G}_B \boldsymbol{s} \end{matrix}, \qquad (13.5d)$$

$$\hat{\boldsymbol{\gamma}} = \boldsymbol{V}_d^{-1} \boldsymbol{d} \equiv \begin{bmatrix} V^{AA} & V^{AB} \\ V^{AB\prime} & V^{BB} \end{bmatrix} \begin{bmatrix} \boldsymbol{d}_A \\ \boldsymbol{d}_B \end{bmatrix} \Rightarrow \begin{matrix} \hat{\boldsymbol{\gamma}}_A = V^{AA} \boldsymbol{d}_A + V^{AB} \boldsymbol{d}_B \\ \hat{\boldsymbol{\gamma}}_B = V^{AB\prime} \boldsymbol{d}_A + V^{BB} \boldsymbol{d}_B \end{matrix}, \qquad (13.5e)$$

$$\hat{\boldsymbol{\theta}} = \boldsymbol{s} + \boldsymbol{V}_e \boldsymbol{G}' \hat{\boldsymbol{\gamma}} \Rightarrow \hat{\boldsymbol{\theta}} = \boldsymbol{s} + \begin{bmatrix} \boldsymbol{V}_e \boldsymbol{G}_A' & \boldsymbol{V}_e \boldsymbol{G}_B' \end{bmatrix} \begin{bmatrix} \hat{\boldsymbol{\gamma}}_A \\ \hat{\boldsymbol{\gamma}}_B \end{bmatrix}$$

$$= \boldsymbol{s} + \boldsymbol{V}_e \boldsymbol{G}_A' \hat{\boldsymbol{\gamma}}_A + \boldsymbol{V}_e \boldsymbol{G}_B' \hat{\boldsymbol{\gamma}}_B, \qquad (13.5f)$$

where $\hat{\boldsymbol{\gamma}}_A = V^{AA} \boldsymbol{d}_A + V^{AB} \boldsymbol{d}_B$, $\hat{\boldsymbol{\gamma}}_B = V^{AB\prime} \boldsymbol{d}_A + V^{BB} \boldsymbol{d}_B$ and $\boldsymbol{d}_A = \boldsymbol{g}_A - \boldsymbol{G}_A \boldsymbol{s}$ and $\boldsymbol{d}_B = \boldsymbol{g}_B - \boldsymbol{G}_B \boldsymbol{s}$.

Substituting (13.5a) and (13.5c) into (13.4b) yields:

$$var[\hat{\boldsymbol{\theta}}] = \boldsymbol{V}_e - \boldsymbol{V}_e \boldsymbol{G}_A' V^{AA} \boldsymbol{G}_A \boldsymbol{V}_e - \boldsymbol{V}_e \boldsymbol{G}_B' V^{AB\prime} \boldsymbol{G}_A \boldsymbol{V}_e -$$

$$\boldsymbol{V}_e \boldsymbol{G}_A' V^{AB} \boldsymbol{G}_B \boldsymbol{V}_e - \boldsymbol{V}_e \boldsymbol{G}_B' V^{BB} \boldsymbol{G}_B \boldsymbol{V}_e. \qquad (13.5h)$$

This completes the development of solution (13.4) in terms of partitions A and B, given by Eqs (13.5f) and (13.5h).

13.2.2 Deriving an Analytical Solution for Each Series

We now develop the formulae at the level of the sub-partitions pertaining to each series. This sub-section may be skipped without little loss of continuity.

Substituting the sub-partitions of $\boldsymbol{V}_e$, $\boldsymbol{G}_A$ and $\boldsymbol{G}_B$ in Eq. (13.5) leads to:

$$\boldsymbol{V}_e \boldsymbol{G}_A{}' = \begin{bmatrix} \boldsymbol{V}_{e_{41}} & \boldsymbol{0} & \boldsymbol{0} & \boldsymbol{0} & \boldsymbol{0} & \boldsymbol{0} \\ \boldsymbol{0} & \boldsymbol{V}_{e_{42}} & \boldsymbol{0} & \boldsymbol{0} & \boldsymbol{0} & \boldsymbol{0} \\ \boldsymbol{0} & \boldsymbol{0} & \boldsymbol{V}_{e_{13}} & \boldsymbol{0} & \boldsymbol{0} & \boldsymbol{0} \\ \boldsymbol{0} & \boldsymbol{0} & \boldsymbol{0} & \boldsymbol{V}_{e_{23}} & \boldsymbol{0} & \boldsymbol{0} \\ \boldsymbol{0} & \boldsymbol{0} & \boldsymbol{0} & \boldsymbol{0} & \boldsymbol{V}_{e_{33}} & \boldsymbol{0} \\ \boldsymbol{0} & \boldsymbol{0} & \boldsymbol{0} & \boldsymbol{0} & \boldsymbol{0} & \boldsymbol{V}_{e_{43}} \end{bmatrix} \begin{bmatrix} \boldsymbol{G}_{41}{}' \\ \boldsymbol{G}_{42}{}' \\ \boldsymbol{0}' \\ \boldsymbol{0}' \\ \boldsymbol{0}' \\ -\boldsymbol{G}_{43}{}' \end{bmatrix} = \begin{bmatrix} \boldsymbol{V}_{e_{41}} \boldsymbol{G}_{41}{}' \\ \boldsymbol{V}_{e_{42}} \boldsymbol{G}_{42}{}' \\ \boldsymbol{0}' \\ \boldsymbol{0}' \\ \boldsymbol{0}' \\ -\boldsymbol{V}_{e_{43}} \boldsymbol{G}_{43}{}' \end{bmatrix}, \quad (13.6a)$$

$$\boldsymbol{V}_e \boldsymbol{G}_B{}' = \begin{bmatrix} \boldsymbol{V}_{e_{41}} & \boldsymbol{0} & \boldsymbol{0} & \boldsymbol{0} & \boldsymbol{0} & \boldsymbol{0} \\ \boldsymbol{0} & \boldsymbol{V}_{e_{42}} & \boldsymbol{0} & \boldsymbol{0} & \boldsymbol{0} & \boldsymbol{0} \\ \boldsymbol{0} & \boldsymbol{0} & \boldsymbol{V}_{e_{13}} & \boldsymbol{0} & \boldsymbol{0} & \boldsymbol{0} \\ \boldsymbol{0} & \boldsymbol{0} & \boldsymbol{0} & \boldsymbol{V}_{e_{23}} & \boldsymbol{0} & \boldsymbol{0} \\ \boldsymbol{0} & \boldsymbol{0} & \boldsymbol{0} & \boldsymbol{0} & \boldsymbol{V}_{e_{33}} & \boldsymbol{0} \\ \boldsymbol{0} & \boldsymbol{0} & \boldsymbol{0} & \boldsymbol{0} & \boldsymbol{0} & \boldsymbol{V}_{e_{43}} \end{bmatrix} \begin{bmatrix} \boldsymbol{0}' \\ \boldsymbol{0}' \\ \boldsymbol{G}_{13}{}' \\ \boldsymbol{G}_{23}{}' \\ \boldsymbol{G}_{33}{}' \\ -\boldsymbol{G}_{43}{}' \end{bmatrix} = \begin{bmatrix} \boldsymbol{0}' \\ \boldsymbol{0}' \\ \boldsymbol{V}_{e_{13}} \boldsymbol{G}_{13}{}' \\ \boldsymbol{V}_{e_{23}} \boldsymbol{G}_{23}{}' \\ \boldsymbol{V}_{e_{33}} \boldsymbol{G}_{33}{}' \\ -\boldsymbol{V}_{e_{43}} \boldsymbol{G}_{43}{}' \end{bmatrix}, \quad (13.6b)$$

$$\boldsymbol{V}_{AA} = \boldsymbol{G}_A(\boldsymbol{V}_e \boldsymbol{G}_A{}') + \boldsymbol{V}_{\varepsilon_A} = (\Sigma_{p=1}^{3} \boldsymbol{G}_{4p} \boldsymbol{V}_{e_{4p}} \boldsymbol{G}_{4p}{}' + \boldsymbol{V}_{\varepsilon_{4\bullet}}), \quad (13.6c)$$

$$\boldsymbol{V}_{AB} = \boldsymbol{G}_A(\boldsymbol{V}_e \boldsymbol{G}_B{}') = \boldsymbol{G}_{43} \boldsymbol{V}_{e_{43}} \boldsymbol{G}_{43}{}', \quad (13.6d)$$

$$\boldsymbol{V}_{BB} = (\Sigma_{n=1}^{4} \boldsymbol{G}_{n3} \boldsymbol{V}_{e_{n3}} \boldsymbol{G}_{n3}{}' + \boldsymbol{V}_{\varepsilon_{\bullet 3}}), \quad (13.6e)$$

$$\boldsymbol{d}_A = \boldsymbol{g}_A + \boldsymbol{G}_{43}\boldsymbol{s}_{43} - \boldsymbol{G}_{41}\boldsymbol{s}_{41} - \boldsymbol{G}_{42}\boldsymbol{s}_{42},$$

$$\boldsymbol{d}_B = \boldsymbol{g}_B + \boldsymbol{G}_{43}\boldsymbol{s}_{43} - \boldsymbol{G}_{13}\boldsymbol{s}_{13} - \boldsymbol{G}_{23}\boldsymbol{s}_{23} - \boldsymbol{G}_{33}\boldsymbol{s}_{33}. \quad (13.6f)$$

Denoting $G_{np} V_{e_{np}}$ by L_{np}, we obtain:

$$V_e G_A' V^{AA} G_A V_e =$$

$$\begin{bmatrix} L_{41}' V^{AA} L_{41} & L_{41}' V^{AA} L_{42} & 0_{T_{41}\times T_{13}} & 0_{T_{41}\times T_{23}} & 0_{T_{41}\times T_{33}} & -L_{41}' V^{AA} L_{43} \\ L_{42}' V^{AA} L_{41} & L_{42}' V^{AA} L_{42} & 0_{T_{42}\times T_{13}} & 0_{T_{42}\times T_{23}} & 0_{T_{42}\times T_{33}} & -L_{42}' V^{AA} L_{43} \\ 0_{T_{13}\times T_{41}} & 0_{T_{13}\times T_{42}} & 0_{T_{13}\times T_{13}} & 0_{T_{13}\times T_{23}} & 0_{T_{13}\times T_{33}} & 0_{T_{13}\times T_{43}} \\ 0_{T_{23}\times T_{41}} & 0_{T_{23}\times T_{42}} & 0_{T_{23}\times T_{13}} & 0_{T_{23}\times T_{23}} & 0_{T_{23}\times T_{33}} & 0_{T_{23}\times T_{43}} \\ 0_{T_{33}\times T_{41}} & 0_{T_{33}\times T_{42}} & 0_{T_{33}\times T_{13}} & 0_{T_{33}\times T_{23}} & 0_{T_{33}\times T_{33}} & 0_{T_{33}\times T_{43}} \\ -L_{43}' V^{AA} L_{41} & -L_{43}' V^{AA} L_{42} & 0_{T_{43}\times T_{13}} & 0_{T_{43}\times T_{23}} & 0_{T_{43}\times T_{33}} & L_{43}' V^{AA} L_{43} \end{bmatrix}, \quad (13.6g)$$

$$V_e G_B' V^{AB'} G_A V_e =$$

$$\begin{bmatrix} 0_{T_{41}\times T_{41}} & 0_{T_{41}\times T_{42}} & 0_{T_{41}\times T_{13}} & 0_{T_{41}\times T_{23}} & 0_{T_{41}\times T_{33}} & 0_{T_{41}\times T_{43}} \\ 0_{T_{42}\times T_{41}} & 0_{T_{42}\times T_{42}} & 0_{T_{42}\times T_{13}} & 0_{T_{42}\times T_{23}} & 0_{T_{42}\times T_{33}} & 0_{T_{42}\times T_{43}} \\ L_{13}' V^{AB'} L_{41} & L_{13}' V^{AB'} L_{42} & 0_{T_{13}\times T_{13}} & 0_{T_{13}\times T_{23}} & 0_{T_{13}\times T_{33}} & -L_{13}' V^{AB'} L_{43} \\ L_{23}' V^{AB'} L_{41} & L_{23}' V^{AB'} L_{42} & 0_{T_{23}\times T_{13}} & 0_{T_{23}\times T_{23}} & 0_{T_{23}\times T_{33}} & -L_{23}' V^{AB'} L_{43} \\ L_{33}' V^{AB'} L_{41} & L_{33}' V^{AB'} L_{42} & 0_{T_{33}\times T_{13}} & 0_{T_{33}\times T_{23}} & 0_{T_{33}\times T_{33}} & -L_{33}' V^{AB'} L_{43} \\ -L_{43}' V^{AB'} L_{41} & -L_{43}' V^{AB'} L_{42} & 0_{T_{43}\times T_{13}} & 0_{T_{43}\times T_{23}} & 0_{T_{43}\times T_{33}} & L_{43}' V^{AB'} L_{43} \end{bmatrix}, \quad (13.6h)$$

$$V_e G_A' V^{AB} G_B V_e = (V_e G_B' V^{AB'} G_A V_e)' =$$

$$\begin{bmatrix} 0_{T_{41}\times T_{41}} & 0_{T_{41}\times T_{42}} & L_{41}' V^{AB} L_{13} & L_{41}' V^{AB} L_{23} & L_{41}' V^{AB} L_{33} & -L_{41}' V^{AB} L_{43} \\ 0_{T_{42}\times T_{41}} & 0_{T_{42}\times T_{42}} & L_{42}' V^{AB} L_{13} & L_{42}' V^{AB} L_{23} & L_{42}' V^{AB} L_{33} & -L_{42}' V^{AB} L_{43} \\ 0_{T_{13}\times T_{41}} & 0_{T_{13}\times T_{42}} & 0_{T_{13}\times T_{13}} & 0_{T_{13}\times T_{23}} & 0_{T_{13}\times T_{33}} & 0_{T_{13}\times T_{43}} \\ 0_{T_{23}\times T_{41}} & 0_{T_{23}\times T_{42}} & 0_{T_{23}\times T_{13}} & 0_{T_{23}\times T_{23}} & 0_{T_{23}\times T_{33}} & 0_{T_{23}\times T_{43}} \\ 0_{T_{33}\times T_{41}} & 0_{T_{33}\times T_{42}} & 0_{T_{33}\times T_{13}} & 0_{T_{33}\times T_{23}} & 0_{T_{33}\times T_{33}} & 0_{T_{33}\times T_{43}} \\ 0_{T_{43}\times T_{41}} & 0_{T_{43}\times T_{42}} & -L_{43}' V^{AB} L_{13} & -L_{43}' V^{AB} L_{23} & -L_{43}' V^{AB} L_{33} & L_{43}' V^{AB} L_{43} \end{bmatrix}, \quad (13.6i)$$

$$\boldsymbol{V}_e\,\boldsymbol{G}_B{}'\,\boldsymbol{V}^{BB}\,\boldsymbol{G}_B\,\boldsymbol{V}_e =$$

$$\begin{bmatrix}
\boldsymbol{0}_{T_{41}\times T_{41}} & \boldsymbol{0}_{T_{41}\times T_{42}} & \boldsymbol{0}_{T_{41}\times T_{13}} & \boldsymbol{0}_{T_{41}\times T_{23}} & \boldsymbol{0}_{T_{41}\times T_{33}} & \boldsymbol{0}_{T_{41}\times T_{43}} \\
\boldsymbol{0}_{T_{42}\times T_{41}} & \boldsymbol{0}_{T_{42}\times T_{42}} & \boldsymbol{0}_{T_{42}\times T_{13}} & \boldsymbol{0}_{T_{42}\times T_{23}} & \boldsymbol{0}_{T_{42}\times T_{33}} & \boldsymbol{0}_{T_{42}\times T_{43}} \\
\boldsymbol{0}_{T_{13}\times T_{41}} & \boldsymbol{0}_{T_{13}\times T_{42}} & \boldsymbol{L}_{13}{}'\boldsymbol{V}^{BB}\boldsymbol{L}_{13} & \boldsymbol{L}_{13}{}'\boldsymbol{V}^{BB}\boldsymbol{L}_{23} & \boldsymbol{L}_{13}{}'\boldsymbol{V}^{BB}\boldsymbol{L}_{33} & -\boldsymbol{L}_{13}{}'\boldsymbol{V}^{BB}\boldsymbol{L}_{43} \\
\boldsymbol{0}_{T_{23}\times T_{41}} & \boldsymbol{0}_{T_{23}\times T_{42}} & \boldsymbol{L}_{23}{}'\boldsymbol{V}^{BB}\boldsymbol{L}_{13} & \boldsymbol{L}_{23}{}'\boldsymbol{V}^{BB}\boldsymbol{L}_{23} & \boldsymbol{L}_{23}{}'\boldsymbol{V}^{BB}\boldsymbol{L}_{33} & -\boldsymbol{L}_{23}{}'\boldsymbol{V}^{BB}\boldsymbol{L}_{43} \\
\boldsymbol{0}_{T_{33}\times T_{41}} & \boldsymbol{0}_{T_{33}\times T_{42}} & \boldsymbol{L}_{33}{}'\boldsymbol{V}^{BB}\boldsymbol{L}_{13} & \boldsymbol{L}_{33}{}'\boldsymbol{V}^{BB}\boldsymbol{L}_{23} & \boldsymbol{L}_{33}{}'\boldsymbol{V}^{BB}\boldsymbol{L}_{33} & -\boldsymbol{L}_{33}{}'\boldsymbol{V}^{BB}\boldsymbol{L}_{43} \\
\boldsymbol{0}_{T_{43}\times T_{41}} & \boldsymbol{0}_{T_{43}\times T_{42}} & -\boldsymbol{L}_{43}{}'\boldsymbol{V}^{BB}\boldsymbol{L}_{13} & -\boldsymbol{L}_{43}{}'\boldsymbol{V}^{BB}\boldsymbol{L}_{23} & -\boldsymbol{L}_{43}{}'\boldsymbol{V}^{BB}\boldsymbol{L}_{33} & \boldsymbol{L}_{43}{}'\boldsymbol{V}^{BB}\boldsymbol{L}_{43}
\end{bmatrix}. \quad (13.6j)$$

Substituting (13.6a) and (13.6b) into (13.5f) yields:

$$\begin{bmatrix} \hat{\boldsymbol{\theta}}_{41} \\ \hat{\boldsymbol{\theta}}_{42} \\ \hat{\boldsymbol{\theta}}_{13} \\ \hat{\boldsymbol{\theta}}_{23} \\ \hat{\boldsymbol{\theta}}_{33} \\ \hat{\boldsymbol{\theta}}_{43} \end{bmatrix} = \begin{bmatrix} \boldsymbol{s}_{41} \\ \boldsymbol{s}_{42} \\ \boldsymbol{s}_{13} \\ \boldsymbol{s}_{23} \\ \boldsymbol{s}_{33} \\ \boldsymbol{s}_{43} \end{bmatrix} + \begin{bmatrix} \boldsymbol{V}_{e_{41}}\boldsymbol{G}_{41}{}' \\ \boldsymbol{V}_{e_{42}}\boldsymbol{G}_{42}{}' \\ \boldsymbol{0}' \\ \boldsymbol{0}' \\ \boldsymbol{0}' \\ -\boldsymbol{V}_{e_{43}}\boldsymbol{G}_{43}{}' \end{bmatrix} \hat{\boldsymbol{\gamma}}_A + \begin{bmatrix} \boldsymbol{0}' \\ \boldsymbol{0}' \\ \boldsymbol{V}_{e_{13}}\boldsymbol{G}_{13}{}' \\ \boldsymbol{V}_{e_{23}}\boldsymbol{G}_{23}{}' \\ \boldsymbol{V}_{e_{33}}\boldsymbol{G}_{33}{}' \\ -\boldsymbol{V}_{e_{43}}\boldsymbol{G}_{43}{}' \end{bmatrix} \hat{\boldsymbol{\gamma}}_B\,. \quad (13.7)$$

This completes the derivation of the analytical solution for each series.

13.2.3 General Analytical Solution of the Marginal Two-Way Reconciliation Model

It is now straightforward to generalize result (13.7) to any number of provinces and industries:

$$\hat{\boldsymbol{\theta}}_{nP} = \boldsymbol{s}_{nP} + \boldsymbol{V}_{e_{nP}}\boldsymbol{G}_{nP}{}'\,\hat{\boldsymbol{\gamma}}_B, \quad n=1,\ldots,N-1,$$

$$\hat{\boldsymbol{\theta}}_{Np} = \boldsymbol{s}_{Np} + \boldsymbol{V}_{e_{Np}}\boldsymbol{G}_{Np}{}'\,\hat{\boldsymbol{\gamma}}_A, \quad p=1,\ldots,P-1,$$

$$\hat{\boldsymbol{\theta}}_{NP} = \boldsymbol{s}_{NP} - \boldsymbol{V}_{e_{NP}}\boldsymbol{G}_{NP}{}'\,(\hat{\boldsymbol{\gamma}}_A + \hat{\boldsymbol{\gamma}}_B), \quad (13.8a)$$

where $\hat{\gamma}_A = V^{AA} d_A + V^{AB} d_B$, $\hat{\gamma}_B = V^{AB'} d_A + V^{BB} d_B$,

$$d_A = g_A - \Sigma_{p=1}^{P} \delta_{pP} G_{Np} s_{Np}\ ,\quad d_B = g_B - \Sigma_{n=1}^{N} \delta_{nN} G_{nP} s_{nP}\ .$$

Scalar δ_{pP} is equal to -1 for $p=P$ and to 1 for $p \neq P$; and δ_{nN}, equal to -1 for $n=N$ and to 1 for $n \neq N$.

Substituting (13.6g) to (13.6j) into (13.5h) and re-arranging, yields the covariance matrices and the cross-covariance matrices of the $N+P-1$ series. The covariance matrices are:

$$var[\hat{\theta}_{nP}] = V_{e_{nP}} - V_{e_{nP}} G_{nP}{'} V^{BB} G_{nP} V_{e_{nP}}\ ,\ n=1,\ldots,N-1,$$

$$var[\hat{\theta}_{Np}] = V_{e_{Np}} - V_{e_{Np}} G_{Np}{'} V^{AA} G_{Np} V_{e_{Np}}\ ,\ p=1,\ldots,P-1,$$

$$var[\hat{\theta}_{NP}] = V_{e_{NP}}$$
$$- V_{e_{NP}} G_{NP}{'} (V^{AA} + V^{AB'} + V^{AB} + V^{BB}) G_{NP} V_{e_{NP}}. \qquad (13.8b)$$

The cross-covariance matrices are given by the four last terms of Eq. (13.5h) namely:

$$- V_e G_A{'} V^{AA} G_A V_e - V_e G_B{'} V^{AB'} G_A V_e -$$
$$V_e G_A{'} V^{AB} G_B V_e - V_e G_B{'} V^{BB} G_B V_e.$$

Substituting the analytical expression of these terms, given by Eqs. (13.6g) to (13.6j) respectively, yields

$$cov[\hat{\theta}_{nP}\ \hat{\theta}_{n'P}] = -V_{e_{nP}} G_{nP}{'} V^{BB} G_{n'P} V_{e_{n'P}},\ n<n'<N, \qquad (13.8c)$$

$$cov[\hat{\theta}_{nP}\ \hat{\theta}_{NP}] = V_{e_{nP}} G_{nP}{'} (V^{BB} + V^{AB'}) G_{NP} V_{e_{NP}},\ n<N, \qquad (13.8d)$$

$$cov[\hat{\theta}_{Np}\ \hat{\theta}_{Np'}] = -V_{e_{Np}} G_{Np}{'} V^{AA} G_{Np'} V_{e_{Np'}},\ p<p'<P, \qquad (13.8e)$$

$$cov[\hat{\theta}_{Np}\ \hat{\theta}_{NP}] = V_{e_{Np}} G_{Np}{'} (V^{AA} + V^{AB}) G_{NP} V_{e_{NP}},\ p<P, \qquad (13.8f)$$

$$cov[\hat{\theta}_{Np}\ \hat{\theta}_{n'P}] = -V_{e_{Np}} G_{Np}{'} V^{AB} G_{n'P} V_{e_{n'P}},\ p<P,\ n'<N. \qquad (13.8g)$$

Eq. (13.8c) provides the covariance matrix between the Industrial Totals; Eq. (13.8d), between the Industrial Totals and the Grand Total; Eq. (13.8e), between the Provincial Totals; Eq. (13.8f), between the Provincial Totals and the Grand Total; and Eq. (13.8g), between the Provincial Totals and the Industrial Totals.

Any series $\boldsymbol{s}_k$ - including the Grand Total - can be specified as exogenous, i.e. pre-determined, by setting its alterability coefficient α_k equal to 0, which implies that $\boldsymbol{V}_{e_k} = \boldsymbol{0}$. Of course, if too many series are un-alterable, there may not be a solution to the reconciliation problem.

The Grand Total can also be made exogenous (i.e. $\boldsymbol{\theta}_{NP} = \boldsymbol{s}_{NP}$) by storing it in vectors $\boldsymbol{g}_A$ and $\boldsymbol{g}_B$ and by ignoring the cases of Eq. (13.8) involving $\boldsymbol{\theta}_{NP}$. This entails having removed $\boldsymbol{I}_{T_{43}}$ and the last row from Eq. (13.3a) and removed $\boldsymbol{\theta}_{43}$ and $-\boldsymbol{G}_{43}$ from Eq. (13.3b). Note that the vectors of cross-sectional discrepancies $\boldsymbol{d}_A$ and $\boldsymbol{d}_B$ remain unchanged.

It is possible to develop a similar model which omits the Grand Total, because the cross-sectional sums over the Provincial Totals is specified to be equal to those of the Industrial Totals. In our framework, the cross-sectional constraints of the model would be

$$\boldsymbol{0} = \begin{bmatrix} -\boldsymbol{G}_{41} & -\boldsymbol{G}_{42} & \boldsymbol{G}_{13} & \boldsymbol{G}_{23} & \boldsymbol{G}_{33} \end{bmatrix} \begin{bmatrix} \boldsymbol{\theta}_{41}' & \boldsymbol{\theta}_{42}' & \boldsymbol{\theta}_{13}' & \boldsymbol{\theta}_{23}' & \boldsymbol{\theta}_{33}' \end{bmatrix}' + \varepsilon,$$

which would replace (13.3b). The Grand Total ($\boldsymbol{\theta}_{43}$), absent from the model, can then be obtained by taking the cross-sectional sums over the Provincial Totals or the Industrial Totals.

This approach can be approximated with model (13.3) by setting the alterability coefficient of the Grand Total to a relatively large value (e.g. 1000) and those of the other series to a low value (e.g. 1). As a result, the Grand Total is very passive in the reconciliation: the marginal totals are largely imposed onto the Grand Total. One advantage of model (13.3) is to provide a comparison of the reconciled Grand Total series to the original one as a by-product.

13.3 Implementation of the Analytical Solution

This section is intended mainly for readers who consider programming the marginal two-way reconciliation model. If the cross-covariance matrices are not required, the implementation of the general analytical solution (13.8) requires two runs of the data. Each series is processed twice.

Note that if the Grand Total is exogenous and contained in vectors $\boldsymbol{g}_A$ and $\boldsymbol{g}_B$, the indices n and p range from 1 to N-1 and from 1 to P-1 respectively.

First Run

The first run performs the following calculations for each series separately.

(1a) The series considered $\boldsymbol{s}_{np}$ is read and its covariance matrix $\boldsymbol{V}_{e_{np}}$ is generated. Alternatively $\boldsymbol{V}_{e_{np}}$ is read from a file, if a previous process generated the covariance matrices. In particular, this process may be benchmarking, in which case the series would also be subject to temporal aggregation constraints.

(1b) To avoid potential numerical problems, the series are rescaled. Each series is divided by

$$\bar{s}=(max(abs(s_{np,t}))-min(abs(s_{np,t}))/2, \qquad (13.9)$$

where $max(abs(s_{np,t}))$ is the maximum absolute value of the various series and $min(abs(s_{np,t}))$ the minimum absolute value. The covariance matrix of the series is divided by the square of $\bar{s}$. The resulting covariance matrix is then multiplied by the alterability coefficient selected for the series. This product becomes the covariance matrix used in the reconciliation model.

(1c) The industry and the province are determined.(This determination governs in which partition and sub-partition the data of the series are stored or accumulated.). The partitions of matrix $\boldsymbol{V}_d$, namely

$\boldsymbol{V}_{AA} = (\Sigma_{p=1}^{P} \boldsymbol{G}_{Np} \boldsymbol{V}_{e_{Np}} \boldsymbol{G}_{Np}{}' + \boldsymbol{V}_{\varepsilon_{N\bullet}})$,

$\boldsymbol{V}_{BB} = (\Sigma_{n=1}^{N} \boldsymbol{G}_{nP} \boldsymbol{V}_{e_{nP}} \boldsymbol{G}_{nP}{}' + \boldsymbol{V}_{\varepsilon_{\bullet P}})$ and $\boldsymbol{V}_{AB} = \boldsymbol{G}_{NP} \boldsymbol{V}_{e_{NP}} \boldsymbol{G}_{NP}{}'$,

are accumulated, each series contributing one term to $\boldsymbol{V}_{AA}$ or $\boldsymbol{V}_{BB}$, or to both for the Grand Total. Similarly, the cross-sectional discrepancies are accumulated in vectors

$\boldsymbol{d}_A = \boldsymbol{g}_A - (\Sigma_{p=1}^{P} \delta_{pP} \boldsymbol{G}_{Np} \boldsymbol{s}_{Np})$ and

$\boldsymbol{d}_B = \boldsymbol{g}_B - (\Sigma_{n=1}^{N} \delta_{nN} \boldsymbol{G}_{nP} \boldsymbol{s}_{nP})$,

where the scalar δ_{pP} is equal to -1 for p=P, and to 1 otherwise; and δ_{nN} is equal to -1 for n=N, and to 1 otherwise.

Steps (1a) to (1c) are repeated for each of the N+P−1 series in the system.

After the first run is completed, the following calculations are performed. In order to deal with potential redundant constraints (discussed in the next section), the Moore-Penrose generalized matrix inversion can be used in all the matrix inversions below.

(a) Matrices V^{AA}, V^{AB} and V^{BB}, i.e. the partitions of matrix V_d^{-1}, are obtained by inverting V_d by parts, which involves the following steps.

  (i) Matrix V_{BB}^{-1} is obtained. At this stage matrix V_{BB} may be discarded.

  (ii) Matrix V^{AA} is set equal to $(V_{AA} - V_{AB} V_{BB}^{-1} V_{AB}^{/})^{-1}$. Matrix V_{AA} may be discarded.

  (iii) Matrix V^{AB} is set equal to $-V^{AA} V_{AB} V_{BB}^{-1}$, where V_{BB}^{-1} originates from (i).

  (iv) Matrix V^{BB} is set equal to $V_{BB}^{-1} - V^{AB/} V_{AB} V_{BB}^{-1}$. Matrices V_{AB} and V_{BB}^{-1} may be discarded.

(b) The Lagrange multipliers are calculated according to (13.8a):

$$\hat{\gamma}_A = V^{AA} d_A + V^{AB} d_B \text{ and}$$

$$\hat{\gamma}_B = V^{AB/} d_A + V^{BB} d_B.$$

Second Run

The second run performs the following calculations for each series separately.

(2a) This step is identical to (1a).

(2b) This step is identical to (1b).

(2c) The reconciled series $\hat{\theta}_{np}$ is calculated according to Eq. (13.8a). If applicable, its covariance matrix $var[\hat{\theta}_{np}]$ is calculated according to Eq. (13.8b).

(2d) The series considered and its covariance matrix are converted to the original scale by multiplying it by $\bar{s}$ of Eq. (13.9) and by the square of this number respectively. Both $\hat{\theta}_{np}$ and $var[\hat{\theta}_{np}]$ (or perhaps the diagonal thereof) are saved on a file.

(2e) The residual cross-sectional discrepancies are accumulated in vectors

$$d_A^{(r)} = g_A - (\Sigma_{p=1}^{P} \delta_{pP} G_{Np} \hat{\theta}_{Np}) \text{ and}$$

$$d_B^{(r)} = g_B - (\Sigma_{n=1}^{N} \delta_{nN} G_{nP} \hat{\theta}_{nP}).$$

These discrepancies may be used to assess whether the reconciliation satisfied the cross-sectional constraints.

Under the solution and implementation proposed, none of the larger matrices in (13.4), (13.5f) and (13.5h), namely $\boldsymbol{V}_e$, $\boldsymbol{G}$, $\boldsymbol{G}_A$ and $\boldsymbol{G}_B$ need be generated. Matrices $\boldsymbol{G}_{np}$ need not be generated either. For example, let the Grand Total $\boldsymbol{s}_{NP}$ start in 2001 and end in 2005 (5 observations); and, the series $\boldsymbol{s}_{N2}$ start in 2002 and end in 2004 (3 observations). The first and last observations of $\boldsymbol{s}_{N2}$ are then $t_{1,N2}=2$ and $t_{L,N2}=4$. Matrix $\boldsymbol{G}_{N2}$ and matrix product $\boldsymbol{G}_{N2}\,\boldsymbol{s}_{N2}$ are then respectively

$$\boldsymbol{G}_{N2} = \begin{bmatrix} 0 & 0 & 0 \\ 1 & 0 & 0 \\ 0 & 1 & 0 \\ 0 & 0 & 1 \\ 0 & 0 & 0 \end{bmatrix}, \quad \boldsymbol{G}_{N2}\,\boldsymbol{s}_{N2} = \begin{bmatrix} 0 & 0 & 0 \\ 1 & 0 & 0 \\ 0 & 1 & 0 \\ 0 & 0 & 1 \\ 0 & 0 & 0 \end{bmatrix} \begin{bmatrix} s_{N2,2} \\ s_{N2,3} \\ s_{N2,4} \end{bmatrix} = \begin{bmatrix} 0 \\ s_{N2,2} \\ s_{N2,3} \\ s_{N2,4} \\ 0 \end{bmatrix}.$$

As a result, subtracting $\boldsymbol{G}_{N2}\,\boldsymbol{s}_{N2}$ (dimension 5×1) from vector $\boldsymbol{d}_B$ (5×1) amounts to subtracting $\boldsymbol{s}_{N2}$ from rows 2 to 4 of $\boldsymbol{d}_B$. Similarly adding $\boldsymbol{G}_{N2}\,\boldsymbol{V}_{e_{N2}}\,\boldsymbol{G}_{N2}{}'$ to $\boldsymbol{V}_{BB}$ (5×5) amounts to adding $\boldsymbol{V}_{e_{N2}}$ (3×3) to rows 2 to 4 and columns 2 to 4 of $\boldsymbol{V}_{BB}$. Operationally, matrix $\boldsymbol{G}_{N2}$ is thus replaced by the use of row and/or column indexes. (The index can contain non-contiguous values, e.g. 1, 2, 4, 5.)

Note that some of the calculations are redundant, namely for series with alterability coefficients equal to 0. We already know that for such series $\hat{\boldsymbol{\theta}}_{np} = \boldsymbol{s}_{np}$ and $var[\hat{\boldsymbol{\theta}}_{np}] = \boldsymbol{0}$. These series contribute to $\boldsymbol{d}_A$ or $\boldsymbol{d}_B$ but not to matrices $\boldsymbol{V}_{AA}$, nor $\boldsymbol{V}_{BB}$ nor $\boldsymbol{V}_{AB}$.

13.4 Redundant Constraints

The marginal two-way model has one set of T_{NP} "industrial" cross-sectional constraints and one set of T_{NP} "provincial" cross-sectional constraints, respectively:

$$\theta_{NP,t} - \Sigma_{p=1}^{P-1}\,\theta_{Np,t} = 0,\ t=1,\ldots,T_{NP},$$

$$\theta_{NP,t} - \Sigma_{n=1}^{N-1}\,\theta_{nP,t} = 0,\ t=1,\ldots,T_{NP}.$$

Similarly to section 12.4, we distinguish two situations regarding the temporal constraints:

(a) the absence of any temporal constraints and

(b) the presence of temporal constraints. In this case, we assume that (i) all series start at the same date and have equal length, (ii) all series have annual benchmarks for each complete "year" and (iii) all the temporal constraints are binding and implicit in the covariance matrices V_{e_k} of the series to be reconciled.

In the absence of temporal constraints, the marginal two-way model (13.3) has no redundant cross-sectional constraints.

In the presence of temporal constraints (case (b) above), the model has one redundant cross-sectional constraint per "year" with benchmark in both sets of cross-sectional constraints. Alternatively, one can say that the set of temporal constraints of any one of the Provincial Totals s_{Np} $(p<P)$ and the set of temporal constraint of any one of the Industrial Totals s_{nP} $(n<N)$ are redundant. In fact choosing which constraints are redundant is arbitrary, because redundancy is an attribute of the whole set of constraints. The problem of redundancy can therefore be approached in a number of different ways:

(1) Omit one constraint per "year" with benchmark in each of the two sets of cross-sectional constraints.

(2) Make one constraint per "year" non-binding in each of the two sets of cross-sectional constraints, by setting the corresponding diagonal element of V_ε equal to 1 (say).

(3) Omit all the temporal constraints for any one of the Provincial Totals and for any one of the Industrial Totals.

(4) Make the temporal constraints non-binding for any one of the Provincial Totals and any one of the Industrial Totals.

(5) Invert V_d by parts using the Moore-Penrose generalized matrix inversion, as described in section 13.3 at the end of the first run.

Under (1) to (4), the omitted or non-binding constraints are satisfied through the remaining ones. If some constraints are not satisfied, it means some of the data is contradictory. For example, the benchmarks do not satisfy the cross-sectional aggregation constraints.

In practical situations some series may have temporal constraints for each "year", and others fewer temporal constraints or none at all. This makes it hard to determine which constraints should be omitted or made non-binding.

The generalized matrix inversion was used by Theil (1971) in a more general context and specifically applied by Di Fonzo (1990) in reconciliation. This approach offers a valid and practical solution. Failure of the resulting reconciled series to satisfy all constraints indicates contradictory data, which should be fixed by the subject matter experts.

13.5 A Real Data Example: the Seasonally Adjusted Canadian Retail Trade Series

We now resume the example of the seasonally adjusted Canadian Retail Trade series (Statistics Canada 2001). The marginal two-way reconciliation model allows more general and thorough discussion of the example than the one-way model used in section 12.6.

Table 13.2. Trade Group Codes for the Canadian Retail Trade Series

n	Code	Trade Group	*n*	Code	Trade Group
1	010	Supermarkets and grocery stores	10	100	Motor vehicle and recreational vehicle dealers
2	020	All other food stores	11	110	Gasoline service stations
3	030	Drugs and patent medicine stores	12	120	Automotive parts, accessories and services
4	040	Shoe stores	13	131	Department stores
5	050	Men's clothing stores	14	132	Other general merchandise stores
6	060	Women's clothing stores	15	140	Other semi-durables stores
7	070	Other clothing stores	16	150	Other durable goods stores
8	080	Household furniture and appliance stores	17	161	Liquor stores
9	090	Household furnishings stores	18	162	Other retail stores
			19	300	"All" Trade Group

Let us recall that the system consists of 31 time series: 18 Trade Group Totals, 12 Provincial Totals and the Grand Total, for the period Jan. 1991 to Nov. 2002. Each series is identified by a Province code displayed in Table 12.6 and by a Trade Group code in Table 13.2. In this case, the source of the cross-sectional aggregation discrepancies is seasonal adjustment. (The seasonal adjustment method used is non linear, and different series required different treatments.) The seasonal adjustment method used produces no covariance matrix.

All the series must satisfy yearly benchmarks which are the yearly sums of the raw (seasonally un-adjusted) series. Each series is benchmarked, and its covariance matrix is read in the reconciliation programme. The model applied is that of proportional benchmarking described in section 3.8, except parameter λ is set to 1/2 instead of 1; parameter φ=0.90 is used. Setting λ equal to 1/2 specifies that the corrections are proportional to the values of the original series.

The marginal two-way reconciliation model is applied to the benchmarked series $\tilde{\theta}_k$ using their covariance matrix $var[\tilde{\theta}_k]$. (In order to alleviate the notation, we sometimes use the subscipt k instead of np.) These covariance matrices are multiplied by the alterability coefficient assigned to the series: $V_{e_k} = \alpha_k var[\tilde{\theta}_k]$. The problem of redundancies in the constraints, due to the presence of binding benchmarks, is addressed using the Moore-Penrose generalized matrix inversion.

Table 13.3a. Cross-sectional aggregation discrepancies in the yearly benchmarks

Industrial		Provincial	
1991	-6	1991	13
1992	4	1992	2
1993	0	1993	10
1994	-3	1994	-2
1995	0	1995	-5
1996	2	1996	0
1997	-2	1997	-4
1998	-5	1998	-1
1999	2	1999	1
2000	-1	2000	1
2001	2	2001	5

We now consider two approaches to seasonal adjustment, the indirect and the direct. These two approaches or strategies call for two different ways of applying the marginal two-way model.

(1) The indirect approach defines the seasonally adjusted Grand Total as the cross-sectional sums of the seasonally adjusted Trade Group Totals. This common practice is often referred to as *indirect seasonal adjustment*. We therefore expect the presence of cross-sectional aggregation discrepancies between the Grand Total and the Provincial Totals, but not between the Grand Total and the Trade Group Totals. However, as observed in section 12.6 there are discrepancies in the latter case as well, because the seasonal adjustment program used at the time performed the calculations in single precision. This implies that the digits beyond the seventh digit are not significant, which causes the emergence of the discrepancies.

The reconciliation process is used to impose the Trade Group Totals (*sic*) onto the Grand Total as intended under indirect seasonal adjustment, and of course to impose the resulting Grand Total onto the Provincial Totals. In the marginal two-way model, this is achieved by setting the coefficients of alterability of the Trade Group Totals to zero and those of the Grand Total and of the Provincial Totals to 1. Under these coefficients of alterability, the Trade Group Totals determine the Grand Total, which in turn determines the Provincial Totals.

(2) The direct approach consists of seasonally adjusting each series separately. This produces cross-sectional discrepancies between the Grand Total and the Trade Group Totals and between the Grand Total and the Provincial Totals. In the marginal two-way model, all series are assigned alterability coefficients equal to 1, and as a result all series more or less equally determine each other.

13.5.1 The Indirect Seasonal Adjustment

The case at hand illustrates a data problem, namely that of contradictory data discussed in section 11.4. The 18 raw Trade Group Totals and the 12 raw Provincial Totals do not cross-sectionally add up to the Grand Total by as much as ±3 units ($,000). As a result, the benchmarks, i.e. the yearly totals of the raw series, do not cross-sectionally add up by amounts larger than ±3. As shown in Table 13.3a, the cross-sectional discrepancies in the yearly benchmarks range from -6 to 13 units. Because the benchmarks are binding, these cross-sectional discrepancies will be forced onto the system of reconciled series. In other words, the system will never satisfy both the temporal and the cross-sectional constraints.

Table 13.3b. Statistics on the temporal discrepancies for the indirectly seasonally adjusted Retail Trade Series, before reconciliation

Prov.	TrGr	Min.	Max.	Prov.	TrGr	Min.	Max.
00	010	-5	0	00	161	-2	1
00	020	-2	2	00	162	-2	1
00	030	-2	0	**00**	**300**	**-27**	**9**
00	040	-2	1	10	300	-2	0
00	050	-1	1	11	300	-1	2
00	060	-2	1	12	300	-2	1
00	070	-1	1	13	300	-1	1
00	080	-3	1	22	300	-4	-1
00	090	-1	3	32	300	-5	1
00	100	-4	-1	42	300	-2	1
00	110	-3	0	47	300	-2	2
00	120	-3	2	48	300	-4	1
00	131	-3	1	52	300	-3	0
00	132	-3	0	60	300	-2	1
00	140	-1	1	64	300	-2	2
00	150	-1	0	.	.	-27	9

Table 13.3b displays statistics on the temporal discrepancies between the yearly benchmarks and the yearly sums of the seasonally adjusted series to be reconciled. In principle these series should already satisfy the yearly benchmarks, because they were benchmarked by the seasonal adjustment program. Some of the series do not satisfy their benchmarks, because the

seasonal adjustment program used at the time is in single precision. This implies the eighth digit is not significant. The Grand Total has more than six digits, exceeding 10,000,000 ($,000) since the beginning of the series in 1991. As a result, the temporal discrepancies of the Grand Total are more noticeable in Table 13.3b (bold line). Note that even if there were no temporal discrepancies, it would still be necessary to benchmark the series, to generate their covariance matrices, for use in the reconciliation process to preserve the temporal constraints.

Table 13.3c. Some statistics on the cross-sectional discrepancies for the indirectly seasonally adjusted Retail Trade series, before reconciliation

	Min. add. discrep.	Max. add. discrep.	Min. prop. discrep.	Max. prop. discrep.
Industrial:	-188,585	333,352	0.9877	1.0163
Provincial:	-10	6	1.0000	1.0000

Table 13.3c displays the minimum and the maximum cross-sectional discrepancies for the 31 series before reconciliation. The first line of the table indicates the presence of relatively large proportional and additive discrepancies between the Grand Total and the cross-sectional sums over the Provincial Totals; the second line, the presence of small industrial additive discrepancies between the Grand Total and the cross-sectional sums over the Trade Groups. The statistics on the proportional cross-sectonal discrepancies indicates that, seasonal adjustment introduces proportional discrepancies between -1.23% and +1.63% in the Canadian Retail Trade series. Fig. 12.1 displays the actual proportional cross-sectional discrepancies between the Grand Total and the Provincial Totals.

Table 13.3d displays the proportional cross-sectional aggregation discrepancies greater than .1.005 or lower than 0.995, before reconciliation. These relatively large discrepancies between the Grand Total and the cross-sectional sums of the Provincial Totals tend to occur more often in May and in December, which seems to indicate the presence of seasonality in the cross-sectional discrepancies. The actual values for the Grand Total appear under the heading "Total Series" for the dates indicated.

Table 13.3d. Larger industrial cross-sectional discrepancies (over the Provincial Totals) for the indirectly seasonally adjusted Retail Trade series, before reconciliation

Starting date	Ending date	Total series	Cross-sect. sum	Addit. discrep.	Propor. discrep.
01/01/91	31/01/91	14,827,980	14,708,092	119,888	1.0082
01/04/91	30/04/91	15,128,850	15,042,784	86,066	1.0057
01/05/91	31/05/91	15,351,340	15,265,438	85,902	1.0056
01/12/91	31/12/91	15,083,340	15,271,925	-188,585	0.9877
01/01/92	31/01/92	15,126,980	15,027,637	99,343	1.0066
01/11/92	30/11/92	15,640,930	15,721,238	-80,308	0.9949
01/06/93	30/06/93	15,959,110	15,838,551	120,559	1.0076
01/12/96	31/12/96	19,030,840	18,879,565	151,275	1.0080
01/05/97	31/05/97	19,659,000	19,770,205	-111,205	0.9944
01/12/97	31/12/97	20,838,910	20,505,558	333,352	1.0163
01/05/99	31/05/99	21,315,770	21,430,666	-114,896	0.9946
01/12/99	31/12/99	22,465,790	22,295,623	170,167	1.0076

Table 13.3e. Some statistics on the residual temporal discrepancies for the indirectly seasonally adjusted Retail Trade series, after reconciliation

Prov	TrGr	Min.	Max.	Prov.	TrGr	Min.	Max.
00	010	-5	0	00	161	-2	1
00	020	-2	2	00	162	-2	1
00	030	-2	0	00	300	-3	4
00	040	-2	1	10	300	-1	1
00	050	-1	1	11	300	-3	1
00	060	-2	1	12	300	-1	1
00	070	-1	1	13	300	-1	1
00	080	-3	1	22	300	-1	1
00	090	-1	3	32	300	-1	2
00	100	-4	-1	42	300	-1	2
00	110	-3	0	47	300	-1	2
00	120	-3	2	48	300	-1	2
00	131	-3	1	52	300	-2	1
00	132	-3	0	60	300	-2	1
00	140	-1	1	64	300	-2	3
00	150	-1	0	.	.	-5	4

Table 13.3e shows some statistics on the residual temporal discrepancies after reconciliation. The discrepancies are now contained in the range -5 to 4, compared to -27 to 9 in Table 13.3b. The remaining additive temporal discrepancies are due to rounding. The reconciled series were rounded to the nearest unit by the reconciliation program before the calculation of the residual temporal and cross-sectional discrepancies. The purpose of rounding at that stage is to anticipate the discrepancies attributable to rounding in the publication.

The problem of constraint redundancies was addressed by using the Moore Penrose matrix inversion. If instead the problem is addressed by making (say) the January cross-sectional constraints non-binding, Table 13.3e remains the same. In other words, the Moore-Penrose approach does not affect the pre-imposed temporal constraints implicit in the input covariance matrices of the original series.

Table 13.3f. Statistics on the cross-sectional discrepancies for the indirectly seasonally adjusted Retail Trade series, after reconciliation

	Min. add. discrep.	Max. add. discrep.	Min. prop. discrep.	Max. prop. discrep.
Industrial:	-2	2	1.0000	1.0000
Provincial:	-2	0	1.0000	1.0000

Table 13.3f displays some statistics on the residual cross-sectional discrepancies after reconciliation. The additive discrepancies are due to rounding to the nearest unit and the fact that the yearly benchmarks do not cross-sectionally add-up. When using the Moore-Penrose generalized matrix inversion, the cross-sectional discrepancies in the benchmarks, i.e. the contradiction in the benchmarks, are distributed evenly over each month of each year with benchmark. When making the January cross-sectional constraints non-binding, the January constraints are not satisfied by larger amounts; the minimum and maximum additive discrepancies are then -8 and 5 (not shown) for the industrial constraints and -8 and 12 for the provincial constraints.

Table 13.3g shows some statistics on the proportional corrections $\hat{\theta}_{k,t}/s_{k,t}$ made to the series to achieve reconciliation. The Trade Group Totals, with alterability coefficients equal to 0, are not corrected by reconciliation. The Grand Total with alterability equal to 1 is corrected by very small amounts to satisfy the cross-sectional sums of the Trade Group Totals. The Provincial Totals, with alterability coefficients equal to 1 display corrections in the range 0.9872 and 1.0169 and with standard deviations equal to 0.0033.

The marginal two-way reconciliation model thus imposes the seasonally adjusted Trade Group Totals to the indirectly adjusted Grand Total and imposes the resulting Grand Total onto the Provincial Totals. This is achieved in one application of the marginal two-way reconciliation model, which minimizes the logistic and the number of files involved.

Table 13.3g. Some statistics on the Proportional Corrections and on the Proportional Movement Preservation

Prov	TrGr	Alter. coef.	Min. cor.	Max. cor.	std.dev. of cor.	Min. change mov.	Max. change mov.	Std.Dev. of ch. mov.
00	010	0	1.0000	1.0000	0.0000	0.0000	0.0000	0.0000
00	020	0	1.0000	1.0000	0.0000	0.0000	0.0000	0.0000
00	030	0	1.0000	1.0000	0.0000	0.0000	0.0000	0.0000
00	040	0	1.0000	1.0000	0.0000	0.0000	0.0000	0.0000
00	050	0	1.0000	1.0000	0.0000	0.0000	0.0000	0.0000
00	060	0	1.0000	1.0000	0.0000	0.0000	0.0000	0.0000
00	070	0	1.0000	1.0000	0.0000	0.0000	0.0000	0.0000
00	080	0	1.0000	1.0000	0.0000	0.0000	0.0000	0.0000
00	090	0	1.0000	1.0000	0.0000	0.0000	0.0000	0.0000
00	100	0	1.0000	1.0000	0.0000	0.0000	0.0000	0.0000
00	110	0	1.0000	1.0000	0.0000	0.0000	0.0000	0.0000
00	120	0	1.0000	1.0000	0.0000	0.0000	0.0000	0.0000
00	131	0	1.0000	1.0000	0.0000	0.0000	0.0000	0.0000
00	132	0	1.0000	1.0000	0.0000	0.0000	0.0000	0.0000
00	140	0	1.0000	1.0000	0.0000	0.0000	0.0000	0.0000
00	150	0	1.0000	1.0000	0.0000	0.0000	0.0000	0.0000
00	161	0	1.0000	1.0000	0.0000	0.0000	0.0000	0.0000
00	162	0	1.0000	1.0000	0.0000	0.0000	0.0000	0.0000
00	300	1	1.0000	1.0000	0.0000	-0.0000	0.0000	0.0000
10	300	1	0.9876	1.0163	0.0033	-0.0154	0.0189	0.0046
11	300	1	0.9878	1.0162	0.0033	-0.0153	0.0190	0.0046
12	300	1	0.9884	1.0162	0.0033	-0.0154	0.0191	0.0046
13	300	1	0.9879	1.0160	0.0033	-0.0154	0.0190	0.0046
22	300	1	0.9872	1.0169	0.0033	-0.0150	0.0188	0.0046
32	300	1	0.9878	1.0161	0.0033	-0.0154	0.0190	0.0046
42	300	1	0.9876	1.0160	0.0033	-0.0154	0.0190	0.0046
47	300	1	0.9881	1.0159	0.0032	-0.0155	0.0190	0.0046
48	300	1	0.9880	1.0160	0.0033	-0.0154	0.0190	0.0046
52	300	1	0.9876	1.0161	0.0033	-0.0154	0.0190	0.0046
60	300	1	0.9886	1.0164	0.0033	-0.0154	0.0191	0.0046
64	300	1	0.9879	1.0164	0.0033	-0.0154	0.0189	0.0046
.	.	.	0.9872	1.0169	0.0020	-0.0155	0.0191	0.0029

13.5.2 The Direct Seasonal Adjustment

Under direct seasonal adjustment, each series is separately seasonally adjusted. In the marginal two-way reconciliation model, we assign alterability coefficients equal to 1 to all series. In other words, the Trade Group Totals no longer have alterability coefficients equal to 0, like in the indirect approach.

The yearly benchmarks used are of course the same for the indirect and the direct seasonal adjustment approach. As discussed in relation to Table 13.3a, the benchmarks display small cross-sectional aggregation discrepancies.

Table 13.4a shows some statistics on the temporal discrepancies between the yearly benchmarks and the yearly sums of the directly seasonally adjusted series to be reconciled. The series do not quite satisfy their yearly benchmarks, because they were benchmarked by the seasonal adjustment program in single precision, as discussed in conjunction with Table 13.3b. The statistics in Table 13.4a and 13.3b are identical, except for the Grand Total (bolded line) which is now directly seasonally adjusted.

Table 13.4a. Some statistics on the temporal discrepancies for the directly seasonally adjusted Retail Trade series, before reconciliation

Prov.	TrGr	Min.	Max.	Prov.	TrGr	Min.	Max.
00	010	-5	0	00	161	-2	1
00	020	-2	2	00	162	-2	1
00	030	-2	0	**00**	**300**	**-11**	**10**
00	040	-2	1	10	300	-2	0
00	050	-1	1	11	300	-1	2
00	060	-2	1	12	300	-2	1
00	070	-1	1	13	300	-1	1
00	080	-3	1	22	300	-4	-1
00	090	-1	3	32	300	-5	1
00	100	-4	-1	42	300	-2	1
00	110	-3	0	47	300	-2	2
00	120	-3	2	48	300	-4	1
00	131	-3	1	52	300	-3	0
00	132	-3	0	60	300	-2	1
00	140	-1	1	64	300	-2	2
00	150	-1	0	.	.	-11	10

Table 13.4b. Some statistics on the cross-sectional discrepancies for the directly seasonally adjusted Retail Trade series, before reconciliation

	Min. add. discrep.	Max. add. discrep.	Min. prop. discrep.	Max. prop. discrep.
Ind.	-76,180	127,198	0.9950	1.0086
Prov .	-234,097	175,913	0.9884	1.0117

Table 13.4b displays the minimum and the maximum cross-sectional aggregation discrepancies for the 31 series before reconciliation. The first line of the table indicates the presence of substantial proportional and additive discrepancies between the Grand Total and the cross-sectional sums over the Provincial Totals; the second line, between the Grand Total and the cross-sectional sums over the Trade Groups.

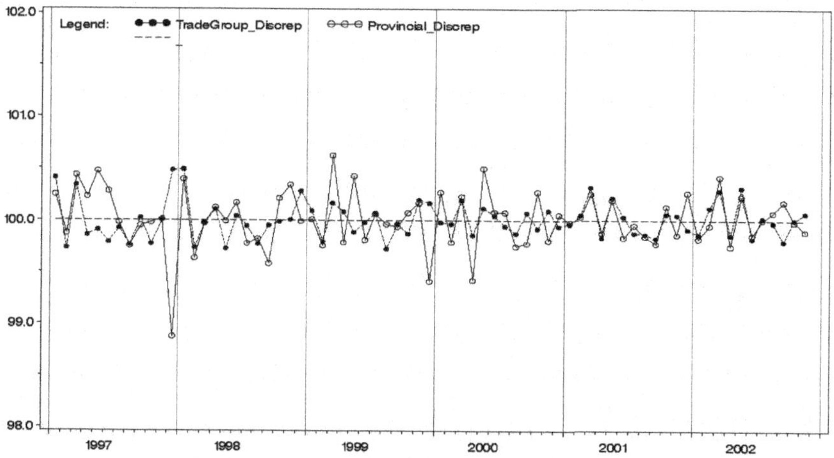

Fig. 13.1. Proportional Industrial (Trade Group) and Provincial cross-sectional aggregation discrepancies for the directly seasonally adjusted Retail Trade series for the period 1997-2002

Table 13.4c. Larger cross-sectional discrepancies of the directly seasonally adjusted Retail Trade series, before Reconciliation

Starting date	Ending date	Total series	Cross-sect. sum	Addit. discrep.	Propor. discrep.
Industrial					
01/01/91	31/01/91	14,835,290	14,708,092	127,198	1.0086
01/06/91	30/06/91	15,071,880	15,148,060	-76,180	0.9950
01/01/92	31/01/92	15,143,840	15,027,637	116,203	1.0077
01/01/93	31/01/93	16,115,180	16,025,345	89,835	1.0056
01/12/94	31/12/94	17,970,260	17,872,886	97,374	1.0054
Provincial					
01/04/91	30/04/91	15,052,790	15,128,847	-76,057	0.9950
01/06/91	30/06/91	15,071,880	15,203,144	-131,264	0.9914
01/11/91	30/11/91	15,403,920	15,296,839	107,081	1.0070
01/12/91	31/12/91	15,259,250	15,083,337	175,913	1.0117
01/06/92	30/06/92	15,342,390	15,466,391	-124,001	0.9920
01/11/92	30/11/92	15,741,560	15,640,928	100,632	1.0064
01/12/92	31/12/92	15,833,000	15,737,966	95,034	1.0060
01/01/93	31/01/93	16,115,180	16,021,038	94,142	1.0059
01/06/93	30/06/93	15,773,930	15,959,116	-185,186	0.9884
01/01/94	31/01/94	16,759,760	16,642,250	117,510	1.0071
01/06/94	30/06/94	17,259,490	17,355,848	-96,358	0.9944
01/04/96	30/04/96	18,206,390	18,115,527	90,863	1.0050
01/12/97	31/12/97	20,604,810	20,838,907	-234,097	0.9888
01/03/99	31/03/99	21,489,020	21,355,302	133,718	1.0063
01/12/99	31/12/99	22,332,760	22,465,794	-133,034	0.9941
01/04/00	30/04/00	22,469,920	22,601,281	-131,361	0.9942
01/05/00	31/05/00	22,915,140	22,801,043	114,097	1.0050

Fig. 13.1 exhibits the actual industrial and provincial proportional aggregation discrepancies for the period 1997 to Nov 2002. Note that the amplitude of the industrial discrepancies is substantially smaller in Fig. 13.1 (direct seasonal adjustment) than in Fig. 12.1 (indirect). This may suggest that indirect seasonal adjustment of the Grand Total, by cross-sectionally summing over the Trade Groups, produce larger discrepancies cross-sectional discrepancies with respect to the Provincial Totals.

Table 13.4c displays the larger proportional cross-sectional aggregation discrepancies outside the range 0.995 to 1.005, before reconciliation. The upper part of the table displays the larger discrepancies between the Grand Total and the Provincial Totals. These discrepancies are now fewer than in Table 13.3d. The lower part of Table 13.4c displays the larger discrepancies between the Grand Total and the the Trade Group Total. There were no such discrepancies outside the range 0.995 and 1.005 in Table 13.3d, because the Grand Total had been obtained by summing over the Trade Group Totals.

Table 13.4d shows some statistics on the residual temporal discrepancies after reconciliation. The remaining additive temporal discrepancies are due to rounding. When making the January cross-sectional constraints not binding, instead of using the Moore-Penrose matrix inversion, the table remains the same. In other words the Moore-Penrose inversion does not affect the temporal constraints.

Table 13.4d. Some statistics on the temporal discrepancies for the directly seasonally adjusted Retail Trade series, after reconciliation

Prov	TrGr	Min.	Max.	Prov.	TrGr	Min.	Max.
00	010	-1	1	00	161	-3	2
00	020	-1	2	00	162	-1	3
00	030	-2	0	00	300	-2	1
00	040	-1	1	10	300	-1	2
00	050	-2	1	11	300	-1	2
00	060	-1	1	12	300	-1	1
00	070	-1	2	13	300	-2	2
00	080	-3	1	22	300	-1	2
00	090	-2	2	32	300	-2	2
00	100	-1	2	42	300	-1	2
00	110	-2	2	47	300	-1	1
00	120	-2	1	48	300	-1	1
00	131	-1	2	52	300	-2	2
00	132	-1	1	60	300	-2	3
00	140	-1	2	64	300	-1	1
00	150	-1	1	.	.	-3	3

Table 13.4e. Some statistics on the cross-sectional discrepancies for the directly seasonally adjusted Retail Trade series, after reconciliation

Trade group	Prov.	Min. add. discrep.	Max. add. discrep.	Min. prop. discrep.	Max. prop. discrep.
300	.	-3	3	1.0000	1.0000
.	00	-3	4	1.0000	1.0000

Table 13.4e shows some statistics on the residual cross-sectional aggregation discrepancies after reconciliation. The additive discrepancies are due to rounding to the nearest unit and the fact that the yearly benchmarks do not cross-sectionally add-up. When addressing the constraint redundancy problem by means of the Moore-Penrose generalized matrix inversion, the cross-sectional discrepancies in the benchmarks are distributed evenly over each month of each year with benchmark. When addressing the problem by making the January cross-sectional constraints non-binding instead, the January constraints are not satisfied by larger amounts, the minimum and maximum additive discrepancies being -6 and 3 (not shown) for the industrial constraints and -6 and 14 for the provincial constraints.

Table 13.4f displays some statistics on the proportional corrections $\hat{\theta}_{k,t}/s_{k,t}$ made to the series to achieve reconciliation. Since all series have alterability coefficients equal to 1, all Trade Group Totals display corrections with standard deviation of 0.0020; and similarly for all Provincial Totals, with standard deviation of 0.0015.

Comparing Table 13.3g to Table 13.4f, the Provincial Totals are more modified in the indirect seasonal adjustment than in the direct one. The reason is that the 18 Trade Group Totals were unalterable in the indirect approach. As a result the 12 Provincial Totals had to absorb all the cross-sectional aggregation discrepancies. In the direct approach of Table 13.4f, the aggregation discrepancies are absorbed by all 31 series; as a result, the corrections are more evenly distributed and therefore smaller. This case provides an example of the following rules of reconciliation:

(1) The more numerous the modifiable series in a system of time series, the smaller are the corrections needed to impose the cross-sectional constraints and the better the proportional movement preservation.

(2) The larger the system of series, the smaller are the corrections needed (*ceteris paribus*) and the better movement preservation.

Table 13.4f. Some statistics on the proportional corrections and on the proportional movement preservation for the directly seasonally adjusted Retail Trade series

Pr.	Trade group	Alter. coef.	Min. cor.	Max. cor.	Std. dev. of cor.	Min. mov.	Max. mov.	Std. of ch. mov.
00	010	1	0.9913	1.0080	0.0020	-0.0099	0.0102	0.0029
00	020	1	0.9913	1.0089	0.0021	-0.0096	0.0101	0.0029
00	030	1	0.9912	1.0080	0.0020	-0.0098	0.0102	0.0029
00	040	1	0.9916	1.0091	0.0021	-0.0096	0.0102	0.0029
00	050	1	0.9913	1.0076	0.0020	-0.0098	0.0102	0.0029
00	060	1	0.9913	1.0080	0.0020	-0.0098	0.0101	0.0029
00	070	1	0.9909	1.0079	0.0020	-0.0098	0.0101	0.0029
00	080	1	0.9911	1.0073	0.0020	-0.0099	0.0101	0.0029
00	090	1	0.9902	1.0070	0.0021	-0.0098	0.0102	0.0029
00	100	1	0.9899	1.0080	0.0021	-0.0099	0.0099	0.0029
00	110	1	0.9914	1.0079	0.0020	-0.0099	0.0101	0.0029
00	120	1	0.9912	1.0083	0.0021	-0.0098	0.0103	0.0029
00	131	1	0.9912	1.0082	0.0020	-0.0098	0.0101	0.0029
00	132	1	0.9914	1.0080	0.0020	-0.0098	0.0101	0.0029
00	140	1	0.9909	1.0077	0.0020	-0.0099	0.0102	0.0029
00	150	1	0.9905	1.0086	0.0021	-0.0097	0.0100	0.0029
00	161	1	0.9913	1.0076	0.0020	-0.0099	0.0102	0.0029
00	162	1	0.9912	1.0078	0.0020	-0.0098	0.0102	0.0029
00	300	1	0.9961	1.0052	0.0016	-0.0058	0.0051	0.0022
10	300	1	0.9955	1.0069	0.0015	-0.0053	0.0092	0.0021
11	300	1	0.9957	1.0069	0.0015	-0.0053	0.0092	0.0021
12	300	1	0.9960	1.0069	0.0015	-0.0053	0.0092	0.0021
13	300	1	0.9957	1.0069	0.0015	-0.0053	0.0092	0.0021
22	300	1	0.9953	1.0070	0.0015	-0.0053	0.0091	0.0021
32	300	1	0.9956	1.0069	0.0015	-0.0053	0.0092	0.0021
42	300	1	0.9955	1.0069	0.0015	-0.0053	0.0092	0.0021
47	300	1	0.9958	1.0069	0.0015	-0.0053	0.0092	0.0021
48	300	1	0.9957	1.0069	0.0015	-0.0053	0.0092	0.0021
52	300	1	0.9955	1.0069	0.0015	-0.0053	0.0092	0.0021
60	300	1	0.9960	1.0071	0.0015	-0.0053	0.0092	0.0021
64	300	1	0.9957	1.0071	0.0015	-0.0053	0.0092	0.0021
.	.	.	0.9899	1.0091	0.0018	-0.0099	0.0103	0.0026

The direct approach opens new possibilities regarding the reconciliation of directly adjusted series:

(a) The Grand Total may be made "doubly indirect", by setting its alterability coefficients to a large number (e.g. 1000) and that of the other series to 1. This makes the Grand Total relatively passive. The resulting estimate of the Grand Total tends to be an optimal trade-off between the cross-sectional sums of the seasonally adjusted Trade Group Totals and cross-sectional sums of the seasonally adjusted Provincial Totals. This approximates the approach where there is no Grand Total (see end of section 13.2). As a quality control measure, it is useful to have an un-reconciled seasonally adjusted Grand Total to compare the reconciled series: they should not be drastically different.

(b) The Grand Total may be imposed onto the other series, by setting its alterability coefficient to 0 and that of the other series to 1.

(c) An intermediate solution may be chosen between the two opposite situations described in (a) and (b).

(d) The non-seasonal series in the system may be kept intact (i.e. seasonally adjusted series equal to the raw series) by setting their alterability coefficients to 0.

In the next chapter we discuss the full two-way reconciliation model.

14 Reconciling Two-Way Classifed Systems of Series

14.1 Introduction

As already discussed in Chapter 11, most time series data produced by statistical agencies are typically part of a system of series classified by attributes. The series of such system must satisfy cross-sectional aggregation constraints, i.e. the elementary series must add up to the marginal totals for each period of time. In many cases, some or all the series must also satisfy temporal aggregation constraints, typically in the form of yearly benchmarks.

This chapter focuses on *two-way classified systems* of series. We shall henceforth refer to the corresponding reconciliation model as the *two-way reconciliation model.*

The two-way classified system contains N sets of *industrial cross-sectional aggregation constraints*, over the Provinces in each Industry, and P sets of *provincial cross-sectional aggregation constraints*, over the Industries in each Province. These sets of constraints are respectively:

$$s_{nP,t} = \Sigma_{p=1}^{P-1} s_{np,t}, \quad t=t_{1,nP},\ldots,t_{L,nP}; \; n=1,\ldots,N; \qquad (14.1a)$$

$$s_{Np,t} = \Sigma_{n=1}^{N-1} s_{np,t}, \quad t=t_{1,Np},\ldots,t_{L,Np}; \; p=1,\ldots,P. \qquad (14.1b)$$

where $t_{1,np}$ and $t_{L,np}$ are respectively the first and last available (time periods of) observations for series $s_{np,t}$. The number of observations for a given series, $s_{np,t}$, is thus $T_{np} = t_{L,np} - t_{1,np} + 1$. That number may vary from series to series; the number of observations may in fact be equal to 0 for some elementary series $n<N$ and $p<P$, because some Industries may not exist in some of the Provinces. Note that the number of observations for the Grand Total, T_{NP}, is greater or equal to T_{np} for all n and p, because the sub-annual periods t covered by the Grand Total must cover all the sub-annual periods covered by all the other series in the system. Similarly, the number of observations for the Industrial Totals T_{nP} is greater or equal to T_{np} for any given Industry n, and the number of observations for the Provincial Totals, T_{Np}

is greater or equal to T_{np} for any given province p. Note that the subscript np should be treated as a single index and not as the product of n by p.

14.2 The Reconciliation Model for Two-Way Classified Systems of Time Series

The reconciliation model for a system of series with N industries and P provinces, including the Marginal Totals and the Grand Total, takes the following form:

$$\begin{bmatrix} \boldsymbol{s}_{11} \\ \vdots \\ \boldsymbol{s}_{1P} \\ \vdots \\ \vdots \\ \boldsymbol{s}_{N1} \\ \vdots \\ \boldsymbol{s}_{NP} \end{bmatrix} = \begin{bmatrix} \boldsymbol{I}_{T_{11}} & \cdots & \boldsymbol{0} & \cdots & \cdots & \boldsymbol{0} & \cdots & \boldsymbol{0} \\ \vdots & \ddots & \vdots & \cdots & \cdots & \vdots & \ddots & \vdots \\ \boldsymbol{0} & \cdots & \boldsymbol{I}_{T_{1P}} & \cdots & \cdots & \boldsymbol{0} & \cdots & \boldsymbol{0} \\ \vdots & \vdots & \vdots & \ddots & \vdots & \vdots & \vdots & \vdots \\ \vdots & \vdots & \vdots & \vdots & \ddots & \vdots & \vdots & \vdots \\ \boldsymbol{0} & \cdots & \boldsymbol{0} & \cdots & \cdots & \boldsymbol{I}_{T_{N1}} & \cdots & \boldsymbol{0} \\ \vdots & \vdots & \vdots & \cdots & \cdots & \vdots & \ddots & \vdots \\ \boldsymbol{0} & \cdots & \boldsymbol{0} & \cdots & \cdots & \boldsymbol{0} & \cdots & \boldsymbol{I}_{T_{NP}} \end{bmatrix} \begin{bmatrix} \boldsymbol{\theta}_{11} \\ \vdots \\ \boldsymbol{\theta}_{1P} \\ \vdots \\ \vdots \\ \boldsymbol{\theta}_{N1} \\ \vdots \\ \boldsymbol{\theta}_{NP} \end{bmatrix} + \begin{bmatrix} \boldsymbol{e}_{11} \\ \vdots \\ \boldsymbol{e}_{1P} \\ \vdots \\ \vdots \\ \boldsymbol{e}_{N1} \\ \vdots \\ \boldsymbol{e}_{NP} \end{bmatrix}, \qquad (14.2.a)$$

where $E(\boldsymbol{e}_{np}) = \boldsymbol{0}$, $E(\boldsymbol{e}_{np}\,\boldsymbol{e}_{np}') = \boldsymbol{V}_{e_{np}}$ and $E(\boldsymbol{e}_{ij}\,\boldsymbol{e}_{np}') = \boldsymbol{0}$ for $i \neq n$ or $j \neq p$, and where $T_{np} \leq T_{NP}$;

$$\begin{bmatrix} \boldsymbol{g}_{A,1} \\ \vdots \\ \boldsymbol{g}_{A,N} \\ \boldsymbol{g}_{B,1} \\ \vdots \\ \boldsymbol{g}_{B,P} \end{bmatrix} = \begin{bmatrix} \boldsymbol{G}_{11} & \boldsymbol{G}_{12} & \cdots & -\boldsymbol{G}_{1P} & \cdots & \cdots & \boldsymbol{0} & \boldsymbol{0} & \cdots & \boldsymbol{0} \\ \vdots & \vdots & \cdots & \vdots & \cdots & \cdots & \vdots & \vdots & \cdots & \vdots \\ \boldsymbol{0} & \boldsymbol{0} & \cdots & \boldsymbol{0} & \cdots & \cdots & \boldsymbol{G}_{N1} & \boldsymbol{G}_{N2} & \cdots & -\boldsymbol{G}_{NP} \\ \boldsymbol{G}_{11} & \boldsymbol{0} & \cdots & \boldsymbol{0} & \cdots & \cdots & -\boldsymbol{G}_{N1} & \boldsymbol{0} & \cdots & \boldsymbol{0} \\ \vdots & \vdots & \ddots & \vdots & \cdots & \cdots & \vdots & \vdots & \ddots & \vdots \\ \boldsymbol{0} & \boldsymbol{0} & \cdots & \boldsymbol{G}_{1P} & \cdots & \cdots & \boldsymbol{0} & \boldsymbol{0} & \cdots & -\boldsymbol{G}_{NP} \end{bmatrix} \begin{bmatrix} \boldsymbol{\theta}_{11} \\ \vdots \\ \boldsymbol{\theta}_{1P} \\ \vdots \\ \vdots \\ \boldsymbol{\theta}_{N1} \\ \vdots \\ \boldsymbol{\theta}_{NP} \end{bmatrix} + \begin{bmatrix} \boldsymbol{\varepsilon}_{1\bullet} \\ \vdots \\ \boldsymbol{\varepsilon}_{N\bullet} \\ \boldsymbol{\varepsilon}_{\bullet 1} \\ \vdots \\ \boldsymbol{\varepsilon}_{\bullet P} \end{bmatrix}, \qquad (14.2b)$$

where $E(\boldsymbol{\varepsilon}_{n\bullet}) = \boldsymbol{0}$, $E(\boldsymbol{\varepsilon}_{\bullet p}) = \boldsymbol{0}$, $E(\boldsymbol{\varepsilon}_{n\bullet}\,\boldsymbol{\varepsilon}_{n\bullet}') = \boldsymbol{V}_{\varepsilon_{n\bullet}}$, $E(\boldsymbol{\varepsilon}_{\bullet p}\,\boldsymbol{\varepsilon}_{\bullet p}') = \boldsymbol{V}_{\varepsilon_{\bullet p}}$ and $E(\boldsymbol{\varepsilon}_{n\bullet}\,\boldsymbol{\varepsilon}_{\bullet p}') = \boldsymbol{0}$. Matrices $\boldsymbol{V}_{\varepsilon_{n\bullet}}$ and $\boldsymbol{V}_{\varepsilon_{\bullet p}}$ are diagonal. Typically these matrices would be equal to zero. However, to make specific constraints non-binding, it is possible to set the corresponding diagonal elements to a value greater than 0. This is a way to deal with redundant constraints.

For the moment, all the series in model (14.2) – including the Totals – are endogenous, that is jointly estimated. As a result, vectors $\boldsymbol{g}_{A,n}$ and $\boldsymbol{g}_{B,n}$ contain zeroes. Specifying all series as endogenous leads to a more general solution. It will then be easy to treat some of the series as exogenous, that is pre-determined.

The matrices $\boldsymbol{G}_{np}$ are usually selector matrices, i.e. identities matrices with $max(T_{Np}, T_{nP})$ rows and with some columns possibly removed; the columns removed correspond to periods t where the series does not exist. These matrices thus allow some of the elementary series ($\boldsymbol{s}_{np}$ for $n < N$ or $p < P$) to begin later and/or end earlier than their marginal totals $\boldsymbol{s}_{Np}$ and $\boldsymbol{s}_{nP}$.

The constraints (14.2b) may be rewritten as:

$$\begin{bmatrix} \boldsymbol{g}_A \\ \boldsymbol{g}_B \end{bmatrix} = \begin{bmatrix} \boldsymbol{G}_A \\ \boldsymbol{G}_B \end{bmatrix} \boldsymbol{\theta} + \begin{bmatrix} \boldsymbol{\varepsilon}_A \\ \boldsymbol{\varepsilon}_B \end{bmatrix}, \tag{14.2b'}$$

where $\boldsymbol{g}_A = \boldsymbol{0}$, $\boldsymbol{g}_B = \boldsymbol{0}$, $E(\boldsymbol{\varepsilon}_A) = \boldsymbol{0}$, $E(\boldsymbol{\varepsilon}_A \boldsymbol{\varepsilon}_A') = \boldsymbol{V}_{\varepsilon_A}$,

$E(\boldsymbol{\varepsilon}_B) = \boldsymbol{0}$, $E(\boldsymbol{\varepsilon}_B \boldsymbol{\varepsilon}_B') = \boldsymbol{V}_{\varepsilon_B}$, $E(\boldsymbol{\varepsilon}_A \boldsymbol{\varepsilon}_B') = \boldsymbol{0}$,

and where matrices $\boldsymbol{G}_A$ and $\boldsymbol{G}_B$ are respectively:

$$\boldsymbol{G}_A = \begin{bmatrix} \boldsymbol{G}_{11} & \boldsymbol{G}_{12} & \cdots & -\boldsymbol{G}_{1P} & \boldsymbol{0} & \boldsymbol{0} & \cdots & \boldsymbol{0} & \cdots\ \cdots & \boldsymbol{0} & \boldsymbol{0} & \cdots & \boldsymbol{0} \\ \boldsymbol{0} & \boldsymbol{0} & \cdots & \boldsymbol{0} & \boldsymbol{G}_{21} & \boldsymbol{G}_{22} & \cdots & -\boldsymbol{G}_{2P} & \cdots\ \cdots & \boldsymbol{0} & \boldsymbol{0} & \cdots & \boldsymbol{0} \\ \vdots & \vdots & \cdots & \vdots & \vdots & \vdots & \cdots & \vdots & \cdots\ \cdots & \vdots & \vdots & \cdots & \vdots \\ \boldsymbol{0} & \boldsymbol{0} & \cdots & \boldsymbol{0} & \boldsymbol{0} & \boldsymbol{0} & \cdots & \boldsymbol{0} & \cdots\ \cdots & \boldsymbol{G}_{N1} & \boldsymbol{G}_{N2} & \cdots & -\boldsymbol{G}_{NP} \end{bmatrix},$$

$$\boldsymbol{G}_B = \begin{bmatrix} \boldsymbol{G}_{11} & \boldsymbol{0} & \cdots & \boldsymbol{0} & \boldsymbol{G}_{21} & \boldsymbol{0} & \cdots & \boldsymbol{0} & \cdots\ \cdots & -\boldsymbol{G}_{N1} & \boldsymbol{0} & \cdots & \boldsymbol{0} \\ \boldsymbol{0} & \boldsymbol{G}_{12} & \cdots & \boldsymbol{0} & \boldsymbol{0} & \boldsymbol{G}_{22} & \cdots & \boldsymbol{0} & \cdots\ \cdots & \boldsymbol{0} & -\boldsymbol{G}_{N2} & \cdots & \boldsymbol{0} \\ \vdots & \vdots & \ddots & \vdots & \vdots & \vdots & \ddots & \vdots & \cdots\ \cdots & \vdots & \vdots & \ddots & \vdots \\ \boldsymbol{0} & \boldsymbol{0} & \cdots & \boldsymbol{G}_{1P} & \boldsymbol{0} & \boldsymbol{0} & \cdots & \boldsymbol{G}_{2P} & \cdots\ \cdots & \boldsymbol{0} & \boldsymbol{0} & \cdots & -\boldsymbol{G}_{NP} \end{bmatrix}.$$

Matrix $\boldsymbol{G}_A$ is the industrial cross-sectional sum operator over the provinces, and $\boldsymbol{G}_B$ is the provincial cross-sectional sum operator over the industries.

Matrix G_A and G_B have dimensions $T_{\bullet P} \times T_{\bullet\bullet}$ and $T_{N\bullet} \times T_{\bullet\bullet}$ respectively, where $T_{\bullet P} = \Sigma_{n=1}^{N} T_{nP}$, $T_{N\bullet} = \Sigma_{p=1}^{P} T_{Np}$ and $T_{\bullet\bullet} = \Sigma_{n=1}^{N} \Sigma_{p=1}^{P} T_{np}$.

If all series start at the same date and have the same number of observations, i.e. $T_{np} = T$, matrices G_A and G_B can be compactly expressed in terms of Kronecker products:

$$G_A = I_N \otimes (\imath_A \otimes I_T) \, , \; G_B = \imath_B \otimes (I_P \otimes I_T) \, , \qquad (14.2d)$$

where $\imath_A$ and $\imath_B$ are row vectors of dimensions $1 \times P$ and $1 \times N$ respectively and containing 1s except for the last element equal to -1. This notation is useful for theoretical analysis.

Our purpose is now to find an analytical solution to the two-way reconciliation problem which is implementable. The general solution (11.8) is repeated here for convenience:

$$\begin{aligned} \hat{\theta} &= s + V_e G' (G V_e G' + V_\varepsilon)^{-1} [g - G s] \\ &= s + V_e G' [V_d^{-1} d] \;\; = s + V_e G' \hat{\gamma} \, , \end{aligned} \qquad (14.3a)$$

$$var[\hat{\theta}] = V_e - V_e G' V_d^{-1} G V_e \, , \qquad (14.3b)$$

where g contains zeroes because all series are considered endogenous, i.e. jointly estimated with the elementary series.

Many of the matrices in (14.3) are large and sparse (i.e. contain many 0s). In order to minimize the number of computations, the matrix multiplications involving partitions known to contain only 0s are avoided. Furthermore, we are typically interested in the block diagonal elements of the covariance matrix $var[\hat{\theta}]$, i.e. in the covariance matrices of each series and not in the cross-covariance matrices between the reconciled series. However we do provide the formulae for both the covariance and the cross-covariance matrices.

The strategy used to arrive at an analytic solution consists of developing the various matrix products in terms of main partitions A and B; and then, in terms of the sub-partitions pertaining to each individual series.

14.2.1 Deriving an Analytical Solution in Terms of the Main Partitions

The development in terms of partitions A and B for the *two-way reconciliation model* (14.2) is very similar to that of the marginal two-way reconciliation model discussed in Chapter 13. Substituting the partitions in the various matrix products yields:

$$\boldsymbol{V}_e\boldsymbol{G}' = \left[\boldsymbol{V}_e\boldsymbol{G}_A' \ \ \boldsymbol{V}_e\boldsymbol{G}_B'\right], \tag{14.4a}$$

$$\boldsymbol{V}_d = \begin{bmatrix} \boldsymbol{G}_A\boldsymbol{V}_e\boldsymbol{G}_A' + \boldsymbol{V}_{\varepsilon_A} & \boldsymbol{G}_A\boldsymbol{V}_e\boldsymbol{G}_B' \\ \boldsymbol{G}_B\boldsymbol{V}_e\boldsymbol{G}_A' & \boldsymbol{G}_B\boldsymbol{V}_e\boldsymbol{G}_B' + \boldsymbol{V}_{\varepsilon_B} \end{bmatrix} = \begin{bmatrix} \boldsymbol{V}_{AA} & \boldsymbol{V}_{AB} \\ \boldsymbol{V}_{AB}' & \boldsymbol{V}_{BB} \end{bmatrix}, \tag{14.4b}$$

where $\boldsymbol{V}_d$ has dimension $(T_{\bullet P}+T_{N\bullet})\times(T_{\bullet P}+T_{N\bullet})$ and where partition $\boldsymbol{V}_{AA}$ has dimension $T_{\bullet P}\times T_{\bullet P}$; $\boldsymbol{V}_{AB}$, dimension $T_{\bullet P}\times T_{N\bullet}$; and $\boldsymbol{V}_{BB}$, dimension $T_{N\bullet}\times T_{N\bullet}$.

We partition $\boldsymbol{V}_d^{-1}$ like matrix $\boldsymbol{V}_d$ of (14.4b):

$$\boldsymbol{V}_d^{-1} = \begin{bmatrix} \boldsymbol{V}^{AA} & \boldsymbol{V}^{AB} \\ \boldsymbol{V}^{AB'} & \boldsymbol{V}^{BB} \end{bmatrix}. \tag{14.4c}$$

Continuing the derivation yields:

$$\boldsymbol{d} = (\boldsymbol{g} - \boldsymbol{G}\boldsymbol{s}) \Rightarrow \begin{bmatrix} \boldsymbol{d}_A \\ \boldsymbol{d}_B \end{bmatrix} = \begin{bmatrix} \boldsymbol{g}_A \\ \boldsymbol{g}_B \end{bmatrix} - \begin{bmatrix} \boldsymbol{G}_A \\ \boldsymbol{G}_B \end{bmatrix}\boldsymbol{s} \Rightarrow \begin{array}{l} \boldsymbol{d}_A = \boldsymbol{g}_A - \boldsymbol{G}_A\boldsymbol{s} \\ \boldsymbol{d}_B = \boldsymbol{g}_B - \boldsymbol{G}_B\boldsymbol{s} \end{array}, \tag{14.4d}$$

$$\hat{\boldsymbol{\gamma}} = \boldsymbol{V}_d^{-1}\boldsymbol{d} \quad \Rightarrow \quad \begin{array}{l} \hat{\boldsymbol{\gamma}}_A = \boldsymbol{V}^{AA}\boldsymbol{d}_A + \boldsymbol{V}^{AB}\boldsymbol{d}_B \\ \hat{\boldsymbol{\gamma}}_B = \boldsymbol{V}^{AB'}\boldsymbol{d}_A + \boldsymbol{V}^{BB}\boldsymbol{d}_B \end{array}, \tag{14.4e}$$

$$\hat{\boldsymbol{\theta}} = \boldsymbol{s} + \boldsymbol{V}_e\boldsymbol{G}'\hat{\boldsymbol{\gamma}} \Rightarrow \hat{\boldsymbol{\theta}} = \boldsymbol{s} + \left[\boldsymbol{V}_e\boldsymbol{G}_A' \ \ \boldsymbol{V}_e\boldsymbol{G}_B'\right]\begin{bmatrix} \hat{\boldsymbol{\gamma}}_A \\ \hat{\boldsymbol{\gamma}}_B \end{bmatrix}$$

$$= \boldsymbol{s} + \boldsymbol{V}_e\boldsymbol{G}_A'\hat{\boldsymbol{\gamma}}_A + \boldsymbol{V}_e\boldsymbol{G}_B'\hat{\boldsymbol{\gamma}}_B, \tag{14.4f}$$

where $\hat{\boldsymbol{\gamma}}_A = \boldsymbol{V}^{AA}\boldsymbol{d}_A + \boldsymbol{V}^{AB}\boldsymbol{d}_B$, $\hat{\boldsymbol{\gamma}}_B = \boldsymbol{V}^{AB'}\boldsymbol{d}_A + \boldsymbol{V}^{BB}\boldsymbol{d}_B$ and $\boldsymbol{d}_A = \boldsymbol{g}_A - \boldsymbol{G}_A\boldsymbol{s}$ and $\boldsymbol{d}_B = \boldsymbol{g}_B - \boldsymbol{G}_B\boldsymbol{s}$.

Substituting (14.4a) and (14.4c) into (14.3b) yields

$$var[\hat{\boldsymbol{\theta}}] = \boldsymbol{V}_e - \boldsymbol{V}_e \boldsymbol{G}_A{}' \boldsymbol{V}^{AA} \boldsymbol{G}_A \boldsymbol{V}_e - \boldsymbol{V}_e \boldsymbol{G}_B{}' \boldsymbol{V}^{AB'} \boldsymbol{G}_A \boldsymbol{V}_e - \boldsymbol{V}_e \boldsymbol{G}_A{}' \boldsymbol{V}^{AB} \boldsymbol{G}_B \boldsymbol{V}_e - \boldsymbol{V}_e \boldsymbol{G}_B{}' \boldsymbol{V}^{BB} \boldsymbol{G}_B \boldsymbol{V}_e . \quad (14.4g)$$

This completes the development of solution (14.3) in terms of the main partitions A and B. We now need to develop the formulae in more details, at the level of the sub-partitions pertaining to each series.

14.2.2 Deriving an Analytical Solution for Each Series

We now develop the formulae in more details, at the level of the sub-partitions pertaining to each series. This sub-section may be skipped without loss of continuity.

Substituting the sub-partitions of $\boldsymbol{V}_e$, $\boldsymbol{G}_A$ and $\boldsymbol{G}_B$ in (14.4) and defining $\boldsymbol{L}_{np} = \boldsymbol{G}_{np} \boldsymbol{V}_{e_{np}}$ and $\boldsymbol{Q}_{np} = \boldsymbol{G}_{np} \boldsymbol{V}_{e_{np}} \boldsymbol{G}_{np}{}' \equiv \boldsymbol{L}_{np} \boldsymbol{G}_{np}{}'$ leads to:

$$\boldsymbol{V}_e \boldsymbol{G}_A{}' = \begin{bmatrix} \boldsymbol{E}_1 & 0 & \cdots & 0 \\ 0 & \boldsymbol{E}_2 & \cdots & 0 \\ \vdots & \vdots & \ddots & \vdots \\ 0 & 0 & \cdots & \boldsymbol{E}_N \end{bmatrix}, \quad \boldsymbol{E}_n = \begin{bmatrix} \delta_{1P} \boldsymbol{L}_{n1}{}' \\ \delta_{2P} \boldsymbol{L}_{n2}{}' \\ \vdots \\ \delta_{PP} \boldsymbol{L}_{nP}{}' \end{bmatrix}; \quad (14.5a)$$

$$\boldsymbol{V}_e \boldsymbol{G}_B{}' = \begin{bmatrix} \boldsymbol{E}_1 \\ \boldsymbol{E}_2 \\ \vdots \\ \boldsymbol{E}_N \end{bmatrix}, \quad \boldsymbol{E}_n = \begin{bmatrix} \delta_{nN} \boldsymbol{L}_{n1}{}' & \boldsymbol{0} & \cdots & \boldsymbol{0} \\ \boldsymbol{0} & \delta_{nN} \boldsymbol{L}_{n2}{}' & \cdots & \boldsymbol{0} \\ \vdots & \vdots & \ddots & \vdots \\ \boldsymbol{0} & \boldsymbol{0} & \cdots & \delta_{nN} \boldsymbol{L}_{nP}{}' \end{bmatrix}, \quad (14.5b)$$

where $\delta_{nN} = 1$ for $n<N$ and $\delta_{nN} = -1$ for $n=N$;

$$\boldsymbol{V}_{AA} = \boldsymbol{G}_A (\boldsymbol{V}_e \boldsymbol{G}_A{}') + \boldsymbol{V}_{\varepsilon_A} = \begin{bmatrix} \boldsymbol{E}_{1,1} & \cdots & \boldsymbol{0} \\ \vdots & \ddots & \vdots \\ \boldsymbol{0} & \cdots & \boldsymbol{E}_{N,N} \end{bmatrix}, \quad (14.5c)$$

where $\boldsymbol{E}_{n,n} = \Sigma_{p=1}^{P} \boldsymbol{Q}_{np} + \boldsymbol{V}_{\varepsilon_{n\bullet}}$, $\boldsymbol{E}_{n,j} = \boldsymbol{0}$ for $j \neq n$;

$$V_{AB} = G_A(V_e G_B{}') = \begin{bmatrix} Q_{11} & Q_{12} & \cdots & -Q_{1P} \\ Q_{21} & Q_{22} & \cdots & -Q_{2P} \\ \vdots & \vdots & \ddots & \vdots \\ -Q_{N1} & -Q_{N2} & \cdots & +Q_{NP} \end{bmatrix}, \quad (14.5d)$$

where $Q_{np} = G_{np} V_{e_{np}} G_{np}{}' \equiv L_{np} G_{np}{}'$;

$$V_{BB} = G_B(V_e G_B{}') + V_{\varepsilon_B} = \begin{bmatrix} E_{1,1} & \cdots & 0 \\ \vdots & \ddots & \vdots \\ 0 & \cdots & E_{P,P} \end{bmatrix}, \quad (14.5e)$$

where $E_{p,p} = \Sigma_{n=1}^{N} Q_{np} + V_{\varepsilon_p}$, $E_{p,j} = 0$ for $j \neq p$;

$$d_A = \begin{bmatrix} d_{A,1} \\ \vdots \\ d_{A,N} \end{bmatrix}, \; d_B = \begin{bmatrix} d_{B,1} \\ \vdots \\ d_{B,P} \end{bmatrix}, \quad (14.5f)$$

where $\qquad d_{A,n} = g_A - \Sigma_{p=1}^{P} \delta_{pP} G_{np} s_{np}$, $d_{B,p} = g_B - \Sigma_{n=1}^{N} \delta_{nN} G_{np} s_{np}$.

We now sub-partition the partitions of V_d^{-1} of Eq. (14.4.c), V^{AA}, V^{AB} and V^{BB}, like V_{AA}, V_{AB} and V_{BB} in (14.5e):

$$V^{AA} = \begin{bmatrix} V^{AA}_{1,1} & \cdots & V^{AA}_{1,N} \\ \vdots & \ddots & \vdots \\ V^{AA}_{N,1} & \cdots & V^{AA}_{N,N} \end{bmatrix}, \; V^{AB} = \begin{bmatrix} V^{AB}_{1,1} & \cdots & V^{AB}_{1,P} \\ \vdots & \ddots & \vdots \\ V^{AB}_{N,1} & \cdots & V^{AB}_{N,P} \end{bmatrix}, \; V^{BB} = \begin{bmatrix} V^{BB}_{1,1} & \cdots & V^{BB}_{1,P} \\ \vdots & \ddots & \vdots \\ V^{BB}_{P,1} & \cdots & V^{BB}_{P,P} \end{bmatrix}, \quad (14.5g)$$

where $V^{AA}_{n,i}$, $V^{AB}_{n,p}$ and $V^{BB}_{p,i}$ have dimensions $T_{nP} \times T_{nP}$, $T_{nP} \times T_{Np}$ and $T_{Np} \times T_{Np}$ respectively. Note that the subscripts of the partitions refer to N sets of industrial aggregation constraints and the P sets of provincial aggregation constraints.

We similarly partition $\hat{\boldsymbol{\gamma}}_A$ and $\hat{\boldsymbol{\gamma}}_B$ from (14.4e)

$$\hat{\boldsymbol{\gamma}}_A = \begin{bmatrix} \hat{\boldsymbol{\gamma}}_{A,1} \\ \hat{\boldsymbol{\gamma}}_{A,2} \\ \vdots \\ \hat{\boldsymbol{\gamma}}_{A,N} \end{bmatrix}, \quad \hat{\boldsymbol{\gamma}}_B = \begin{bmatrix} \hat{\boldsymbol{\gamma}}_{B,1} \\ \hat{\boldsymbol{\gamma}}_{B,2} \\ \vdots \\ \hat{\boldsymbol{\gamma}}_{B,P} \end{bmatrix}, \tag{14.5h}$$

where $\hat{\boldsymbol{\gamma}}_{A,n}$ and $\hat{\boldsymbol{\gamma}}_{B,p}$ have dimensions $T_{nP} \times 1$ and $T_{Np} \times 1$ respectively.

Continuing the above approach yields:

$$\boldsymbol{V}_e \boldsymbol{G}_A{}' \boldsymbol{V}^{AA} = \begin{bmatrix} \boldsymbol{E}_1 \\ \vdots \\ \boldsymbol{E}_N \end{bmatrix}, \quad \boldsymbol{E}_n = \begin{bmatrix} \delta_{1P} \boldsymbol{L}_{n1}{}' \boldsymbol{V}^{AA}_{1,1} & \cdots & \delta_{1P} \boldsymbol{L}_{n1}{}' \boldsymbol{V}^{AA}_{1,N} \\ \delta_{2P} \boldsymbol{L}_{n1}{}' \boldsymbol{V}^{AA}_{1,1} & \cdots & \delta_{2P} \boldsymbol{L}_{n1}{}' \boldsymbol{V}^{AA}_{1,N} \\ \vdots & \ddots & \vdots \\ \delta_{PP} \boldsymbol{L}_{nP}{}' \boldsymbol{V}^{AA}_{1,1} & \cdots & \delta_{PP} \boldsymbol{L}_{nP}{}' \boldsymbol{V}^{AA}_{1,N} \end{bmatrix}; \tag{14.5i}$$

$$\boldsymbol{V}_e \boldsymbol{G}_B{}' \boldsymbol{V}^{BB} = \begin{bmatrix} \boldsymbol{E}_1 \\ \vdots \\ \boldsymbol{E}_N \end{bmatrix}, \boldsymbol{E}_n = \begin{bmatrix} \delta_{nN} \boldsymbol{L}_{n1}{}' \boldsymbol{V}^{BB}_{1,1} & \cdots & \delta_{nN} \boldsymbol{L}_{n1}{}' \boldsymbol{V}^{BB}_{1,P} \\ \delta_{nN} \boldsymbol{L}_{n2}{}' \boldsymbol{V}^{BB}_{2,1} & \cdots & \delta_{nN} \boldsymbol{L}_{n2}{}' \boldsymbol{V}^{BB}_{2,P} \\ \vdots & \ddots & \vdots \\ \delta_{nN} \boldsymbol{L}_{nP}{}' \boldsymbol{V}^{BB}_{P,1} & \cdots & \delta_{nN} \boldsymbol{L}_{nP}{}' \boldsymbol{V}^{BB}_{P,P} \end{bmatrix}, \tag{14.5j}$$

where $\delta_{nN}=1$ for $n<N$ and $\delta_{nN}=-1$ for $n=N$;

$$\boldsymbol{V}_e \boldsymbol{G}_A{}' \boldsymbol{V}^{AB} = \begin{bmatrix} \boldsymbol{E}_1 \\ \vdots \\ \boldsymbol{E}_N \end{bmatrix}, \boldsymbol{E}_n = \begin{bmatrix} \delta_{1P} \boldsymbol{L}_{n1}{}' \boldsymbol{V}^{AB}_{n,1} & \cdots & \delta_{1P} \boldsymbol{L}_{n1}{}' \boldsymbol{V}^{AB}_{n,P} \\ \delta_{2P} \boldsymbol{L}_{n2}{}' \boldsymbol{V}^{AB}_{n,1} & \cdots & \delta_{2P} \boldsymbol{L}_{n2}{}' \boldsymbol{V}^{AB}_{n,P} \\ \vdots & \ddots & \vdots \\ \delta_{PP} \boldsymbol{L}_{nP}{}' \boldsymbol{V}^{AB}_{n,1} & \cdots & \delta_{PP} \boldsymbol{L}_{nP}{}' \boldsymbol{V}^{AB}_{n,P} \end{bmatrix}; \tag{14.5k}$$

$$\boldsymbol{V}_e \boldsymbol{G}_A{}' \boldsymbol{V}^{AA} \boldsymbol{G}_A \boldsymbol{V}_e = \begin{bmatrix} \boldsymbol{E}_{1,1} & \cdots & \boldsymbol{E}_{1,N} \\ \vdots & \ddots & \vdots \\ \boldsymbol{E}_{N,1} & \cdots & \boldsymbol{E}_{N,N} \end{bmatrix}, \tag{14.5l}$$

where $$E_{i,j} = \begin{bmatrix} \delta_{1P}\,\delta_{1P}\,L_{i1}'V_{i,j}^{AA}L_{j1} & \cdots & \delta_{1P}\,\delta_{PP}\,L_{i1}'V_{i,j}^{AA}L_{jP} \\ \delta_{2P}\,\delta_{1P}\,L_{i2}'V_{i,j}^{AA}L_{j1} & \cdots & \delta_{2P}\,\delta_{PP}\,L_{i2}'V_{i,j}^{AA}L_{jP} \\ \vdots & \ddots & \vdots \\ \delta_{PP}\,\delta_{1P}\,L_{iP}'V_{i,j}^{AA}L_{j1} & \cdots & \delta_{PP}\,\delta_{PP}\,L_{iP}'V_{i,j}^{AA}L_{jP} \end{bmatrix};$$

$$V_e G_B' V^{BB} G_B V_e = \begin{bmatrix} E_{1,1} & \cdots & E_{1,N} \\ \vdots & \ddots & \vdots \\ E_{N,1} & \cdots & E_{N,N} \end{bmatrix}, \tag{14.5m}$$

where $$E_{i,j} = \begin{bmatrix} \delta_{iN}\delta_{jN}\,L_{i1}'V_{1,1}^{BB}L_{j1} & \cdots & \delta_{iN}\delta_{jN}\,L_{i1}'V_{1,P}^{BB}L_{jP} \\ \delta_{iN}\delta_{jN}\,L_{i2}'V_{2,1}^{BB}L_{j1} & \cdots & \delta_{iN}\delta_{jN}\,L_{i2}'V_{2,P}^{BB}L_{jP} \\ \vdots & \ddots & \vdots \\ \delta_{iN}\delta_{jN}\,L_{iP}'V_{P,1}^{BB}L_{j1} & \cdots & \delta_{iN}\delta_{jN}\,L_{iP}'V_{P,P}^{BB}L_{jP} \end{bmatrix},$$

and $\delta_{iN}=1$ for $i<N$ and $\delta_{iN}=-1$ for $i=N$; and similarly for δ_{jN};

$$V_e G_A' V^{AB} G_B V_e = \begin{bmatrix} E_{1,1} & \cdots & E_{1,N} \\ \vdots & \ddots & \vdots \\ E_{N,1} & \cdots & E_{N,N} \end{bmatrix}, \tag{14.5n}$$

where

$$E_{i,j} = \begin{bmatrix} \delta_{jN}\,\delta_{1P}\,L_{i1}'V_{i,1}^{AB}L_{j1} & \cdots & \delta_{jN}\,\delta_{1P}\,L_{i1}'V_{i,P}^{AB}L_{jP} \\ \delta_{jN}\,\delta_{2P}\,L_{i2}'V_{i,1}^{AB}L_{j1} & \cdots & \delta_{jN}\,\delta_{2P}\,L_{i2}'V_{i,P}^{AB}L_{jP} \\ \vdots & \ddots & \vdots \\ \delta_{jN}\,\delta_{PP}\,L_{iP}'V_{i,1}^{AB}L_{j1} & \cdots & \delta_{jN}\,\delta_{PP}\,L_{iP}'V_{i,P}^{AB}L_{jP} \end{bmatrix},$$

where $\delta_{jN}=1$ for $j<N$ and $\delta_{jN}=-1$ for $j=N$.

14.2.3 Analytical Solution of the Two-Way Reconciliation Model

Substituting (14.5a) and (14.5b) into (14.4f) and remembering that $\boldsymbol{L}_{np} = \boldsymbol{G}_{np} \boldsymbol{V}_{e_{np}}$ yields

$$\hat{\boldsymbol{\theta}}_{np} = \boldsymbol{s}_{np} + \boldsymbol{V}_{e_{np}} \boldsymbol{G}_{np}{}' (\delta_{pP} \hat{\boldsymbol{\gamma}}_{A,n} + \delta_{nN} \hat{\boldsymbol{\gamma}}_{B,p}),\ n=1,...,N;\ p=1,...,P, \quad (14.6a)$$

where $\delta_{nN}=1$ for $n<N$ and $\delta_{nN}=-1$ for $n=N$, and similarly for δ_{pP}. Vectors $\hat{\boldsymbol{\gamma}}_{A,n}$ and $\hat{\boldsymbol{\gamma}}_{B,p}$ originate from (14.5h) and (14.4e). Substituting (14.5l), (14.5m) and (14.5n) into (14.4g) and re-arranging, yields the covariance matrices of the series:

$$var[\hat{\boldsymbol{\theta}}_{np}] = \boldsymbol{V}_{e_{np}} - \boldsymbol{V}_{e_{np}} \boldsymbol{G}_{np}{}' \\ [\delta_{pP}^2 \boldsymbol{V}_{n,n}^{AA} + \delta_{nN} \delta_{pP} (\boldsymbol{V}_{n,p}^{AB\,\prime} + \boldsymbol{V}_{p,n}^{AB}) + \boldsymbol{V}_{p,p}^{BB} \delta_{nN}^2] \\ \boldsymbol{G}_{np} \boldsymbol{V}_{e_{np}},\ n=1,...,N;\ p=1,...,P. \quad (14.6b)$$

The cross-covariance matrices between the reconciled series are given by

$$cov[\hat{\boldsymbol{\theta}}_{np} \hat{\boldsymbol{\theta}}_{n'p'}] = -\boldsymbol{V}_{e_{np}} \boldsymbol{G}_{np}{}' \\ [\delta_{pP} \delta_{p'P} \boldsymbol{V}_{n,n'}^{AA} + \delta_{pP} \delta_{n'N} (\boldsymbol{V}_{n,p'}^{AB\,\prime} + \boldsymbol{V}_{p',n}^{AB}) + \delta_{nN} \delta_{n'N} \boldsymbol{V}_{p,p'}^{BB}] \\ \boldsymbol{G}_{n'p'} \boldsymbol{V}_{e_{n'p'}},\ n \neq n' \text{ or } p \neq p'. \quad (14.6c)$$

If $n=n'$ and $p \neq p'$, the cross-covariance matrix is between two provinces within an industry; if $n \neq n'$ and $p=p'$, between two industries within a province; if $n \neq n'$ and $p \neq p'$, between two different industries in different provinces. In the vast majority of cases, there would be no need to calculate the cross-covariance matrices.

Any series $\boldsymbol{s}_k$ - including the Totals - can be specified as exogenous, i.e. pre-determined, by setting its alterability coefficient α_k equal to 0, which implies that $\boldsymbol{V}_{e_k} = \boldsymbol{0}$. Of course, if too many series are un-alterable, there may be no solution to the reconciliation problem. This would be the case if all series in an Industry (a row of Table 11.1) or all series in a Province (column) are assigned alterability coefficients equal to 0.

Note that if the Industrial Totals and Provincial Totals cross-sectionally add up to the Grand Total, the Totals can be made exogenous by setting their coefficients of alterability to 0 except for the Grand Total. The Grand Total will be satisfied since the Industrial and Provincial Totals remain unaltered.

14.3 Particular Cases of the Two-Way Reconciliation Model

The solution of the two-way reconciliation model (14.6) holds other simpler models as particular cases, namely the one-way model, the marginal two-way model and the two-way model without the Grand Total.

14.3.1 The One-Way Model as a Particular Case

In the one-way reconciliation model discussed in Chapter 12, there is only one set of constraints. Solution (14.6) is valid if the terms pertaining to the inapplicable constraints are ignored.

For a model classified by Province, solution (14.6) is applicable if only one Industry $n=\bar{n}$ is considered and $\delta_{nN}=0$ for all n, which leads to

$$\hat{\boldsymbol{\theta}}_{\bar{n}p} = \boldsymbol{s}_{\bar{n}p} + \delta_{pP}\, \boldsymbol{V}_{e_{\bar{n}p}} \boldsymbol{G}_{\bar{n}p}{}' \hat{\boldsymbol{\gamma}}_A,\; p=1,\ldots,P,$$

$$var[\hat{\boldsymbol{\theta}}_{\bar{n}p}] = \boldsymbol{V}_{e_{\bar{n}p}} - \delta_{pP}^2\, \boldsymbol{V}_{e_{\bar{n}p}} \boldsymbol{G}_{\bar{n}p}{}'\, \boldsymbol{V}^{AA}\, \boldsymbol{G}_{\bar{n}p} \boldsymbol{V}_{e_{\bar{n}p}},\; p=1,\ldots,P,$$

where $\boldsymbol{V}^{AA}$ and $\hat{\boldsymbol{\gamma}}_A$ respectively coincide with $\boldsymbol{V}_d^{-1}$ and $\hat{\boldsymbol{\gamma}}$ of Eq. (12.7) of Chapter 12.

Similarly, for a model classified by Industry, solution (14.6) applies, if only one Province $p=\bar{p}$ is considered and $\delta_{pP}=0$ for all p.

$$\hat{\boldsymbol{\theta}}_{n\bar{p}} = \boldsymbol{s}_{n\bar{p}} + \delta_{nN}\, \boldsymbol{V}_{e_{n\bar{p}}} \boldsymbol{G}_{n\bar{p}}{}' \hat{\boldsymbol{\gamma}}_{B,\bar{p}},\; n=1,\ldots,N,$$

$$var[\hat{\boldsymbol{\theta}}_{n\bar{p}}] = \boldsymbol{V}_{e_{n\bar{p}}} - \delta_{nN}^2\, \boldsymbol{V}_{e_{n\bar{p}}} \boldsymbol{G}_{n\bar{p}}{}'\, \boldsymbol{V}^{BB}\, \boldsymbol{G}_{n\bar{p}} \boldsymbol{V}_{e_{n\bar{p}}},\; n=1,\ldots,N,$$

where $\boldsymbol{V}^{BB}$ and $\hat{\boldsymbol{\gamma}}_B$ respectively coincide with $\boldsymbol{V}_d^{-1}$ and $\hat{\boldsymbol{\gamma}}$ of Eq. (12.7).

14.3.2 The Marginal Two-Way Model as a Particular Case

In the marginal two-way model discussed in Chapter 13, only the N-th set industrial cross-sectional aggregation constraints and the P-th set of provincial constraints are present, namely

$$\Sigma_{p=1}^{P} \delta_{pP} \boldsymbol{G}_{Np} \boldsymbol{\theta}_{Np} = \boldsymbol{0}, \tag{14.7a}$$

$$\Sigma_{n=1}^{N} \delta_{nN} \boldsymbol{G}_{Np} \boldsymbol{\theta}_{Np} = \boldsymbol{0}, \tag{14.7b}$$

As a result, the Industrial Totals, $\boldsymbol{\theta}_{nP}$, n=1,...,N−1, are subject to the provincial constraints (14.7b) only; the Provincial Totals $\boldsymbol{\theta}_{Np}$, p=1,...,P−1, to the industrial constraints (14.7a) only; and the Grand Total $\boldsymbol{\theta}_{NP}$ to both sets of constraints.

Solution (14.6) is valid if the inapplicable constraints are ignored. This can be achieved by setting δ_{pP} to 0 for $n < N$ and δ_{nN} equal to 0 for $p < P$, which leads to solution

$$\hat{\boldsymbol{\theta}}_{nP} = \boldsymbol{s}_{nP} + \boldsymbol{V}_{e_{nP}} \boldsymbol{G}_{nP}{}' (\delta_{nN} \hat{\boldsymbol{\gamma}}_B), \; n=1,...,N-1,$$

$$\hat{\boldsymbol{\theta}}_{Np} = \boldsymbol{s}_{Np} + \boldsymbol{V}_{e_{Np}} \boldsymbol{G}_{Np}{}' (\delta_{pP} \hat{\boldsymbol{\gamma}}_A), \; p=1,...,P-1,$$

$$\hat{\boldsymbol{\theta}}_{NP} = \boldsymbol{s}_{NP} + \boldsymbol{V}_{e_{NP}} \boldsymbol{G}_{NP}{}' (\delta_{PP} \hat{\boldsymbol{\gamma}}_A + \delta_{NN} \hat{\boldsymbol{\gamma}}_B), \tag{14.8a}$$

$$var[\hat{\theta}_{nP}] = \boldsymbol{V}_{e_{nP}} - \boldsymbol{V}_{e_{nP}} \boldsymbol{G}_{nP}{}' \; [\boldsymbol{V}^{BB} \delta_{nN}^2] \, \boldsymbol{G}_{nP} \boldsymbol{V}_{e_{nP}}, \; n=1,...,N-1,$$

$$var[\hat{\theta}_{Np}] = \boldsymbol{V}_{e_{Np}} - \boldsymbol{V}_{e_{Np}} \boldsymbol{G}_{Np}{}' \; [\delta_{pP}^2 \boldsymbol{V}^{AA}] \, \boldsymbol{G}_{Np} \boldsymbol{V}_{e_{Np}}, \; p=1,...,P-1,$$

$$var[\hat{\theta}_{NP}] = \boldsymbol{V}_{e_{NP}} - \boldsymbol{V}_{e_{NP}} \boldsymbol{G}_{NP}{}'$$
$$[\delta_{PP}^2 \boldsymbol{V}^{AA} + \delta_{NN} \delta_{PP} (\boldsymbol{V}^{AB\prime} + \boldsymbol{V}^{AB}) + \boldsymbol{V}^{BB} \delta_{NN}^2] \, \boldsymbol{G}_{np} \boldsymbol{V}_{e_{np}}, \tag{14.8b}$$

where $\boldsymbol{V}^{AA}$, $\boldsymbol{V}^{BB}$ and $\boldsymbol{V}^{AB}$, $\hat{\boldsymbol{\gamma}}_A$ and $\hat{\boldsymbol{\gamma}}_B$ now coincide with the corresponding matrices and vectors in Eq. (13.8) of Chapter 13.

14.3.3 The Two-Way Model Without the Grand Total

In some applications, the Grand Total is not part of the reconciliation problem, because the Industrial Totals already cross-sectionally add up to the Grand Total and similarly for the Provincial Totals. Furthermore in some models, the row totals do not have to sum to the column totals. In other words, there is no Grand Total.

In the absence of the Grand Total $\boldsymbol{\theta}_{NP}$, the N-th set and the P-th set of cross-sectional aggregation constraints (14.7a) and (14.7b) are excluded from the two-way reconciliation model. As a result, the Industrial Totals are no longer subject to (the P-th) set of provincial cross-sectional aggregation constraints; and the Provincial Totals, no longer subject to (the N-th) set of industrial constraints. In other words, in the absence of the Grand Total, solution (14.6) is valid, if the inapplicable constraints are ignored. This is achieved by setting δ_{pP} equal to 0 for $n = N$ and δ_{nN} equal to 0 for $p = P$. The solution for the elementary series is still given by Eq. (14.6), and the solution of Industrial Totals and the Provincial Totals simplifies to

$$\hat{\boldsymbol{\theta}}_{nP} = \boldsymbol{s}_{nP} + \delta_{PP} \boldsymbol{V}_{e_{nP}} \boldsymbol{G}_{nP}{}' \hat{\boldsymbol{\gamma}}_{A,n}, \; n=1,...,N-1,$$
$$\hat{\boldsymbol{\theta}}_{Np} = \boldsymbol{s}_{Np} + \delta_{NN} \boldsymbol{V}_{e_{Np}} \boldsymbol{G}_{Np}{}' \hat{\boldsymbol{\gamma}}_{B,p}, \; p=1,...,P-1, \qquad (14.9a)$$

$$var[\hat{\theta}_{nP}] = \boldsymbol{V}_{e_{nP}} - \delta_{PP}^2 \boldsymbol{V}_{e_{nP}} \boldsymbol{G}_{nP}{}' \boldsymbol{V}_{n,n}^{AA} \boldsymbol{G}_{nP} \boldsymbol{V}_{e_{nP}}, \; n=1,...,N-1,$$
$$var[\hat{\theta}_{Np}] = \boldsymbol{V}_{e_{Np}} - \delta_{NN}^2 \boldsymbol{V}_{e_{Np}} \boldsymbol{G}_{Np}{}' \boldsymbol{V}_{p,p}^{BB} \boldsymbol{G}_{Np} \boldsymbol{V}_{e_{Np}}, \; p=1,...,P-1, \qquad (14.9b)$$

where $\delta_{PP} = -1$ and $\delta_{NN} = -1$.

In order to maintain the Industrial and Provincial Totals intact and - if applicable - to preserve their cross-sectional additivity to the Grand Total, they can be specified as exogenous by setting their alterability coefficients 0.

The Industrial Totals and the Provincial Totals can also be made exogenous by storing them in vectors $\boldsymbol{g}_A$ and $\boldsymbol{g}_B$. Solution (14.6) is valid if n and p respectively vary from 1 to $N-1$ and 1 to $P-1$.

Note that there are advantages to keep the Grand Total (if applicable) in the reconciliation process, as this provides statistics on the cross-sectional aggregation discrepancies between the Grand Total and the Industrial Totals and the Provincial Totals.

The two-way reconciliation model can be used to reconcile pseudo two-way systems of series, such as in Tables 11.4 and 11.5 of Chapter 11. These tables contained no column totals at all or no row totals.

An important special case of the two-way model without Grand Total is that of Input-Output models.

14.4 Input-Output Models

Input-Output models were conceived by Leontief (1951). The model accounts for all the input "commodities" (goods and services) purchased by each industry in the economy and all the output commodities sold to other industries (output). The models were heavily used by socialist countries to plan economic activity. They are still widely used to integrate various economic statistics in an accounting framework.

In developing countries, the Input-Output model is often the point of departure to build a system of national statistics. The model is used as a skeleton of the system on which to add statistics as they are developed.

Contrary to the National Accounts which deals with final goods and services, Input-Output models account for both intermediate and final goods and services.

Mathematically, the closed Input-Output model takes the following form

$$\begin{bmatrix} a_{11} & \cdots & a_{1P} \\ \vdots & \ddots & \vdots \\ a_{N1} & \cdots & a_{NP} \end{bmatrix} \begin{bmatrix} x_1 \\ \vdots \\ x_P \end{bmatrix} = \begin{bmatrix} y_1 \\ \vdots \\ y_N \end{bmatrix} \quad \text{or } \tilde{\boldsymbol{A}} \boldsymbol{x} = \boldsymbol{y} \, , \tag{14.10a}$$

where vector $\boldsymbol{x}$ stands for the inputs and vector $\boldsymbol{y}$ for the outputs. Matrix $\tilde{\boldsymbol{A}}$ contains *input coefficients* a_{np} which take values between 0 and 1 inclusively, where the tilde is used to avoid any confusion with the main partitions of some matrices and vectors, namely $\boldsymbol{G}_A$ and $\boldsymbol{\gamma}_A$. The coefficients a_{np} reflect the proportion of *input commodity* x_p needed to produce one unit of *output commodity* y_n. For example, in order to produce one unit of commodity 5 (say), a_{51} units of commodity 1 are needed, a_{52} units of commodity 2, a_{53} units of commodity 3, and so forth. The column sums of matrix $\tilde{\boldsymbol{A}}$ must be equal to 1: $\Sigma_{n=1}^{N} a_{np} = 1, \, p=1,...,P$:

$$\tilde{\boldsymbol{A}}' \boldsymbol{1}_{N \times 1} = \boldsymbol{1}_{P \times 1} \, . \tag{14.10b}$$

Input-Output models usually pertain to one period of time, typically a year. Given vectors $\boldsymbol{x}$ and $\boldsymbol{y}$, the content of matrix $\tilde{\boldsymbol{A}}$ has to be reconciled so that the identities (14.10a) and (14.10b) are fulfilled.

When vector $\boldsymbol{x}$ is equal to $\boldsymbol{y}$, the model is called a Social Account Model or SAM (Golan et al. 1994), which implies that matrix $\tilde{\boldsymbol{A}}$ is square instead of rectangular. SAM models expand Input-Output Accounts to include the final output, i.e. the demand of final goods and services; and the primary inputs such as labour.

The two-way reconciliation model without Grand Total discussed in section 14.3.3 can be applied to Input-Output matrices. Let vector $\boldsymbol{s}$ of dimension $NP\times 1$ contain preliminary input coefficients a^0_{np} not satisfying constraints (14.10a) and (14.10b). Let vector $\boldsymbol{\theta}$ contain the true coefficients a_{np} satisfying the same constraints. An appropriate variant of the model (14.2) consists of Eq. (14.11a) and (14.11b):

$$\boldsymbol{s} = \boldsymbol{I}\,\boldsymbol{\theta} + \boldsymbol{e}, \quad E(\boldsymbol{e}) = \boldsymbol{0}, \quad E(\boldsymbol{e}\,\boldsymbol{e}') = \boldsymbol{V}_e, \tag{14.11a}$$

where the covariance of the original input coefficients are set equal to $V_{e_{np}} = \alpha_{np}\,\Xi^{\lambda}_{np}\,\Omega_{np}\,\Xi^{\lambda}_{np}$, where $\boldsymbol{\Xi}_{np} = s_{np} = a^0_{np}$, $\lambda = 1/2$, $\Omega_{np} = 1$ with alterability coefficients α_{np} reflecting the relative reliability of the initial estimate a^0_{np}.

Appropriate constraints alalogous to (14.2b),

$$\begin{bmatrix} \boldsymbol{g}_A \\ \boldsymbol{g}_B \end{bmatrix} = \begin{bmatrix} \boldsymbol{G}_A \\ \boldsymbol{G}_B \end{bmatrix} \boldsymbol{\theta} + \begin{bmatrix} \boldsymbol{\varepsilon}_A \\ \boldsymbol{\varepsilon}_B \end{bmatrix}, \tag{14.11b}$$

are a reformulation of constraints (14.10a) and (14.10b), with $\boldsymbol{g}_A = \boldsymbol{y}$ and $\boldsymbol{g}_B = \boldsymbol{1}_{P\times 1}$, $E(\boldsymbol{\varepsilon}_A) = \boldsymbol{0}$, $E(\boldsymbol{\varepsilon}_A\boldsymbol{\varepsilon}'_A) = \boldsymbol{V}_{\varepsilon_A}$, $E(\boldsymbol{\varepsilon}_B) = \boldsymbol{0}$, $E(\boldsymbol{\varepsilon}_B\boldsymbol{\varepsilon}'_B) = \boldsymbol{V}_{\varepsilon_B}$, $E(\boldsymbol{\varepsilon}_A\boldsymbol{\varepsilon}'_B) = \boldsymbol{0}$. Matrices $\boldsymbol{G}_A$ and $\boldsymbol{G}_B$ have the following content:

$$\boldsymbol{G}_A = \boldsymbol{I}_N \otimes \boldsymbol{x}' = \begin{bmatrix} x_1 & x_2 & \dots & x_P & 0 & 0 & \dots & 0 & \dots & \dots \\ 0 & 0 & \dots & 0 & x_1 & x_2 & \dots & x_P & \dots & \dots \\ \vdots & \vdots & \vdots & \vdots & \vdots & \vdots & \vdots & \vdots & \vdots & \vdots \end{bmatrix}, \tag{14.12a}$$

$$\boldsymbol{G}_B = \boldsymbol{1}_{1\times N} \otimes \boldsymbol{I}_P = \begin{bmatrix} 1 & 0 & \dots & 0 & 1 & 0 & \dots & 0 & \dots & \dots \\ 0 & 1 & \dots & 0 & 0 & 1 & \dots & 0 & \dots & \dots \\ \vdots & \vdots & \ddots & \vdots & \vdots & \vdots & \ddots & \vdots & \vdots & \vdots \\ 0 & 0 & \dots & 1 & 0 & 0 & \dots & 1 & \dots & \dots \end{bmatrix}. \tag{14.12b}$$

Note that there is no "Grand Total" and that the Output "Totals" $\boldsymbol{y}$ and the Input "Totals" $\boldsymbol{x}$ are now formally exogenous, i.e. on the left side of Eqs. (14.11b) in vectors $\boldsymbol{g}_A$ and $\boldsymbol{g}_B$ respectively.

If the Input-Output matrix is very large, the rows of vector $\boldsymbol{s}$ containing zeroes and the corresponding rows of $\boldsymbol{\theta}$ should be removed as well as the corresponding columns of matrices $\boldsymbol{G}_A$ and $\boldsymbol{G}_B$. The reason is that these entries do not contribute to the aggregation discrepancies; furthermore the solution for these entries is known to be zero with certainty. The removed zeroes are then re-inserted in the reconciled estimates afterwards.

The solution of model (14.11) is given by (14.4f) and (14.4g). That solution replicates the cross-entropy estimates $\hat{\theta}_{np}$ provided by Golan et al. (1994, p. 546) in their experimental example.

The analytical solution (14.6) is not applicable, because matrix $\boldsymbol{G}_A$ does not contain partitions $\boldsymbol{G}_{np}$ like in (14.2c), where the partitions appear in both matrices and the same column (of partitions). In other words, our algebraic results depending on the commonality of partitions $\boldsymbol{G}_{np}$ in $\boldsymbol{G}_A$ and $\boldsymbol{G}_B$.

However, the analytical solution (14.6) of the two-way reconciliation model becomes applicable if the problem is reformulated in a slightly different manner. Let vector $\boldsymbol{s}$ of dimension $NP\times 1$ contain $\overset{0}{a}_{np} x_p$ instead of $\overset{0}{a}_{np}$, and let vector $\boldsymbol{\theta}$ contain $a_{np} x_p$ instead of a_{np}. The variance of the original input coefficients are set equal to $V_{e_{np}} = \alpha_{np}\, \Xi_{np}^{\lambda}\, \Omega_{np}\, \Xi_{np}^{\lambda}$, where $\Xi_{np} = s_{np} = \overset{0}{a}_{np} x_p$, $\lambda = 1/2$, $\Omega_{np} = 1$, with appropriate alterability coefficients α_{np} reflecting the relative reliability of $\overset{0}{a}_{np}$. As a result, $\boldsymbol{V}_e$ is a diagonal matrix containing the values of $\alpha_{np}\, s_{np} = \alpha_{np}\, \overset{0}{a}_{np} x_p$.

Furthermore, set $\boldsymbol{g}_A = \boldsymbol{y}$ and $\boldsymbol{g}_B = \boldsymbol{x}$ and

$$\boldsymbol{G}_A = \boldsymbol{I}_N \otimes \boldsymbol{1}_{1\times P} = \begin{bmatrix} 1 & 1 & \dots & 1 & 0 & 0 & \dots & 0 & \dots & \dots \\ 0 & 0 & \dots & 0 & 1 & 1 & \dots & 1 & \dots & \dots \\ \vdots & \vdots & \vdots & \vdots & \vdots & \vdots & \vdots & \vdots & \vdots & \vdots \end{bmatrix}, \quad (14.13a)$$

$$G_B = 1_{1\times N} \otimes I_P = \begin{bmatrix} 1 & 0 & \ldots & 0 & 1 & 0 & \ldots & 0 & \ldots & \ldots \\ 0 & 1 & \ldots & 0 & 0 & 1 & \ldots & 0 & \ldots & \ldots \\ \vdots & \vdots & \ddots & \vdots & \vdots & \vdots & \ddots & \vdots & \vdots & \vdots \\ 0 & 0 & \ldots & 1 & 0 & 0 & \ldots & 1 & \ldots & \ldots \end{bmatrix}. \quad (14.13b)$$

like in Eq. (14.2d). Note that $g_B = G_B\, \theta$ is equivalent to $x_p = \Sigma_{n=1}^{N} a_{np}\, x_p$, which on dividing both sides by x_p implies (14.10b), i.e. $1 = \Sigma_{n=1}^{N} a_{np}$ for all p.

Again, if the matrix $\tilde{A}$ is very large, the rows of vector s containing zeroes and the corresponding rows of θ should removed as well as the corresponding columns of matrices G_A and G_B in Eq. (14.13). These zeroes are re-inserted in $\hat{\theta}$ after reconciliation.

The solution (14.4f) and (14.4g) is applicable to the modified model. The analytical solution (14.6) also applies, provided n and p vary from 1 to $N-1$ and from 1 to $P-1$; as a result all the Kronecker deltas (δ_{pP} and δ_{nN}) are equal to 1.

The estimated input coefficients are then given by $\hat{a}_{np} = \hat{\theta}_{np} / x_p$ and their variances by $var(\hat{\theta}_{np}) / x_p^2$. The estimates $\hat{a}_{np}$ also coincide with the cross-entropy estimates of Golan et al. (1994 p. 546).

14.5 Implementation of the Two-Way Reconciliation Model

This section is intended mainly for readers who consider programming the two-way reconciliation. If the cross-covariance matrices are not required, the implementation of the general analytical solution (14.6) requires two runs of the data. Each series is processed twice.

It is useful to identify each set of aggregation constraints. For a set of Industrial constraints to exist, the Industrial Total must be present and at least one Province in the Industry considered. Similarly, for a set of Provincial constraints to exist, the Provincial Total must be present and at least one Industry in the Province .

Each set of industrial aggregation constraints of Eq. (14.1a) should be identified by the industrial code (or n) and the starting dates (or t) of the

Industrial Total; and each set of provincial constraints, by the provincial codes (or p) and the starting dates (or t). The resulting *industrial constraint identifiers* also identify the rows of matrices and vectors V_{AA}, V_{AB}, d_A and γ_A and the columns of V_{AA}. Similarly the *provincial constraint identifiers* also identify the rows of matrices or vectors V_{BB}, d_B and γ_B and the columns of V_{BB} and V_{AB}. These constraint identifiers determine the sizes of these matrices and vectors and governs the rows and columns in which the information from each series has to be stored.

First Run

The first pass performs the following calculations for each series separately. The series can be read in any order.

(1a) The series considered s_k is read and its covariance matrix V_{e_k} is generated[1]. Alternatively V_{e_k} is read from a file, if a previous process generated the covariance matrices. In particular, this process may be benchmarking, in which case the series would also be subject to temporal aggregation constraints.

(1b) To avoid potential numerical problems, the series is re-scaled. Each series is divided by

$$\bar{s}=(max(abs(s_{k,t}))-min(abs(s_{k,t}))/2 \qquad (14.14)$$

where $max(abs(s_{k,t}))$ is the maximum absolute value of all the series and $min(abs(s_{k,t}))$ the minimum absolute value. The covariance matrix of the series is divided by the square of $\bar{s}$. The resulting covariance matrix is then multiplied by the alterability coefficient selected for the series. This product becomes the covariance matrix used in the reconciliation model.

(1c) The Industry n and the Province p of the series s_k are determined. If the series is an Industrial Total, δ_{nN} is set equal to -1 and to 1 otherwise. If the series is a Provincial Total, δ_{pP} is set equal to -1 and to 1 otherwise. The partitions of V_{AA}, V_{AB} and V_{BB}, given by Eqs. (14.5c) to (14.5e), and the partitions of d_A and d_B, given by Eq. (14.5f), are accumulated. The constraint identifiers determine in which rows and/or

[1] In order to alleviate the notation, the series are sometimes identifies by index k.

columns of matrices V_{AA}, V_{AB} and V_{BB}, and vectors d_A and d_B, vectors and matrices should be stored.

Steps (1a) to (1c) are repeated for each series in the system. Each series contributes terms to some sub-partitions.

After the first run, the following calculations are performed.

(a) In the *full* two-way model, one set of cross-sectional constraints is redundant. In order to eliminate that redundancy, add an identity matrix to one set of constraints with maximum number of individual constraints as described in section 14.6 (if applicable). The identity matrix would be added to the block diagonal partitions of V^{AA} or V^{BB}.

(b) Matrices V^{AA}, V^{AB} and V^{BB}, i.e. the partitions of matrix V_d^{-1}, are obtained by inverting V_d by parts and by exploiting the fact that V_{AA} and V_{BB} are block diagonal. In order to deal with potential redundant constraints (discussed in the next section), the Moore-Penrose generalized matrix inversion can be used in all the matrix inversions.

If matrix V_{BB} is larger than V_{AA}, the following steps are performed.

(i) Matrix V_{BB}^{-1} is obtained by inverting V_{BB} of Eq. (14.5e) by diagonal blocks, which minimizes both calculations and loss of accuracy. Matrix V_{BB} may then be discarded.

(ii) Matrix V^{AA} is set equal to $(V_{AA} - V_{AB} V_{BB}^{-1} V_{AB}')^{-1}$. Matrix V_{AA} may be discarded.

(iii) Matrix V^{AB} is set equal to $-V^{AA} V_{AB} V_{BB}^{-1}$.

(iv) Matrix V^{BB} is set equal to $V_{BB}^{-1} - V^{AB\prime} V_{AB} V_{BB}^{-1}$. Matrices V_{AB} and V_{BB}^{-1} may then be discarded.

If matrix V_{BB} is *not* larger than V_{AA}, the following steps are performed.

(i) Matrix V_{AA}^{-1} is obtained by inverting V_{AA} of Eq. (14.5c) by diagonal blocks. Matrix V_{AA} may be discard.

(ii) Matrix V^{BB} is set equal to $(V_{BB} - V_{AB}' V_{AA}^{-1} V_{AB})^{-1}$. Matrix V_{BB} may now be discarded.

(iii) Matrix V^{AB} is set equal to $-V_{AA}^{-1} V_{AB} V^{BB}$.

(iv) Matrix V^{AA} is set equal to $V_{AA}^{-1} - V^{AB} V_{AB}{}' V_{AA}^{-1}$. Matrices V_{AB} and V_{AA}^{-1} may be discarded.

(c) The Lagrange multipliers $\hat{\gamma}_A$ and $\hat{\gamma}_B$ are calculated according to (14.4e).

Second Run

The second run performs the following calculations for each series separately.

(2a) This step is identical to (1a).

(2b) This step is identical to (1b).

(2c) The reconciled series $\hat{\boldsymbol{\theta}}_{np}$ is calculated according to the last expression of (14.6a). If applicable, its covariance matrix, $var[\hat{\theta}_{np}]$, is calculated according to (14.6b). (The cross-covariance matrices of (14.6c) would not be required.)

(2d) The series considered and its covariance matrix are converted into the original scale by multiplying it by $\bar{s}$ of Eq. (14.14) and by the square of this number respectively. Both $\hat{\boldsymbol{\theta}}_{np}$ and $var[\hat{\theta}_{np}]$ (or more probably the diagonal thereof) are written to file.

(2e) The residual cross-sectional aggregation discrepancies $\boldsymbol{d}_A^{(r)}$ and $\boldsymbol{d}_B^{(r)}$ are accumulated. These discrepancies may be used to assess whether the reconciliation satisfied the cross-sectional constraints.

Under the solution and implementation proposed, none of the larger matrices in (14.3) and (14.4), namely $\boldsymbol{V}_e$, $\boldsymbol{G}$, $\boldsymbol{G}_A$ and $\boldsymbol{G}_B$ and $\boldsymbol{G}_{np}$, need to be generated, if the industrial and provincial constraint identifiers described above are used. For example, subtracting $\boldsymbol{G}_{32}\boldsymbol{s}_{32}$ from vector $\boldsymbol{d}_A$ and $\boldsymbol{d}_B$ amounts to subtracting $\boldsymbol{s}_{32}$ from to the appropriate rows of vectors $\boldsymbol{d}_A$ (e.g. rows 101 to 120) and from the appropriate rows of $\boldsymbol{d}_B$ (e.g. 61 to 80). Similarly adding $\boldsymbol{G}_{32}\boldsymbol{V}_{e_{32}}\boldsymbol{G}_{32}{}'$ to $\boldsymbol{V}_{AB}$ amounts to adding $\boldsymbol{V}_{e_{32}}$ to the appropriate rows and columns of V_{AB} (e.g. rows 101 to 120 and columns 61 to 80).

Note that if the "all" Industry were the first δ_{n1} would be equal to -1 for n=1 and 1 for $n \neq 1$. Similarly, if the "all" Province were the second, δ_{p2} would be equal to -1 for p=2 and 1 for $p \neq 2$.

The following recursive algorithm has been proposed which incorporates the cross-sectional aggregations constraints one at the time:

$$\hat{\boldsymbol{\theta}}^{(i)} = \hat{\boldsymbol{\theta}}^{(i-1)} + var[\hat{\boldsymbol{\theta}}^{(i-1)}] \, \boldsymbol{G}_i' (\boldsymbol{V}_d^{(i)})^{-1} [g_i - \boldsymbol{G}_i \hat{\boldsymbol{\theta}}^{(i-1)}], \tag{14.15a}$$

$$var[\hat{\boldsymbol{\theta}}^{(i)}] = var[\hat{\boldsymbol{\theta}}^{(i-1)}] -$$

$$var[\hat{\boldsymbol{\theta}}^{(i-1)}] \, \boldsymbol{G}_i' (\boldsymbol{V}_d^{(i)})^{-1} \boldsymbol{G}_i \, var[\hat{\boldsymbol{\theta}}^{(i-1)}], \quad i=1,2...,(T_{\bullet P}+T_{N\bullet}), \tag{14.15b}$$

where $\boldsymbol{G}_i$ is the i-th row of matrix $\boldsymbol{G}$ of (14.3) and $\boldsymbol{V}_d^{(i)} = (\boldsymbol{G}_i \; var[\hat{\boldsymbol{\theta}}^{(i-1)}] \, \boldsymbol{G}_i' + \boldsymbol{V}_{\varepsilon_i})$. The vector $\hat{\boldsymbol{\theta}}^{(i)}$ and the matrix $var[\hat{\boldsymbol{\theta}}^{(i)}]$ are respectively the vector of reconciled series and their covariance matrix, estimated on the basis on the first i constraints incorporated. The recursion is initialized by $\hat{\boldsymbol{\theta}}^{(0)} = \boldsymbol{s}$ and by $var[\hat{\boldsymbol{\theta}}^{(0)}] = \boldsymbol{V}_e$. The advantage of this approach is that matrix $\boldsymbol{V}_d^{(i)}$ to be inverted is in fact a scalar.

The disadvantage, on the other hand, is that it requires the repeated updating of matrix $var[\hat{\boldsymbol{\theta}}^{(i)}]$ which has the same dimensions as $var[\hat{\boldsymbol{\theta}}]$ of (14.3b). That matrix can have a prohibitive size, because it includes the cross-covariances between the series which are of little or no interest. For example, with P=53 provinces and N=1 industry and series of 40 observations, the dimensions of matrices $var[\hat{\boldsymbol{\theta}}^{(i)}]$ and $var[\hat{\boldsymbol{\theta}}^{(i-1)}]$ are 2120×2120. With N=1, each matrix requires more then 34 megabytes of computer memory;, with N=19, 12,378 megabytes. The problem is now one of storing, when inverting matrix $\boldsymbol{V}_d$ is generally not a problem under our proposed implementation.[2]

The recursive strategy (14.15) can also be applied in such a way to incorporate one *set* of industrial or provincial constraints at the time. This

[2] Reading and writing to disk especially is very time consuming compared to calculations.

would translate into fewer updates of $var[\hat{\boldsymbol{\theta}}^{(m)}]$, where $m=1,2,...,N+P$ identifies the set of constraints incorporated.

14.6 Redundant Constraints in the Two-Way Reconciliation Model

The *full* two-way reconciliation model (with Grand Total) has N set of "industrial" cross-sectional constraints and P set of "provincial" cross-sectional constraints respectively:

$$\theta_{nP,t} - \Sigma_{p=1}^{P-1}\,\theta_{Np,t} = 0, \quad n=1,...,N;\ t=1,...,T_{nP},$$
$$\theta_{Np,t} - \Sigma_{n=1}^{N-1}\,\theta_{np,t} = 0, \quad p=1,...,P;\ t=1,...,T_{Np}.$$

where T_{nP} and T_{Np} are respectively the number of individual constraints in each set.

We distinguish two situations regarding the temporal constraints:

(a) the absence of any temporal constraints and

(b) the presence of temporal constraints. In this case, we assume that (i) all series have equal length, (ii) all series have annual benchmarks for each complete "year" and (iii) all the temporal constraints are binding and implicit in the covariance matrices $\boldsymbol{V}_{e_k}$ of the series to be reconciled.

In the absence of temporal constraints, the two-way reconciliation model has one set of redundant cross-sectional constraints. This applies to both the full model and the model without the Grand Total. The number of *non-redundant* constraints is the rank of matrix product $\boldsymbol{G}\boldsymbol{G}'$. Our implementation deals with the problem by making the set of industrial constraints with maximum number of individual constraints non-binding. Since several sets of constraints have the same maximum number of individual constraints, the first one is chosen.

In the presence of temporal constraints (case (b) above), the model has one redundant cross-sectional constraint per "year" with benchmark in both sets of cross-sectional constraints. Alternatively, one can say that the set of temporal constraints of any one Province $\bar{p}$ (say) for each industry $\boldsymbol{s}_{n\bar{p}}$, $(n=1,...,N-1)$ and the set of temporal constraints in any one Industry $\bar{n}$ for each Province $s_{\bar{n}p}$ $(p=1,...,P-1)$ are redundant. In fact, identifying redundant constraint is arbitrary, because redundancy is an attribute of the whole set of constraints. The problem of redundancy can therefore be addressed in a number of different ways:

(1) Omit one constraint per "year" with benchmark in each sets of cross-sectional constraints.
(2) Make the constraints identified in (1) non-binding by setting the corresponding diagonal element of V_ε equal to 1 (say).
(3) Omit all the temporal constraints for any one of the Province for all Industries and for any one Industry for all Provinces.
(4) Make the constraints identified in (3) non-binding.
(5) Invert V_d by parts using the Moore-Penrose generalized matrix inversion, as described in section (14.5) at the end of the first run.

Under (1) to (4), the omitted or non-binding constraints are satisfied through the remaining ones. If some constraints are not satisfied, it means some of the data is contradictory. For example the benchmarks do not satisfy the cross-sectional aggregation constraints.

In practical situations some series may have temporal constraints for each "year", and others fewer temporal constraints or none at all. This makes it hard to determine which constraints should be omitted or made non-binding. The generalized matrix inversion offers a valid and practical solution. If the resulting reconciled series fail to satisfy all constraints, the system contains contradictory data, which should be fixed by the subject matter experts. Chen and Dagum (1997) propose a theoretical approach to the problem.

14.7 A Real Data Example of a Large Two-Way System of Series

The system of time series now exemplified pertains to yearly Exports of Commodity by Province from 1997 to 1999. The Exports are classified into 679 Commodities and 13 Provinces, plus the "all" Commodity and the "all" Province, respectively identified by "Call " and "CA" in Table 14.1a. Province "OC" consists of Canadian diplomatic missions abroad. The system of series thus comprises 9,520 (680×14) series and entails 680 sets of Commodity cross-sectional constraints (over the Provinces) and 14 sets of Provincial cross-sectional constraints (over the Commodities). Each of the 694 sets of constraints contains three constraints (one per year), for a total of 2,082 constraints. The marginal totals originate from the Input-Output model of Statistics Canada and must be imposed onto the sub-provincial series. There are no temporal aggregation constraints.

Out of the 28,560 observations, 8,025 (28%) have zero values. We excluded these values, because (a) their reconciled values are known to be zero, (b) the zeroes cannot affect the cross-sectional sums and (c) the presence of zeroes consumes memory and complicates the calculations (e.g. division by zero). We therefore set the zero values aside for later reintegration into the reconciled system of series. Without the zeroes, the system of series now comprises 591 Commodities and 14 Provinces, for a total of 7,106 series (some cells now being empty) instead of 9,520. The number of constraints thus drops from 2,082 to 1,805, which significantly reduces the processing time.

Table 14.1a. Some statistics on the cross-sectional aggregation discrepancies for the Exports of Commodities by Province, before reconciliation

Commod.	Prov.	Min. add. discrep.	Max. add. discrep.	Min. prop. discrep.	Max. prop. discrep.
C91		-1.00	0.00	0.9997	1.0000
C92		0.00	2.00	1.0000	1.0000
C93		0.00	1.00	1.0000	1.0001
C96		-1.00	-1.00	1.0000	1.0000
C97		0.00	1.00	1.0000	1.0000
C98		-1.00	1.00	1.0000	1.0000
C99		0.00	0.00	1.0000	1.0000
Call		***-2.00***	***0.00***	***1.0000***	***1.0000***
	AB	-416,990.00	-231,909.00	0.9904	0.9940
	BC	268,579.00	447,122.00	1.0070	1.0127
	CA	**0.00**	**0.00**	**1.0000**	**1.0000**
	MB	342,678.00	466,246.00	1.0389	1.0507
	NB	-78,714.00	-55,760.00	0.9875	0.9920
	NF	-177,630.00	-114,003.00	0.9504	0.9741
	NS	12,965.00	19,977.00	1.0024	1.0039
	NT	-31,716.00	6,808.00	0.9563	1.0134
	OC	0.00	3.00	1.0000	1.0005
	ON	-715,304.00	-274,462.00	0.9963	0.9984
	PE	-2,239.00	15,201.00	0.9977	1.0245
	PQ	-60,208.00	-28,280.00	0.9993	0.9996
	SK	131,336.00	586,739.00	1.0109	1.0515
	YT	11,949.00	21,493.00	1.0433	1.1033
.	.	-715,304.00	586,739.00	0.9504	1.1033

Table 14.1a displays the additive and proportional cross-sectional aggregation discrepancies for a few of the Commodity constraints; except for rounding errors, these discrepancies are equal to zero at the outset.

The lower part of Table 14.1a exhibits the Provincial cross-sectional discrepancies, which are not equal to zero, except for the Grand Total (in bold). The goal of reconciliation is to impose the provincial aggregation constraints and to keep the Commodity Totals un-altered.

The two-way reconciliation model described above is applied. The Totals are specified as un-alterable by means of alterability coefficients equal to 0, except for the Grand Total with coefficient equal to 1. Since the Totals already cross-sectionally add up, the Grand Total remain un-altered as well.

The covariances matrices of the original series are set equal to

$$\boldsymbol{V}_{e_{np}} = \alpha_{np} \, \boldsymbol{\Xi}_{np}^{\lambda} \, \boldsymbol{\Omega}_{np} \, \boldsymbol{\Xi}_{np}^{\lambda},$$

where $\boldsymbol{\Xi}_{np} = diag(0.01 \times \boldsymbol{s}_{np})$, where $\lambda = 1/2$, $\boldsymbol{\Omega}_{np} = \boldsymbol{I}_{T_{np}}$, $\alpha_{np} = 0$ for the Totals and $\alpha_{np} = 1$ otherwise. The resulting objective function minimizes the proportional errors,

$$(1/0.01) \times min \{ \Sigma_{n=1}^{N} \, \Sigma_{p=1}^{P} \, \alpha_{np} \, \Sigma_{t=t_{1,np}}^{t_{L,np}} \, e_{np,t}^{2} \, / s_{np,t} \}$$

where $e_{np,t} = s_{np,t} - \theta_{np,t}$. As a result, reconciliation minimizes the size of the proportional corrections θ_{kt} / s_{kt} made to the series to satisfy the constraints.

Table 14.1b shows that after reconciliation the cross-section aggregation constraints are satisfied, except for rounding errors. In order to anticipate the impact of rounding for publication purposes, the reconciled series are rounded to the nearest integer immediately after reconciliation, before the calculation of the residual aggregation discrepancies. In the absence of such rounding, the residual aggregation discrepancies are equal to 0 (not shown).

Note that rounding affects the residual provincial cross-sectional discrepancies much more than the industrial ones. The reason is that each province contains up to 680 industries, which entails adding up 680 rounding errors; and, each industry on the other hand contains at most 14 provinces.

Reconciling this system of series takes less than 12 minutes on a computer operating at a speed of one megahertz with 256 megabytes of random access memory; and, less than 2 1/2 minutes on a computer operating at three megahertz with 1024 megabytes of memory. The program used is SAS (Statistical Analysis System); the reconciliation algorithm is in SAS/IML which is the matrix algebra procedure of SAS.

Table 14.1b. Some tatistics on the residual cross-sectional aggregation discrepancies for the Exports of Commodities by Province, after reconciliation

Commod.	Prov.	Min. add. discrep.	Max. add. discrep.	Min. prop. discrep.	Max. prop. discrep.
C91		0.00	1.00	1.0000	1.0003
C92		-1.00	1.00	1.0000	1.0000
C93		0.00	1.00	1.0000	1.0001
C96		0.00	1.00	1.0000	1.0000
C97		0.00	0.00	1.0000	1.0000
C98		0.00	1.00	1.0000	1.0000
C99		0.00	1.00	1.0000	1.0000
Call		-2.00	0.00	1.0000	1.0000
	AB	-2.00	5.00	1.0000	1.0000
	BC	0.00	6.00	1.0000	1.0000
	CA	0.00	0.00	1.0000	1.0000
	MB	-9.00	1.00	1.0000	1.0000
	NB	-3.00	2.00	1.0000	1.0000
	NF	-5.00	4.00	1.0000	1.0000
	NS	2.00	17.00	1.0000	1.0000
	NT	-10.00	28.00	1.0000	1.0001
	OC	0.00	0.00	1.0000	1.0000
	ON	-6.00	5.00	1.0000	1.0000
	PE	17.00	22.00	1.0000	1.0000
	PQ	-6.00	5.00	1.0000	1.0000
	SK	-2.00	-1.00	1.0000	1.0000
	YT	10.00	20.00	1.0000	1.0001
.	.	-10.00	28.00	0.9972	1.0027

Appendix A: Extended Gauss-Markov Theorem [1]

Let the mixed linear model consist of the following two equations:

$$y = X_1 \beta + X_2 \theta + e, \quad E(e) = 0, \; E(e e') = V_e \tag{1a}$$

$$D \theta = \eta, \quad E(\eta) = 0, \; E(\eta \eta') = V_\eta \tag{1b}$$

where β is a vector of p unknown fixed parameters and θ is a vector of T non-stationary stochastic (random) parameters. After applying an appropriate linear operator D of dimension $q \times T$ ($q \leq T$), the stochastic parameters have mean 0 and positive definite covariance matrix V_η. In our case, θ is a time series, D is often a regular and/or seasonal differencing operator, and V_η is typically the covariance matrix corresponding to an Autoregressive Moving Average (ARMA) process.

Model (1) can be written in a condensed form as

$$\begin{bmatrix} y \\ r \end{bmatrix} = \begin{bmatrix} X_1 & X_2 \\ 0 & D \end{bmatrix} \begin{bmatrix} \beta \\ \theta \end{bmatrix} + \begin{bmatrix} e \\ -\eta \end{bmatrix}, \quad \begin{matrix} E(e\, e') = V_e \\ E(\eta\, \eta') = V_\eta \end{matrix}, \quad E(e\, \eta') = 0 \tag{2a}$$

or

$$\tilde{y} = Z a + u, \quad E(u) = 0, \; E(u u') = V_u = \begin{bmatrix} V_e & 0 \\ 0 & V_\eta \end{bmatrix} \tag{2b}$$

where r in $\tilde{y}$ is 0 for the non-stationary case, but not necessarily for other variants discussed in the book.

[1] Original version published in Slottje DJ ed. (1999), Advances in Econometrics, Income Distribution and Scientific Methodology, Essays in Honor of Professor Camilo Dagum, Physica-Verlag, pp. 27-39.

Theorem: The Generalized Least Square (GLS) solution of model (2),

$$\hat{\boldsymbol{a}} = (\boldsymbol{Z}'\boldsymbol{V}_u^{-1}\boldsymbol{Z})^{-1}(\boldsymbol{Z}'\boldsymbol{V}_u^{-1}\tilde{\boldsymbol{y}})$$

$$= \begin{bmatrix} \boldsymbol{X}_1'\boldsymbol{V}_e^{-1}\boldsymbol{X}_1 & \boldsymbol{X}_1'\boldsymbol{V}_e^{-1}\boldsymbol{X}_2 \\ \boldsymbol{X}_2'\boldsymbol{V}_e^{-1}\boldsymbol{X}_1 & \boldsymbol{X}_2'\boldsymbol{V}_e^{-1}\boldsymbol{X}_2 + \boldsymbol{D}'\boldsymbol{V}_\eta^{-1}\boldsymbol{D} \end{bmatrix}^{-1} \begin{bmatrix} \boldsymbol{X}_1'\boldsymbol{V}_e^{-1}\boldsymbol{y} \\ \boldsymbol{X}_2'\boldsymbol{V}_e^{-1}\boldsymbol{y} + \boldsymbol{D}'\boldsymbol{V}_\eta^{-1}\boldsymbol{r} \end{bmatrix}, \qquad (3a)$$

$$\boldsymbol{V}_{\hat{a}} = (\boldsymbol{Z}'\boldsymbol{V}_u^{-1}\boldsymbol{Z})^{-1}$$

$$= \begin{bmatrix} \boldsymbol{X}_1'\boldsymbol{V}_e^{-1}\boldsymbol{X}_1 & \boldsymbol{X}_1'\boldsymbol{V}_e^{-1}\boldsymbol{X}_2 \\ \boldsymbol{X}_2'\boldsymbol{V}_e^{-1}\boldsymbol{X}_1 & \boldsymbol{X}_2'\boldsymbol{V}_e^{-1}\boldsymbol{X}_2 + \boldsymbol{D}'\boldsymbol{V}_\eta^{-1}\boldsymbol{D} \end{bmatrix}^{-1} \equiv \begin{bmatrix} \boldsymbol{V}_{\hat{\beta}} & \boldsymbol{V}_{\hat{\beta},\hat{\theta}} \\ \boldsymbol{V}_{\hat{\theta},\hat{\beta}} & \boldsymbol{V}_{\hat{\theta}} \end{bmatrix}, \qquad (3b)$$

is the Minimum Variance Linear Unbiased Estimator of both fixed and stochastic parameters in $\boldsymbol{a}$.

To prove the theorem we start from the well known result of Whittle (1963): For any two random vectors $\boldsymbol{a}$ and $\boldsymbol{y}$, the minimum variance unbiased linear prediction of $\boldsymbol{a}$ using data $\boldsymbol{y}$ is given by:

$$\hat{\boldsymbol{a}} = E(\boldsymbol{a}) + \boldsymbol{V}_{a,y}\boldsymbol{V}_y^{-1}(\boldsymbol{y} - E(\boldsymbol{y})), \qquad (4a)$$

with covariance matrix

$$\boldsymbol{V}_{\hat{a}} = E(\hat{\boldsymbol{a}} - \boldsymbol{a})(\hat{\boldsymbol{a}} - \boldsymbol{a})' = \boldsymbol{V}_a - \boldsymbol{V}_{a,y}\boldsymbol{V}_y^{-1}\boldsymbol{V}_{y,a} \qquad (4b)$$

where $\boldsymbol{V}_y$ is the covariance matrix of $\boldsymbol{y}$, $\boldsymbol{V}_a$ is the covariance matrix of $\boldsymbol{a}$ and $\boldsymbol{V}_{a,y}$ is the covariance matrix of $\boldsymbol{a}$ and $\boldsymbol{y}$. For stationary stochastic parameters $\boldsymbol{a}$ this result applies directly. For non-stationary parameters, the covariance matrix $\boldsymbol{V}_a$ is not uniquely defined, which is one motivation for this theorem.

First we establish the particular form of (4) under model (1a), which can be written as

$$\boldsymbol{y} = \boldsymbol{X}\boldsymbol{a} + \boldsymbol{e}, \quad E(\boldsymbol{e}) = \boldsymbol{0}, \; E(\boldsymbol{e}\boldsymbol{e}') = \boldsymbol{V}_e \qquad (5)$$

where $\boldsymbol{X} = \begin{bmatrix} X_1 & X_2 \end{bmatrix}$ and $\boldsymbol{\alpha} = \begin{bmatrix} \boldsymbol{\beta}' & \boldsymbol{\theta}' \end{bmatrix}'$. Under a model such as (5), we have $E(\boldsymbol{y}) = \boldsymbol{X}\,E(\boldsymbol{\alpha})$, $\boldsymbol{V}_y = (\boldsymbol{X}\boldsymbol{V}_\alpha\boldsymbol{X}' + \boldsymbol{V}_e)$, $\boldsymbol{V}_{\alpha,y} = \boldsymbol{V}_\alpha\boldsymbol{X}'$. Substituting these in (4) yields:

$$\hat{\boldsymbol{\alpha}} = E(\boldsymbol{\alpha}) + \boldsymbol{V}_\alpha\boldsymbol{X}'(\boldsymbol{X}\boldsymbol{V}_\alpha\boldsymbol{X}' + \boldsymbol{V}_e)^{-1}(\boldsymbol{y} - \boldsymbol{X}\,E(\boldsymbol{\alpha}))$$
$$= \boldsymbol{V}_{\hat{\alpha}}\,\boldsymbol{V}_\alpha^{-1}\,E(\boldsymbol{\alpha}) + \boldsymbol{V}_{\hat{\alpha}}\boldsymbol{X}'\boldsymbol{V}_e^{-1}\boldsymbol{y}\,, \tag{6a}$$

$$\boldsymbol{V}_{\hat{\alpha}} = \boldsymbol{V}_\alpha - \boldsymbol{V}_\alpha\boldsymbol{X}'(\boldsymbol{X}\boldsymbol{V}_\alpha\boldsymbol{X}' + \boldsymbol{V}_e)^{-1}\boldsymbol{X}\boldsymbol{V}_\alpha = (\boldsymbol{V}_\alpha^{-1} + \boldsymbol{X}'\boldsymbol{V}_e^{-1}\boldsymbol{X})^{-1}$$

$$= \left[\boldsymbol{V}_\alpha^{-1} + \begin{bmatrix} \boldsymbol{X}_1'\boldsymbol{V}_e^{-1}\boldsymbol{X}_1 & \boldsymbol{X}_1'\boldsymbol{V}_e^{-1}\boldsymbol{X}_2 \\ \boldsymbol{X}_2'\boldsymbol{V}_e^{-1}\boldsymbol{X}_1 & \boldsymbol{X}_2'\boldsymbol{V}_e^{-1}\boldsymbol{X}_2 \end{bmatrix} \right]^{-1}. \tag{6b}$$

In order to derive (6b), it is sufficient to apply the identity $(\boldsymbol{A}^{-1} + \boldsymbol{B}\boldsymbol{C}^{-1}\boldsymbol{B}')^{-1} \equiv \boldsymbol{A} - \boldsymbol{A}\boldsymbol{B}(\boldsymbol{B}'\boldsymbol{A}\boldsymbol{B} + \boldsymbol{C})^{-1}\boldsymbol{B}'\boldsymbol{A}$ to the third member of the equation.

In order to derive (6a), we will proceed backwards by first showing

(a) $\boldsymbol{V}_{\hat{\alpha}}\boldsymbol{V}_\alpha^{-1}E(\boldsymbol{\alpha}) = E(\boldsymbol{\alpha}) - [\boldsymbol{V}_\alpha\boldsymbol{X}'(\boldsymbol{X}\boldsymbol{V}_\alpha\boldsymbol{X}' + \boldsymbol{V}_e)^{-1}]\,\boldsymbol{X}\,E(\boldsymbol{\alpha})$,

(b) $\boldsymbol{V}_{\hat{\alpha}}\boldsymbol{X}'\boldsymbol{V}_e^{-1}\boldsymbol{y} = [\boldsymbol{V}_\alpha\boldsymbol{X}'(\boldsymbol{X}\boldsymbol{V}_\alpha\boldsymbol{X}' + \boldsymbol{V}_e)^{-1}]\,\boldsymbol{y}$.

These two result imply the first expression of $\hat{\boldsymbol{\alpha}}$ in (6a).

Proposition (a) is obtained by applying result (6b) to left-hand member of the equation.

$$\boldsymbol{V}_{\hat{\alpha}}\boldsymbol{V}_\alpha^{-1}E(\boldsymbol{\alpha}) = (\boldsymbol{V}_\alpha - \boldsymbol{V}_\alpha\boldsymbol{X}'(\boldsymbol{X}\boldsymbol{V}_\alpha\boldsymbol{X}' + \boldsymbol{V}_e)^{-1}\boldsymbol{X}\boldsymbol{V}_\alpha)\,\boldsymbol{V}_\alpha^{-1}E(\boldsymbol{\alpha})$$
$$= E(\boldsymbol{\alpha}) - [\boldsymbol{V}_\alpha\boldsymbol{X}'(\boldsymbol{X}\boldsymbol{V}_\alpha\boldsymbol{X}' + \boldsymbol{V}_e)^{-1}]\,\boldsymbol{X}\,E(\boldsymbol{\alpha}).$$

Proposition (b) is obtained by substituting (6b), which yields:

$$\boldsymbol{V}_{\hat{\alpha}}\boldsymbol{X}'\boldsymbol{V}_e^{-1}\boldsymbol{y} = (\boldsymbol{V}_\alpha - \boldsymbol{V}_\alpha\boldsymbol{X}'(\boldsymbol{X}\boldsymbol{V}_\alpha\boldsymbol{X}' + \boldsymbol{V}_e)^{-1}\boldsymbol{X}\boldsymbol{V}_\alpha)\boldsymbol{X}'\boldsymbol{V}_e^{-1}\boldsymbol{y}$$
$$= \{\boldsymbol{V}_\alpha\boldsymbol{X}'\boldsymbol{V}_e^{-1} - \boldsymbol{V}_\alpha\boldsymbol{X}'[(\boldsymbol{X}\boldsymbol{V}_\alpha\boldsymbol{X}' + \boldsymbol{V}_e)^{-1}(\boldsymbol{X}\boldsymbol{V}_\alpha\boldsymbol{X}' + \boldsymbol{V}_e - \boldsymbol{V}_e)]\,\boldsymbol{V}_e^{-1}\}\,\boldsymbol{y}$$
$$= \{\boldsymbol{V}_\alpha\boldsymbol{X}'\boldsymbol{V}_e^{-1} - \boldsymbol{V}_\alpha\boldsymbol{X}'[(\boldsymbol{X}\boldsymbol{V}_\alpha\boldsymbol{X}' + \boldsymbol{V}_e)^{-1}(\boldsymbol{X}\boldsymbol{V}_\alpha\boldsymbol{X}' + \boldsymbol{V}_e)$$
$$- (\boldsymbol{X}\boldsymbol{V}_\alpha\boldsymbol{X}' + \boldsymbol{V}_e)^{-1}\boldsymbol{V}_e]\,\boldsymbol{V}_e^{-1}\}\,\boldsymbol{y}$$
$$= \{\boldsymbol{V}_\alpha\boldsymbol{X}'\boldsymbol{V}_e^{-1} - \boldsymbol{V}_\alpha\boldsymbol{X}'[\boldsymbol{I} - (\boldsymbol{X}\boldsymbol{V}_\alpha\boldsymbol{X}' + \boldsymbol{V}_e)^{-1}\boldsymbol{V}_e]\,\boldsymbol{V}_e^{-1}\}\,\boldsymbol{y}$$

$$= \{ V_\alpha X' V_e^{-1} - V_\alpha X' V_e^{-1} + V_\alpha X' (X V_\alpha X' + V_e)^{-1} V_e V_e^{-1} \} y$$
$$= [V_\alpha X' (X V_\alpha X' + V_e)^{-1}] y.$$

Next, under commonly used assumptions, we prove the following three propositions:

(c) $V_{\hat{\alpha}}$ of (6b) is equal to $(Z' V_u^{-1} Z)^{-1}$ of (3b),

(d) the first term in $E(\alpha)$ of (6a) is zero and

(e) the second term of (6a) can be expressed as $(Z' V_u^{-1} Z)^{-1} (Z' V_u^{-1} \tilde{y})$ of (3a).

This will complete the proof.

Let us define, $$\zeta = \begin{bmatrix} \beta \\ \theta_0 \\ \eta \end{bmatrix} = \begin{bmatrix} I_p & 0 & 0 \\ 0 & I_d & 0 \\ 0 & D_1 & D_2 \end{bmatrix} \begin{bmatrix} \beta \\ \theta_0 \\ \theta^* \end{bmatrix} = K \begin{bmatrix} \beta \\ \theta \end{bmatrix} = K\alpha , \qquad (7)$$

where θ_0 contains the d initial conditions of θ, where θ^* contains the last $q{=}T{-}d$ values of θ, and where D_1 and D_2 are respectively the first d and the last q columns of D of (1b). Note that Eq. (7) implies $V_\zeta = K V_\alpha K'$.

Because the stochastic parameters are non-stationary, their variance is not uniquely defined; the common treatment in the literature is to assume diffuse and orthogonal initial conditions (e.g. Ansley and Kohn 1985). We assume diffuse and orthogonal priors for β and θ_0 as follows:

$$E(\begin{bmatrix} \beta \\ \theta_0 \end{bmatrix} \begin{bmatrix} \beta' & \theta_0' \end{bmatrix}) = h\, I_{p+d},\ h \to \infty ; \quad E(\beta\, \eta') = 0,\ E(\theta_0\, \eta') = 0. \quad (8)$$

For a given value $h > 0$, the covariance matrix of ζ is given by $V_\zeta(h) =$ $block(h\, I_{p+d}, V_\eta)$ Since Eq. (7) implies $V_\zeta = K V_\alpha K'$, it is legitimate to state $V_\zeta(h) = K V_\alpha(h) K'$. The identity $(ABC)^{-1} = C^{-1} B^{-1} A^{-1}$ implies $V_\zeta^{-1}(h) = (K V_\alpha(h) K')^{-1} = (K')^{-1} V_\alpha^{-1}(h)\, K^{-1}$. Pre- and post-multiplying the latter expression of $V_\zeta^{-1}(h)$ by K' and K, and substitution of K defined in (7) yields:

$$V_\alpha^{-1}(h) = K' V_\zeta^{-1}(h) K = \begin{bmatrix} h^{-1} I_p & 0 & 0 \\ 0 & h^{-1} I_d & 0 \\ 0 & 0 & 0_q \end{bmatrix} + \begin{bmatrix} 0_p & 0 \\ 0 & D' V_\eta^{-1} D \end{bmatrix}. \qquad (9)$$

where $D' V_\eta^{-1} D$ has dimension T by T.

Replacing V_α^{-1} by $V_\alpha^{-1}(h)$ in the last expression of (6b), and a few rearrangements yield

$$V_{\hat{\alpha}}(h) =$$

$$\left\{ \begin{bmatrix} h^{-1} I_p & 0 & 0 \\ 0 & h^{-1} I_d & 0 \\ 0 & 0 & 0_q \end{bmatrix} + \begin{bmatrix} 0_p & 0 \\ 0 & D' V_\eta^{-1} D \end{bmatrix} + \begin{bmatrix} X_1' V_e^{-1} X_1 & X_1' V_e^{-1} X_2 \\ X_2' V_e^{-1} X_1 & X_2' V_e^{-1} X_2 \end{bmatrix} \right\}^{-1} \qquad (10)$$

After letting $h \to \infty$ Eq. (10) becomes identical to $V_{\hat{\alpha}} = (Z' V_u^{-1} Z)^{-1}$ of (3b). This proves proposition (c).

Replacing V_α^{-1} by $V_\alpha^{-1}(h)$, the first term, $V_{\hat{\alpha}} V_\alpha^{-1} E(\boldsymbol{\alpha})$, of (6a) can be written as

$$V_{\hat{\alpha}} K V_\zeta^{-1}(h) K' \begin{bmatrix} E(\boldsymbol{\beta}) \\ E(\boldsymbol{\theta}) \end{bmatrix} = V_{\hat{\alpha}} K \begin{bmatrix} h^{-1} I_p & 0 & 0 \\ 0 & h^{-1} I_d & 0 \\ 0 & 0 & V_\eta^{-1} \end{bmatrix} \begin{bmatrix} E(\boldsymbol{\beta}) \\ E(\boldsymbol{\theta}_0) \\ E(\boldsymbol{\eta}) \end{bmatrix}. \qquad (11)$$

Letting $h \to \infty$ and noting that $E(\boldsymbol{\eta}) = 0$, Eq. (11) converges to zero. This proves proposition (d) and also shows that the generalized least square solution does not have the problem of the initial conditions.

The last term $V_{\hat{\alpha}} X' V_e^{-1} y$ of (6a) is equal to (3a) because $V_{\hat{\alpha}} = (Z' V_u^{-1} Z)^{-1}$ as demonstrated by (10) and $X' V_e^{-1} y$ is equal to $Z' V_u^{-1} \tilde{y}$ for $r = 0$:

$$X' V_e^{-1} y = \begin{bmatrix} X_1' V_e^{-1} y \\ X_2' V_e^{-1} y \end{bmatrix} = Z' V_u^{-1} \tilde{y} = \begin{bmatrix} X_1' V_e^{-1} y \\ X_2' V_e^{-1} y + D' V_\eta^{-1} 0 \end{bmatrix}, \quad (12)$$

which proves proposition (e) and completes the proof.

Solution (3) can be applied as such but depending on the size of the matrices involved, it may be appropriate to invert matrix $(Z' V_u^{-1} Z)$ by parts, under *any* convenient partitioning. If matrix $V_{\hat{\alpha}} = (Z' V_u^{-1} Z)^{-1}$ is partitioned according to (3b), a generic expression for (3) is then:

$$\hat{\beta} = (V_{\hat{\beta}} X_1' + V_{\hat{\beta},\hat{\theta}} X_2')\, V_e^{-1} y + V_{\hat{\beta},\hat{\theta}}\, D' V_\eta^{-1} r, \quad (13a)$$

$$\hat{\theta} = (V_{\hat{\theta},\hat{\beta}} X_1' + V_{\hat{\theta}} X_2')\, V_e^{-1} y + V_{\hat{\theta}}\, D' V_\eta^{-1} r. \quad (13b)$$

where r is 0 in the non-stationary cases.

The usual corollary to the Gauss-Markov theorem holds: the Minimum Variance Linear Unbiased Predictor of any linear combination of the parameters, $w = W\alpha$, is given by:

$$\hat{w} = W\hat{\alpha} = [\, W_1 \;\; W_2 \,]\, \hat{\alpha} = W_1 \hat{\beta} + W_2 \hat{\theta} \quad (13c)$$

$$\begin{aligned} V_{\hat{w}} &= W V_{\hat{\alpha}} W' \\ &= W_1 V_{\hat{\beta}} W_1' + W_2 V_{\hat{\theta},\hat{\beta}} W_1' + W_1 V_{\hat{\beta},\hat{\theta}} W_2' + W_2 V_{\hat{\theta}} W_2' \end{aligned} \quad (13d)$$

When V_e does not have not full rank, the above formulae containing its inverse are no longer valid. This problem can be solved by applying the same derivation but where V_e is replaced by $V_e + \delta I$ ($\delta>0$). Result (6) for $\hat{\alpha}$ and $V_{\hat{\alpha}}$ can be interpreted as the limits to which the resulting equations converge when $\delta \to 0$. Note however that only the first expression of $V_{\hat{\alpha}}$ in (6b) and the first expression of $\hat{\alpha}$ in (6a) (where $E(\alpha)$ was shown to be 0) remain valid for $\delta = 0$.

Appendix B: An Alternative Solution for the Cholette-Dagum Model for Binding Benchmarks

The purpose of this Appendix is to provide an alternative solution of the Cholette-Dagum benchmarking model, in order to allow for binding benchmarks. The alternative solution also applies to interpolation model of Eq. (7.4), and to reconciliation model (11.4), to allow binding cross-sectional aggregation constraints.

The generalized least square solution of model (3.18) is

$$\hat{a} = (X' V_u^{-1} X)^{-1} X' V_u^{-1} y, \tag{1a}$$

$$var[\hat{a}] = (X' V_u^{-1} X)^{-1}. \tag{1b}$$

Given model (3.18), the estimator (1) can be written as

$$\begin{bmatrix} \hat{\beta} \\ \hat{\theta} \end{bmatrix} = \begin{bmatrix} R' V_e^{-1} R & R' V_e^{-1} \\ V_e^{-1} R & (V_e^{-1} + J' V_\varepsilon^{-1} J) \end{bmatrix}^{-1} \begin{bmatrix} R' V_e^{-1} s \\ V_e^{-1} s + J' V_\varepsilon^{-1} a \end{bmatrix}, \tag{2a}$$

$$var[\hat{a}] = (X' V_u^{-1} X)^{-1} \equiv \begin{bmatrix} var[\hat{\beta}] & cov[\hat{\beta}, \hat{\theta}] \\ cov[\hat{\theta}, \hat{\beta}] & var[\hat{\theta}] \end{bmatrix} \tag{2b}$$

$$= \begin{bmatrix} R' V_e^{-1} R & R' V_e^{-1} \\ V_e^{-1} R & (V_e^{-1} + J' V_\varepsilon^{-1} J) \end{bmatrix}^{-1} \tag{2c}$$

$$= \begin{bmatrix} A_{11} & A_{12} \\ A_{12}' & A_{22} \end{bmatrix}^{-1} = \begin{bmatrix} A^{11} & A^{12} \\ A^{12\prime} & A^{22} \end{bmatrix}. \tag{2d}$$

Eq. (2a) and (2b) imply:

$$\hat{\boldsymbol{\beta}} = var[\hat{\boldsymbol{\beta}}]\, \boldsymbol{R}' \boldsymbol{V}_e^{-1} \boldsymbol{s} + cov[\hat{\boldsymbol{\beta}}, \hat{\boldsymbol{\theta}}]\, (\boldsymbol{V}_e^{-1} \boldsymbol{s} + \boldsymbol{J}' \boldsymbol{V}_\varepsilon^{-1} \boldsymbol{a}), \tag{3a}$$

$$\hat{\boldsymbol{\theta}} = cov[\hat{\boldsymbol{\theta}}, \hat{\boldsymbol{\beta}}]\, \boldsymbol{R}' \boldsymbol{V}_e^{-1} \boldsymbol{s} + var[\hat{\boldsymbol{\theta}}]\, (\boldsymbol{V}_e^{-1} \boldsymbol{s} + \boldsymbol{J}' \boldsymbol{V}_\varepsilon^{-1} \boldsymbol{a}). \tag{3b}$$

Solution (1) requires that $\boldsymbol{V}_\varepsilon$ be positive definite, which excludes the possibility of binding benchmarks.

We now show that the estimator (1) can also be expressed as:

$$\hat{\boldsymbol{\beta}} = -(\boldsymbol{R}' \boldsymbol{J}' \boldsymbol{V}_d^{-1} \boldsymbol{J} \boldsymbol{R})^{-1}\, \boldsymbol{R}' \boldsymbol{J}' \boldsymbol{V}_d^{-1}\, [\boldsymbol{a} - \boldsymbol{J} \boldsymbol{s}], \tag{4a}$$

$$var[\hat{\boldsymbol{\beta}}] = (\boldsymbol{R}' \boldsymbol{J}' \boldsymbol{V}_d^{-1} \boldsymbol{J} \boldsymbol{R})^{-1} \tag{4b}$$

$$\hat{\boldsymbol{\theta}} = \boldsymbol{s}^{\dagger} + \boldsymbol{V}_e\, \boldsymbol{J}' \boldsymbol{V}_d^{-1}\, [\boldsymbol{a} - \boldsymbol{J} \boldsymbol{s}^{\dagger}], \tag{4c}$$

$$var[\hat{\boldsymbol{\theta}}] = \boldsymbol{V}_e - \boldsymbol{V}_e \boldsymbol{J}' \boldsymbol{V}_d^{-1} \boldsymbol{J} \boldsymbol{V}_e + (\boldsymbol{R} - \boldsymbol{V}_e \boldsymbol{J}' \boldsymbol{V}_d^{-1} \boldsymbol{J} \boldsymbol{R})\, var[\hat{\boldsymbol{\beta}}]\, (\boldsymbol{R}' - \boldsymbol{R}' \boldsymbol{J}' \boldsymbol{V}_d^{-1} \boldsymbol{J} \boldsymbol{V}_e) \tag{4d}$$

where $\boldsymbol{s}^{\dagger} = \boldsymbol{s} - \boldsymbol{R}\hat{\boldsymbol{\beta}}$ and $\boldsymbol{V}_d = [\boldsymbol{J} \boldsymbol{V}_e \boldsymbol{J}' + \boldsymbol{V}_\varepsilon]$.

The derivation assumes that both $\boldsymbol{V}_e$ and $\boldsymbol{V}_\varepsilon$ and invertible. For $\boldsymbol{V}_e$, this is not a problem. For $\boldsymbol{V}_\varepsilon$ this is a problem because that matrix is typically not invertible and often equal to $\boldsymbol{0}$. When $\boldsymbol{V}_\varepsilon$ is not invertible, it can be replaced by a matrix $\boldsymbol{V}_\delta$ equal to $\boldsymbol{V}_\varepsilon + \delta \boldsymbol{I}_M$ with δ close to zero. Parameter δ is then set to 0 after the derivation.

The lower-right partition of (2c) corresponding to partition $\boldsymbol{A}_{22}$ in (2d) may be inverted using the identity $(\boldsymbol{D} + \boldsymbol{B}\boldsymbol{C}\boldsymbol{B}')^{-1} \equiv \boldsymbol{D}^{-1} - \boldsymbol{D}^{-1}\boldsymbol{B}\, (\boldsymbol{B}'\boldsymbol{D}^{-1}\boldsymbol{B} + \boldsymbol{C}^{-1})^{-1}\boldsymbol{B}'\boldsymbol{D}^{-1}$:

$$(\boldsymbol{V}_e^{-1} + \boldsymbol{J}' \boldsymbol{V}_\varepsilon^{-1} \boldsymbol{J})^{-1} \equiv \boldsymbol{V}_e - \boldsymbol{V}_e \boldsymbol{J}' (\boldsymbol{J} \boldsymbol{V}_e \boldsymbol{J}' + \boldsymbol{V}_\varepsilon)^{-1} \boldsymbol{J} \boldsymbol{V}_e$$

$$\equiv \boldsymbol{V}_e - \boldsymbol{V}_e \boldsymbol{J}' \boldsymbol{V}_d^{-1} \boldsymbol{J} \boldsymbol{V}_e = \boldsymbol{A}_{22}^{-1}. \tag{5}$$

Given $\boldsymbol{A}_{22}^{-1}$ from (5), the matrix inversion in (2d) is performed by parts using the following formulae:

$$A^{11} = (A_{11} - A_{12}A_{22}^{-1}A_{12}^{/})^{-1}, \tag{6a}$$
$$A^{12} = -A^{11}A_{12}A_{22}^{-1} = -(A_{11} - A_{12}A_{22}^{-1}A_{12}^{/})^{-1}A_{12}A_{22}^{-1}, \tag{6b}$$
$$A^{22} = A_{22}^{-1} - A^{12/}A_{12}A_{22}^{-1} = A_{22}^{-1} + A_{22}^{-1}A_{12}^{/}A^{11}A_{12}A_{22}^{-1}, \tag{6c}$$
$$= A_{22}^{-1} + A_{22}^{-1}A_{12}^{/}(A_{11} - A_{12}A_{22}^{-1}A_{12}^{/})^{-1}A_{12}A_{22}^{-1}. \tag{6c}$$

Substituting (6a) in the upper left partition of (2b) yields:

$$var[\hat{\beta}] = A^{11} = [(R^{/}V_e^{-1}R) - (R^{/}V_e^{-1})(V_e - V_eJ^{/}V_d^{-1}JV_e)(V_e^{-1}R)]^{-1}$$
$$= (R^{/}J^{/}V_d^{-1}JR)^{-1}, \tag{7a}$$

which proves (4b);

Substituting (6b) into the upper right partion of (2b) yields:

$$cov[\hat{\beta},\hat{\theta}] = A^{12} = -var[\hat{\beta}]\,(R^{/}V_e^{-1})\,(V_e - V_eJ^{/}V_d^{-1}JV_e)$$
$$= -var[\hat{\beta}]\,R^{/}\,(I - J^{/}V_d^{-1}JV_e), \tag{7b}$$

$$cov[\hat{\theta},\hat{\beta}] = A^{12/} = -(I - V_eJ^{/}V_d^{-1}J)\,R\,var[\hat{\beta}]. \tag{7c}$$

Substituting (6d) into the lower-right partition of (2b) yields:

$$var[\hat{\theta}] = A^{22} = (V_e - V_e\,J^{/}\,V_d^{-1}J\,V_e)$$

$$- cov[\hat{\theta},\hat{\beta}]\,(R^{/}V_e^{-1})(V_e - V_e\,J^{/}\,V_d^{-1}J\,V_e)$$
$$= (V_e - V_e\,J^{/}\,V_d^{-1}J\,V_e)$$
$$+ (I - V_eJ^{/}V_d^{-1}J)\,R\,var[\hat{\beta}]\,R^{/}\,(I - J^{/}\,V_d^{-1}J\,V_e) \tag{7d}$$

which proves (4d).

Substituting (7) in (3a) and expanding leads to cancellations and simplifications

$$\hat{\beta} = var[\hat{\beta}]\,R^{/}\,V_e^{-1}s \;-\; var[\hat{\beta}]\,R^{/}\,(I - J^{/}V_d^{-1}JV_e)(V_e^{-1}s + J^{/}V_\varepsilon^{-1}a)$$
$$= var[\hat{\beta}]\,R^{/}J^{/}V_d^{-1}Js \;+\; var[\hat{\beta}]\,R^{/}J^{/}\,V_d^{-1}(JV_eJ^{/} - V_d)\,V_\varepsilon^{-1}a\,.$$

Replacing the inner occurrences of V_d in the last equation by $J V_e J' + V_\varepsilon$ yields:

$$\begin{aligned}\hat{\beta} &= var[\hat{\beta}]\, R' J' V_d^{-1} J s \\ &\qquad + var[\hat{\beta}]\, R' J'\, V_d^{-1}(J V_e J' - J V_e J' - V_\varepsilon)\, V_\varepsilon^{-1} a \\ &= -(R' J' V_d^{-1} J R)^{-1} R' J' V_d^{-1}(a - J s),\end{aligned} \tag{8}$$

which proves (4a).

Substituting (7) in (3b) and expanding yields

$$\begin{aligned}\hat{\theta} &= -(I - B J) R\, var[\hat{\beta}]\;\; R' V_e^{-1} s \;+ \{(V_e - B J V_e) \\ &\qquad + (I - B J) R\, var[\hat{\beta}]\;\; R' (I - J' B')\}\, (V_e^{-1} s + J' V_\varepsilon^{-1} a),\end{aligned}$$

where $B = V_e J' V_d^{-1}$.

After lengthy algebraic transformations and collecting the terms in $V_\varepsilon^{-1} a$ and s we obtain:

$$\begin{aligned}\hat{\theta} &= s - V_e J' V_d^{-1} J s \\ &\quad + V_e J' V_d^{-1} J R\, var[\hat{\beta}]\;\; R' J' V_d^{-1} J s \;- R\, var[\hat{\beta}]\, R' J' V_d^{-1} J s \\ &\quad - V_e J' [V_d^{-1} J V_e J' - I]\, V_\varepsilon^{-1} a \\ &\quad - [R\, var[\hat{\beta}]\;\; R' J'][V_d^{-1} J V_e J' - I]\, V_\varepsilon^{-1} a \\ &\quad + [V_e J' V_d^{-1} J R\, var[\hat{\beta}]\;\; R' J'][V_d^{-1} J V_e J' - I]\, V_\varepsilon^{-1} a\,.\end{aligned}$$

Replacing $[V_d^{-1} J V_e J' - I]$ by $V_d^{-1}[J V_e J' - V_d]$, then replacing the inner V_d by $J V_e J' + V_\varepsilon$ and rearranging yields

$$\begin{aligned}\hat{\theta} &= (s - R\hat{\beta}) + [V_e J' V_d^{-1}][a - J(s - R\hat{\beta})] \\ &= s^{\dagger} + V_e J' V_d^{-1} s^{\dagger},\end{aligned} \tag{9}$$

which proves (4c) and completes the proof.

Appendix C: Formulae for Some Recurring Matrix Products

This appendix provides algebraic formulae to directly generate the elements of matrix products involved in the solution of the Cholette-Dagum benchmarking and the interpolation models. These formulae remove the need to generate some of the matrices involved. They simplify and reduce the calculations required, especially when the benchmarked or interpolated series is daily.

We provide formulae for the additive and the multiplicative models.

Formulae for the C-D Additive Model

In the case of benchmarking, these products appear in equations (3.20) and (3.21) of Chapter 3.

The elements q_{tm} of matrix product $\boldsymbol{Q} = \boldsymbol{V}_e \boldsymbol{J}'$ are given by

$$q_{tm} = \Sigma_{i=t_{1m}}^{t_{2m}} j_{mi}\, \omega(|t-i|)\, \sigma_t \sigma_i, \quad t=1,...,T, \; m=1,...,M, \tag{1}$$

where $\omega(|\ell|)$ are the autocorrelations of the standardized error for lags $\ell = 0, ..., T-1$. For an autoregressive error model of order 1 with parameter φ, $\omega(|\ell|) = \varphi^{|\ell|}$. For an autoregressive model of order 2, the autocorrelations are given by $\omega_0 = 1$, $\omega_1 = \varphi_1 / (1 - \varphi_2)$, $\omega_\ell = \varphi_1\, \omega_{\ell-1} + \varphi_2\, \omega_{\ell-2}$, $\ell = 2, ..., T-1$.

The elements p_{mi} of matrix product $\boldsymbol{P} = \boldsymbol{V}_d = [\boldsymbol{J} \boldsymbol{V}_e \boldsymbol{J}' + \boldsymbol{V}_\varepsilon]$ are given by

$$p_{mi} = (\Sigma_{t=t_{1m}}^{t_{2m}} j_{mt}\, q_{ti}) + \xi_{mi}, \quad m=1,...,M, \; i=m,...,M, \tag{2a}$$

$$p_{im} = p_{mi}, \quad m=1,...,M-1, \; i=m+1,...,M, \tag{2b}$$

where $\xi_{mi} = \sigma^2_{\varepsilon_m}$ for i=m and 0 otherwise.

The elements z_{mh} of matrix product $\boldsymbol{Z} = \boldsymbol{J}\boldsymbol{R}$ are given by

$$z_{mh} = \Sigma_{t=t_{1m}}^{t_{2m}} j_{mt}\, r_{th}, \quad m=1,...,M, \; h=1,...,H. \tag{3}$$

The elements of vectors $\boldsymbol{d} = [\boldsymbol{a} - \boldsymbol{J}\boldsymbol{s}]$ and $\boldsymbol{d}^{\dagger} = [\boldsymbol{a} - \boldsymbol{J}\boldsymbol{s}^{\dagger}]$ are given by:

$$d_m = a_m - \Sigma_{t=t_{1m}}^{t_{2m}} j_{mt}\, s_t,\ m=1,...,M, \tag{4a}$$

$$d_m^{\dagger} = a_m - \Sigma_{t=t_{1m}}^{t_{2m}} j_{mt}\, s_t^{\dagger},\ m=1,...,M. \tag{4b}$$

With these formulae, solution (3.20) can be implemented without generating the T by T matrix $\boldsymbol{V}_e$ nor the M by T matrix $\boldsymbol{J}$. In the case of $\boldsymbol{V}_e$, it is sufficient to store the T standard deviations, σ_t, and the T autocorrelations, $\omega(\ell)$ for lags $\ell = 0, 1,..., T-1$, in vectors. In the case of $\boldsymbol{J}$, it is sufficient to store the first and last sub-annual periods covered by each benchmark, t_{1m} and t_{2m}, and the positive coverage fractions j_{mt}, $t=t_{1m},..., t_{2m}$; $m=1,..., M$.
As a result of the above formulae, the solution (3.20) can be written as:

$$var[\hat{\boldsymbol{\beta}}] = (\boldsymbol{Z}'(\boldsymbol{V}_d^{-1}\boldsymbol{Z}))^{-1}, \tag{5a}$$

$$\hat{\boldsymbol{\beta}} = -\,var[\hat{\boldsymbol{\beta}}]\,(\boldsymbol{Z}'(\boldsymbol{V}_d^{-1}\boldsymbol{d})), \tag{5b}$$

$$\hat{\boldsymbol{\theta}} = \boldsymbol{s}^{\dagger} + \boldsymbol{Q}(\boldsymbol{V}_d^{-1}\boldsymbol{d}^{\dagger}), \tag{5c}$$

$$var[\hat{\boldsymbol{\theta}}] = \boldsymbol{V}_e - \boldsymbol{Q}\boldsymbol{V}_d^{-1}\boldsymbol{Q}' + \boldsymbol{W}\,var[\hat{\boldsymbol{\beta}}]\,\boldsymbol{W}', \tag{5d}$$

where $\boldsymbol{s}^{\dagger} = \boldsymbol{s} - \boldsymbol{R}\hat{\boldsymbol{\beta}}$, $\boldsymbol{W} = \boldsymbol{R} - \boldsymbol{Q}(\boldsymbol{V}_d^{-1}\boldsymbol{Z})$ and $\boldsymbol{V}_d = \boldsymbol{P}$ from (1b). The superfluous inner pairs of brackets suggest which calculations should be performed first, in order to minimize computations and to recycle as many matrix products as possible.

Note that matrix $\boldsymbol{V}_e$ need not be generated, especially if $var[\hat{\boldsymbol{\theta}}]$ is not required; in many cases, only the diagonal elements of $var[\hat{\boldsymbol{\theta}}]$ are required.

In the absence of $\boldsymbol{R}\boldsymbol{\beta}$, the solution (3.21) simplifies to:

$$\hat{\boldsymbol{\theta}} = \boldsymbol{s} + \boldsymbol{Q}(\boldsymbol{V}_d^{-1}\boldsymbol{d}), \tag{6a}$$

$$var[\hat{\boldsymbol{\theta}}] = \boldsymbol{V}_e - \boldsymbol{Q}\boldsymbol{V}_d^{-1}\boldsymbol{Q}'. \tag{6b}$$

Note that in most cases only the diagonal elements of $var[\,\hat{\boldsymbol{\theta}}\,]$ are needed. The formula to directly calculate the diagonal elements of matrix product $\boldsymbol{C} = \boldsymbol{Q}\,\boldsymbol{G}\,\boldsymbol{Q}'$ is

$$c_{tt} = \Sigma_{k=1}^{M}\, \Sigma_{m=1}^{M}\, q_{tm}\, g_{mk}\, q_{tk},\; t=1,...,T \tag{6c}$$

where matrices $\boldsymbol{Q}$ and $\boldsymbol{G} = \boldsymbol{V}_d^{-1}$ are of dimensions T by M and M by M respectively. Formula (6c) can also be applied to calculate the diagonal elements of $\boldsymbol{W}\, var[\,\hat{\boldsymbol{\beta}}\,]\, \boldsymbol{W}'$, *mutatis mutandis*. The result may of course be stored in a vector rather than a diagonal matrix.

Formulae for the C-D Multiplicative Model

In the case of multiplicative benchmarking discussed in Chapter 5, the matrix products appear in Eqs (5.9) and (5.10). It is advisable to read the above section on the additive benchmarking first.

The elements of the following matrix products: $\boldsymbol{Q}_0 = \boldsymbol{V}_{e*}\boldsymbol{J}_0'$, are given by

$$q_{tm} = \Sigma_{i=t_{1m}}^{t_{2m}}\, j_{mi}\, \theta_{0i}\, \omega(|t-i|)\, c_t c_i,\;\; t=1,...,T,\; m=1,...,M, \tag{7}$$

where the c_ts are the coefficients of variations of the s_ts and $\omega(|\ell|)$ are the autocorrelations of the standardized error described after Eq. (1).

The elements p_{mi} of matrix product $\boldsymbol{P}_0 = \boldsymbol{V}_d = (\boldsymbol{J}_0\,\boldsymbol{V}_{e*}\boldsymbol{J}_0' + \boldsymbol{V}_\varepsilon)$ are given by ,

$$p_{mi} = (\Sigma_{t=t_{1m}}^{t_{2m}} j_{mt}\, \theta_{0t}\, q_{ti}) + \xi_{mi},\; m=1,...,M,\; i=m,...,M, \tag{8a}$$

$$p_{im} = p_{mi},\; m=1,...,M-1,\; i=m+1,...,M, \tag{8b}$$

where $\xi_{mi} = \sigma_{\varepsilon_m}^2$ for i=m and 0 otherwise.

The elements z_{mh} of matrix product $\boldsymbol{Z}_0 = \boldsymbol{J}_0\boldsymbol{R}$ are given by

$$z_{mh} = \Sigma_{t=t_{1m}}^{t_{2m}} j_{mt}\, \theta_{0t}\, r_{th},\;\; m=1,...,M,\; h=1,...,H, \tag{9}$$

The elements of vectors $\boldsymbol{d}_0 = [\boldsymbol{a}_0 - \boldsymbol{J}_0\,\boldsymbol{s}]$ and $\boldsymbol{d}_0^{\dagger} = [\boldsymbol{a}_0 - \boldsymbol{J}_0\,\boldsymbol{s}^{\dagger}]$ are given by:

$$d_m = a_{0\,m} - \Sigma_{t=t_{1m}}^{t_{2m}} j_{m\,t}\,\theta_{0\,t}\,s_t^{\,*}, \tag{10a}$$

$$d_m^{\dagger} = a_{0\,m} - \Sigma_{t=t_{1m}}^{t_{2m}} j_{m\,t}\,\theta_{0\,t}\,s_t^{\,\dagger}, \quad m=1,...,M, \tag{10b}$$

where $a_{0\,m} = a_m - \Sigma_{t=t_{1m}}^{t_{2m}} j_{m\,t}\,\theta_{0\,t} + \Sigma_{t=t_{1m}}^{t_{2m}} j_{m\,t}\,\theta_{0\,t}\,\theta_{0\,t}^{*}$ and $s_t^{\dagger} = s_t^{\,*} - \Sigma_{h=1}^{H} r_{th}\,\hat{\beta}_h$ where the asterisks indicates the logarithmic scale.

As a result of the above formulae, solution (5.9) can be written as

$$var[\hat{\boldsymbol{\beta}}^*] = (\boldsymbol{Z}_0'(\boldsymbol{V}_d^{-1}\,\boldsymbol{Z}_0))^{-1}, \tag{11a}$$

$$\hat{\boldsymbol{\beta}}^* = -var[\hat{\boldsymbol{\beta}}^*]\,\boldsymbol{Z}_0'(\boldsymbol{V}_d^{-1}\,\boldsymbol{d}_0), \tag{11b}$$

$$\hat{\boldsymbol{\theta}}^* = \boldsymbol{s}^{\dagger} + \boldsymbol{Q}_0\,(\boldsymbol{V}_d^{-1}\,\boldsymbol{d}_0^{\dagger}), \tag{11c}$$

$$var[\,\hat{\boldsymbol{\theta}}^*] = \boldsymbol{V}_{e*} - \boldsymbol{Q}_0\,\boldsymbol{V}_d^{-1}\boldsymbol{Q}_0' + \boldsymbol{W}\,var[\hat{\boldsymbol{\beta}}^*]\,\boldsymbol{W}', \tag{11d}$$

where $\boldsymbol{s}^{\dagger} = \boldsymbol{s}^* - \boldsymbol{R}\hat{\boldsymbol{\beta}}^*$, $\boldsymbol{W} = \boldsymbol{R} - \boldsymbol{Q}_0\,(\boldsymbol{V}_d^{-1}\,\boldsymbol{Z}_0)$, and $\boldsymbol{V}_d = \boldsymbol{P}_0$.

In the absence of the regression part $\boldsymbol{R\beta}$, the solution (5.10) simplifies to:

$$\hat{\boldsymbol{\theta}}^* = \boldsymbol{s}^* + \boldsymbol{Q}_0\,(\boldsymbol{V}_d^{-1}\,\boldsymbol{d}_0), \tag{12a}$$

$$var[\,\hat{\boldsymbol{\theta}}^*] = \boldsymbol{V}_{e*} - \boldsymbol{Q}_0\,\boldsymbol{V}_d^{-1}\,\boldsymbol{Q}_0'. \tag{12b}$$

If only the diagonal elements of $var[\,\hat{\boldsymbol{\theta}}\,]$ are needed, formula (6c) can be used to directly calculate the diagonal elements of matrix product $\boldsymbol{Q}_0\,\boldsymbol{V}_d^{-1}\,\boldsymbol{Q}_0'$ and $\boldsymbol{W}\,var[\,\hat{\boldsymbol{\beta}}^*]\,\boldsymbol{W}'$, *mutatis mutandis.*

As explained after Eq. (4), solution (5.9) and (5.10) can be implemented without generating the larger matrices, namely $\boldsymbol{V}_{e*}$ and $\boldsymbol{J}_0$.

Appendix D: Seasonal Regressors

This appendix discusses various approaches to model seasonality in regression-based interpolation methods and in regression analysis in general.

Seasonal variations represent the effects of climatic and institutional factors which repeat regularly from year to year. They are related to weather, social and religious events, namely Christmas, the school year, the taxation year.

The seasonal pattern gives the relative importance of the months (say) of the year for certain variables, in terms of activity (flow) or level (stock). For example, Total Retail Trade Sales in Canada are about 25% higher in December than on an average month; and about 15% lower in January and February. The variations of series within years are primarily due to seasonality, which usually surpasses that of the business-cycle. Seasonality occurs in most sub-annual series, namely in quarterly, monthly, weekly data, but also in data with irregular sub-annual reporting periods, 4-weeks periods, 5-weeks periods, etc.

The econometric literature usually models seasonality by means of dummy variables. In the case of quarterly series, the first quarter is modelled by a dummy variable, which takes value 1 for the first quarter and 0 otherwise; and similarly, for the second and third quarters. For the fourth quarter, the three dummy variables take the value -1; this specification ensures that the seasonal component cancels out to 0 over any four consecutive quarters. The dummy variable approach is illustrated by the following regression

$$\begin{bmatrix} y_1 \\ y_2 \\ y_3 \\ y_4 \\ y_5 \\ y_6 \\ y_7 \\ y_8 \\ \vdots \end{bmatrix} = \begin{bmatrix} 1 & 1 \\ 1 & 2 \\ 1 & 3 \\ 1 & 4 \\ 1 & 5 \\ 1 & 6 \\ 1 & 7 \\ 1 & 8 \\ \vdots & \vdots \end{bmatrix} \begin{bmatrix} \alpha_1 \\ \alpha_2 \end{bmatrix} + \begin{bmatrix} 1 & 0 & 0 \\ 0 & 1 & 0 \\ 0 & 0 & 1 \\ -1 & -1 & -1 \\ 1 & 0 & 0 \\ 0 & 1 & 0 \\ 0 & 0 & 1 \\ -1 & -1 & -1 \\ \vdots & \vdots & \vdots \end{bmatrix} \begin{bmatrix} \beta_1 \\ \beta_2 \\ \beta_3 \end{bmatrix} + \begin{bmatrix} e_1 \\ e_2 \\ e_3 \\ e_4 \\ e_5 \\ e_6 \\ e_7 \\ e_8 \\ \vdots \end{bmatrix}, \qquad (1a)$$

$$y = Z\alpha + X\beta + e, \tag{1b}$$

where series y is assumed to start in a first quarter and the disturbance to follow a seasonal ARMA model. The matrix product $Z\alpha$ accounts for a linear trend, with constant α_1 and slope α_2. The matrix product $X\beta$ accounts for the seasonal component, where matrix X contains the three seasonal dummy variables, and vector β contains the additive seasonal coefficients of the three first quarters. The seasonal coefficient of the fourth quarter β_4 is implicitly defined as $-\Sigma_{h=1}^{3}\beta_h$.

Table D.1. Seasonal dummy regressors to estimate seasonality in quarterly series

q	(1)	(2)	(3)
1	1.000	0.000	0.000
2	0.000	1.000	0.000
3	0.000	0.000	1.000
4	-1.000	-1.000	-1.000

Table D.1 displays the seasonal dummy variables to be used to estimate seasonality in quarterly series. The regressors of Table D.1 also account for the length of each quarter, except for the first, which contains 91 days in leap years and 90 otherwise. In order to account for the length of the first quarter, the dummy variable in column (1) should be multiplied by 91/90.25 in leap years; and by 90/90.25 for non-leap year. The entries of the Table D.1 thus become

1	p	0.000	0.000
2	0.000	1.000	0.000
3	0.000	0.000	1.000
4	$-p$	-1.000	-1.000

where p is equal to 91/90.25 in leap years and 90/90.25 otherwise. The resulting estimated coefficients standardize the length of the first quarter.

For monthly series, the seasonal dummy regressors are similar to the quarterly regressors, with eleven dummy variables instead of three and with the December regressor equal to -1. These regressors account for the length of each month except for February, which can contains 29 or 28 days. In order account for the length of the month, the dummy variable for February should be multiplied by 29/28.25 in leap years; and by 28/28.25 other years. The resulting estimated coefficients standardize the length of February. This is equivalent to multiplying the monthly dummy variables by the following operator

$$\begin{bmatrix} 1 & 0 & 0 & \dots & 0 & 0 \\ 0 & p & 0 & \dots & 0 & 0 \\ 0 & 0 & 1 & \dots & 0 & 0 \\ \vdots & \vdots & \vdots & \ddots & \vdots & \vdots \\ 0 & 0 & 0 & \dots & 1 & 0 \\ -1 & -p & -1 & \dots & -1 & 0 \end{bmatrix}, \tag{2}$$

where p is equal to 29/28.25 in leap years and 28/28.25 otherwise.

The seasonal component can also be specified by means of sine and cosine functions (e.g. Pierce et al. 1984, Dagum 1988b, Harvey 1989). The quarterly effects η_t are then given by

$$\eta_t = \Sigma_{j=1}^{2} [\delta_j cos(\lambda_j t) + \gamma_j sin(\lambda_j t)], \tag{3}$$

where $\lambda_j = 2\pi j/4$ stands for the seasonal frequencies corresponding to cycles occurring once per year and twice per year.[1] Note that $sin(\lambda_2 t) = sin(\pi t)$ is never used because it is always equal to 0. The three remaining trigonometric functions appear in the columns of matrix $\boldsymbol{X}$ of equation (4).

$$\begin{bmatrix} y_1 \\ y_2 \\ y_3 \\ y_4 \\ y_5 \\ y_6 \\ y_7 \\ y_8 \\ \vdots \end{bmatrix} = \begin{bmatrix} 1 & 1 \\ 1 & 2 \\ 1 & 3 \\ 1 & 4 \\ 1 & 5 \\ 1 & 6 \\ 1 & 7 \\ 1 & 8 \\ \vdots & \vdots \end{bmatrix} \begin{bmatrix} \alpha_1 \\ \alpha_2 \end{bmatrix} + \begin{bmatrix} 0 & 1 & -1 \\ -1 & 0 & 1 \\ 0 & -1 & -1 \\ 1 & 0 & 1 \\ 0 & 1 & -1 \\ -1 & 0 & 1 \\ 0 & -1 & -1 \\ 1 & 0 & 1 \\ \vdots & \vdots & \vdots \end{bmatrix} \begin{bmatrix} \beta_1 \\ \beta_2 \\ \beta_3 \end{bmatrix} + \begin{bmatrix} e_1 \\ e_2 \\ e_3 \\ e_4 \\ e_5 \\ e_6 \\ e_7 \\ e_8 \\ \vdots \end{bmatrix}, \tag{4a}$$

$$\boldsymbol{y} = \boldsymbol{Z\alpha} + \boldsymbol{X\beta} + \boldsymbol{e}, \tag{4b}$$

In structural time series modelling, seasonality is often specified by means of these trigonometric functions. One advantage of the trigonometric specification is that regressors with non-significant parameters can be

[1] Sines and cosines repeat every 2π. A good approximation of π is 3.1415926536.

dropped from the regression, thus achieving a more parsimonious specification. The resulting seasonal pattern always sums to zero for any four consecutive quarters, because each regressor does so as can be verified in Eq. (4a).

Despite dropping the non-significant regressors, the seasonal pattern is still defined for all four quarters, because each trigonometric function affects all quarters. For example, if only the first column of $\boldsymbol{X}$ is kept in the model (i.e. $cos(\pi t/2)$), the seasonal factors for the four quarters are respectively equal to 0, $-\hat{\beta}_1$, 0 and $\hat{\beta}_1$. If the first two columns are kept, the seasonal factors are respectively equal to $\hat{\beta}_2$, $-\hat{\beta}_1$, $-\hat{\beta}_2$ and $\hat{\beta}_1$. If all columns of $\boldsymbol{X}$ are kept, the seasonal factors are equal to $\hat{\beta}_2-\hat{\beta}_3$, $-\hat{\beta}_1+\hat{\beta}_3$, $-\hat{\beta}_2-\hat{\beta}_3$ and $\hat{\beta}_1+\hat{\beta}_3$. In the latter case, the trigonometric seasonal component $\boldsymbol{X\beta}$ in (4) produces the same estimates as the usual specification in (1).

Table D.2. Trigonometric regressors to estimate seasonality in quarterly series

q	$cos(\lambda_1 q)$	$sin(\lambda_1 q)$	$cos(\lambda_2 q)$
1	0.000	1.000	-1.000
2	-1.000	0.000	1.000
3	0.000	-1.000	-1.000
4	1.000	0.000	1.000

q : quarter; $\lambda_j = 2\pi j/4$

In order to standardize time across series starting at different dates, the quarter q is used instead of time t as shown in Table D.2.

In order to standardize for the varying length of the first quarter, the first row of Table D.2 is multiplied by 91/90.25 in leap years and by 90/(90.25×3) for non-leap year. The last row of the table is redefined as minus the sum of the first three rows. The entries of Table D.2 thus become:

1	0.000	p	$-p$
2	-1.000	0.000	1.000
3	0.000	-1.000	-1.000
4	1.000	$1-p$	p

This transformation is equivalent to multiplying the entries of Table D.2 by the following matrix

$$\begin{bmatrix} p & 0 & 0 & 0 \\ 0 & 1 & 0 & 0 \\ 0 & 0 & 1 & 0 \\ -p & -1 & -1 & 0 \end{bmatrix}, \tag{5}$$

where p is equal to 91/90.25 in leap year and 90/90.25 otherwise. In principle, this transformation is not required for stock series.

Table D.3. Trigonometric regressors to estimate seasonal patterns in monthly series

q	(1)	(2)	(3)	(4)	(5)	(6)	(7)	(8)	(9)	(10)	(11)
1	0.87	0.50	0.50	0.87	-0.00	1.00	-0.50	0.87	-0.87	0.50	-1.00
2	0.50	0.87	-0.50	0.87	-1.00	-0.00	-0.50	-0.87	0.50	-0.87	1.00
3	-0.00	1.00	-1.00	-0.00	0.00	-1.00	1.00	0.00	-0.00	1.00	-1.00
4	-0.50	0.87	-0.50	-0.87	1.00	0.00	-0.50	0.87	-0.50	-0.87	1.00
5	-0.87	0.50	0.50	-0.87	-0.00	1.00	-0.50	-0.87	0.87	0.50	-1.00
6	-1.00	-0.00	1.00	0.00	-1.00	-0.00	1.00	0.00	-1.00	-0.00	1.00
7	-0.87	-0.50	0.50	0.87	0.00	-1.00	-0.50	0.87	0.87	-0.50	-1.00
8	-0.50	-0.87	-0.50	0.87	1.00	0.00	-0.50	-0.87	-0.50	0.87	1.00
9	0.00	-1.00	-1.00	-0.00	-0.00	1.00	1.00	0.00	0.00	-1.00	-1.00
10	0.50	-0.87	-0.50	-0.87	-1.00	-0.00	-0.50	0.87	0.50	0.87	1.00
11	0.87	-0.50	0.50	-0.87	0.00	-1.00	-0.50	-0.87	-0.87	-0.50	-1.00
12	1.00	0.00	1.00	0.00	1.00	0.00	1.00	0.00	1.00	0.00	1.00

q: month; (1) and (2): $cos(\lambda_1 q)$, $sin(\lambda_1 q)$;
(3) and (4): $cos(\lambda_2 q)$, $sin(\lambda_2 q)$, etc.; $\lambda_j = 2\pi j/12$

The monthly seasonal component can also be specified by means of sine and cosine functions. The monthly seasonal effects η_t are then given by

$$\eta_t = \Sigma_{j=1}^{6} [\delta_j cos(\lambda_j t), + \gamma_j sin(\lambda_j t)], \tag{6}$$

where $\lambda_j = 2\pi j/12$, $j=1,2,...,6$, stand for the seasonal frequencies. These frequencies correspond to cycles lasting $12/j$ months, i.e. 12, 6, 4, 3, 2.4 and 2 months respectively. The function $sin(2\pi 6t/12) = sin(\pi t)$ is always null and never included in Eq. (6). The eleven first functions take values displayed in Table D.3. Similarly to quarterly series, time is standardized by replacing t by the month, denoted by q in the table. This standardizes the trigonometric functions across series starting in different months.

In order to standardize for the varying length of February, the second row of Table D.3 is multiplied by 29/28.25 in leap years and by 28/28.25 in other years. The last row is re-defined as minus the sum of the first eleven rows. This is equivalent to multiplying the entries of Table D.3 by the following matrix of dimension 12×12

$$\begin{bmatrix} 1 & 0 & 0 & \dots & 0 & 0 \\ 0 & p & 0 & \dots & 0 & 0 \\ 0 & 0 & 1 & \dots & 0 & 0 \\ \vdots & \vdots & \vdots & \ddots & \vdots & \vdots \\ 0 & 0 & 0 & \dots & 1 & 0 \\ -1 & -p & -1 & \dots & -1 & 0 \end{bmatrix}, \tag{7}$$

where p is equal to 29/28.25 in leap years and 28/28.25 otherwise.

The seasonal component in weekly data can also be specified trigonometric regressors as follows:

$$\eta_t = \Sigma_{j=1}^{26} \delta_j cos(\lambda_j t), + \gamma_j sin(\lambda_j t), \tag{8}$$

where $\lambda_j = 2\pi j/52$, 1, 2, ..., 26 stand for the seasonal frequency corresponding to one cycle per 52 weeks, two cycles per 52 weeks, and so forth. The function $sin(\lambda_{26} t)$ is never used because it is always equal to 0.

One problem with these regressors is that 52 weeks covers 364 days, ignoring one or two days per year. For a given week of the year (e.g. the third complete week of the first year), the seasonal effect is measured one or two days earlier from year to year, depending on the configuration of the leap years.

A simple solution is to redefine the frequency as $\lambda_j = 2\pi j/(52+1.25/7)$. This specification provides a seasonal pattern lasting 52 and 1.25/7 weeks. Weekly data would provide contiguous measurements everywhere along the seasonal pattern without omitting any day.

However, the last frequency $\lambda_{26} = 26/(52+1.25/7)$ is equal to 0.498288 with a duration of 2.006868 weeks. Perhaps it would be preferable to make the last frequency equal to $\lambda_{26} = 2\pi 26/52$, as this two-week cycle is meaningful to capture important phenomena like bi-weekly pay-days. Usually a small subset of the trigonometric functions is required (Pierce et al. 1984, Harvey et al. 1997).

Note that adding one cosine and one sine functions of the same periodicity defines a new cosine wave with new amplitude and phase-shift. Fig. D.1a exhibits one pair of *monthly* cosine and sine, namely $10cos(\lambda_1 q)$ and $7sin(\lambda_1 q)$, and their sum. Adding the sine wave amplifies the original cosine and shifts its peaks and troughs to the right. This feature is useful to adjust the timing of the cosine to the data.

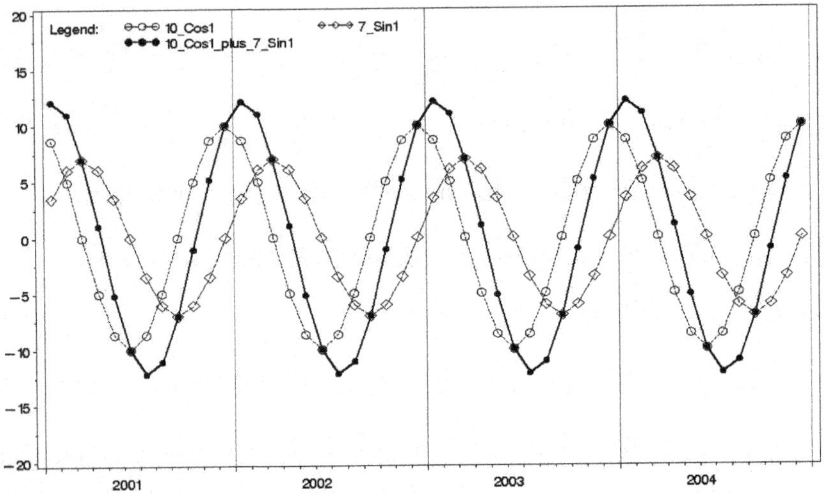

Fig. D.1a. Fundamental (12-month) cosine and sine functions of amplitudes 10 and 7 respectively, and their sum

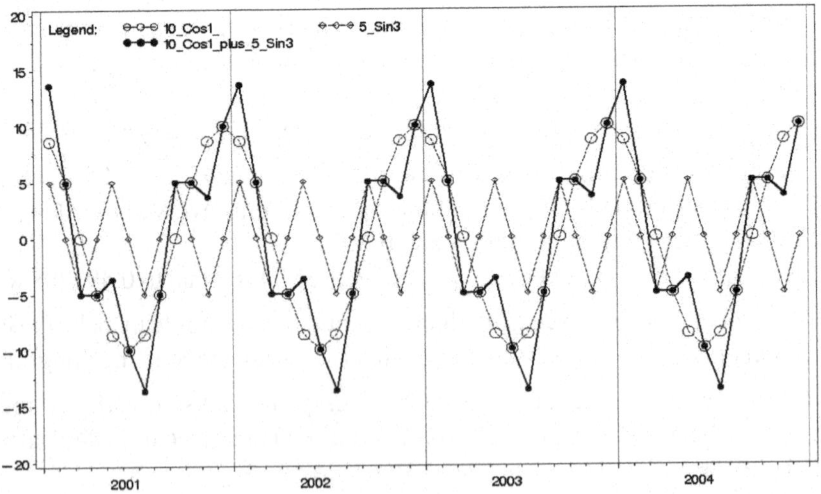

Fig. D.1b. Fundamental (12-month) cosine of amplitude 10 and 4-month sine of amplitude 5 and their sum

Adding a few of these trigonometric functions can produce intricate seasonal patterns. Fig. D.1b illustrates this point by adding $5 sin(\lambda_3 t)$ of 4-month to the cosine $10 cos(\lambda_1 t)$ of 12-month duration. The resulting seasonal pattern is more complex than the two original functions. Each trigonometric term affects the whole seasonal pattern, i.e. the twelve seasonal factors. When all eleven trigonometric regressors of Eq. (6) are used, the resulting daily pattern coincides with that obtained with the six dummy regressors.

For both the dummy and the trigonometric regressors, the seasonal pattern is given by $\boldsymbol{X}\hat{\boldsymbol{\beta}}$, where vector $\hat{\boldsymbol{\beta}} = \left[\hat{\beta}_1 \dots \hat{\beta}_H\right]'$ contains the H estimated seasonal parameters and matrix $\boldsymbol{X}$ contains the regressors as exemplified by Eqs. (1) and (4) respectively.

There are cases where the dummy approach may be more parsimonious, namely when two or more quarters (say) share the same seasonal coefficient. The left portion of Table D4 displays the quarterly seasonal dummy regressors when quarters 1 and 2 share the same coefficient; and the right part, when quarters 1 and 2 share the same coefficient and similarly for quarters 3 and 4. Note that the column totals are equal to 0. It is possible to derive similar dummy variables for monthly series.

Table D.4. Seasonal dummy regressors to estimate seasonality in quarterly series when some quarters share the same seasonal coefficient

Quarters 1 and 2 sharing same coefficient			Quarters 1 and 2 and quarters 3 and 4 sharing same coefficient
q	(1)	(2)	(1)
1	1.000	0.000	1.000
2	1.000	0.000	1.000
3	0.000	1.000	-1.000
4	-2.000	-1.000	-1.000

A commonly used specification to account for seasonality in state-space models is the following. The current monthly seasonal factor η_t is equal to minus the sum of the previous eleven factors

$$\eta_t = -\Sigma_{j=1}^{11} \eta_{t-j} + \zeta_t, \quad t=12,13,\dots,T, \tag{9}$$

where ζ_t is an innovation of mean 0 and constant variance. If the variance of ζ_t is zero, the seasonal effects are the same from year to year. Otherwise, this stochastic model is unacceptable to describe seasonal behaviour. Indeed, the seasonal effect of a month fluctuates up and down from year to year. This even happens when the innovations are autocorrelated. It is already known that seasonality usually evolves gradually and in a smooth manner and not erratically.

Appendix E: Trading-Day Regressors

Trading-day variations consist of fluctuations attributed to the relative importance of the days for some socio-economic variables, in terms of activity, e.g. sales, births; or in terms of level, e.g. prices, inventories. The trading-day component implies the existence of a daily pattern analogous to the seasonal pattern. However, these "daily factors" are usually referred to as the *daily coefficients*. Depending on the socio-economic variable (e.g. sales of food, sales of gasoline, births, deaths, green-house gaz emissions), some days may be 50% more important than an average day and other days, 80% less important.

If the more important days of the week appear 5 times in a month (instead of 4), the month may record an excess of activity in the order of 4% (say). If the less important days appear 5 times, the month may record a short-fall in the order of 4%. As a result, the monthly *trading-day component* can cause increase of +8% or −8% between neighbouring months and between same-months of neighbouring years.

Table E.1. Dummy regressors to estimate daily patterns over one week in daily series

	τ	(1)	(2)	(3)	(4)	(5)	(6)
M	1	1.000	0.000	0.000	0.000	0.000	0.000
T	2	0.000	1.000	0.000	0.000	0.000	0.000
W	3	0.000	0.000	1.000	0.000	0.000	0.000
T	4	0.000	0.000	0.000	1.000	0.000	0.000
F	5	0.000	0.000	0.000	0.000	1.000	0.000
S	6	0.000	0.000	0.000	0.000	0.000	1.000
S	7	-1.000	-1.000	-1.000	-1.000	-1.000	-1.000

Note: τ is the standardized daily time index

Like for the seasonal pattern, the daily pattern defined over one week may be specified either by means of dummy variables or trigonometric functions (e.g. Pierce et al. 1984; Harvey 1989). The dummy variables to estimate daily patterns in daily series are displayed in Table E.1, where τ stands for *daily time*: 1 for Monday (M), 2 for Tuesday (T), etc. They are very similar to the dummy variables for seasonality.

The trigonometric functions describing the daily pattern are,

$$cos(\lambda_j \tau),\ sin(\lambda_j \tau),\ i=1,...,3, \tag{1}$$

where $\lambda_j = 2\pi j/7$ stand for the frequency corresponding to one cycle per week, two and three cycles per week. The values of the functions of Eq. (1) are displayed in Table E.2. A sufficient standardization is given by $\tau=1$ for Mondays (say), $\tau=2$ for Tuesdays, etc.

Table E.2. Trigonometric regressors to estimate daily patterns over one week in daily series

	τ	$cos(\lambda_1\tau)$	$sin(\lambda_1\tau)$	$cos(\lambda_2\tau)$	$sin(\lambda_2\tau)$	$cos(\lambda_3\tau)$	$sin(\lambda_3\tau)$
M	1	0.623	0.782	-0.223	0.975	-0.901	0.434
T	2	-0.223	0.975	-0.901	-0.434	0.623	-0.782
W	3	-0.901	0.434	0.623	-0.782	-0.223	0.975
T	4	-0.901	-0.434	0.623	0.782	-0.223	-0.975
F	5	-0.223	-0.975	-0.901	0.434	0.623	0.782
S	6	0.623	-0.782	-0.223	-0.975	-0.901	-0.434
S	7	1.000	0.000	1.000	0.000	1.000	0.000

Note: τ is the standardized daily time index; $\lambda_j = 2\pi j/7$

The trigonometric approach usually describes trading-day patterns with fewer parameters than the widely applied dummy variable approach. The daily effects η_t are then given by

$$\eta_t = \Sigma_{j=1}^{3} \delta_j cos(\lambda_j \tau), + \gamma_j sin(\lambda_j \tau), \tag{2}$$

where $\lambda_j = 2\pi j/7$ is the frequency: 1, 2, and 3 cycles per 7 days. The six functions take values displayed in Table E.2.

In order to estimate trading-day variations in monthly flow series, the monthly sums of the regressors in Tables E.1 or E.2 are used in additive models. The monthly sums of the dummy variables of Table E.1 take the following values for the months indicated:

Jan/2004	0.000	0.000	0.000	1.000	1.000	1.000
Feb/2004	-1.000	-1.000	-1.000	-1.000	-1.000	-1.000
Mar/2004	1.000	1.000	1.000	0.000	0.000	0.000

In Jan. 2004, Thursday, Friday and Saturday occurred five times (instead of four); in Feb., Sunday occurred five times (leap-year); and in March, Monday, Tuesday and Wednesday.

The corresponding monthly sums for the trigonometric regressors of Table E.2 are:

Jan/2004	-0.500	-2.191	-0.500	0.241	-0.500	-0.627
Feb/2004	1.000	0.000	1.000	0.000	1.000	0.000
Mar/2004	-0.500	2.191	-0.500	-0.241	-0.500	0.627

In our experience, daily patterns for the monthly Canadian Retail Trade series can be described by as few as three parameters. The reason may be that statistically it not easy to estimate the daily pattern even in monthly series. Indeed monthly data record the sums of two or three extra days (since the 28 other days cancel out to 0). This reduces the variability of the monthly data caused by of the various days, especially for the cosines and sines of frequency 2/7 and 3/7, i.e. of duration 3.5 days and 2 1/3 days. If all months contained 29 days, it would be statistically easier to estimated trading-day variations: there would be not doubt as to which day caused the excess or short-fall of activity. Ideally, the methods to estimate daily patterns should also rely on prior daily patterns based on extra information.

Trading-day variations are rarely identified in quarterly series. However, there are cases where an extra day in a quarter can influence not so much the level but movement between quarters.

In order to estimate trading-day variations in quarterly flow series, the quarterly sums of the regressors in Tables E.1 or E.2 are used in additive models. The quarterly sums of the dummy variables of Tables E.1 for year 2004 take the following values:

Q1/2004	0.000	0.000	0.000	0.000	0.000	0.000
Q2/2004	0.000	0.000	0.000	0.000	0.000	0.000
Q3/2004	0.000	0.000	0.000	1.000	0.000	0.000
Q4/2004	0.000	0.000	0.000	0.000	1.000	0.000

The four quarters of 2004 respectively contain 91, 91, 92 and 92 days. In the first and second quarters of 2004, the seven days of the week occur 13 times. As a result, no days appear an extra time to produce trading-day variations. In the third quarter, Thursday occurs 14 times instead of 13; and in the fourth quarter Friday, occurs 14 times. Quarterly series thus provide very few degrees of freedom to estimate trading-day variations, especially in leap years like 2004. In non-leap years, the first quarter contains 90 days, as a result six of the days appearing 13 times and one day appears 12 times.

The quarterly sums of the trigonometric regressors of Table E.2 for year 2004 take the following values:

Q1/2004	0.000	0.000	0.000	0.000	0.000	0.000
Q2/2004	0.000	0.000	0.000	0.000	0.000	0.000
Q3/2004	-0.901	-0.434	0.623	0.782	-0.223	-0.975
Q4/2004	-0.223	-0.975	-0.901	0.434	0.623	0.782

Note that the first two quarters of 2004 contains only zeroes, both under the dummy and the trigonometric specification, meaning the absence of trading-day variations.

In the case multiplicative or logarithmic models, the monthly or quarterly sums are divided by the number of days in the month or quarter.

The monthly and quarterly trading-day regressors are then

$$(\Sigma_{\tau \in t} cos(\lambda_j \tau))/n_t, \quad (\Sigma_{\tau \in t} sin(\lambda_j \tau))/n_t, \quad j=1,...,3, \tag{3}$$

where $\lambda_j = 2\pi j/7$, τ stands for the days in the month or quarter. In additive decomposition models, parameter n_t is equal to 1; and in multiplicative (logarithmic) models, to the number of days in the month or quarter t. The reason for this denominator is to conform with the standard interpretation of the daily pattern in multiplicative models, namely that the daily pattern should sum to 7 over each week (Young, 1965).

When all six trigonometric functions are used, the trigonometric specification of trading-day variations produces the same daily patterns as the classical dummy variable approach. When the non-significant parameters are excluded, a more parsimonious specification is achieved.

The trigonometric specification of trading-day variations also addresses the problem of non-significant parameters for some of the days, encountered in the dummy approach. Quenneville et al. (1999) used this specification to estimate abrupt changes in the daily pattern, resulting from Sunday opening of stores in the 1990s.

There are cases where the dummy variable approach to trading-day variations is more parsimonious, namely when some days of the weeks have the same importance (coefficient). Table E.3 displays on the left side the daily regressors to be used when the first five days of the week (Monday to Friday) share the same coefficient and Saturday and Sunday have their own coefficient. On the right, the table displays the daily regressors to be used when the first five days of the week share the same coefficient and Saturday and Sunday share the same coefficient. In the latter case, the trigonometric specification requires all six trigonometric regressors.

Table E.3. Dummy regressors to estimate daily patterns in daily series some days share the same coefficients

	τ	Monday to Friday		Monday to Friday Saturday and Sunday
M	1	1.000	0.000	1.000
T	2	1.000	0.000	1.000
W	3	1.000	0.000	1.000
T	4	1.000	0.000	1.000
F	5	1.000	0.000	1.000
S	6	0.000	1.000	-2.500
S	7	-5.000	-1.000	-2.500

Note: τ is the standardized daily time index

In order to estimate trading-day variations in monthly flow series, the monthly sums of the regressors in Tables E.3 are used for additive models. The monthly sums of the dummy variables of Table E.3 take the following values for the months indicated:

	(1)	(2)	(1)
Jan/2004	2.000	1.000	-0.500
Feb/2004	-5.000	-1.000	-2.500
Mar/2004	3.000	0.000	3.000

In Jan. 2004, Thursday, Friday and Saturday occur five times (instead of four); in Feb., Sunday occur five times; and in March, Monday, Tuesday and Wednesday.

Bi-Weekly Daily Pattern

In some cases the daily pattern is defined over two weeks instead of one, because many employers pay their employees every second week, typically on Wednesday, Thursday or Friday. The existence of such a daily pattern is more prevalent in industrial sectors characterized by few firms (e.g. mining). As a result, two or three times per year, some months contain three bi-weekly pay-days instead of two, which can causes an excess of "activity" in order of 50%. Similarly, some quarters may contain from six to eight pay-days instead of six, which can cause an excess of 17% or 25%. Indeed, depending on the distribution of the days, 90-day quarters may contain six or seven bi-weekly paydays; 91-day quarter, seven pay-days; and 92-day quarters, seven or eight paydays. Note that for a monthly value to exhibit lamented excesses of 50% requires that six weeks of data was assigned to that specific month.

The trigonometric approach should be particularly efficient in dealing with such situations, because of its parsimony. The appropriate daily regressors analogous to those in Table E.2 are then

$$cos(\lambda_j \tau),\ sin(\lambda_j \tau)\ ,\ j=1,...,7, \tag{4}$$

where $\lambda_j = 2\pi j\tau/14$ the frequency: 1, 2, ...7 cycles per 14 days. The last sine is never used because always equal to 0.

For flow series, the monthly and quarterly regressors would then be

$$(\Sigma_{\tau\in t}\, cos(\lambda_j \tau))/n_t,\ (\Sigma_{\tau\in t}\, sin(\lambda_j \tau))/n_t\ ,\ i=1,...,7, \tag{5}$$

where τ stands for the days in the month or quarter. The number n_t is equal to 1 in additive decomposition models and to the number of days in the month or quarter t in multiplicative (logarithmic) models.

In order to use these regressors on monthly or quarterly data, the data must cover only the days in each month or quarter. If this were the case, the data could be corrected for bi-weekly trading-day variations.

References

Aadland DM (2000) Distribution and interpolation using transformed data. Journal of Applied Statistics 27:141-156.

Alba E de (1988) Disaggregation and forecasting: a bayesian analysis. Journal of Business and Economic Statistics 6:197-206.

Ansley CF, Kohn R (1985) Estimation, filtering and smoothing in state space models with incompletely specified initial conditions. Annals of Statistics 13:1286-1316.

Bacharach M (1965) Estimating nonnegative matrices from marginal data. International Economic Review 6:294-310.

Bacharach M (1971) Biproportional matrices and input-output change. Cambridge University Press, Cambridge.

Barker TF, Ploeg van der F, Weale MR (1984) A balance system of national accounts for the United Kingdom. Review of Income and Wealth, Series 30 4:451-486.

Bell WR (1984) Signal extraction for nonstationary time series. The Annals of Statistics 12:646-664.

Bell WR, Hillmer SC (1984) Issues involved with the seasonal adjustment of economic time series. Journal of Business and Economic Statistics 2:291-320.

Bell WR, Hillmer SC (1990) The time series approach to estimation for repeated surveys. Survey Methodology 16:195-215.

Bell WR and Wilcox DW (1993) The effect of sampling errors on the time series behavior of consumption data. Journal of Econometrics 55:235-265.

Binder DA and Dick JP (1990) Analysis of seasonal ARIMA models from survey data. In: Sing AC and Withridge P (eds) Analysis of Data in Time, Proceedings of the 1989 International Symposium, Statistics Canada, Ottawa, pp 57-65.

Binder DA, Bleuer SR, Dick JP (1993) Time series methods applied to survey data. Bulletin of the International Statistical Institute, Proceedings of the 49th Session, Book 1, pp 327-344.

Bloem AM, Dippelsman RJ, Mæhle ON (2001) Quarterly national accounts manual, concepts, data sources, and compilation. Washington, DC: International Monetary Fund (//www.imf.org/external/pubs).

Boot JCG, Feibes W, Lisman JHC (1967) Further methods of derivation of quarterly figures from annual data. Applied Statistics 16:65-75

Bournay J, Laroque G (1979) Réflexions sur la méthode d'élaboration des comptes trimestriels. Annales de l'I.N.S.E.E. 36:3-30.

Box GEP, Jenkins GM (1970, 1976) Time series analysis, forecasting and control. Holden-Day, San Francisco.

Box G, Cox D (1964) An Analysis of Transformations. Journal of the Royal Statistical Society, Series B, pp 211-264.

Box GEP, Tiao GC (1975) Intervention analysis with applications to economic and environmental problems. Journal of the American Statistical Association 70:70-79.

Burman JP (1980) Seasonal adjustment by signal extraction. Journal of the Royal Statistical Society, Series A 143:321-337.

Byron RP (1978) The estimation of large social account matrices. Journal of the Royal Statistical Society, Series A 141:359-367.

Chen ZG, Cholette PA, Dagum EB (1993) A nonparametric time series approach for benchmarking survey data. Proceedings of the American Statistical Association, Business and Economic Statistics Section, pp 181-186.

Chen ZG, Cholette PA, Dagum EB (1997) A nonparametric method for benchmarking survey data via signal extraction. Journal of the American Statistical Association 92:1563-1571.

Chen ZG, Dagum EB (1997) A recursive method for predicting variables with temporal and contemporaneous constraints. Proceedings of the American Statistical Association, Business and Economic Statistics Section, pp 229-233.

Chhab N, Morry M, Dagum EB (1999) Further results on alternative trend-cycle estimation for current economic analysis. Estadistica 51:231-257.

Cholette PA, Dagum EB (1994) Benchmarking time series with autocorrelated sampling errors. International Statistical Review 62:365-377.

Cholette PA (1979) Adjustment methods of sub-annual series to yearly benchmarks. In: Gentleman JF (ed) Proceedings of the Computer Science and Statistics, 12th Annual Symposium on the Interface, University of Waterloo, pp 358-366.

Cholette PA (1982) Prior information and ARIMA forecasting. Journal of Forecasting 1:375-383.

Cholette PA (1990) L'annualisation des chiffres d'exercices financiers. L'actualité économique 66:218-230.

Cholette PA (1990) Transforming fiscal quarter data into calendar quarter values. In: Singh AC and Withridge P (eds) Analysis of data in time, Proceedings of the 1989 International Symposium, Statistics Canada, Ottawa, pp 131-141.

Cholette PA, Chhab N (1991) Converting aggregates of weekly data into monthly values. Applied Statistics 40:411-422.

Chow GC, Lin AL (1971) Best linear unbiased interpolation, distribution and extrapolation of time series by related series. The Review of Economics and Statistics 53:372-375.

Chow GC, Lin AL (1976) Best linear unbiased estimation of missing observations in an economic time series. Journal of the American Statistical Association 71:719-721.

Cleveland WP, Tiao GC (1976) Decomposition of seasonal time series: a model for the Census X-11 program. Journal of the American Statistical Association 71:581-587.

Cleveland RB, Cleveland WS, McRae JE, Terpenning IJ (1990) STL a seasonal trend decomposition procedure based on LOESS. Journal of Official Statistics 6:3-33.

Cohen KJ, Müller W, Padberg MW (1971) Autoregressive approaches to the disaggregation of time series data. Applied Statistics 20:119-129.

Dagum EB (1980) The X-11-ARIMA seasonal adjustment program. Statistics Canada, Cat. 12-564E.

Dagum EB (1988a) The X-11-ARIMA seasonal adjustment method, foundations and users manual. Time Series Research and Analysis Centre, Statistics Canada, Ottawa, Canada.

Dagum EB (1988b) Seasonality. In: Kotz S and Johnson N (eds) Encyclopedia of Statistical Sciences, John Wiley and Sons, New York, 8, pp 321-326.

Dagum EB (2001) Time series: seasonal adjustment. In: Smelser NJ Baltes PB (eds), International Encyclopedia of Social and Behavioral Science, Vo. II - Methodology Elsevier Sciences Publishers, Amsterdam, pp 15739-15746.

Dagum EB, Cholette PA, Chen ZG (1998) A unified view of signal extraction, benchmarking, interpolation and extrapolation of time series. International Statistical Review 66:245-269.

Dagum EB, Cholette PA (1999) An extension of the Gauss-Markov theorem for mixed linear regression models with non-stationary stochastic parameters. In: Slottje DJ (ed) Advances in Econometrics, Income Distribution and Scientific Methodology Essays in Honor of Professor Camilo Dagum, Physica-Verlag, pp 27-39.

Dagum C, Dagum EB (1988) Trend. In: Kotz S, Jonhson NL (eds) Encyclopedia of statistical sciences, John Wiley and sons, New York, 9, pp 321-324.

Dagum EB, Huot G, Morry M (1988) Seasonal adjustment in the 80's: some problems and solutions. The Canadian Journal of Statistics 16:109-126.

Dagum EB, Luati A (2000) Predictive performances of some nonparametric linear and nonlinear smoothers for noisy data. Statistica 4:635-654.

Dagum EB, Quenneville B, Sutradhar B (1992) Trading-day variations multiple regression models with random parameters. International Statistical Review 60:57-73.

Deming WE and Stephan DF (1940) On the least squares adjustment of a sampled frequency table when the expected marginal totals are known. Annals of Mathematical Statistics 11:427-444.

Denton FT (1971) Adjustment of monthly or quarterly series to annual totals: an approach based on quadratic minimization. Journal of the American Statistical Association 66:99-102.

Di Fonzo T (1990) The estimation of M disaggregated time series when contemporaneous and temporal aggregates are known. The Review of Economics and Statistics 72:178-182.

Di Fonzo T (2003) Temporal disaggregation of economic time series: towards a dynamic extension. Dipartimento di Scienze Statische, Università di Padova.

Durbin J, Quenneville B (1997) Benchmarking by state space models. International Statistical Review 65:23-48.

Fernandez RB (1981) A methodological note on the estimation of time series. Review of Economic and Statistics 63:471-476.

Findley DF, Monsell BC, Bell WR, Otto MC, Chen BC (1998) New capabilities and methods of the X-12-ARIMA seasonal-adjustment program. Journal of Business and Economic Statistics 16:127-177.

Ginsburgh VA (1973) A further note on the derivation of quarterly figures consistent with annual data. Applied Statistics 22:368-374.

Golan A, Judge G, Robinson S (1994) Recovering information from incomplete or partial multisectoral economic data. The Review of Economics and Statistics 76:541-549.

Golan A, Judge G, Robinson S (1996) Recovering information from incomplete or partial multisectoral economic data. Environment and Planning, A 9:687-701.

Gómez V, Maravall A (1996) Programs Tramo and Seats, instructions for the user. Banco de España, Servicio de Estudios, documento de Trabajo No. 9628.

Gray A, Thomson P (1996) Design of moving-average trend filters using fidelity and smoothness criteria. In: Robinson PM and Rosemblatt M (eds) Time series analysis (in memory of EJ Hannan), Springer Lecture Notes in Statistics No 115, New York, 2, pp 205-219,

Grégoir S (1995) Propositions pour une désagrégation temporelle basée sur des modèles dynamiques simples. I.N.S.E.E, Département des Comptes nationaux, document No 22/G430.

Gudmundson G (1999) Disaggregation of annual flow data with multiplicative trends. Journal of Forecasting 18:33-37

Guerrero VM (1989) Optimal conditional ARIMA forecast. Journal of Forecasting 8:215-229.

Guerrero VM (1990) Temporal disaggregation of time series: an ARIMA-based approach. International Statistical Review 58:29-46.

Guerrero VM, Peña D (2000) Linear combination of restrictions and forecasts in time series analysis. Journal of Forecasting 19:103-122.

Hannan EJ (1967) Measurement of a wandering signal amid noise. Journal of Applied Probability 4:90-102.

Harvey AC (1981) Time series models. Oxford: Phillip Allan Publishers Limited.

Harvey AC (1985) Trends and cycles in macroeconomic time series. Journal of Business and Economic Statistics 3:216-227.

Harvey AC (1989) Forecasting, structural time series models and the Kalman filter. Cambridge University Press, New York, Melbourne.

Harvey AC, Koopman JS, Riani M (1997) The modelling and seasonal adjustment of weekly observations. Journal of Business and Economic Statistics 15:354-368.

Harvey AC, Pierse RG (1984) Estimating missing observations in econcomic time series. Journal of the American Statistical Association 79:125-131.

Harville DA (1976) Extension of the Gauss-Markov theorem to include the estimation of random effects. Annals of Statistics 4:384-395.

Helfand SD, Monsour NJ, Trager ML (1977) Historical revision of current business survey estimates. Proceedings of the Business and Economic Statistics Section, American Statistical Association, pp 246-250.

Henderson R (1916) Note on graduation by adjusted average. Transactions of the Actuarial Society of America 17:43-48.

Henderson CR (1975) Best linear unbiased estimation and prediction under a selection model. Biometrics 31:423-447.

Hillmer SC, Tiao GC (1982) An ARIMA based approach to seasonal adjustment. Journal of the American Statistical Association 77:63-70.

Hillmer SC, Trabelsi A (1987) Benchmarking of economic time series. Journal of the American Statistical Association 82:1064-1071.

Hotta LK, Vasconellos KL (1999) Aggregation and disaggregation of structural time series models. Journal of Time Series Analysis 20:155-171.

Jones RG (1980) Best linear unbiased estimators for repeated surveys. Journal of the Royal Statistical Society, Series B 42:221-226.

Junius T, Oosterhaven J (2003) The solution of updating or regionalizing a matrix with both positive and negative entries. Economic Systems Research 15:87-96.

Kapur JN, Kesavan HK (1992) Entropy optimization principles with applications. Academic Press, San Diego, California.

Kitagawa G, Gersch W (1984) A smoothness priors state-space modelling of time series with trend and seasonality. Journal of the American Statistical Society 79:378-389.

Kohn R, Ansley CF (1987) Signal extraction for finite nonstationary time series. Biometrika 74:411-421.

Kolmogorov AN (1939) Sur l'interpolation et extrapolation des suites stationnaires C. R Acad. Sci. Paris 208:2043-2045.

Kolmogorov AN (1941) Stationary sequences in a Hilbert space. Bulletin Moscow University 2:1-40.

Koopman SJ, Harvey AC, Doornik JA, Shephard N (1998) Structural time series analyser, STAMP (5.0). International Thomson Business Press, London, England.

Koopmans LH (1974) The spectral analysis of time series. Academic Press, New York.

Leontief WW (1951) The structure of the American economy 1919-1939: an empirical application of equilibrium analysis. New York: Oxford University Press.

Litterman RB (1983) A random walk Markov model for the distribution of time series. Journal of Business and Economic Statistics 1:169-173.

Ljung GM, Box GEP (1978) On a Measure of Lack of Fit in Time Series Models, Biometrika, 65:297-303.

Macaulay FR (1931) The smoothing of time series. National bureau of Economic Research.

MacGill SM (1977) Theoretical properties of biproportional matrix adjustment. Environment and Planning, A 9: 687-701.

Malthus TR (1798) An essay on the principle of population. Johnson J, in St-Paul's church courtyard; reproduced by MacMillan Publications, London, 1926.

Maravall A (1993) Stochastic and linear trend models and estimators. Journal of Econometrics 56:5-37.

McLeod IA (1975) Derivation of the theoretical autocovariance function of autoregressive-moving average time series. Applied Statistics 24:255-256.

Mian IUH, Laniel N (1993) Maximum likelihood estimation of constant multiplicative bias benchmarking model with application. Survey Methodology 19:165-172.

Morris ND, Pfeffermann D (1984) A Kalman filter approach to the forecasting of monthly time series affected by moving festivals. Journal of Time Series 5:225-268.

Nelson CR, Plosser CI (1982) Trends and random walks in macroeconomic time series. Journal of Monetary Economics 10:139-162.

Pankratz A (1989) Time series forecasts and extra-model information. Journal of Forecasting 8:75-83.

Pearl R, Reid LJ (1920) On the rate of growth of the population of the United States. Proceedings of the National Academy of Science, U.S.A. 6, pp 275-288.

Persons WM (1919) Indices of business conditions. Review of Economic Statistics 1:5-107.

Pfeffermann D, Bleuer SR (1993) Robust joint modelling of labour force series of small areas. Survey Methodology 19:149-163.

Pierce DA, Grupe MR, Cleveland WP (1984) Seasonal adjustment of the weekly monetary aggregates: a model-based approach. Journal of Business and Economic Statistics 2:260-270.

Ploeg F van der (1982) Reliability and the adjustment of sequences of large economic accounting matrices. Journal of the Royal Statistical Society, Series A 145:169-194.

Proietti T (1999) Distribution and interpolation revisited: a structural approach. Statistica 58:411-432.

Quenneville B, Cholette PA, Morry M (1999) Should stores be open on sunday? The impact of Sunday opening on the retail trade sector in New Brunswick. Journal of Official Statistics 15:449-463.

Quenneville B, Rancourt E (2005) Simple methods to restore additivity of a system of time series. Proceedings of the workshop on frontiers in benchmarking techniques and their application to official statistics, Eurostat, Luxembourg.

Rao CR (1965) Linear statistical inference and its applications. John Wiley and Sons: New York.

Rao JNK (1999) Some recent advances in model-based small area estimation. Survey Methodology 25:175-187.

Rao JNK (2003) Small area estimation. Wiley, New York.

Robinson GK (1991) That BLUP is a good thing: the estimation of random effects. Statistical Science 6:15-51.

Rossana RJ, Seater JJ (1995) Temporal aggregation and economic time series. Journal of Business and Economic Statistics 13:441-451.

Santos Silva JMC, Cardoso FN (2001) The Chow-Lin method using dynamic models. Economic Modelling 18:269-280.

Särndal CE, Swensson B, Wretman J (1992) Model Assisted Survey Sampling. Springer-Verlag, New York.

Shiskin J, Young AH, Musgrave JC (1967) The X-11 variant of the census method II seasonal adjustment program. U.S. Bureau of the Census, Technical Paper No. 15.

Sobel EL (1967) Prediction of a noise-distorted multivariate non-stationary signal. Journal of applied Probability 4:330-342.

Solomou S, Weale M (1993) Balanced estimates of national accounts when measurement errors are autocorrelated: the UK, 1920-38. Journal of the Royal Statistical Society, Series A 156:89-105.

Statistics Canada (2001) Retail Trade. Catalogue No. 63-005-XPB, Ottawa, Canada.

Stone JRN, Champernowne DG, Meade JE (1942) The precision of national income estimates. Review of Economic Studies 9:111-125.

Stone JRN (1980) The adjustment of observations. Mimeo, Department of Applied Economics, Cambidge University, Cambridge, England.

Stone JRN (1982) Balancing the national accounts. In: Ingham A and Ulph A (eds) Demand, Equilibrium and Trade, Macmillan, London.

Stram DO, Wei WW (1986) Temporal aggregation in the ARIMA process. Journal of Time Series Analysis 7:279-292.

Stram DO, Wei WW (1986) A methodological note on the disaggregation of time series totals. Journal of Time Series Analysis 7:293-302.

Theil H (1967) Economics and information theory. North-Holland Publishing Company, Amsterdam; Rand McNally and Company, Chicago.

Theil H (1971) Principle of econometrics. John Wiley and Sons, New York London Sydney Toronto.

Trabelsi A, Hillmer SC (1989) A benchmarking approach to forecast combination. Journal of Business and Economic Statistics 7:353-362.

Tsimikas J, Ledolter J (1994) REML and best linear unbiased prediction in state space models. Communications in Statistics, Theory and Methods 23:2253-2268.

Verhulst PF (1838) Notice sur la loi que la population suit dans son accroissement. In: Evetelet A (ed) Correspondance mathématique et physique, Tome X, pp 113-121.

Weale M (1988) The reconciliation of values, volumes and prices in the national accounts. Journal of the Royal Statistical Society, Series A 151:211-221.

Wei WW, Stram DO (1990) Disaggregation of time series models. Journal of the Royal Statistical Society, Series B 52:453-467.

Wiener N (1949) Extrapolations, interpolations and smoothing of stationary time series. Wiley, New York.

Whittle P (1963) Prediction and regulation. D. Van Nostrand, New York.

Whittaker ET (1923) On a new method of graduation. Proceeding of the Edinburgh Mathematical Association, 78, pp 81-89.

Whittaker ET, Robinson G (1924) Calculus of observations: a treasure on numerical calculations. Blackie and Son, London.

Wold HO (1938) A study in the analysis of stationary time series. Almquist and Wacksell, Uppsala, Sweden, second edition 1954.

Young AH (1965) Estimating trading-day variation in monthly economic time series. U.S. Bureau of the Census, Technical Paper No. 12.

Zellner A, Hong CH, Min C (1991) Forecasting turning points in international output growth rates using Bayesian exponentially weighted autoregressive time varying parameters and pooling techniques. Journal of Econometrics 48:275-304

Index

Lecture Notes in Statistics (*Continued from p. ii*)

162: Constantine Gatsonis, Bradley Carlin, Alicia Carriquiry, Andrew Gelman, Robert E. Kass, Isabella Verdinelli, and Mike West (Editors), Case Studies in Bayesian Statistics, Volume V. xiv, 448 pp., 2001.

163: Erkki P. Liski, Nripes K. Mandal, Kirti R. Shah, and Bikas K. Sinha, Topics in Optimal Design. xii, 164 pp., 2002.

164: Peter Goos, The Optimal Design of Blocked and Split-Plot Experiments. xiv, 244 pp., 2002.

165: Karl Mosler, Multivariate Dispersion, Central Regions and Depth: The Lift Zonoid Approach. xii, 280 pp., 2002.

166: Hira L. Koul, Weighted Empirical Processes in Dynamic Nonlinear Models, Second Edition. xiii, 425 pp., 2002.

167: Constantine Gatsonis, Alicia Carriquiry, Andrew Gelman, David Higdon, Robert E. Kass, Donna Pauler, and Isabella Verdinelli (Editors), Case Studies in Bayesian Statistics, Volume VI. xiv, 376 pp., 2002.

168: Susanne Rässler, Statistical Matching: A Frequentist Theory, Practical Applications and Alternative Bayesian Approaches. xviii, 238 pp., 2002.

169: Yu. I. Ingster and Irina A. Suslina, Nonparametric Goodness-of-Fit Testing Under Gaussian Models. xiv, 453 pp., 2003.

170: Tadeusz Caliński and Sanpei Kageyama, Block Designs: A Randomization Approach, Volume II: Design. xii, 351 pp., 2003.

171: D.D. Denison, M.H. Hansen, C.C. Holmes, B. Mallick, B. Yu (Editors), Nonlinear Estimation and Classification. x, 474 pp., 2002.

172: Sneh Gulati, William J. Padgett, Parametric and Nonparametric Inference from Record-Breaking Data. ix, 112 pp., 2002.

173: Jesper Møller (Editor), Spatial Statistics and Computational Methods. xi, 214 pp., 2002.

174: Yasuko Chikuse, Statistics on Special Manifolds. xi, 418 pp., 2002.

175: Jürgen Gross, Linear Regression. xiv, 394 pp., 2003.

176: Zehua Chen, Zhidong Bai, Bimal K. Sinha, Ranked Set Sampling: Theory and Applications. xii, 224 pp., 2003.

177: Caitlin Buck and Andrew Millard (Editors), Tools for Constructing Chronologies: Crossing Disciplinary Boundaries, xvi, 263 pp., 2004.

178: Gauri Sankar Datta and Rahul Mukerjee, Probability Matching Priors: Higher Order Asymptotics, x, 144 pp., 2004.

179: D.Y. Lin and P.J. Heagerty (Editors), Proceedings of the Second Seattle Symposium in Biostatistics: Analysis of Correlated Data, vii, 336 pp., 2004.

180: Yanhong Wu, Inference for Change-Point and Post-Change Means After a CUSUM Test, xiv, 176 pp., 2004.

181: Daniel Straumann, Estimation in Conditionally Heteroscedastic Time Series Models, x, 250 pp., 2004.

182: Lixing Zhu, Nonparametric Monte Carlo Tests and Their Applications, xi, 192 pp., 2005.

183: Michel Bilodeau, Fernand Meyer, and Michel Schmitt (Editors), Space, Structure and Randomness, xiv, 416 pp., 2005.

184: Viatcheslav B. Melas, Functional Approach to Optimal Experimental Design, vii., 352 pp., 2005.

185: Adrian Baddeley, Pablo Gregori, Jorge Mateu, Radu Stoica, and Dietrich Stoyan, (Editors), Case Studies in Spatial Point Process Modeling, xiii., 324 pp., 2005.

186: Estela Bee Dagum and Pierre A. Cholette, Benchmarking, Temporal Distribution, and Reconciliation Methods for Time Series, xiv., 410 pp., 2006.

187: Patrice Bertail, Paul Doukhan and Philippe Soulier, Dependence in Probability and Statistics, viii., 504 pp., 2006.